Joining of Advanced Materials

Joining of
Advanced Materials

Robert W. Messler, Jr.
Rensselaer Polytechnic Institute
Troy, New York

Butterworth–Heinemann

Boston   London   Oxford   Singapore   Sydney   Toronto   Wellington

Library of Congress Cataloging-in-Publication Data
Messler, Robert W., 1942–
  Joining of advanced materials / Robert W. Messler, Jr.
    p. cm.
  Includes bibliographical references and index.
  ISBN 0–7506–9008–9
  1. Joints (Engineering)   I. Title.
  TA660.J64M47 1993
  621.8´8--dc20   93-22442
                                                        CIP

British Library Cataloguing in Publication Data
A catalogue record for this book is available from the British Library.

Butterworth-Heinemann
80 Montvale Avenue
Stoneham, MA 02180

10 9 8 7 6 5 4 3 2 1

Printed in the United States of America

Contents

4 Adhesive Bonding 107

7 The Metallurgy of Welding   235

8 Brazing   283

10 Other Joining Processes 365

II JOINING OF SPECIFIC MATERIAL TYPES AND COMBINATIONS

11 Joining Advanced Metals, Alloys, and Intermetallics 389

13　Joining of Polymers　457

14　Joining Composite Materials　477

15   Joining Dissimilar Material Combinations   509

Preface

Since the dawn of humankind, the joining of similar and different materials to produce tools, devices, and structures has been a critical technology. Lashing a pointed stone to a split stick of wood to produce an ax or spear, attaching a wood or stone wheel to an axle to produce a cart, forge welding iron to produce tools or weapons, mortaring and cementing hewn stones to create the great cathedrals—all extended human strength, mobility, comfort, and well-being.

As time advances, so does the need for material technology and the parallel need for joining technology. How does one join thermo-mechanically processed, microalloyed, high-strength, low-alloy steels into ship structures; or, powder-metallurgically produced, elevated temperature, aluminum alloys into advanced air- and spacecraft structures? How does one join structural ceramics into advanced internal combustion or gas turbine engines for automobiles? How does one join highly refractory intermetallics into advanced air-breathing and rocket engines for hypersonic transports? How does one join composites into high strength, lightweight vehicles of all sorts? This book addresses the joining of these advanced materials.

Materials and materials processing have entered a new era. Where metals and alloys once dominated, advanced structural and electronic ceramics, new glasses, engineered polymers, intermetallic compounds, long-range ordered alloys, and composites with polymeric, metallic, ceramic, intermetallic and carbon matrices have entered to extend and expand applications. The increasing use of "engineered" compositions and microstructures and complex hybrid structures and assemblies places increasing demands on engineers responsible for selecting and processing strong, reliable, efficient, and economical joints. To meet the challenges posed by these evermore diverse and sophisticated materials, design and process engineers must look to evermore diverse and sophisticated methods of mechanical fastening, adhesive bonding, welding, brazing, soldering, weldbonding, weldbrazing, braze welding, rivet-bonding, and thermal spraying.

There are many ongoing evolutions, and even revolutions, in modern industry. New challenges face the design and process engineer—whether employing conventional materials in unconventional ways or advanced materials in conventional or novel ways. The increasing use of composites in aircraft manufacture is shifting joining from mechanical fastening to adhesive bonding, welding, brazing, and hybrid weldbonding or weldbrazing. The use of coated steels, aluminum alloys, and polymers in automobile manufacture demands new welding and adhesive bonding processes. The dramatic and continuing shift to automated assembly places new demands on mechanical fastening, welding, and bonding. And, as a final example, the cry for increased quality, reliability, and serviceability demands stronger, more durable permanent joints, on one hand, and easily disassembled joints, on the other.

This book is the result of years of contemplation, but, most immediately, the result of a course at Rensselaer Polytechnic Institute developed and presented by the author after years of observing the rapid evolution of materials and the sometimes agonizingly slow parallel evolution of joining processes. The course was developed to introduce engineers to the challenges and options of joining, to dispel misinformation and unwarranted prejudice, and to enlighten. This book is the result—or the extension—of the course. Too many material engineers, if they know anything about joining, know only of welding or brazing and possibly something about soldering. They may know little about adhesive bonding and essentially nothing about mechanical fastening. Mechanical and civil engineers who know mechanical fastening well, often know too little about welding, and may know almost nothing about adhesive bonding. There are few sources that discuss all of the major issues and options for joining both conventional and advanced materials, and none that does so primarily from the material perspective.

This book is intended for all engineering disciplines, including mechanical, civil, electrical, industrial, and materials engineering. It is intended to be a comprehensive primer as opposed to a comprehensive handbook and a primary textbook or a collateral source for undergraduate and graduate engineering students. It is also intended for both recently graduated and seasoned practicing engineers to broaden their knowledge of joining issues and options. Hopefully, it will become a reference they return to over and over to refresh, reflect, and refine their knowledge and understanding.

Joining of Advanced Materials approaches the subject of joining from the material perspective, but without ignoring issues of joint design, structural performance, practical productivity, economics, and service reliability. Part I addresses the process of joining in general (Chapter 1), mechanical fastening (Chapters 2 and 3), adhesive bonding (Chapters 4 and 5), welding (Chapters 6 and 7), brazing (Chapter 8), soldering (Chapter 9), and other processes such as weldbonding, weldbrazing, braze welding, rivet-bonding and thermal spraying (Chapter 10). For each process the principles of operation, relative advantages and disadvantages, and key technical issues are addressed, along with an overview of design considerations and performance. Part II considers the challenges posed by specific material types, including metals, alloys and intermetallics (Chapter 11); ceramics and glasses (Chapter 12); polymers (Chapter 13); composites with polymeric, metallic, ceramic, intermetallic, or carbonaceous matrices (Chapter 14). The particular challenges associated with forming joints between dissimilar materials to produce increasingly important hybrid structures is also addressed (Chapter 15).

While written by one person, this book is the compilation, over time, of the knowledge, ideas, views, and suggestions of many. The absence of specific names is not meant to detract from any one person's contributions but, rather, is simply the result of a continuous process of learning through professional interaction. To all the people with whom I have had the good fortune and pleasure of interacting professionally in industry and in the university over the last twenty-odd years, I say thank you.

In hindsight, I thank my editor Alex Greene at Butterworth-Heinemann. He visited me at RPI just over two years ago to talk about new ideas for technical books and, while doing so, he excited me with the prospect of putting down on paper, in one place, something meaningful and informative and, hopefully, interesting about the joining of advanced materials. There were times during the writing of this book that I wasn't sure I would thank Alex. But now I do—deeply.

I want to give special thanks to two people at Rensselaer who helped me with the production of this book. First, I thank Mary Haughney for helping get everything together for my publisher and for being so patient. Second, I thank Katherine Miller for acting as a sounding-board, for proofreading my original manuscript, for innumerable corrections and suggestions, for always being tough but always caring, and for showing me what strength and human spirit really is.

Most of all, I thank my wife, Joan, and my two daughters, Vicki and Kerri, for giving me up to this time-, energy-, and emotion-consuming project, but for always encouraging me to complete a dream. Without their understanding, their encouragement, and their love, this project would have been impossible.

Finally, I thank my mom and my dad—for nurturing my intellectual curiosity, for helping me appreciate other's views, for sacrificing to enable me to become an engineer, and for always encouraging me to reach for a goal. My dad taught me to love engineering and to always remember that it is the process of building and creating that really contributes to humankind. I remember dad. Thanks.

Robert W. Messler, Jr.

PART I

Joining Processes

1

Introduction to Joining

1.1 JOINING DEFINED

From the beginning of time, the ability to join similar or different materials has been important in the creation or manufacture of useful tools, products, or structures. In fact, joining was clearly one of the very first manufacturing technologies, beginning when a stone was first joined to a stick with a vine lashing to produce a hammer or ax or spear, creating a functional tool from the assembly of simple components. With the passage of time, the need for joining has not abated but, in fact, has grown. More diverse materials were fabricated into more sophisticated components, and these materials were joined in more diverse and effective ways to produce ever more sophisticated assemblies. Today, from a Wheatstone bridge to a suspension bridge, joining is a critically important aspect of both design and manufacturing. Thus, an understanding of what joining is, how it can be accomplished, and what properties can be expected is vitally important to all engineers.

In the most general sense, *joining* is the act or process of putting or bringing things together to make them continuous or to form a unit. As it applies to fabrication, joining is the process of attaching one component or structural element to another to create an assembly, where the assembly of component parts is required to perform some function or combination of functions that are needed or desired. It is the joining of structural elements or components into assemblies, and the materials from which they are composed, into assemblies that is of interest here.

An *assembly* is a collection of manufactured parts, brought together by joining, to perform any one of several primary functions. These primary functions can be broadly divided into three categories: structural, mechanical, and electrical. In *structural assemblies,* the primary function is to carry static or dynamic loads, or both. Examples are buildings, bridges, and the airframes of commercial airliners or space shuttles. In *mechanical assemblies,* the primary function, while often seemingly structural, is to create, enable, or permit some desired motion or series of motions through the interaction of the component parts. Examples are engines, linkages, actuators, and gear trains. Without question, such assemblies must be capable of carrying loads and, therefore, must be structurally sound, but the loads being carried are secondary to the purpose of creating or permitting motion. Finally, in *electrical assemblies,* the primary purpose is to create, transmit, or process some desired electromagnetic signal to perform some function. The most notable examples are microelectronic packages and printed circuit boards, but also included are power generators and transformers.

Usually, assemblies perform multiple functions, with one function being primary and the others secondary. Thus, the joints in an assembly also perform multiple functions. For example, the solder joints in a printed wire board have the primary function of providing electrical connectivity, but they also sustain mechanical forces in physically holding the assembly of electrical components together in proper arrangement under acceleration or differential thermal expansion and contraction. Regardless of the primary or secondary functions of an assembly and its component joints, joints are an extremely important part of any assembly or structure and are found in almost every structure. In fact, joints make complex structures, machines, and devices possible, so joining is a critically important process.

1.2 REASONS FOR JOINING MATERIALS

Ideally, a structure would be designed without joints, since joints are generally a source of weakness or excess weight. In practice, limitations on component size imposed by manufacturing processes and the requirements of inspection, accessibility, repair, portability and assembly, mean that some load-carrying joints are inevitable in all large structures.

There are three predominant reasons for joining materials or parts into assemblies or structures: (1) to achieve function, (2) to achieve structural efficiency, (3) to minimize costs. Where relative motion is required between parts, those parts cannot be made in one piece and so must be joined together. This is most apparent in certain mechanical assemblies. In fact, the joining required in some mechanical assemblies is special just because relative motion between components of the assembly must be possible. Component parts must be held in proper arrangement and proximity, but the joints must be capable of allowing motion, while, at the same time, having the ability to carry the various loads generated by and/or imposed on the assembly. An example is a transmission gear box comprised of an assembly of gears, shafts, bearings and a housing used to transmit power in various ratios and directions relative to the primary power source.

Sometimes to achieve function, diverse properties are required, and multiple parts each of different materials must be employed. When this is the case, joints must be created between those different components and different materials. There are many examples, including optically transparent glass windows in opaque wood or brick houses or metal or plastic automobiles and electrical devices to produce motion in mechanical components. Figure 1.1 shows a glass windshield assembly being installed into a truck cab. The joining of different materials to achieve function is often the most challenging aspect of joining, as the joining of optically transparent, brittle glass into a tough optically opaque structural metal frame.

One obvious need for joining is in the creation of extremely large structures: structures so large that they could not reasonably be fabricated from one piece. In such

Figure 1.1   The glass windshield of a truck cab is being installed into a metal frame using a combination of mechanical interlocking and adhesive bonding to accomplish attachment. (Photo courtesy of General Motors.)

instances, joining is clearly needed for manufacturing efficiency, including cost efficiency. Figure 1.2 schematically illustrates the impracticality of a one-piece versus a
built-up truss for a bridge.

Often, one of the principal demands made in joining is that the resulting joints
provide high structural efficiency. That is, the joints in a structural assembly should be
capable of carrying the loads imposed on them without adding to the bulk (i.e., mass,
weight, volume) of the structure. This usually means that the joint should be able to
operate at stress levels comparable to those being sustained in the structural elements
making up the assembly. Although structural efficiency would seemingly be greater if
the entire structure could indeed be fabricated in one piece, this is usually not the case.
Often, different properties are needed in different portions of a structure, so different
materials should be employed in each portion. To enable such material and structural
optimization, the ability to join is critical and may provide additional benefits in terms
of damage tolerance by changing properties along a potential crack path, disrupting and
arresting crack propagation.

If a designer has the freedom to select the most appropriate material to meet the
load and environmental requirements of an application, as well as the manufacturability, quality, and cost demands and constraints, a design is said to be *optimized*. In large,
complex structures and assemblies, this often means that different materials should be
used in different areas or functional elements of the structure or assembly. The result of
such freedom, from a structural standpoint, is that the overall weight of the structure is

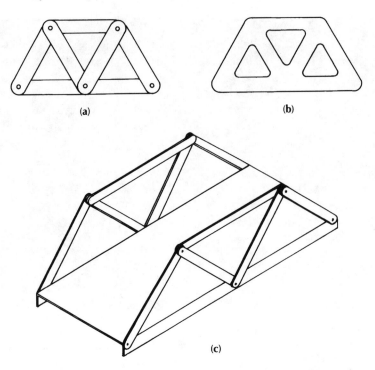

Figure 1.2   One-piece versus built-up truss assembly. (a) Built-up truss; (b) one-piece truss; (c) built-up truss bridge.

minimized. Clearly, to allow the designer such freedom of choice, however, imposes added demands on manufacturing, with the greatest demand being the need and ability to produce sound joints.

The actual process is an iterative one between design and manufacturing. To optimize the design, the designer must have the freedom to select the most appropriate material for each functional element of the assembly. An associated benefit is that the manufacture of the individual components from individual materials is often easier than the manufacture of a one-piece structure from a single material. The need to join the individual components, however, complicates both the design and the manufacturing processes. It complicates design by requiring added analysis of the joint, and it complicates manufacturing by requiring the production of the joint.

The ability to manufacture large or complex structures or assemblies by joining individual, optimized functional elements or components can also result in improved structural efficiency by improving the tolerance of the structure or assembly to damage, especially catastrophic damage. If a large structure is made in one piece, any flaw that initiates a crack anywhere in the structure can propagate, without interruption, until the integrity of the entire structure is lost. On the other hand, if the structure is built from individual structural elements that are joined together, a flaw that initiates a crack in one of those elements may only propagate until it causes that element to fail, without proceeding into some other element of the structural assembly. Naturally, to arrest the propagation of a crack in one element, that element must be joined to other elements in such a way that there is no continuous elastic path. This can be accomplished by changing the material at the joint, to change the elastic properties, or by joining in such a way that the path is interrupted. Several approaches will be described in later chapters, but Figure 1.3 shows how mechanical fastening of details is used to create large complex, but damage-tolerant, assemblies in aircraft.

Finally, the cost of a structure or assembly is strongly affected by the material(s) used in its fabrication and the ease or complexity of fabricating and assembling its functional elements. Later, once the structure or assembly is in use, its cost is also measured and affected by the ease of conducting routine maintenance or needed repair, which may be facilitated by selecting a method of joining that allows disassembly. Through joining, it is possible and practical to allow individual parts or structural elements to be fabricated from optimum materials, matching material properties to design or functional requirements. Expensive or difficult-to-fabricate high-performance materials need only be used where they are absolutely necessary. Manufacturability is improved, and cost is reduced, by making the fabrication of individual components or structural elements as simple as possible. What must be considered carefully is that the added cost of labor for assembling and joining the details does not offset the savings of fabricating the overall structure from details in the first place. Another cost benefit afforded by joining is that it often improves material utilization and reduces scrap loss. By joining small pieces efficiently and economically into larger pieces, it is frequently possible to

Figure 1.3 Built-up aircraft fuselage to provide damage tolerance. (Photo courtesy of Grumman Aircraft Systems Division of Grumman Aerospace Corp.)

use more of the material from certain product forms such as sheets or plates by nesting, or from rolled, drawn, or extruded shapes to build up complex shapes. Figure 1.4 shows how nesting and build-up of details optimizes material utilization and reduces scrap for sheet materials (e.g., metals and composites).

A fourth important reason for joining materials together into structural elements or components, and those elements and components into structures and assemblies, is practicality, which, often relates to cost. Practical reasons for joining include limited availability of large or complex product forms or detailed parts in certain materials, difficulty in handling extremely large parts or structures, and the need for portability in the finished product.

Every material has a limit as to how big a product form or part can be produced in a single piece. This relates both to the limits set by the basic material source (e.g., heat or batch size during the pour of an ingot or casting) and the primary and secondary working processes (e.g., mold sizes, forging press tonnage, rolling mill or cutting machine widths). Consequently, once it is necessary to produce a structure that is larger than can be produced with given processing facilities, it becomes necessary to join pieces together to build up the desired size. An example is the joining of rolled steel plates to produce very large, flat aircraft carrier decks. For some materials, such as ceramics, the inherent susceptibility of the material to process-induced flaws (e.g., shrinkage cracks) limits both the size and the complexity of the product form or part that can be produced. For these materials, joining is principally necessary to build up the shape required. Sometimes the timely availability of material is most important.

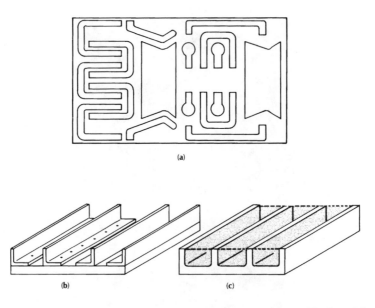

Figure 1.4   Part nesting and build-up to improve material utilization. (a) Nesting of details; (b) built-up; (c) machined showing wasted material removed as chips.

Here, joining can be used to build up a shape from details (e.g., extrusions or machined parts) rather than wait for a net shape produced by some primary processing method such as casting or forging. Figure 1.5 shows photograph of the aft closure beam for the F-14 Tomcat. To overcome an unacceptably long lead time for tapered titanium alloy forgings, simple, machined L-extrusions were welded longitudinally to produce the required tapered channel. In such cases, joining becomes the means for producing near-net shapes so highly desired in many industries.

The ability to handle the material forms to be fabricated into a structure or assembly (and, eventually, to handle the finished structure or assembly) is obvious. There is always a limit set by either the sheer size and weight of the items being handled or the susceptibility of the item to damage during handling due to its own mass (e.g., a very large ceramic part might not be able to support its own weight or tolerate the forces applied to move and manipulate it during processing) or both. Usually the size that can be handled as a finished product is much larger than the size of details that can be handled, because the details must be manipulated with more precision into and out of processing machines and because the built up structure is inherently more rigid. The need for and benefit of joining to facilitate handling during manufacturing is clear.

Portability of the final product may or may not be an important consideration, and, so, the design must include provisions for assembly, disassembly, and reassembly. Although most bridges, for example, are not intended to be moved once they are

Figure 1.5   Weld built-up F-14 wing closure beam to shorten lead time. (Photo courtesy of Grumman Aircraft Systems Division of Grumman Aerospace Corp.)

erected, for military operations, portable bridges are often desirable. As another example, it may be preferable to move a large new machine into an existing building in pieces, joining those pieces together on site to create the needed structure, rather than cutting a hole in the building. Thus, portability may be an important design consideration, and joining may provide practical approaches.

1.3 SPECIAL CHALLENGES OF JOINING ADVANCED MATERIALS

It can safely be said that fewer, simpler, lower-performance materials require less elaborate joining processes and procedures. Not surprisingly, then, advanced materials require special attention and more elaborate joining processes and procedures. Advanced materials include such things as

- high-strength, high-temperature, corrosion-resistant or specialty metals and alloys
- intermetallic compounds and long-range ordered alloys
- high-performance structural or electrical ceramics or glasses
- engineered polymers
- advanced composites with polymeric, metallic, ceramic, intermetallic, or carbon matrices.
- Electronic materials, although not primarily structural, could also be included in this list.

The reason the joining of advanced materials requires special attention becomes obvious from this list, that is, the sheer variety of fundamentally different material types and the number of different compositions within a type. More different types mean more possible combinations and greater differences in basic properties and, therefore, greater chemical, physical, or mechanical incompatibility. Ceramics and engineered polymers really have become practical candidates for structural applications that were once almost the exclusive domain of metals and alloys. Further, modern designs tend to place higher demands on the individual materials used. Modern designs demand, and modern, sophisticated analysis techniques allow, higher operating stresses, permit combined or complex loading, and enable combined properties for severe environments, all at minimal weight. Often, to meet these demands, designers combine diverse materials used in individual functional elements, creating hybrid structures that optimize overall function, performance, and cost. One example is the use of polymeric outer body panels over metal structural spaceframes in automobiles (as in GM's Pontiac Fiero shown in Figure 1.6). Finally, it tends to be the nature of advanced materials that they are more sensitive to secondary processing, such as heat treatment or welding, often because their compositions have been carefully balanced and their microstructures carefully developed to achieve optimum properties.

Figure 1.6 GM Fiero, showing joining of plastic body panels to a metal space frame to enable hybrid structures. (Photo courtesy of General Motors.)

The net result is that to join advanced materials, one must consider their variety, diversity, and sensitivity.

1.4 JOINING OPTIONS

The three fundamental options for joining are (1) mechanical fastening, (2) adhesive bonding, and (3) welding. Welding is usually subdivided into brazing and soldering, to bring the total number of distinct joining options to five. In addition, several additional options may correctly be considered either variations or combinations (or hybrids), including thermal spraying, rivet-bonding, weld-bonding, weld-brazing, and braze welding. Each of the five principal options, as well as the other variations or hybrids, is described briefly in the following sections.

1.4.1 Mechanical Fastening

In *mechanical fastening*, the attachment of components or structural elements in an assembly or structure is accomplished simply through a design feature of the component or element or with a supplemental device (called a *fastener*) that transfers loads from one component to another without relying on any primary or secondary atomic or molecular bonding forces. Attachment or joining is achieved completely through mechanical forces, from either physical interference (or interlocking) or friction, or both. Physical interference, or interlocking, develops forces at the macroscopic level, while friction develops forces at the microscopic level, largely by interference or inter-

Figure 1.7   Typical large bolted structural assembly. (Photo courtesy of Chicago Bridge & Iron Company.)

locking at this level. Figure 1.7 shows a typical use of mechanical fastening in a large bolted structure. The forces that join the structural components in this assembly arise from the macroscopic interference produced by the bolts themselves and by the friction arising from microscopic interference at the mating part faces (or *faying surfaces*). The principal advantage of mechanical fastening over all other forms of joining is the ease of disassembly it affords. Disassembly is often essential for maintenance and repair.

As mentioned, mechanical fastening can employ (1) design features that are integral to the component(s) or (2) supplemental devices called fasteners. Examples of *integral design features* include flanges, interlocking members or protrusions (e.g., tabs, tongues-and-grooves, teeth), roughened (e.g., knurled) gripping surfaces, and deformed features (e.g., crimps, hems, punch marks, or stakes). Examples of *fasteners* include both unthreaded and threaded types such as nails, pins or dowels, rivets, keys, retaining rings, eyelets and grommets, clips, or washers, and bolts, screws, nuts, or tapping screws. Some special forms of mechanical fastening include stapling, stitching, and snap-fit fasteners.

Besides allowing disassembly, mechanical fastening has the distinct advantage of not altering either the chemical composition or the microstructure of the materials of the parts or structural elements being joined. Despite this important and valuable advantage, however, mechanical fastening does introduce severe stress concentrations to the joint at the point of fastening. This can be, and often is, the cause of premature failure in improperly designed joints.

Mechanical fastening can be used with any material type but is best for metals and, to a lesser extent, composites. Problems arise in materials that are susceptible to damage through deformation (such as certain polymers with certain types of highly loaded fasteners or fastening features) or fracture by stress concentrations as a result of poor inherent damage tolerance (such as brittle ceramics or glasses) or with severe reductions of strength or damage tolerance in some particular direction due to anisotropy (such as directionally reinforced, laminated composites). Dissimilar material types or compositions can be readily joined since no interatomic or intermolecular bonding is required to develop joint strength and, thus, material intermixing and chemical compatibility are not issues. Electrochemical and thermal expansion mismatch must be dealth with, however.

1.4.2 Adhesive Bonding

In *adhesive bonding,* materials are joined with the aid of a substance capable of holding those materials together by surface attachment attraction forces. This attachment relies on a combination of varying degrees of microscopic mechanical locking and chemical bonding through primary or (usually more importantly) secondary atomic or molecular forces. The substance used to cause the attachment is called an *adhesive,* which is usually, but not always, added to the parts to be joined at the joint mating or faying surfaces. Figure 1.8 shows a typical adhesively bonded structure. Depending on the nature of the

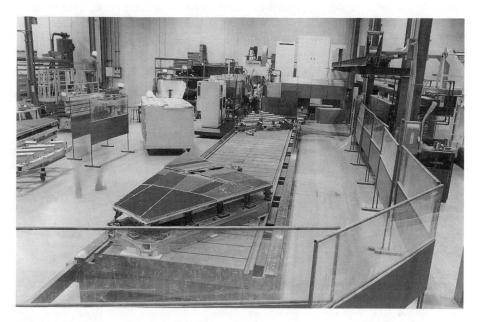

Figure 1.8   Typical adhesively bonded structure, i.e., a horizontal stabilizer for a fighter aircraft. (Photo courtesy of Grumman Aircraft Systems Division of Grumman Aerospace Corp.)

adhesive and the substrate being joined, adhesive bonding usually causes little or no disruption of the microstructure of the parts being joined but may cause varying degrees of chemical alteration or disruption. Because attachment occurs over the surfaces of the joint, loads are spread, and stress concentrations are minimized. The greatest shortcoming of adhesive bonding is susceptibility of the adhesive to environmental degradation.

Metals, ceramics, glasses, polymers, and composites, as well as dissimilar combinations of these, can be successfully adhesively bonded. Disassembly can occasionally be accomplished but never without great difficulty.

1.4.3 Welding

Welding is the process of uniting two or more materials through the application of heat or pressure or both to produce primary atomic or molecular bonds across the interface. Figure 1.9 shows a large welded structure from metal components. The process may or may not involve the need for a filler material of the same or different composition but of the same basic type[1] as the material(s) being joined.

The terms *welding* and *welding processes* commonly pertain to metallic materials, but it is possible to produce welds in polymers and glasses and, to a lesser extent, in some ceramics. Welding of composites can be accomplished, but only in the matrix. By the definition given above, welds cannot be produced between dissimilar types or classes of materials.

The relative amount of heat or pressure or both required to produce a weld can vary greatly. There can be enough heat to cause melting of the substrate or base material(s), with little or no pressure, except to hold the joint elements in contact. When this is the case, the process is known as *fusion welding*. Alternatively, there can be little or no heating but with enough pressure to cause some macroscopic plastic deformation (commonly called *upsetting* in welding), or microscopic plastic deformation and/or creep. In either case, melting or fusion is not required to establish bonds; only pressure is required to cause physical contact between the atoms or molecules of mating joint components. Such processes are known as *solid-phase* or *nonfusion welding* or, if the pressure is significant, *pressure welding*. In the extreme case, where no heating is employed, the process is commonly referred to as cold welding. One special form of solid-phase welding is called *diffusion bonding,* which uses varying degrees of pressure and temperature in appropriate combinations to accelerate diffusion kinetics in the solid state.

There are two subdivisions of welding in which the base materials are heated but not melted, a filler is added and melted, and little or no pressure is applied. These two, known as *brazing* and *soldering,* are described separately.

1.4.4 Brazing

Brazing is a subdivision of welding in which the joint is heated to a suitable temperature in the presence of a filler material having a liquidus above 450°C (840°F) and

---

[1] Composites can be welded to base materials that are of the same type, and, preferably, the same composition as their matrix.

Figure 1.9 Typical large welded structure. (Photo courtesy of Chicago Bridge & Iron Company.)

below the solidus of the base material(s). Bonding is accomplished without melting the substrate(s). The filler material (usually a metal but possibly a ceramic) is distributed between the close-fitting faying surfaces of the joint by wetting of the substrate and capillary flow. Bonding occurs by the formation of primary bonds: metallic in metals, ionic or covalent or mixed in ceramics. Bonding in brazing usually depends fairly significantly on interdiffusion between the filler and the substrate(s).

1.4.5 Soldering

Soldering is a subdivision of welding (or even brazing) that, like brazing, requires a filler that melts and substrates that do not. It is distinguished from brazing by the fact that the filler's liquidus is below 450°C (840°F). As in brazing, the filler material (which is almost always a metal but can be a glass for some joining applications), or *solder*, is distributed between close-fitting components making up the joint by wetting and capillary flow.

In soldering, the joining force is usually some combination of some primary (i.e., metallic or covalent) bonding and mechanical locking. The mechanical locking is often a combination of macroscopic interlocking of some feature of the component being soldered (such as a lead) as well as microscopic interlocking between the solder and substrate asperities. Figure 1.10 shows typical soldered joints in a microelectronic

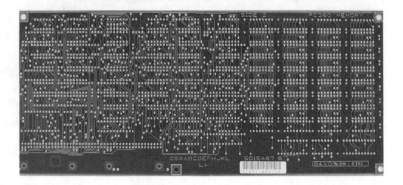

Figure 1.10   Typical soldered microelectronic printed wire board assembly. (Photo courtesy of Digital Equipment Corp.)

assembly or printed wire board. As in brazing, successful soldering requires some dissolution and interdiffusion between the filler and the substrate(s).

1.4.6 Other Joining Processes

Some other processes for joining materials are actually either variations or hybrids of the fundamental processes of mechanical fastening, adhesive bonding, welding, and brazing.

Variations

- *Thermal spraying,* which can be considered a variation of welding, is a special means of applying solid or fused (i.e., melted and solidified) material to a solid substrate and can be used for joining
- *Braze welding* is a variation of welding that uses a low-melting braze filler to fill a joint without relying on capillary action

Hybrid Processes

- *Rivet-bonding,* combines mechanical fastening by riveting and adhesive bonding
- *Weld-bonding,* combines welding and adhesive bonding
- *Weld-brazing,* combines welding and brazing.

In addition to these variations and hybrids of the fundamental joining processes, some special hybrids within welding can be included, such as laser/gas-tungsten arc welding and plasma/gas-metal arc welding.

1.5 JOINT LOAD-CARRYING CAPACITY VERSUS JOINT EFFICIENCY

Normally, when design of a joint in a structural assembly begins, the designer first thinks about how much *load* that joint must carry or transfer. The load that the joint must carry is the same as the load that is being carried by the structural elements on each side of the joint for simple joints but can be higher for more complex joints composed of more than two elements. After thinking about the load, the designer considers the *stress,* or the load per unit of cross-sectional area, in each structural element to be sure that that stress does not exceed the allowable stress for the material used in each of those elements. At this point, the designer must also consider the stress in the joint. So, first, the designer considers a joint's load-carrying capacity and then its stress.

The joint stress[2] is determined by dividing the load in the joint by the effective cross-sectional or load-bearing area of the joint. This effective joint area depends on the type of joint or joint design (e.g., straight- or scarf-butt, single- or double-lap), the size or dimensions of the joint, and the joining method, since the method of joining directly determines how much of the joint is really carrying the load. For welded, brazed, soldered, or adhesively bonded joints, the effective joint area is almost the same as the area of the faying surfaces (assuming continuous, full-penetration welding or continuous full-area brazing, soldering, or adhesive bonding). For mechanically fastened joints, the effective joint area is almost always much less than the area of the faying surfaces and is given by the cross-sectional area of all of the fasteners used in making the joint (assuming fasteners are being used). The actual points of joining are never continuous. Brazing, soldering, and adhesive bonding, on the other hand, are almost always continuous. Welding can be continuous or discontinuous (or intermittent), depending on the method chosen. Figure 1.11 shows the effective area of various joints, including continuous and intermittent welds, brazes, solders, adhesives, and rivets or bolts.

---

[2] The *joint stress* as defined here is different from the stress in the joint section. The stress in the joint section is simply the load carried by the joint divided by the effective area of the structure at the joint. For fastened joints, this is the area of the structural element minus the area of fastener holes along some plane cutting through the joint. The joint stress, on the other hand, would be the load carried by the joint divided by the load-bearing, cross-sectional area of fasteners along some plane cutting through the joint.

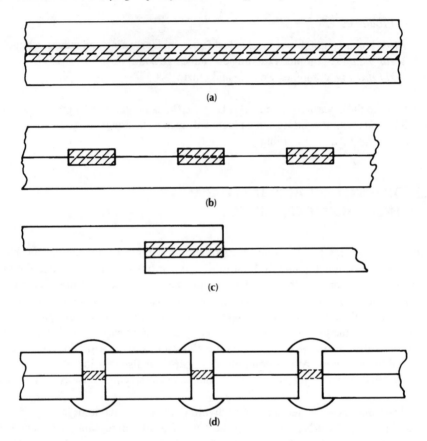

Figure 1.11   The effective joint area. (a) Continuous welded; (b) intermittent welded; (c) brazed, soldered, or adhesively bonded; (d) bolted (or riveted) joints.

Example 1.1: Calculation of Joint Stress

For the joint shown in Figure 1.12, this example compares the use of two ¼-in. diameter rivets to the use of a structural adhesive for a single-lap joint arrangement with a 1 ½ in. overlap. The actual stress in the joint for an 1,800 lb. load differs dramatically from the stress in the structural elements adjacent to the joint when it is bolted or riveted as opposed to when it is brazed or soldered or adhesively bonded. Stress (σ) in the structural elements, at point A and C:

$\sigma_{A(or\ C)}$ = 1,800 lb./ (3.0 in.)(0.125 in.) = 4,800 lb. per sq. in. (psi)

σ_{rivets} for a riveted joint = 1,800 lb./ (2)π(0.250 in./ 2)^2 = 18,335 psi

σ_{bond} for a bonded joint = 1,800 lb./ (1.5 in.)(3.0 in.) = 400 psi.

The much lower stress in the bond (whether adhesive, braze, or solder) than in the structural elements is due to the much greater area carrying the load. The much higher stress

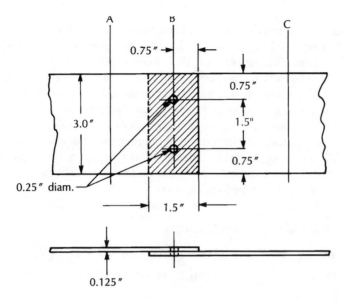

Figure 1.12   Fastened versus adhesively bonded joint. Sections at A and C pass through the joint elements with cross-sectional area = 3.0 in. × 0.125 in. Section B passes through the overlap area and the fastener holes, so the cross-sectional area of each joint plate is reduced by the total area of all holes.

in the rivets than in the structural elements or bond is due to the much smaller area of fastener carrying the load. Because of the effect of effective joint or load-carrying area, lower strength adhesives can compare favorably to much higher strength fasteners in total load-carrying capacity.

The true measure of a joint's structural effectiveness is whether it can safely carry the load or loads imposed. The ultimate efficiency of the structure, in terms of its overall load-carrying capacity, its size, and its weight, however, is often dependent on the efficiency of the joints making up that structural assembly. *Joint efficiency* is a measure of the effectiveness of the joint compared to the rest of the structure for carrying the design or service load and is defined by:

$$\text{Joint efficiency} = \frac{\text{Joint stress}}{\text{Stress in the structure}} \times 100\% \qquad (1.1)$$

Joint efficiency varies widely depending on the joining process or method used and can range from very low values (for example, 10%) to over 100%. Some examples will help illustrate the point.

In welding two pieces of the same metal with a continuous, full-penetration straight butt weld using a filler metal with the same composition as the base metal, the joint efficiency is typically 100%. That is, the strength of or stress developed in the weld filler metal is typically equal to the strength of or stress developed in the base metal. A 100% joint efficiency can be considered to be a perfect joint. While it is possible to have the joint efficiency exceed 100% when welding certain metals, it is usually of limited benefit. For example, using a filler metal that is different from the base metal and with a higher strength will guarantee that the weld will not fail but will force yielding upon overloading to occur in the base metal adjacent to the weld rather than in the weld. The joint in this case, will be stronger than the surrounding structure, but to do that will be meaningless if the integrity of the surrounding structure is compromised.

The efficiency of mechanically fastened joints is typically very high and often greater than 100%. This is so because the mechanical fasteners are typically stressed to fairly high levels compared to the joint elements.

Example 1.2: Calculation of Joint Efficiency

The stress in the joint elements or splice plates at point A or C in Example 1.1 is:

$$\sigma_{plate} = \text{Load} / \text{Area}_{splice\ plate} = 1{,}800\ \text{lb.} / (3.0\ \text{in.})(0.125\ \text{in.}) = 4{,}800\ \text{psi}$$

The stress in the rivets is

$$\sigma_{rivets} = \text{Load} / \text{Area}_{rivets} = 1{,}800\ \text{lb.} / (2)\pi(0.250/2)^2 = 18{,}335\ \text{psi}$$

The stress in the adhesive is

$$\sigma_{adhesive} = \text{Load} / \text{Area}_{adhesive} = 1{,}800\ \text{lb.} / (3.0\ \text{in.})(1.5\ \text{in.}) = 400\ \text{psi}$$

So the joint efficiency for the riveted configuration versus the adhesively bonded configuration is

$$\text{Joint efficiency for rivet} = 18{,}335\ \text{psi} / 4{,}800\ \text{psi} \times 100\% = 375\%$$

while

$$\text{Joint efficiency for adhesive} = 400\ \text{psi} / 4{,}800\ \text{psi} \times 100\% = 8.3\%.$$

Obviously, the effective load-bearing area in a joint has an extremely strong effect on the joint's efficiency, with small effective areas leading to high efficiencies and large effective areas leading to low efficiencies. Because of this, joints that are fastened with rivets or bolts tend to have high efficiencies, while joints that are bonded, brazed, or soldered tend to have low efficiencies.

Welded joints can exhibit efficiencies of less than 100% in three ways:

1. The weld is not continuous but, rather, is intermittent.
2. The weld is not full penetration.
3. The weld filler metal is lower in strength than the base metal.

The reasons for electing to use intermittent versus continuous or partial- versus full-penetration welds will be discussed later (Chapter 6, Section 6.6) but usually relate to cost considerations. Very often, the filler metal is intentionally chosen to have a strength lower than the base metal, as well as a higher ductility than the base metal, so that the weld itself yields before fracturing. In this way, the weld acts as a safety valve to the rest of the structure.

Welded joint efficiencies typically range from approximately 50% to 100% but, as has been explained, can exceed 100%. Brazed and soldered joints usually exhibit joint efficiencies that are lower than 100% and, often, much lower than welds because the area of braze or solder over which loading occurs is typically high (at least for lap-type joints). Brazed joint efficiencies typically range from 40 to 80% (or higher) in metals but only 10 to 50% in ceramics. Soldered joint efficiencies typically range from 5 to 25% of the metals they are used to join but from 50 to 100% when solder glasses are used to join glasses. Larger effective joint areas are needed with brazes and solders because these fillers tend to be lower strength than most weld fillers. Adhesive bonded joints can exhibit widely varying joint efficiencies, largely depending on the substrate, or *adherend*. In polymers, joint efficiencies typically are very high and near 100%, while in metals, such as titanium, joint efficiencies are typically quite low, say less than 10%. What makes these low joint efficiencies tolerable, however, is that the joints can still have high load-carrying capacity as a result of their large effective bonded area.

Mechanically fastened joints also exhibit fairly high efficiencies, often exceeding 100%. In these joints, high efficiency is obtained by using fasteners with higher strengths than the joint material(s), to compensate for the relatively low effective joint area. In addition, the fasteners are often worked to higher stress levels (as a percentage of their yield strength). It should be recognized that we have been referring to the efficiency of joints under static loads, and that dynamic (e.g., fatigue) loads can reduce the strength of a joint significantly, especially if the method of joining gives rise to stress concentrations (e.g., mechanical fastening).

Table 1.1 summarizes the joint efficiencies typically observed for various joining methods in various materials.

Table 1.1   Typical Maximum Joint Efficiencies (in percent)

Jointing Method	Metals	Ceramics	Glass	Polymers	Composites
Mechanically fastened	75–100+	N/A	N/A	100	50–100
Adhesively bonded	<20	<20	<20	40–100	20–50
Inorganic adhesive	N/A	50–100	UNK	N/A	UNK
Welded	100	30–80	100	100	50+
Brazed	40–90	20–50	N/A	N/A	UNK
Soldered	5–20	<20	10–40	N/A	N/A
Weld-bonded	25–50	N/A	N/A	UNK	UNK
Weld-brazed	50–90	N/A	N/A	N/A	UNK
Rivet-bonded	75–100	N/A	N/A	100	UNK

N/A, not applicable (generally).
UNK, unknown for process.

1.6 LOAD- OR STRESS-STATE COMPLEXITY

The state of stress or the complexity of loading in a structure is critically important in selecting a joining method. As a general rule, the more complex the loading or stress state, the poorer the performance of a given joint. Biaxial loading is more severe than uniaxial loading, and triaxial loading is more severe than biaxial loading. However, the effect of stress state complexity is much greater for some forms of joining than for others. Figure 1.13 illustrates the states of stress complexity in simple elements as well as in a more complex element subjected to multiple loads.

For welding, the stress state generally does not matter very much, as the weld filler metal used exhibits fairly isotropic properties and usually constitutes a fair volume over which the applied stresses can operate. For adhesively bonded joints, on the other hand, it is critically important to keep the state of stress as near to perfect shear as possible, since the thin layer of adhesive usually performs badly under out-of-plane loading. Brazed and soldered joints exhibit similar behavior, since they too usually employ thin layers of filler, with little ability to tolerate strain through the filler layer thickness. For brazes and solders, however, out-of-plane tensile strengths are usually better than for adhesives. In mechanical fastening, loading should be in accordance with the type of fastener or method of fastening selected, but, in general, stress state complexity can be dealt with by proper design and fastener selection, and is not a major issue.

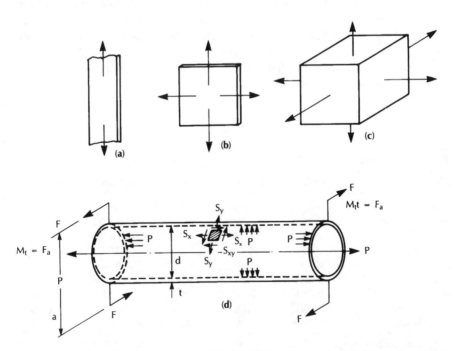

Figure 1.13   Stress-state complexity. (a) Uniaxial; (b) biaxial; (c) triaxial; (d) combined loading from internal pressure, axial loading, and torsion lending to a complex stress state.

SUMMARY

It is often desirable or necessary to fabricate structures or devices by assembling structural elements or components to achieve function, provide structural efficiency, and minimize costs. Such assembly relies on the formation of joints between similar or different materials, and the method of joining becomes a critical consideration. The challenge is greater with advanced materials, such as advanced alloys, ceramics, glasses, intermetallics, engineered polymers, and composites, since these materials are available in more basic types and compositions within a type, they are often selected for their higher performance capabilities, they tend to be used in combinations to optimize designs, and they tend to be inherently more susceptible to degradation of their highly developed and controlled microstructures and properties than conventional materials. Five basic methods are available for joining materials: mechanical fastening, adhesive bonding, welding, brazing, and soldering, with several variations and hybrids of these as well. Each has distinct advantages, as well as shortcomings, including differing joint efficiencies and sensitivity to loading or stress state complexity.

PRACTICE QUESTIONS AND PROBLEMS

1. Define the term *joining* in your own words.
2. What are the three major types of assemblies and what is the primary function of each? Give two examples of each type of assembly, in which there is essentially no function other than the primary function.
3. Give some examples of assemblies with definite multiple functions. What are these functions? Which are primary, and which are secondary? Can an assembly have more than one primary function? Explain and give an example.
4. What are the three predominant reasons for joining materials into parts or parts into assemblies or structures? What is a fourth major reason for joining?
5. Modern manufacturing speaks of the desirability of net-shape processing methods. The most common examples of net-shape processing methods are casting, forging, and machining. Explain how joining methods can be used for net-shape processing. Give at least one example for each of mechanical fastening, adhesive bonding, and welding.
6. How can joining render a structure damage tolerant even if the materials composing the structural elements are not inherently damage tolerant?
7. What are three special challenges associated with joining advanced materials? Give an example of each.
8. Define a *hybrid structure*. Give some examples. What is a *hybrid joining process*? Give some examples.
9. Each specific joining process has relative advantages and disadvantages. What is the predominant advantage and what is the predominant disadvantage of each of the following:
   a. mechanical fastening relative to all other processes
   b. adhesive bonding relative to all other processes

    c. welding relative to brazing and soldering
    d. brazing or soldering relative to welding
    e. brazing relative to adhesive bonding.
10. Explain what limits the load-carrying capacity of a joint.
11. A single-lap joint between two pieces of 2-in. wide by 0.1-in. thick aluminum has a 2-in. overlap and contains four ⅛-in. diameter rivets arranged in a square pattern (in two parallel rows of two rivets each) with centers ½-in. from each edge and with 1 in. between rivet centers. Calculate the stress in the rivets (i.e., the joint stress) for a unidirectional 2,000 lb. load applied along the longitudinal centerline of the aluminum joint elements. What is the joint efficiency for this joint?

    Extra: What is the net stress in the joint elements at the plane of either row of rivets?
12. For the joint in Problem 11, calculate the stress in the joint if, instead of being riveted, the joint was brazed over the entire area of overlap. What is the joint efficiency for this case?

    Extra: What is the net stress in the joint elements at the mid-plane of the overlap?
13. What would be the stress in the brazed joint if the joint was a straight butt rather than a single lap (i.e., if the two pieces of aluminum were simply butted end to end)? What would be the joint efficiency for this case?

    Extra: What would the net stress be at the abutment?
14. Of simple tension, simple compression, simple shear, bending, and torsion, which would cause the least problem with adhesively bonded joints? Which would cause the most problem? Rank order all loading types from least problematic to worst.
15. How do you think riveted joints might respond to the types of loading given in Problem 14? Would you change your answer for joints that were bolted?

REFERENCES

1. Charles, James A., and Crane, Frederick A.A., *Selection and Use of Engineering Materials,* 2nd ed., London: Butterworths, 1989.
2. Faupel, Joseph H., and Fisher, Franklin E., *Engineering Design,* 2nd ed., New York: John Wiley, 1981.
3. Lindberg, Roy A., *Processes and Materials of Manufacture,* 4th ed., Boston: Allyn and Bacon, 1990.
4. Parmley, Robert O., *Standard Handbook of Fastening & Joining,* 2nd ed., New York: McGraw-Hill, 1989.

Mechanical Fastening

2.1 INTRODUCTION

Undoubtedly, the oldest method used by humankind to join materials together was mechanical means, i.e., mechanical fastening. Mechanical fastening has been the principal method used to hold things together throughout recorded history and, no doubt, even before recorded history. From the first tool or weapon, when a stone was wedged into a forked stick and, later, lashed with a strip of plant or animal fiber, through the first time a stone or wooden wheel was locked onto a wooden axle by a wooden peg, to the assembly of modern jet airliners with nearly a million rivets and nuts and bolts in each, mechanical fastening has been critical to engineering and manufacturing.

In this chapter we define *mechanical fastening* as a joining process, look at the advantages and disadvantages of this method of joining versus other methods, and then consider the sources and types of loads that must be carried by mechanical joints. Next, the two predominant joint types, *shear-loaded joints* and *tension-loaded joints,* and their subtypes, are briefly described in terms of the design analysis required. No attempt is made in this elementary overview to delve into the intricacies of joint design and analysis, however, so the reader is referred to other sources (see Bibliography). Some attention is given to the subject of bolt preloading, including determination of the appropriate level, methods of achieving the desired level, and causes for its loss in service. Finally, various factors that affect fastener and joint performance in service are presented. Actual fasteners and fastening methods are described in Chapter 3.

2.2 MECHANICAL FASTENING AS A JOINING PROCESS

2.2.1 General Description

Mechanical fastening is the attachment of components in an assembly (or elements in a structure) through an integral design feature of the components or through the use of a supplemental device called a *fastener.* Loads are transferred from one component (or element) to another strictly through the development of mechanical forces from the physical interference or interlocking of component or fastener features at the macro-scopic level or through friction arising from interlocking of material asperities at the microscopic level, or both. In mechanical fastening, there is no dependence on the development of any primary or secondary atomic or molecular bonds between the components' materials, although some bonding may occur locally (giving rise to some of the friction). Like other joining processes, mechanical fastening is used to create assemblies or structures from detail parts or structural elements. Also, like any process, mechanical fastening has both advantages and disadvantages when compared with other methods of joining.

2.2.2 Advantages and Disadvantages of Mechanical Fastening

Mechanical fastening offers many advantages as a joining method, some of which make it unique. These advantages are summarized in Table 2.1 and are described in more detail below.

First and foremost, mechanical fastening uniquely allows simple and practical disassembly without component damage. No bonds need to be formed to accomplish joining, nor do any need to be broken to accomplish disassembly. Disassembly is useful and, often, essential for the purposes of portability, repair of damage, access for maintenance or servicing, modification or reconfiguration, addition of accessories or new structure (i.e., expansion), and so on. Second, and related to this, mechanical fastening, if done appropriately, permits relative motion between parts, while maintaining the needed geometric arrangement to achieve fit and function (recall the gear transmission box example). This is again a result of the fact that actual bonds are not created between the materials making up the parts being brought together. Allowing relative motion between components in an assembly, while providing mechanical alignment and structural integrity, is often vital to the assembly being able to perform its intended function and is another unique capability of mechanical fastening. Third, and possibly most important for many applications, mechanical fastening causes no change to the chemi-

Table 2.1 Advantages and Disadvantages of Mechanical Fastening

Advantages	*Disadvantages*
1. Uniquely allows disassembly	1. Creates significant stress concentrations
2. Can permit relative motion	2. Utility limited with some materials
3. Causes no change to chemical composition or microstructure	3. Allows fluid intrusion or leakage
4. Allows joining of dissimilar material types	4. Installation labor is high
5. Provides damage tolerance to assembly	5. Often a weight penalty
6. Simple; no special joint preparation	6. Joints can loosen
7. Low cost	
8. Sometimes automatable	

cal composition or microstructure of the materials composing the parts being joined. This is because the forces needed to hold the joint components together are purely mechanical. No atomic or molecular bonds are created; thus, no chemical interaction, pressure-induced plastic deformation, or heat to facilitate plastic deformation or melting is necessary.

Related to the fact that mechanical fastening causes no change in composition or microstructure are a fourth and fifth advantage, namely, it allows the joining of dissimilar materials and provides a simple means of achieving structural damage tolerance. The joining of dissimilar materials is again made possible since bonding is not necessary, and, thus, chemical (other than electrochemical or galvanic) compatibility is of little consequence. Materials are not mixed, merely brought into contact. The provision of damage tolerance is achieved by being able to restrict the propagation of a growing crack to the structural element in which it initiated, with no possibility of extension into an adjoining element. A sixth advantage is that mechanical fastening is simple and requires little or no special preparation of the joint, other than reasonable finishing to achieve proper fit and cleaning to avoid unwanted contamination. Besides being simple, mechanical fastening is usually relatively low in cost and, normally, requires only limited operator skill compared with other processes, like welding. Finally, the process, with only a few exceptions such as bolting and keying, is generally amenable to simple automation for most types of fastening (e.g., crimping, sewing) or fasteners (e.g., stapling, riveting, nailing, and screwing).

Despite all of these advantages, there are some disadvantages. First and foremost, mechanical fastening creates significant stress concentration at the point of fastening, primarily because of the hole required for or produced by any fastener.[1] This is a particular concern in fatigue-critical structures. Second, and related to the first, the utility of mechanical fastening, with or without fasteners, is limited with certain materials. This is true largely because of the effect of the aforementioned stress concentration. In materials (e.g., polymers) that exhibit viscoelastic (i.e., time-dependent strain) behavior under an applied stress, the stress concentration associated with fasteners or some integral design features can lead to distortion of the fastener hole or some fastening feature. This is often referred to as "cold flow". In inherently brittle materials, such as ceramics or glasses, the concentration of stress around fastener holes or at sites of point loading can cause intolerable strain, leading to fracture. Finally, utility can be limited by a material's anisotropy, especially if that anisotropy leads to weakness through the thickness direction, as in most laminated composites. The result can be fastener pull-through or pull-out.

A third disadvantage of mechanical fastening can be the open nature of the joint between points of fastening and around fasteners. Such a joint permits moisture, water, or fluid intrusion, permits leaks, and can accelerate corrosion due to the entrapment of the corrosive agent, often aggravated by the dissimilar electrochemical natures of the joined materials or any residual tensile stresses associated with the fastener. Other disadvantages of mechanical fastening are that the labor intensity can be high, especially

---

[1]In an attempt to reduce the stress concentration associated with fastener holes, the fastener is occasionally designed or sized to produce an offsetting compressive residual stress through interference.

for high-performance systems (e.g., preloaded bolts, upset rivets); there can be a weight penalty when compared with other joining processes such as welding, brazing, soldering, or adhesive bonding (if for no other reason than the fasteners themselves add extra weight), and joints can loosen in service as a result of vibration, mechanical flexing, thermal cycling, fastener or joint relaxation, and so on.

As a final point, the efficiency of mechanically fastened joints can vary considerably, depending on the materials making up the joint elements, the design, size, number, and composition of the fasteners, and the geometric factors involved in the joint design. Welding, on the other hand, almost always offers higher joint efficiencies than mechanical fastening, while brazing can result in higher or lower efficiencies, depending on the particular filler's strength. Soldering and adhesive bonding always result in less efficient joints than fastening.

2.3 SOURCES AND TYPES OF JOINT LOADING

As stated earlier, most joints are critical elements of a structure and are, therefore, the most likely areas of the structure to fail. Thus, joints demand careful design for all forms of joining, including mechanical fastening. One critical aspect in the design of all joints is identifying the sources and estimating the magnitudes of applied loads. Applied loads can be static (i.e., unchanging), dynamic (i.e., changing randomly or periodically), or impact (i.e., sudden and, often, unexpected and nonrepeating or highly irregular). The sources of loads can be from weights (e.g., snow, water, wind, or other parts of the structure) or from inertial forces, shock, vibration, temperature changes, fluid pressures, prime movers, and so on.

Regardless of the source of the applied loads, mechanically fastened joints are of two principal types, based on the direction of the primary loading versus the fastener's axis:

1. *shear-loaded joints,* in which the primary loads are applied at right angles to the axes of the fasteners
2. *tension-loaded joints,* in which the primary loads are applied more or less parallel to the axes of the fasteners.

The design procedures for shear-loaded and tension-loaded joints are quite different.

2.4 SHEAR-LOADED JOINTS

2.4.1 Types of Shear-Loaded Joints

Two basic types of joints can be loaded predominantly in shear: lap joints and butt joints. Figure 2.1 shows lap- and butt-type (or double-lap) shear-loaded joints schematically. These types are further defined as being either *friction-type joints,* where the fas-

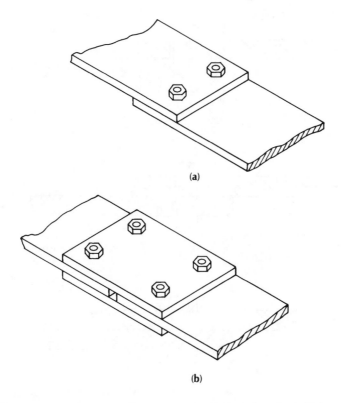

Figure 2.1 (a) Lap- and (b) butt-type (or double-lap) shear-loaded joints.

teners create a significant amount of clamping force on the joint, holding the joint elements together, and with the resulting friction created between joint members preventing joint slip, or *bearing-type joints,* where the fasteners, in effect, act as points to prevent slip, with the pieces being held together by shear in the fastener and bearing in the joint elements.

Figure 2.2 illustrates friction- and bearing-type joints, including loading. Only bolts (see Chapter 3, Section 3.3.3) can be used in friction-type joints, because only bolts can be counted on to develop the high clamping forces required to produce the necessary frictional forces to resist slip. Friction-type connections are generally used for alternating (i.e., fatigue), impact, or vibration loads. Rivets (see Chapter 3, Section 3.4.2), bolts (Section 3.3.3), pins (Section 3.4.4), or other fasteners can be used in bearing-type joints, where the loading is generally static.

The shear stress, in bearing-type connections, or the friction, in friction-type connections, is related to the arrangement of the pieces (or elements) composing the joint, as shown in Figure 2.3. This figure shows single shear and double shear arrangements in lap joints.

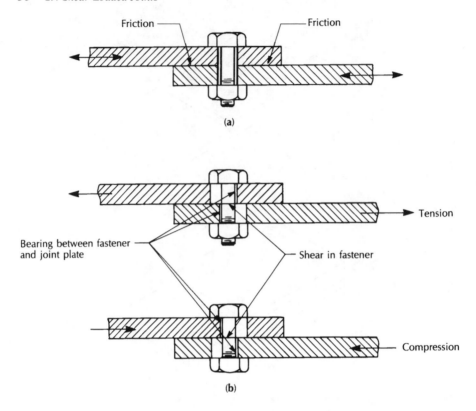

Figure 2.2 Loading in (a) friction and (b) bearing-type joints.

2.4.2 Fastener Spacing and Edge Distance

Because mechanical fasteners impose a stress and a stress concentration on the joint elements at each fastener under the application of a load, they must be carefully located both relative to the edges of the joint elements (i.e., their edge distance) and to each other (i.e., their spacing). This is true for both shear- and tension-loaded types of fasteners, but is especially a concern with bearing-type shear-loaded fasteners. Bearing forces imposed by the fastener on the joint element under static shear loading can cause elongation of the fastener hole and tear-out of a slug of material from the hole to the element's edge if edge distances are insufficient. Dynamic loading can cause even more problems, both with edge distance and between fasteners.

Recommended distances for locating fasteners from one another and from edges are given in various specifications, such as American Institute of Steel Construction (AISC) Spec J3.8 and J3.9. As shown in Figure 2.4, these specifications require spacing between fasteners of ≥2⅔ times the diameter of the fastener (3 d preferred), and edge distances ≤1¾ times the diameter of the fastener for sheared edges or 1¼ times the diameter of the fastener for rolled edges. Edge distances and fastener spacings can vary from this specification in other specifications, but careful consideration should always

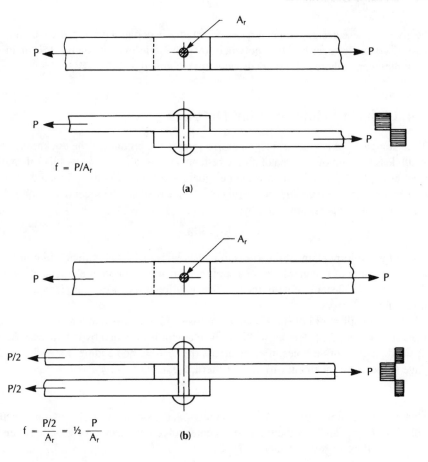

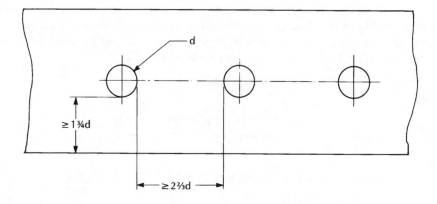

Figure 2.3 (a) Single- and (b) double-shear arrangements in lap joints.

Figure 2.4 Fastener spacings and edge distances.

be given to these design parameters. Edge distance becomes a particular concern in certain viscoelastic materials like polymers or polymeric-matrix composites, or in thin joint members, where bearing loads cannot be tolerated.

2.4.3 Effect of Holes on Joint Net Area

The net area A_n of a structural member in a joint is the product of the thickness and the width less the sum of the area of all the fasteners holes taken as the width of the bolt or rivet hole plus $\frac{1}{16}$ in. (although some designers add $\frac{1}{8}$ in.). Figure 2.5 (a and b) shows two patterns of holes, the first consisting of a single row and the second of a double row, for which the net areas are given by:

$$A_n = t\,[w - n(d + \tfrac{1}{16} \text{ in.}] \tag{2.1}$$

where t is the joint element thickness in inches, w is the joint element width in inches, n is the number of fastener holes in a straight path across the joint element, and d is the diameter of the fastener hole in inches. The additional factor of $\frac{1}{16}$ in. is to account for hole tolerance errors.

For a pattern of holes that extends across a joint element in a diagonal or zigzag line, as in Figure 2.5c, the net width of the element is obtained by subtracting the sum of the diameters of all holes in a line across the section, and adding, for each so-called gage space or line segment (m) in the pattern, the quantity

$$s^2/\,4g \tag{2.2}$$

where s is the longitudinal center-to-center spacing or pitch of any two consecutive holes, in inches, and g is the transverse center-to-center spacing or gage between fastener gage lines or rows, in inches. This gives

$$A_n = t[w - 2(d + \tfrac{1}{16} \text{ in.}) + m(s^2/4g)] \tag{2.3}.$$

Example 2.1: Determining Net Area for a Joint

For a $\frac{3}{4}$ in. thick joint with a staggered pattern of nominal $\frac{5}{8}$ in. diameter bolt holes as shown in Figure 2.6, there is the gross area (where there are no holes) and two different net areas; one along the straight transverse path AC and the other along the zigzag path ABC. These various areas are calculated by:

$$A_{gross} = (\tfrac{3}{4} \text{ in.})\,(12 \text{ in.}) = 9.0 \text{ in.}^2$$

while

$$A_{net\,AC} = \tfrac{3}{4} \text{ in.}[12 \text{ in.} - 2(\tfrac{5}{8} \text{ in.} + \tfrac{1}{16} \text{ in.})]$$

$$= 7.97 \text{ in.}^2$$

and

$$A_{netABC} = \tfrac{3}{4} \text{ in.}[12 \text{ in.} - 3(\tfrac{5}{8} \text{ in.} + \tfrac{1}{16} \text{ in.}) + 2\,(1\tfrac{3}{4} \text{ in.})^2 / 4(3 \text{ in.})]$$

$$= 7.83 \text{ in.}^2$$

Obviously, the smaller net area governs the design.

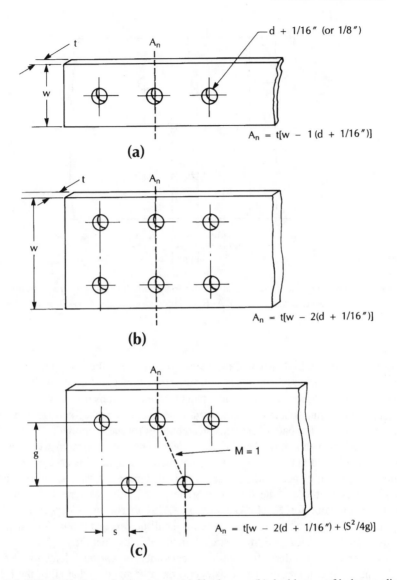

Figure 2.5 Effect of holes on joint net area. (a) Single row; (b) double row of holes on aligned centers; (c) double row of holes on staggered centers.

2.4.4 Allowable-Stress Design Procedure

The predominant procedure for designing shear-loaded fastened joints is the *allowable-stress design procedure,* in which all fasteners in a joint are assumed to carry an equal share of the applied loads. In fact, this assumption is only valid for joints that are composed of perfectly rigid materials, which is not the real case, and is least valid for long

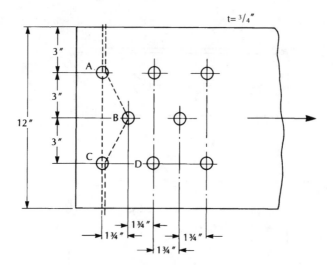

Figure 2.6 Net joint area. A, B, C, and D refer to specific holes along three different possible joint failure paths: A–C, straight across; A–B–C, zig-zag shear; and A–B–D, an alternative shear path.

joints containing multiple rows of fasteners. Nevertheless, the assumption is generally accepted and is perfectly safe when conservative allowables are used. In fact, a joint can reasonably be assumed to be rigid when (1) plane sections in the structural member, e.g., a gusset, remain plane and do not warp and (2) a straight line from the center of gravity of the fastener pattern remains straight after torque is applied.

Empirical means have been used to determine the maximum working stresses that can be allowed in the fasteners and joint members under this assumption. Various professional society specifications use slightly different allowables, but two common specifications are those of the American Society for Testing Materials (ASTM) and the American Institute for Steel Construction (AISC). The AISC specifies allowable stresses as follows: "The allowable stress σ_t shall not exceed $0.60\sigma_y$ [yield stress] on the gross area nor $0.50F_u$ [ultimate stress] on the effective net area."

Under this procedure, for bearing-type joints the various elements of the joint (including structural members and fasteners) must be sized so that all of the following conditions are satisfied: (1) the fasteners will not shear, (2) the joint plates will not fail in tension by overload nor (3) be deformed by bearing loads, and (4) the fasteners will not tear loose from the joint plates at edges. These various modes of failure are shown schematically in Figure 2.7. None of these things will happen if the appropriate allowable stresses are not exceeded in the fasteners (i.e., shear allowable) or in the joint plates (i.e, tension or bearing allowables) as given in design tables, such as those in the *Standard Handbook of Machine Design*. A sampling of the data from this book is shown in Table 2.2.

The allowable-stress design procedure is probably best understood through an example, as follows.

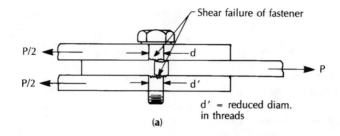

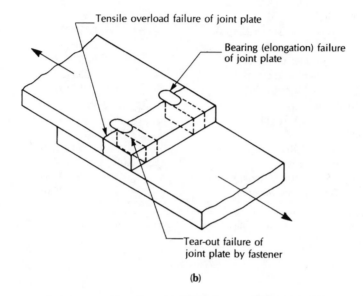

Figure 2.7 The various failure modes in bearing-type joints.

Example 2.2: Allowable-Stress Design Procedure Applied to a Bearing-Type Shear-Loaded Joint

Given that the double-lap shear joint shown in Figure 2.8 is composed of ASTM A36 steel, contains five 22 mm (nominally, ⅞ in.) diameter ASTM A325 steel bolts arranged as shown (although the particular pattern does not matter for symmetrical loading), that the bolts have a pitch of 2.12 mm per thread (or 12 threads per inch), that one of the shear planes in the double-lap passes through the unthreaded portion and one passes through the threaded portion of each bolt, the problem is to determine the various loads produced in the fastener and the joint plates by a load of 300 kN (67,000 lb.).

Table 2.2   Allowable Stresses in Some Representative Fastener and Joint Plate Materials

Material/Condition		*Allowable Stress MPA (kpsi)*		
		Tension	*Shear*	*Bearing*
ASTM A325 Bolts				
Bearing-type	— threads in shear plane	—	145 (21.0)	[a]
	— no threads in shear plane	—	207 (30.0)	[a]
Friction-type	— clean mill scale	—	52 (17.5)	[a]
	— blasted clean	—	190 (27.5)	[a]
	— blasted + Zn paint	—	203 (29.5)	[a]
ASTM A490 Bolts				
Bearing-type	— threads in shear plane	—	193 (28.0)	[a]
	— no threads in shear plane	—	276 (40.0)	[a]
Friction-type	— clean mill scale	—	152 (22.0)	[a]
	— blasted clean	—	238 (34.5)	[a]
	— blasted + Zn paint	—	255 (37.0)	[a]
ASTM SA31 Rivets				
	— SA 515 plate	—	62 (9.0)	124 (18.0)
ASTM A502-1 Rivets				
	— A36 plate	—	93 (13.0)	276 (40.1)
ASTM A36 Joint Material		152 (22.0)	100 (14.5)	335 (48.6)
58 kpsi UTS Steel Joint Material		—	160 (23.2)	—
100 kpsi UTS Steel Joint Material		—	200 (29.0)	—

[a] L S_u/2d or 1.5 S_u, whichever is least.
Reprinted with permission from J. Shigley and C. Mischke, *Standard Handbook of Machine Design*, McGraw-Hill, 1977.

1. Determine the shear stress in the fastener.

The shear stress produced in a fastener by a given load depends on the actual cross section of the fastener, and this is affected by whether the fastener is threaded or unthreaded in the region through which a shear plane passes. For an unthreaded rivet, the shear stress τ within each fastener is

$$\tau = F / bmA_r \qquad (2.4)$$

where F is the force in kilonewtons (or pounds), b is the number of shear planes that pass through the fastener and/or the number of slip planes in a shear joint (see Figure 2.9), m is the number of fasteners in the joint, and A_r is the cross-sectional area of the body of the fastener in square millimeters (or square inches).

For a threaded fastener such as a bolt, the shear stress τ within each is

$$\tau = F / A_T \qquad (2.5)$$

where A_T is the total cross-sectional area of all bolts in square millimeters (or square inches).

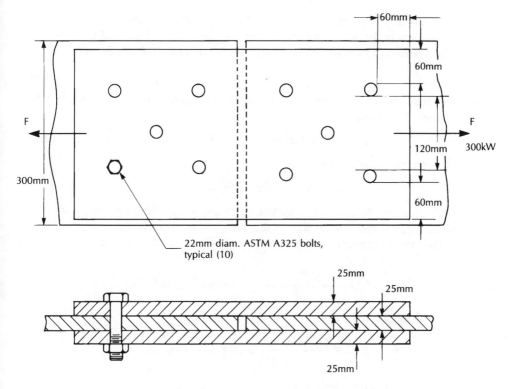

Figure 2.8 Double-lap shear joint.

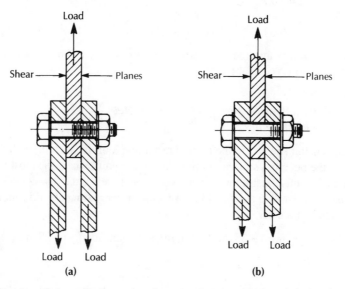

Figure 2.9 Number of shear planes passing through a fastener. (a) Threads in bearing in one of two shear planes; (b) preferred design.

A bolt can have different cross-sectional areas, however, depending where the shear planes involved in the particular joint pass through the bolt (see Figure 2.9). If the shear plane passes through the unthreaded portion of the body of the bolt, the area of a bolt is

$$A_B = \pi d^2 / 4 \qquad (2.6)$$

while if the shear plane passes through the threaded portion of the bolt, for unified threads, the area of a bolt is

$$A_s = \pi / 4(d - 0.9743 / n)^2 \qquad (2.7)$$

while for metric threads, the area per bolt is

$$A_s = \pi / 4(d - 0.9382P)^2 \qquad (2.8)$$

where d is the bolt diameter (in millimeters or inches), n is the number of threads per millimeter or inch, and P is the pitch of the threads in millimeters (or inches).

The total cross-sectional area through the bodies of all five bolts in this example is thus

$$5A_s = \frac{5\pi(22 \text{ mm})^2}{4} = 1900 \text{ mm}^2 (2.94 \text{ in.}^2)$$

and through the threads is

$$5A_s = \frac{5\pi(22 \text{ mm})^2}{4} = 1900 \text{ mm}^2 (2.94 \text{ in.}^2)$$

The shear stress in each bolt is thus

$$\tau = F / A_T \qquad (2.9)$$

$$= 300 \text{ kN} / 1{,}900 \text{ mm}^2 + 1{,}572 \text{ mm}^2 = 86.4 \text{ MPa } (12{,}513 \text{ psi})$$

This value is well within the shear stress of 145 MPa (21,000 psi) allowed for A325 steel bolts and so is acceptable.

2. Determine the tension stress in the joint plates within the joint itself.

The tension stress σ in the joint plates at the joint is taken to be the applied load divided by the net cross section of joint plate at the point where the most fastener holes are located. For this illustration, the cross-sectional area of the row containing the most bolts (i.e., the least A36 steel), adding a factor of approximately 1.5 mm (¹⁄₁₆ in.) extra for each hole's diameter, is

$$A = 25 \text{ mm}(60 \text{ mm}) + 25 \text{ mm}(120 \text{ mm}) + 25 \text{ mm}(60 \text{ mm}) - 25 \text{ mm}(2)(1.5 \text{ mm})$$

$$= 5{,}925 \text{ mm}^2 (9.2 \text{ in.}^2).$$

Thus, the stress in two such cross sections (given that there are two splice plates making up the joint) is

$$\sigma = F / A = 300 \text{ kN} / (5,925 \text{ mm}^2)2 = 25.3 \text{ MPa} (3,661 \text{ psi})$$

This is well within the allowable tension stress value of 152 MPa (22 ksi) for A36 steel joint material, and, therefore, is acceptable.

3. Determine the bearing stresses on the joint plate.

If the fasteners exert too great a load on the plates making up a joint, the joint can be deformed if the bolt holes elongate. From Figure 2.9, the bearing stress on the joint plate σ_B is

$$\sigma_B = F/mdl_G \qquad (2.10)$$

where here $l_G = 75$ mm (2.95 in.), $m = 5$, $d = 22$ mm (nominally ⅞ in.), so

$$\sigma_B = 300 \text{ kN} / (5)(22 \text{ mm})(75 \text{ mm}) = 36.4 \text{ MPa} (5,230 \text{ psi}).$$

This falls well within the allowable bearing stress of 335 MPa (48.6 ksi) for A36 steel, and, therefore, is acceptable.

4. Determine the fastener tearout stress.

Finally, the designer must assure that the fasteners will not tear out of the joint plate as shown in Figure 2.10. This can only occur if the fasteners are located too close to the edge of the plate, such that the shear stress developed by bearing exceeds the ability of the material to sustain that stress over the cross sections from the fastener to the edge. (In fact, tear-out cannot occur if there are multiple rows of fasteners, given the

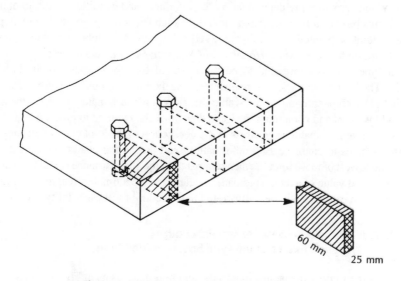

Figure 2.10 Effect of bearing stress on fastener holes.

assumption of equal sharing of load by all fasteners, since a slug of material near the edge could only tear out if the amount of elongation of all fastener holes due to bearing equaled or exceeded the edge distance.)

In this example, there are multiple rows of fasteners, so tear-out could not occur. If there were, however, three bolts in a single row, there would be six shear areas, so the shear stress in the tearout sections would be

$$\tau = F \,/\, mtH \tag{2.11}$$

where t is the thickness of the plate, and H is the distance from the fastener centerline to the edge of the plate. So, here

$$t = 300 \text{ kN} \,/\, 6(22 \text{ mm})(60 \text{ mm}) = 37.8 \text{ MPa } (5,480 \text{ psi})$$

The value of 37.8 MPa (5,480 psi) is well within the allowable shear stress of 100 MPa (14.5 ksi) for A36 steel so would be acceptable for this arrangement of fasteners.

Thus, the joint shown in Figure 2.8 meets all of the design requirements for the given loading.

For a bearing-type joint, the design is limited by whichever of the four stresses calculated by the allowable-stress design procedure exceeds the appropriate allowable stress.

For friction-type joints, the design analysis is slightly different than for bearing-type joints. Recall that in a friction-type joint, the purpose of the fastener is to develop a sufficiently high friction force at the joint interface that the joint elements will not slip relative to one another under the applied load. This slip resistance depends strongly on the surface condition of the materials at the joint interface, and, especially, the coefficient of friction at the interface. Typical slip coefficients (μ_s) can be found in tables in various handbooks, such as Table 2.3, which is from the *Standard Handbook of Machine Design*. Values are strongly dependent on the treatment and condition of the joint surface and must be carefully defined and limited for such friction-type joints. For example, surfaces cannot be painted, unless an approved paint is used; no lubricants can be used.

According to the *Standard Handbook of Machine Design*, "the ultimate strength of a friction-type joint is considered to be the lower of its slip resistance or its bearing strength." The bearing strength is computed by using the same formulas as before (Eqs. 2.4 through 2.11), except entering the allowable shear stress for each material used in the joint plates and fasteners and computing the force that would be required to produce that shear stress in the fasteners, that tension stress in the net section of the joint plates, that stress in the fastener holes to cause bearing deformation, and that in the plate to cause fastener tearout. The lowest of these forces is then compared to the slip resistance calculated earlier (for the assumed value of average preload) to determine the ultimate strength of the joint.

Again, an example will best illustrate the procedure for these joint types.

Example 2.3: Allowable-Stress Design Procedure
Applied to a Friction-Type Shear-Loaded Joint

Using the same dimensions, materials, and bolt patterns as in Figure 2.8, but this time preloading the bolts high enough in tension to create frictional forces between the

Table 2.3 Typical Slip Coefficients in Friction-Type Joints Prepared by Various Means

Typical Slip Surfaces	*Coefficient,* μ_s
Free of paint, applied finish, oil dirt, loose rust or scale, etc.	0.45
Clean mill scale	0.35
Hot dip galvanized	0.16
Hot dip galvanized, wire brushed	0.3–0.4
Grit blasted	0.331–0.527
Sand blasted	0.47
Metallized Zn sprayed onto grit blasted surface	0.422

Reprinted with permission from J. Shigley and C. Mischke, *Standard Handbook of Machine Design*, McGraw-Hill, 1977.

joint members (i.e., at the faying surfaces) that slip is prevented under the design load of 300 kN (67,500 lb.), the procedure first involves computing the slip resistance.

1. Computing slip resistance.

The slip resistance of a joint under shear loading is given by the *Standard Handbook of Machine Design* as:

$$R_s = \mu_s F_{PA} bm \tag{2.12}$$

where μ_s is the slip coefficient of a friction-type joint, F_{PA} is the average preload in a group of bolts in kilonewtons or pounds, b is the number of shear planes (as before), and m is the number of fasteners in the joint.

Assuming that joint surfaces are sand blasted and $\mu_s = 0.47$, the task at hand is to estimate the average preload, F_{PA}, in the bolts required to develop the needed slip resistance. This is an iterative process. To estimate the average preload, F_{PA}, the designer assumes some average pressure is created in each of the bolts involved in the joint and then compares the various loads created in the joint that tend to cause shear in the fasteners, tensile overload in the joint plates, elongation of the fastener holes under bearing, or fastener tearout to the load needed to cause slip in the joint under that assumed value of preload.

In this example, if the average preload is assumed to be 77.5 kN (17,400 lb.) in each of the five bolts in the joint, then the slip resistance is:

$$R_s = \mu_s F_{PA} bm = 0.47(77.5 \text{ kN})(2)(5) = 364 \text{ kN} \ (81,780 \text{ lb.}).$$

2. Comparing slip resistance to strength in bearing.

For the example given earlier, with ASTM 325 steel bolts in ASTM A36 steel joint plates, where the assumed average preload is 77.5 kN (17,400 lb.), the following loads result for the various allowables:

1. Load to exceed bolt shear allowable, 660 kN
2. Load to exceed splice plate tension allowable, 912 kN
3. Load to exceed splice plate bearing allowable, 2,764 kN
4. Load to exceed splice plate shear allowable, 792 kN.

Thus, bolt shear determines the ultimate strength of this friction-type joint.

2.4.5 Axial Shear versus Eccentric Shear

A joint is not always loaded in such a way that all of the fasteners see the same load. In fact, joints are only loaded in such a way when the resultant of all of the externally applied loads passes through the centroid (or center of mass or gravity) of the joint's fastener pattern. When this occurs, the joint is said to be *axially loaded* and the joint is called an *axial shear joint*.[2] When the resultant of the externally applied loads does not pass through the centroid of the fastener pattern, as in Figure 2.11, a net moment acts on the pattern, and each of the fasteners in the pattern resists this moment differently. In such a case, the joint is said to be *eccentrically loaded,* and the joint is called an *eccentric shear joint.*

Shear loaded fasteners, of both the bearing-type and friction-type, are commonly encountered in structural steel construction (e.g., bridges, building, and ships) and in riveted airframe structures, and many times loading is eccentric. The analysis of an eccentrically loaded joint is much more involved than for an axially loaded joint, although it is similar in most respects. Without going into great detail, the procedure involves the following steps.

1. Determine the location of the centroid of the fastener pattern within a reference Cartesian coordinate system, i.e., $(\bar{x}, \bar{y})$, where

$$\bar{x} = A_1 x_1 + A_2 x_2 + . . . + A_n x_n / A_1 + A_2 + . . . + A_n \qquad (2.13)$$

$$\bar{y} = A_1 y_1 + A_2 y_2 + . . . + A_n y_n / A_1 + A_2 + . . . + A_n \qquad (2.14)$$

2. Determine the primary shear forces in the fasteners by dividing the applied load by the number of fasteners

$$F_{\text{fastener}} = F/m = \text{applied load / no. of fasteners} \qquad (2.15)$$

3. Determine the secondary shear forces or reaction moment forces in each fastener

$$M = F_1 r_1 + F_2 r_2 + . . . + F_n r_n \qquad (2.16)$$

and

$$F_1 / r_1 = F_2 / r_2 = . . . = F_n / r_n \qquad (2.17)$$

---

[2] In fact, even an axially loaded joint has some variation in the loading seen by each fastener due to the fastener's fit in the fastener hole, its tightness or clamping load, or other factors.

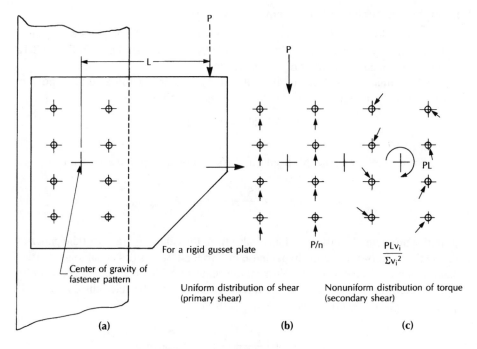

For a rigid gusset plate

Uniform distribution of shear
(primary shear)

Nonuniform distribution of torque
(secondary shear)

Center of gravity of
fastener pattern

P/n $\dfrac{PLv_i}{\Sigma v_i^2}$

(a)   (b)   (c)

Figure 2.11 Eccentrically loaded shear joint.

where r_i is the radial distance of a fastener i from the centroid of the fastener pattern (in millimeters or inches) and combine

$$F_n = Mr_n / r_1^2 + r_2^2 + \ldots + r_n^2 \tag{2.18}$$

where n is a given fastener.

4. Combine the primary and secondary shear forces for each bolt by vectorial means (with secondary shears perpendicular to a line drawn from the centroid to any particular fastener)

5. Compute the shear stress in each fastener by dividing the reaction moment force in each fastener by the area of that fastener,

$$\tau = F_n / A_n \tag{2.19}$$

and complete these values to the maximum allowable shear stress for the fastener material.

Viewed another way, an eccentric load on a fastener group can be analyzed as a combination of a shear force (primary shear force, above) plus a torsional moment (secondary shear force, above). This approach is shown in Figure 2.11b and c. Example 2.4 clarifies this latter approach.

Example 2.4: Analysis of an Eccentric Shear Joint

Given a joint consisting of eight ⅞ in. diameter A490 bolts in A36 steel arranged and eccentrically loaded with a 40 kip (i.e., 40,000 lb.) force as shown in Figure 2.12, determine the load on the most highly loaded bolt.

The primary shear loads on the bolts of this example act in a vertical direction down, as shown in Figure 2.13a. The shear stress per bolt is

$$40 \text{ kips} / 8 \text{ bolts} = 5 \text{ kips per bolt},$$

from Eq. 2.15.

The torque or moment M, from Eq. 2.16, is shown in Figure 2.13 and is

$$40 \text{ kips} \left(8 \text{ in.} + \frac{5 \frac{1}{2} \text{ in.}}{2} \right) = 430 \text{ kip-in.}$$

If, in this problem, the distance of any bolt i from the center of gravity of the pattern is r_i, and this is known to be the hypotenuse of a triangle with sides equal to the vertical and horizontal distances of the bolt from the centroid, then values of r_i^2 needed in Eq. 2.18 to determine the secondary shear force on each bolt can be determined for all bolts as follows. For the four bolts close to the centroid,

$$r_1^2 = 2^2 + \left(\frac{5 \frac{1}{2} \text{ in.}}{2} \right)^2 = 11.56 \text{ in.}^2$$

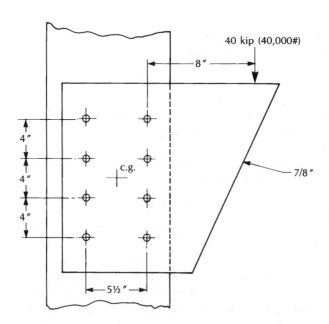

Figure 2.12 Eccentrically loaded joint.

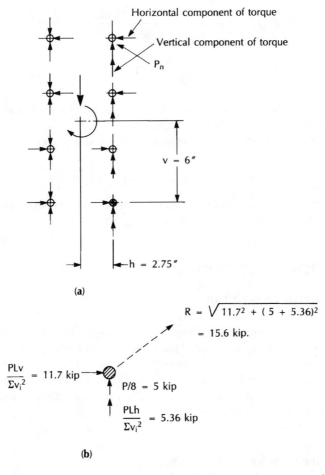

Figure 2.13 (a) Primary and (b) secondary shear forces and vector summing.

so the sum of r_1^2's for these four bolts is 46.25 in.², while for the four bolts far from the centroid,

$$r_2^2 = 6^2 + \left(\frac{5\,^1\!/_2 \text{ in.}}{2}\right)^2 = 43.56 \text{ in.}^2$$

so the sum of r_2^2's for these four bolts is 174.25 in.²

Thus, the vertical and horizontal components of the secondary shear force for the most highly loaded bolt are from Eq. 2.18:

$$F_{\text{vert.}} = \frac{Mh}{r_i^2} = \frac{430\,\text{k}-\text{in.}(2.75\,\text{in.})}{220.5\,\text{in.}^2} = 5.36\,\text{kip}$$

$$F_{\text{hor.}} = \frac{Mv}{r_i^2} = \frac{430(6)}{220.5} = 11.7\,\text{kip}$$

where h and v are the horizontal and vertical distances of the most highly loaded fastener from the centroid, as shown in Figure 2.13.

The vector sum of all of the primary and secondary shear components, from Figure 2.13 is:

$$R = \left[11.72 + (5 + 5.36)^2\right]^{1/2} = 15.6\,\text{kips}$$

This value compares favorably to the shear allowable for A490, which is 24.1 kips, and, therefore, is acceptable.

2.5 TENSION-LOADED JOINTS

2.5.1 General Description

Many mechanically fastened joints must withstand applied forces that load the fasteners in tension and not in shear. Bolted joints are the best example. In bolted joints, the applied load is usually more or less parallel to the axes of the bolts, and the clamping force developed by the bolts is critical to the joint's performance. The development of a clamping force is called *preloading*. This section discusses these important joints and considers, the critical role of preloading.

2.5.2 Procedure for Determining Appropriate (Target) Preload

The key to strong bolted connections with long service life is a thorough analysis of the joint forces. The analysis of tension-loaded joints usually focuses on an analysis of the tension or preloading in the fasteners. It first considers the initial or preload in the fasteners when they are initially installed and tightened and, then, considers the working loads that exist in the fasteners and in the joint members when external loads are applied in service. These working loads consist of the preload plus or minus some portion of the external load.

In theory, if the tensile preload induced in a bolt or bolt system equals or exceeds the applied tensile loading on the joint, the fastener will not sense any increased tensile load and the joint will remain tight or rigid. In practice, loads tend to fluctuate and tightened fasteners may actually sense tensile loadings in excess of the induced preload. An increase in the tensile load on the fastener is accompanied by a decrease in the compressive load in the joint and a resultant loss of the clamping load effect. Where joints are designed to be rigid, excessive loss of clamp load can pose a serious structural problem (see Section 2.5.6 for sources of preload loss).

The initial clamping load and the working load carried by the bolt largely establish the performance of the joint. Under ideal conditions, the initial clamping force, usually called the *preload,* must be high enough to compress the joined parts, thereby improving the resistance to external tensile loads. This clamping force also creates friction between joined parts, providing resistance to shear loads (as in friction-type shear-loaded joints). The amount of clamping force, however, must be limited so that the combination of preload and bolt working load does not exceed the yield-point load limit of the fastener.

Normally, bolted joints are designed so that the clamped parts are much stiffer than is the bolt itself. This is usually accomplished by keeping the bolt diameters small relative to their length (as a rule, six times the diameter), allowing them to stretch. If the joint is highly preloaded, the bolt working load is, then, relatively small. Such a low working load is desirable in dynamic applications to reduce the possibility of fatigue and to prevent fastener loosening.

For strong, reliable joints, the initial clamping load and the bolt working load must meet four basic requirements:

1. The maximum clamping load must be greater than the required nominal preload to allow for embedding the fastener and for offsetting tensioning scatter.
2. The magnitude of the maximum clamping load must be no more than 70% of the load at the 0.2% offset yield point of the fastener material.
3. The bolt working load must be no greater than 10% of the 0.2% offset load.
4. The bolt working load must also not exceed the endurance limit of the bolt, if fatigue is a consideration.

The analysis required to determine the appropriate level of preload is complicated but, without going into great detail, involves

1. Estimating the external loads seen by each bolted joint, including: static, dynamic, and impact loads from weights and from inertial sources.
2. Computing the stiffness or spring rate of the fasteners from an empirical relationship

$$k_b = A_s A_B E / L_s A_B + L_B A_s \qquad (2.20)$$

where the various parameters relate to the areas of the threaded and unthreaded portions of the bolt, the threaded and overall lengths of the bolt, and the bolt material's modulus of elasticity.
3. Determining the joint stiffness experimentally (versus analytically) by applying an external tension load to the fastener and measuring the tension load in the bolt with a strain gauge or ultrasonically. Then, when the stiffness of the bolt (from above) is known, a technique known as joint-diagram, which relates to the force-elongation diagrams for the bolt and the joint materials, is used to estimate the stiffness of the joint.

4. Selecting a target preload. In general, this is the greatest load, known as the *allowable upper limit,* that the bolt can stand without yielding, accounting for the torque (e.g., from nuts) and other loads (e.g., shear) but considering the *acceptable lower limit*[3] to prevent each possible failure mode (e.g., leaking, vibrating loose, shortened fatigue life).
5. Determining whether or not the tension actually developed in any bolt will exceed the maximum allowable tension for the particular bolt material. Here, assembly tool (e.g., torque wrench) errors and operator problems (e.g., skill level, worker fatigue, bolt accessibility) must be taken into account by adding to the preliminary target preload.
6. Considering actual lower limits on clamping force due to relaxation effects (e.g., plastic flow in the bolt threads of new parts), often called *embedment relaxation,* and/or elastic interactions between bolts (e.g., effect of tightening sequence). These effects must be subtracted from the preliminary target preload.

From the foregoing, it should be obvious that the process of determining a target preload is involved and imprecise, relying on a combination of analytical and empirical means. A more complete description of the analysis procedure, as well as numerous examples, are given in references by Bickford [1983], Shigley and Mischke [1986], and Shigley [1977].

2.5.3 Bolt Torque

Although torque is the most common way to tighten a fastener, it is not a very good way to control the preload developed, typically causing ±30% or more scatter (see Table 2.4). The torque-preload interaction is affected by many things, including: (1) the finish of the nuts, bolts, or joint members; (2) the fit between male and female threads; (3) the size (i.e., tightness) of holes and their perpendicularity to the joint; (4) the hardness of parts; (5) the speed of tightening; and (6) the age, temperature, quantity, condition, and type of lubricant used, if any.

In general, such things cannot be controlled perfectly in most applications, and, further, the economics of the situation just will not permit enough control to guarantee the desired results. The objective of many modern tightening devices is to develop a technique that either directly measures the preload or yields a better correlation between the applied torque and the preload.

When torque is used to tighten a bolt, the correct torque level is selected from the so called *short-form torque equation,* given in the *Standard Handbook of Machine Design* as

$$T = kdF_{PT} \tag{2.21}$$

---

[3]The acceptable lower limit can be particularly difficult to estimate or evaluate. It can be zero but is usually 60% of the acceptable upper load limit.

Table 2.4   Typical Scatter for Various Fastener Tightening Methods

Control Parameter and Type of Tool	*Reported Scatter in Preload, %*
Torque Control with	
1. Hand wrench	±21 to 81
2. Hand wrench + multiplier	±70 to 150
3. Dial or click wrench	±60 to 80
4. Wrench with electronic readout	±40 to 60
5. Power wrench (air)	±23 to 28
6. Air tool with one-shot clutch	±30
7. Air tool with feedback	±20
8. Stall torque air tool	±35
Torque-turn control with	
9. Yield control system	±8
10. Turn-of-nut procedure	±15
11. Logarithmic rate method	±2.2 to 2.6
Miscellaneous methods	
12. Strain-gauged bolts	±1
13. Strain-gauged load washers	±15
14. Swaged lockbolts	±5
15. Bolt headers	—
16. Air-powered impact wrench	−300 to +150
17. Hydraulic tensioner	±20
18. Operator feel	±35
19. Fastener elongation	±3 to 5
20. Ultrasonic control of preload	±1 to 10

Reprinted with permission from J. Shigley and C. Mischke, *Standard Handbook of Machine Design*, McGraw-Hill, 1977.

where k is an experimentally determined constant, known as a "bugger factor," which defines the relationship between the applied torque and the achieved preload in a given situation, d is the nominal diameter of the bolt (in millimeters or inches), and F_{PT} is the target preload (in kilonewtons or pounds).

2.5.4 Achieving a Desired Preload in Bolts

In most cases, the target preload or tension selected for a bolt is achieved by using a wrench of some sort to apply torque to the nut or the head of the bolt, on the premise that this applied torque correlates to induced tension. The problem is there is considerable scatter in the value of preload obtained for a given torque value. Table 2.4 shows how severe this scatter can be for various tightening methods. Because of the importance of preload in the performance of a bolted tension-loaded joint, methods are sought for tightening that reduce scatter in the preload values.

Preferred methods for achieving precise values of preload do so by inducing the preload by literally and directly stretching the bolt or by directly measuring the stretch induced by some method of torsional loading or tightening. Some other methods still rely on developing preload indirectly by torquing, without making any direct measurements. Four methods are commonly used:

1. turn-of-the-nut control
2. microprocessor-controlled torque-turn tools
3. hydraulic tensioners and bolt headers
4. ultrasonic control devices

The turn-of-the-nut control technique, employed in structural steel, tightens the nut to produce a preload value that is 60 to 80% of the yield stress, using an air-powered impact wrench, then marking the nut, and, finally, turning another half turn to cause yielding. The technique is rather imprecise since it is still subject to the same sources of scatter in preload as other torque tightening techniques, relying on a predetermined correlation between bolt tension and torque. Scatter is reduced, however, by going to very high levels of torque.

Microprocessor-controlled torque-turn tools again rely on predetermined correlations between induced tension and applied torque but reduce scatter by more precisely controlling the applied torque. These devices measure both the applied torque and the turn of the nut to monitor and control fastener preload.

Hydraulic tensioners or bolt headers and ultrasonic control devices directly control or measure the stretch induced in the bolt by tension, respectively. Hydraulic tensioners or bolt headers stretch a large bolt from the threads to a desired preload based on extension. Then, having achieved the desired tension, they run down (i.e., snug) the nut. Ultrasonic devices control tension in the bolt by directly measuring the effect of tightening on the time-of-flight of an ultrasonic signal caused by the increased length of the tensioned bolt shank and/or velocity changes caused by the presence of a residual stress.

In addition to these methods, some other interesting techniques are used for installing high-strength steel bolts in steel construction. These include so-called calibrated washers and bolts. Calibrated washers have diametrically opposed bumps on each face that deform under the appropriate bolt preload, while calibrated bolts have a small diameter protrusion at their shank end that breaks off when the proper preload value is reached.

2.5.5 Measuring Residual Preload

Estimating the initial preload by measuring the amount of torque applied to a fastener has never been particularly easy or reliable because of the variables involved, as described previously. Measuring residual preload some time after tightening, however, is even more difficult, yet is important since loss of preload can occur from any of several causes to be described in the next Section 2.5.6.

For all intents and purposes, it is impossible to make such measurements by measuring the torque required to restart fastener motion (so-called breakaway torqu*e*) on a previously tightened fastener because of the large uncertainties of the static friction at these load levels and because it is essentially impossible to tell exactly when breakaway or restart has truly occurred. Visible motion and noise have been shown (by strain gauge studies) to be unreliable indicators. Thus, better methods are needed. Ultrasonics seems to be one such method.

There are two basic ways to measure preload ultrasonically. The first, best, and easiest is to use ultrasonics to control the initial tightening, keep a permanent record of the initial acoustic length of each fastener after tightening, and remeasure each fastener's length after some time in service. By comparing these initial and current lengths, one can make an estimate of the stress levels in the fastener regardless of how much time has passed since initial tightening. The second method is for assessing the preload in fasteners that were not initially tightened ultrasonically. Here, ultrasonics is used to measure the length before and after loosening. The decrease in length is a good measure of the level of preload that was in the fastener before loosening. By doing one fastener at a time and restoring the preload (or retightening) before moving on, one can make such measurements without disturbing the overall joint. The obvious shortcoming of this approach is not knowing what the initial preload was and, thus, how much loss of preload may have occurred.

2.5.6 Loss of Preload in Service

The initial preload in a tightened fastener can be, and often is, lost in many ways. At least five mechanisms for such loss have been identified.

The first mechanism is *embedment relaxation.* Upon initial tightening of a new threaded fastener, only the high points of the mating male and female threads and of the joint surfaces contact. These high points are, therefore, overloaded well past their yield points, and so plastic deformation by creep occurs until a large enough percentage of the available contact surfaces have been engaged to stabilize the process. Such plastic flow occurs in thread roots, in the innermost bolt threads to engage the nut threads, under the face of the bolt head or nut (if the bolt is not exactly perpendicular to the joint), and at the corners of the bolt heads and nuts. As a result, losses of 5 to 10% of the initial preload are not uncommon, and losses of up to 25% are possible. Embedment cannot be avoided in most new commercial parts but can be at least partially overcome by retightening.

The second source of loss is *gasket creep.* When gaskets are used to seal a joint against fluid leakage or infiltration, they inevitably relax over time under loading as a result of viscoelastic flow of the plastic or semiplastic materials from which they are generally made. In fact, such flow is necessary for the gasket to work, that is, to seat itself. Nothing can be done to avoid gasket creep or to compensate for it initially, but it can be compensated for later by retightening.

The third source of preload loss is *elastic interactions.* In the process of tightening a group of fasteners, there are inevitable elastic interactions among the various fasteners because of joint compression. After the first bolts are tightened, the joint compresses, and the fastener feels or measures tight. When an adjacent fastener is installed and tightened, however, the joint is often further compressed, and the preload in the first fastener is partially or completely lost. This process goes on throughout the fastener installation and tightening process. Losses of 40 to 50% are common and up to 100% are known. Tightening fasteners in a sequence that attempts to balance joint compression, retightening during a second (or subsequent) pass, or even overtightening the first fasteners are all methods that can be used to compensate for elastic relaxation.

Embedment relaxation, gasket creep, and elastic relaxation are all short-term sources of loss of preload. Two other sources take longer to occur. The first is *vibration loosening.* Under vibrational loads, threaded fasteners often loosen, unless special precautions are taken. Under severe vibration, the fastener at first loses preload gradually over time. Once the preload in the fastener has fallen so far that it is no longer able to prevent transverse slip between male and female threads or nut and joint surfaces, then the loosening action accelerates rapidly and can result in the loss of the nut or the fastener or both. For vibration that is parallel to the axis of the fastener, 20 to 40% loss of initial preload is typical, but for transverse vibration, complete loss can readily occur. Unlike the short-term loss mechanisms, vibrational loosening can be prevented by properly designing the joint, choosing a fastener that can be preloaded high enough to prevent transverse slip, or choosing a fastener design that inherently resists, or a device that prevents, vibration. Special thread forms, adhesives on the threads, or spring nuts, lock washers, and so on, can all help.

The second source of long-term loss of preload is *stress relaxation.* When fasteners are used in applications where the temperature is extreme or where there is nuclear radiation, the atoms in the fastener can realign by a process related to creep, and stress relaxation occurs. The only way to avoid this long-term effect is to use fasteners made from materials that have a resistance to it.

2.6 FATIGUE LOADING OF FASTENED JOINTS

2.6.1 Sources and Signs of Fatigue

Dynamic loading that is repetitive or cyclic in nature leads to fatigue in materials and structures. Fatigue can arise from highly repetitive loading sources such as vibration, aeroelastic flutter, resonance, or repetitive motions in the assembly (e.g., the loads associated with the various stages of combustion in a four-cycle internal combustion engine). It can also arise from more erratic but repetitive loading such as various *g*-loadings during high-speed turns during maneuvers, or from takeoffs and landings in aircraft.

Regardless of the source, fatigue leads to sudden and unexpected joint failure, usually by breakage, and is the most common cause of failure. The performance of a structure subjected to fatigue (often measured by its life, or number of loading cycles until failure) is highly sensitive to part design and inherent stress concentrations and is dependent on the level of stress, or *stress intensity*. Fatigue life increases as the stress level or stress intensity is lowered. The nature of fatigue is such that under dynamic loading, failure can occur at strength levels well below the accepted static strength level.

The same basic principles of design employed for static tension-loaded joints apply to tension-fatigue-loaded joints. Adequate preloading is the prime factor necessary to meet or exceed anticipated dynamic or cyclic loading on the joint. A properly tightened and preloaded bolt will experience only minimal external tensile forces imposed by repetitive tensile loadings. The stresses expected during service must be carefully defined in proper designs and may involve detailed experimental stress analysis and simulated service operation, which can be difficult and expensive. In any case, as a result of the typical stress-cycle (*S-N*) curve relationship, the allowable design stresses that have to be used for joints of this type are substantially lower than those for comparable static-loaded joints, especially where long service life is required. This is illustrated in Figure 2.14.

While the actual failure caused by fatigue is difficult to predict in time, there are several clues that a failure may have been caused by fatigue, including (1) cyclic (especially tensile) loads were present; (2) there was no warning of the onset of failure (by necking or obvious wear); (3) the rough appearance of the fracture surface where overload occurred contains a smooth (almost polished) appearance where fatigue was causing slow crack propagation, possibly with telltale striations or crack arrest marks; and

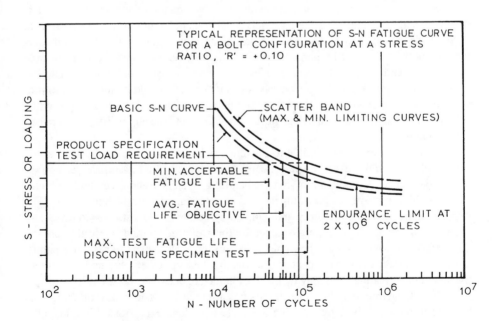

Figure 2.14 Typical representation of the *S-N* curve for a fatigue-rated fastener. (Reprinted with permission from Robert O. Parmley, *Standard Handbook of Fastening and Joining*, 2nd ed., McGraw-Hill, 1989, p. 1-14.)

(4) failure appears to have initiated at points of high stress concentration such as sharp radii on joint elements or fasteners (e.g., under heads), machining marks or gouges on the fastener or joint elements, at thread run-out areas on bolts or where bolt threads first engage a nut, in areas of wear or fretting in friction-type joints, or where a joint splice plate ends.

2.6.2 Reducing the Tendency for Fatigue

While in many materials it is impossible to completely avoid or prevent fatigue, it is possible to reduce the likelihood of fatigue within the needed service life of a structure or assembly by proper design, proper analysis, and proper manufacturing. This section discusses some of the things one can do to reduce the tendency for fatigue.

First, select materials carefully. Materials with higher tensile strength have longer fatigue lives than materials with lower tensile strength, because it is easier to operate these materials at lower proportions of their ultimate fatigue strength. In addition, some materials (e.g., face-centered cubic metals and polymers below their glass transition temperatures) have an inherently lower notch sensitivity (as measured by crack growth rate, da/dt, that is, the incremental growth of a crack, a, per increment of time, t). During processing, materials should be treated to avoid potentially damaging reactions such as decarburization and untempered martensite formation (e.g., during welding of steels).

Second, treat the parts or joint with care. No matter how carefully a part is designed, and no matter what the material, abusive manufacture or operation can lead to problems in fatigue. In fastened assemblies or structures, nut faces and the undersurfaces of bolt heads must be perpendicular to the fastener axis, and the fastener hole must be perpendicular to the joint. A ±2% error in angularity can reduce fatigue life 25%. In bolted joints, the threads on the bolt (and/or nut) should be lubricated to prevent corrosion as a source of cracks. For friction-type joints, faying surfaces should be grit blasted to increase the slip resistance, since increasing the slip resistance increases the fatigue life.

Third, do everything possible to prevent crack initiation. Fatigue initiates at the surface of a material or part, so special care should be given to the surface during fabrication and in service. Machining marks, handling scratches, and other surface blemishes should be removed by polishing in areas of potential fatigue and high loading. The surface can be protected further by inducing compressive residual stresses to offset any cyclic tensile loads. Shot peening is generally the preferred method. To further protect the surface, corrosion should be avoided by any possible means (e.g., using paints, lubricants, surface conversion coatings, or platings) except by employing hard protective coatings or platings that may themselves induce residual tensile stresses.

Fourth, attempt to reduce load excursions as much as possible. Several things can be done to reduce fatigue by reducing the severity of load excursions. First, keep the ratio of the minimum to maximum load as near to unity as possible. Then, keep preload high compared to the worst case external load and keep the ratio between the stiffness of the joint and the stiffness of the bolt high, so that the joint absorbs the larger portion of the applied loads. Finally, make every effort to maintain the preload against losses (see Section 2.5.6).

Finally, among the ways of improving the fatigue life of mechanically fastened joints are to reduce the stress and, especially, the stress concentrations in the fasteners themselves. Methods include ensuring that there are at least three threads above and three below the nut faces; avoiding having the thread run-out coincide with a shear plane in the joint; using large head-to-body fillets; using a large thread root radius; using rolled rather than cut threads; using collars between the head and the joint plates and between the nut and the joint plates to increase the length-to-diameter ratio of the bolts; turning down the diameter of the bolt body just below the head to reduce the bolt's stiffness; using long nuts; using spherical washers to help a bolt adjust to bending loads; and using so-called tension nuts to reduce the level of stress in the threads.

In summary, consider fatigue carefully in the design of all mechanically fastened joints.

2.7 OTHER FACTORS AFFECTING FASTENERS AND FASTENED JOINTS

Many factors besides fatigue need to be considered when employing fasteners for joining. These include bending loads, vibration, and corrosion, to name a few of the most important.

2.7.1 Bending Loading

Structural joints are not always loaded in pure tension or pure shear. Often, structures are subjected to bending forces that create combined tensile and shear loading. For applications where bending-load conditions are expected, the design limits for the fasteners used in joints must be established from interaction curves, which are developed from actual tests where a number of bolts are evaluated at different combinations of applied tensile and shear loadings. From such a curve (as shown in Figure 2.15), it is possible to calculate the maximum allowable design stresses for tension and for shear for any degree of bending.

Even when carefully developed interaction curves are used in the design analysis, additional caution must be exercised when bending loading is expected. Bending loads increase stress concentration effects at notches, such as thread roots, run-outs, and head-to-shank radii.

2.7.2 Vibration

While vibration in a structure is a source of fatigue, the loading produced by vibration is typically low compared to the strength of the fastener but may be associated with various ranges of operating frequencies. Unless the frequency, loading, and amplitude of

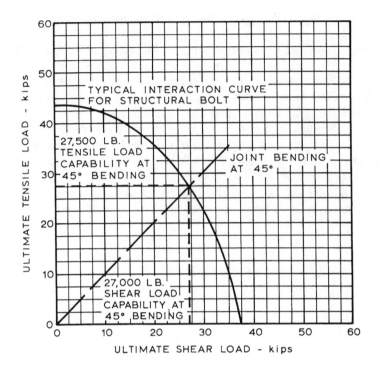

Figure 2.15 Typical interaction curve for fasteners subjected to bending loading, combining tension, and shear loading. (Reprinted with permission from Robert O. Parmley, *Standard Handbook of Fastening and Joining*, 2nd ed., McGraw-Hill, 1989, p. 1-14.)

vibrations cause resonance in the structure, catastrophic failure by fatigue due to vibration rarely occurs. Nevertheless, vibration can cause serious problems.

Extensive or continued vibrations affect threaded fasteners in a joint by causing loosening (see Section 2.5.6). The mechanism is complex, and the results can be serious. Nuts can literally walk off a bolt thread, or preloading can be totally lost.

2.7.3 Corrosion

Corrosion can be the most serious problem and the overriding concern with some design applications. Corrosion can lead to the complete failure of the fastener or the joint and can occur by any one of several mechanisms, including galvanic corrosion, stress corrosion, fretting corrosion, pitting, and oxidation (or rust).

The most common form of corrosion is oxidation, typified by the rusting of many steel structures and fasteners. The next most common form is galvanic corrosion. Galvanic corrosion arises from the combination of two dissimilar metals with an electrolyte, which forms an electrochemical cell. This leads to the sacrificial loss of the more electropositive or anodic metal in the electrochemical couple. The electrolyte can be something as aggressive as an acidic chemical, less aggressive ocean salt spray or acid rain, or seemingly benign rain, dew, snow, or high humidity. Galvanic corrosion is a distinct possibility with any dissimilar metal structure and with metals near some electronegative nonmetals like graphite.

Stress corrosion is a particular condition where cracks are initiated and propagated under the combined effects of imposed stress and a corrosive environment. Any structure or component, fabricated from susceptible metals or alloys and subjected to high stress concentrations (e.g., threaded fasteners) is prone to this type of attack. Corrosion generally starts at a point of high stress concentration, a crack forms, and propagation occurs under continued exposure to the corrosive agent and the stress state. Eventually, damage can be severe, and catastrophic failure can occur.

Pitting or concentration-cell corrosion occurs in some metals when the metal is exposed to certain corrosive agents that act as electrolytes. Corrosion occurs electrochemically in the single metal species because of the effective localized battery established between the metal and an oxygen gradient established in the metal in the presence of the electrolyte.

Fretting corrosion is more closely related to wear and occurs as the result of surface damage that results when two surfaces in contact experience slight periodic relative motion. Fretting corrosion differs from wear, however, in that the relative velocity of the two surfaces is usually much lower than that encountered in wear and since the two surfaces are never brought out of contact there is no chance for the corrosion products to be removed. Initially, surface pitting and deterioration occur, then fatigue cracks originate and grow.

To prevent galvanic corrosion, fasteners should be as compatible as possible with the materials in the joints. To prevent most other forms of corrosion, protective coatings or finishes should be employed, such as primers, paints, inhibitors, platings, oils, or greases. Fretting corrosion is especially difficult to prevent, unless the relative motion can be prevented or the coefficient of friction is reduced through the use of lubricants.

SUMMARY

One especially attractive method for joining materials or parts into structures or assemblies is *mechanical fastening*. In mechanical fastening, the forces that allow parts to be joined together are purely physical, relying on the simple phenomena of mechanical interlocking at the macroscopic level and/or friction, which is mechanical interlocking at the microscopic level. No atomic or molecular bond formation is needed or involved in mechanical fastening. The method uniquely allows relative motion between fastened parts and facilitates disassembly for maintenance, repair, or portability. Further, the process causes no chemical or microstructural changes in the materials being fastened, although it does give rise to *stress concentrations* at the points of fastening. When fasteners are employed, there are two fundamental types of joints, depending on the direction of primary loading relative to the fastener axis: *shear-loaded joints* (when loading is at right angles) and *tension-loaded joints* (when loading is parallel). Shear-loaded joints are further classified as *friction* or *bearing types*, depending on whether the shear loads are resisted by the interfacial friction generated at joint faying surfaces or by local pinning of the joint by the fastener. The *allowable-stress design procedure* for mechanically fastened shear-type joints assures that allowable stresses are not exceeded in the fastener (by shear), in the joint plates (by tension overload), in the fastener holes (by bearing deformation), or because of fastener tear-out from the joint plates. For friction-type joints, the *slip resistance* must be compared to the loads that will just cause these various modes of failure. The analysis of tension-loaded joints focuses on determining the amount of *preloading* that must be given to a fastener to allow the fastener to function properly. Methods of achieving preload, relating preload to the torque used for tightening, measuring preload, and the various mechanisms by which preload can be lost are all important in a tension-loaded joint's performance. When dynamic, cyclic loading is involved, special precautions must be taken to reduce the tendency for the fasteners or the joint to fail in fatigue. Other factors can affect the performance of fasteners and fastened joints, such as bending-loading, vibration, and corrosion.

PRACTICE QUESTIONS AND PROBLEMS

1. Define mechanical fastening in your own words. Are actual fasteners always required? If so, why? If not, why not?
2. What two related features of mechanical fastening are unique among all joining methods? What other major advantage does mechanical fastening offer relative to its effect on the materials it is used to join? How does this compare to adhesive bonding? To welding? To brazing or soldering?
3. Describe what you think would be involved in trying to automate mechanical fastening for certain types of fasteners. *Hint:* How about automating riveting versus bolting when nuts are used?
4. What is the single biggest shortcoming of mechanical fastening as a joining method? How does this aspect of mechanical fastening affect the methods used with various materials such as metals, ceramics, glasses and polymers?
5. Distinguish between shear- and tension-loaded fastened joints. Distinguish between the two subtypes of shear-loaded joints.
6. Calculate the net area for the $1/2$ in. thick joint shown below.
    Hint: Do not forget to use the extra $1/16$ in. factor for each fastener hole.

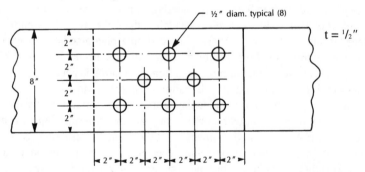

7. If the joint in Problem 6 were composed of ASTM A36 steel, for which the yield strength is 36 ksi and the ultimate tensile strength is 58 ksi, what would be the

maximum tensile load that can imposed under AISC specifications. *Hint:* See the reference to AISC specification in Section 2.4.4.

8. Apply the allowable-stress design procedure to the single-lap, bearing-type, shear-loaded joint below. Assume the bolts in the joint are made from ASTM 325 steel and have 12 threads per inch, and that the joint plate material is ASTM A36 steel.

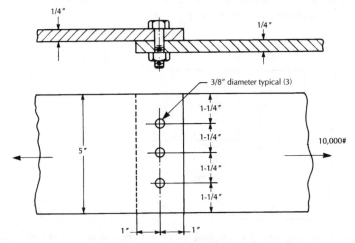

9. Suppose the bolts in Problem 8 were threaded for their entire length (from just under the head to the end of the shank), how would the values for each stress change? Calculate the new values.

10. For the joint shown in Problem 8, what would be the maximum load that could be applied to the joint to just cause failure? *Hint:* Check each possible failure mode for the allowables for the bolt and joint materials.

11. Apply the allowable-stress design procedure to the butt or double-lap, bearing-type, shear-loaded joint below, using the same bolt and joint plate materials as in Problem 8. Assume the bolts have a pitch of 2 mm per thread.

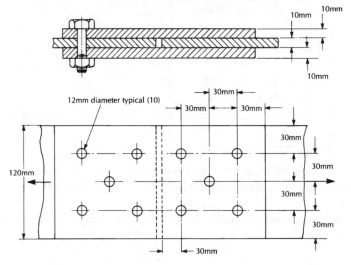

12.   For the joint shown in Problem 11, what would be the maximum load that could be applied to the joint to just cause failure?

13.   Using the allowable-stress design procedure for the joint in Problem 11, what value of preload would be needed to assure there would be no slip at the joint interface, assuming the coefficient of friction at the interface (μ_s) was 0.35?

14.   For the joint shown below, what is the force on each rivet in the pattern, in terms of the general load and dimensions?

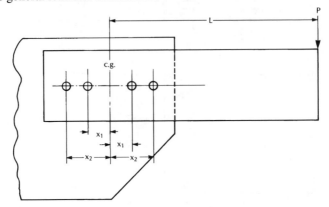

15.   Suppose the load were 200 kN, and the dimensions were $x_1 = 25$ mm, $x_2 = 75$ mm, and $L = 200$ mm, what would be the load on the most highly loaded fastener if the fasteners were 10 mm in diameter? What if the fasteners were 15 mm in diameter?

16.   For a joint like the one shown here, what is the load in the most highly loaded fastener?

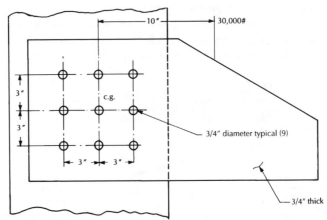

17.   For the joint in Problem 16, if the fasteners were ASTM A490 steel bolts, what would be the maximum allowable load? *Hint:* Use allowables for A490 shear and bearing.

18.   Describe the general procedure for determining a target preload for a tension-loaded joint.

19. Why is torque not a good measure of preload? How can preload be measured accurately during fastener installation? How can preload in bolts in bolt groups in a structure be checked in service, without disrupting the structure?
20. Describe the four sources of loss of bolt preload for a non-gasketed joint? Differentiate between short-term and long-term sources.
21. Why is fatigue such an insidious potential mode of failure for mechanically fastened structures?
22. What are several things to watch out for in mechanically fastened joint design if the structure is fatigue critical? What things in the joint elements and their arrangement? What things in the fasteners themselves and in how they are installed?
23. How does vibration cause loosening of a bolted joint? What types of things can be done to prevent loosening?
24. Describe the difference between various mechanisms of corrosion that can occur in mechanically fastened metal joints.

REFERENCES

1. Aaronson, Stephen F., "Analyzing Critical Joints," *Machine Design,* Jan. 21, 1982, pp. 95–101.
2. Bickford, John H., *An Introduction to the Design and Behavior of Bolted Joints,* 2nd ed., New York: Marcel Dekker, 1990.
3. Bickford, John H., "That Initial Preload—What Happens to It?" *Mechanical Engineering,* Oct. 1983, pp. 57–61.
4. Donald, Eric P., "A Practical Guide to Bolt Analysis," *Machine Design,* April 1981, pp. 225–231.
5. "Fastening and Joining," *Machine Design,* Reference Issue, Vol. 14, 1967.
6. Faupel, Joseph H., and Fisher, Franklin E., Appendix D: Joints and Connections, *Engineering Design,* 2nd ed., New York: John Wiley, 1981.
7. Fisher, John W., and Struik, John H.A., *Guide to Design Criteria for Bolted and Riveted Joints,* New York: John Wiley, 1974.
8. Griffith, Howard T., "Torque Tensioning—Part I," *Fastener Technology,* Vol 10, No. 5, Oct. 1987, pp. 62–63.
9. Shigley, Joseph E., and Mischke, Charles R., *Standard Handbook of Machine Design,* New York: McGraw-Hill, 1986, pp. 23.1–23.39.
10. Shigley, Joseph E., *Mechanical Engineering Design,* 3rd ed., New York: McGraw-Hill, 1977, pp. 227–273.
11. Trilling, Jack, "Torque Tensioning—Part III," *Fastener Technology,* Vol. 11, No. 1, Feb. 1988, pp. 48–49.
12. Trucks, H.E., "Torque Tensioning—Part II," *Fastener Technology,* Vol. 10, No. 6, Dec. 1987, pp. 38–39.

Mechanical Fasteners and Joining Methods

3.1 INTRODUCTION

Mechanical fastening of elements or parts into structures or assemblies requires some means for developing interference forces or interlocking between those elements or parts at their mating or faying surfaces, on either a macroscopic level or a microscopic level or both. No atomic- or molecular-level bonding occurs or needs to occur to effect joining. The interference forces may need to resist joint separation in shear or in tension or both, depending on the principal direction of the applied load or loads relative to the joint. Further, the joint may resist separation through some pinning action (i.e., bearing) or friction under shear loading or through a clamping action under tension loading. Whatever type of interference force is needed (i.e., to resist shear or tension), some physical, macroscopic structure or structural feature is necessary to provide that force. When the interference force is provided by a device or structural entity that is separate from the structural elements being joined, a *mechanical fastener* is said to be involved. Alternatively, the interference force may be provided purely by the interaction between some integral design features of the structural elements themselves or be developed in those elements by some physical means (e.g., plastic deformation). In this case, the fastening method does not rely on a fastener but, instead, relies on mechanical interlocking.

This chapter looks at the various types of mechanical fasteners and joining methods. Fasteners are divided into two broad categories: those that use threads to develop the needed interference forces, usually through clamping (threaded fasteners) and those that develop the needed interference forces principally by a pinning action without the need for threads (unthreaded fasteners). Methods of joining without fasteners that rely on interlocking are divided into those that rely on integral features in the design of the joint elements and those that rely on some other means such as elastic or plastic deformation. In addition, some novel means of mechanically fastening are difficult to classify one way or the other.

3.2 INTERLOCKS VERSUS FASTENERS

3.2.1 Mechanical Interlocking

In mechanical fastening, the forces needed to keep the elements and materials making up a joint in proper juxtaposition, and to enable the transfer of loads across the joint, arise from purely physical sources, as opposed to chemical sources dependent on atomic or molecular bonding. In mechanical fastening, the forces are more macroscopic and arise from the physical interference of the elements to prevent any undesired motions. These macroscopic forces may arise from obvious sources, such as the physical interlocking, meshing, or engaging of features of the design of the elements, or from less obvious, but still macroscopic (relative to atomic dimensions), sources such as the microscopic asperities of the mating surfaces that give rise to friction (see Figure 3.1). In both cases, the real source of the forces that oppose relative motion of the parts in certain directions is the physical interference of macroscopic (or relatively macroscopic) geometric or physical features.

Mechanical interlocking can be accomplished through the use of design features that are integral to the elements involved in the joint, such as tongues-and-grooves or dovetails, or that are developed in the joint elements by some secondary processing operation, such as plastic deformation or coforming of interlocking features or structures. Integral design features achieve interlocking from the inherent geometric fit of the elements. When the inherent geometry of the joint elements does not produce interlocking, interlocking may be achieved by altering the geometry, possibly through local plastic deformation (as in folding and hemming, or in crimping, or by staking and setting), through the use of snap features that rely on elastic deformation, or through the use of supplemental devices called mechanical fasteners.

3.2.2 Mechanical Fasteners

Mechanical fasteners are supplemental devices that can be used to create interlocking between abutting or mating joint elements, thereby preventing relative motion and physical separation. Joint elements tend to be acted upon by loads that act primarily parallel to the joint faying surfaces, producing sliding or shear forces, or perpendicular to the faying surfaces, producing opening or tensile forces.[1] Either type of force can ultimately lead to the unacceptable separation and/or misalignment or misorientation of the joint elements. When fasteners are used, the type of loading in the joint is referenced to the fastener's axis. Joints are said to be shear-loaded when the applied loads are perpendicular to the fastener's axis and parallel to the joint faying surface. Joints are said to be tension-loaded when the applied loads are parallel to the fastener's axis and perpendicular to the joint faying surface. Chapter 2, Sections 2.4 and 2.5, describe shear- and tension-loaded joints, respectively.

---

[1] It is possible for either simple or combined loads to act on a joint in such a way that both shear and tension forces are produced.

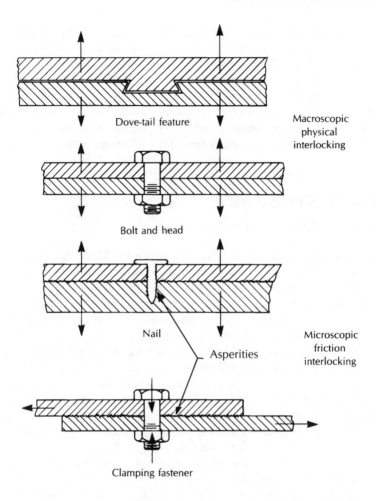

Figure 3.1 Physical interference from macroscopic mechanical interlocking and microscopic friction.

The way in which mechanical fasteners need to achieve interlocking between joint elements depends on the type of applied loading. When joints are shear-loaded, the fasteners resist the load either by pinning the joint against motion using bearing forces between the fastener and the joint elements or by resisting the shear through the action of friction at the joint faying surfaces. Since the magnitude of a frictional force (F) depends on the product of the coefficient of friction (μ) and the applied normal force (N), part of the function of a fastener in a shear-loaded friction-type joint is to develop a normal force through a clamping action on the joint elements. When joints are tension-loaded, the fasteners resist the load entirely by developing a preload or clamping force.

Given these requirements, mechanical fasteners are broadly classified into two types: (1) those that are designed primarily to develop clamping forces or preload through the use of threads, called *threaded fasteners,* and (2) those that are designed primarily to resist shear through bearing with a pinning action, called *unthreaded fasteners* (although threaded fasteners can also be used to resist shear through bearing). Threaded fasteners, in particular, facilitate disassembly, which is one of mechanical fastening's principal advantages over other joining processes, although some unthreaded fasteners, and most interlocking integral design features, also allow disassembly.

3.3 THREADED FASTENERS

3.3.1 General Description

One large category of mechanical fasteners uses threads. These are called threaded fasteners. Threads are a helical or inclined ramp around a cylindrical shaft that allow the development of a clamping force on a joint and/or the development of a preload in the fastener through the principle of a screw. Rotation of the fastener causes a translation parallel to the axis of the fastener and introduces tensile elastic strain in the fastener shank that creates a clamping force by trying to relax.

Some threaded fasteners develop the clamping force in a joint by themselves, simply through the friction between the thread and the material into which the fastener is forced, often creating a path as it goes. Others develop the clamping force through the opposing motion of another, previously threaded piece, e.g., an internally threaded part or an internally threaded backing piece called a *nut.* In either case, the clamping force arises from the opposing translation of the joint element (or the nut against the joint element) and the fastener's head. This is shown in Figure 3.2).

Usually, threaded fasteners, through the clamping action produced by the thread, are used to develop friction in a friction-type shear-loaded joint or are used to develop preloads in tension-loaded joints. The four major types of threaded fasteners are (1) bolts, (2) screws, (3) nuts, and (4) self-tapping screws. In addition to these major types, two special forms of what are usually threaded fasteners are known as integral and self-clinching fasteners.

3.3.2 Threads

The threads on threaded fasteners, and on or in threaded parts, are of two main types, based on the units of dimensions:

- Unified Inch Series (or UN[2] profile), using English units
- Metric Series (or M or MJ[3] profile), using metric units.

---

[2] The UNR profile has a rounded fillet at the root of the external thread.
[3] The MJ profile has a rounded fillet at the root of the external thread and a larger diameter of both the internal and external threads for use under high fatigue loading (e.g., for aerospace).

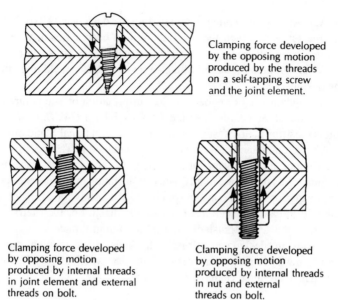

Clamping force developed
by the opposing motion
produced by the threads
on a self-tapping screw
and the joint element.

Clamping force developed
by opposing motion
produced by internal threads
in joint element and external
threads on bolt.

Clamping force developed
by opposing motion
produced by internal threads
in nut and external
threads on bolt.

Figure 3.2   Source of clamping action created by a threaded fastener, with a part or with a nut.

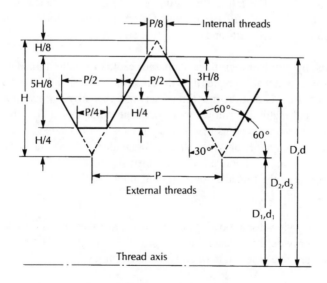

Figure 3.3   Basic thread profiles.

Figure 3.3 shows the basic profile of threads for both thread types. These threads can be external, that is, on the shaft of a fastener or on the outside of a cylindrical part, or internal, that is, in the body of a nut or in a hole in the joint part. External threads are desinated UNA, MA, or MJA, while internal threads are designated UNB, MB, or MJB, depending on the series to which they belong. Within each series or type, the actual shape of the thread root and crown can vary, with either rounded or sharp (vee) profiles. The method of fabrication has a major influence on thread shape, although the shape is defined in the design.

There are standards that provide complete information on screw threads. One is the American National Standards Institute (ANSI) B1.1-1974, *Unified Inch Screw Threads*. Others exist for metric fasteners under the International Standards Organization, or ISO. A more complete list of standards is contained in Table 3.1.

The most important parameter for a threaded fastener or part is the *pitch* of the threads. The pitch refers to the rate of advance or translation of the fastener per rotation and is almost always constant. A constant-pitch Unified Series consisting of 4, 6, 8, 12, 16, 20, 28, and 32 threads per inch has been standardized. For the Metric Series, the pitch is specified by the actual translation per rotation in millimeters.

While threads are designated primarily by their series, nominal diameter, and pitch, they can also vary in their coarseness or fineness, which affects the looseness or tightness of fit. There are three categories as follows:

- UNC (or MC or MJC), coarse: Offers the most resistance to thread damage by stripping and is best for aluminum, brass, cast iron, and other lower-strength materials
- UNF (or MF or MJF), fine: Offers the best vibration resistance
- UNEF (or MEF or MJEF), extra fine: Offers the best performance in thin nuts or internally threaded parts and enables fine adjustment.

Finer threads, by virtue of their finer pitch and resultant smaller thread depth, produce a tensile loading area that is greater than the area for a corresponding coarse-thread fastener. The net effect is that the fine-thread series is typically stronger than the coarse-thread series and, naturally, allows more accurate adjustment. Fine threads are also easier to tap into hard materials than coarse threads, while coarse threads, because of their additional clearance, provide latitude for plating or other finishes and are less plagued by the presence of contaminants or burrs during assembly.

Regardless of the thread's coarseness or fineness, there are tolerances on the fit that are fixed by class.

- Class 1 provides for a loose fit and affords easy assembly or disassembly.
- Class 2 provides for a general fit.
- Class 3 provides for a tight fit and affords accuracy.

A typical designation or call-out for a threaded fastener, say a bolt, includes the series, diameter, category, pitch, and fit class. An example is:

¼ – 20 UNC or ¼ – 20 UNRC, Class 1

where the diameter is ¼ in. in the Unified Inch Series; the threads are coarse, have a pitch of 20 per inch, and have a round profile; and where the overall fit can be expected to be loose. The pitch of a thread can also be right- or left-handed. For right-handed

Table 3.1 Standards for Threaded Fasteners

Type Specification	Source
AAR specifications	Association of American Railroads
AMS specifications[a]	Society of Automotive Engineers
ANSI standards	American National Standards Institute
API specifications	American Petroleum Institute
ASTM specifications	American Society for Testing Materials
EEI specifications	Edison Electric Institute
Federal specification	Naval Publication and Forms Center
Handbook H28, National	National Institute of Standards
Bureau of Standards	and Technology
IFI documents	Industrial Fasteners Institute
IOS[b] recommendations	American National Standards Institute
Military specifications	Naval Publication and Forms Center
NAS[c] specifications	National Standards Association

[a] Aeronautical Materials Specifications
[b] International Organization for Standardization
[c] National Aerospace Standards Committee
Reprinted with permission from Robert O. Parmley, *Standard Handbook of Fastening and Joining,* 2nd ed., McGraw-Hill, 1989, p. 1-16.

threads, which are most common, tightening occurs when the fastener is turned clockwise. For left-handed threads, tightening occurs when the fastener is turned counterclockwise. Left-handed threads are often used to prevent or hinder loosening accidentally.

3.3.3 Bolts

Bolts are usually two-piece threaded fasteners that develop the clamping force in a joint using a second, back-up piece called a nut on an externally threaded shank (see Figure 3.4). Bolts can also be used without a nut, where the last (most distant) joint element is internally threaded.

The critical dimensions on a bolt are the nominal diameter of its threaded shaft and its length. The unthreaded portion of the shank is usually machined or ground in one or more places or over its entire length to a diameter approximately that of its thread-root diameter. The basic thread length in a bolt is (as shown in Figure 3.4):

$$L_T = 2D + 0.25 \text{ in.} \qquad \text{for } L < 6 \text{ in.}$$
$$= 2D + 0.50 \text{ in.} \qquad \text{for } L > 6 \text{ in.}$$
$$= 2D + 6 \text{ mm} \qquad \text{for } L < 125 \text{ mm and } D < 48 \text{ mm}$$
$$= 2D + 12 \text{ mm} \qquad \text{for } 125 < L < 200 \text{ mm}$$
$$= 2D + 25 \text{ mm} \qquad \text{for } L > 200 \text{ mm,}$$

where L is the length of the bolt under the head, and L_T is the threaded length.

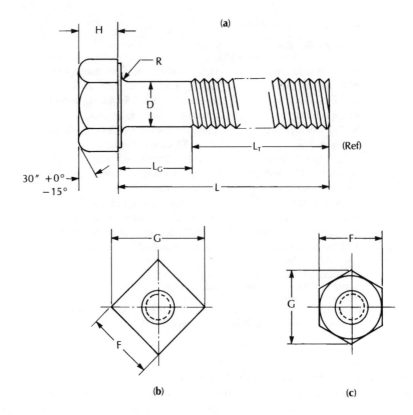

Figure 3.4   Typical bolt and nut, showing principal dimensions. (a) Hexagonal bolt; (b) square nut; (c) hexagonal nut.

Bolts are classified by their head type, with three types predominating as shown in Figure 3.5: (1) square-head, (2) hex(agonal) head, and (3) round-head, or carriage. Hex-heads can be either (a) heavy structural or (b) finished, and carriage bolts can have different neck shapes as shown in Figure 3.6. Dimensions of square-head, hex-head, and various round-head bolts are contained in ANSI B18.5-1971. There are also other less commonly used head types, including countersunk, T-, askew, elliptical, or oval, as well as eyebolts and bent or hook bolts (see Figure 3.7).

Grade markings are found on the heads of most bolts and indicate the material from which the bolt is made and the strength level to which the material is treated. A complete listing of grade markings is available in the ANSI B18.2.1-1972 published by the American Society of Mechanical Engineers. A summary of some of the more important grade markings is given in Table 3.2.

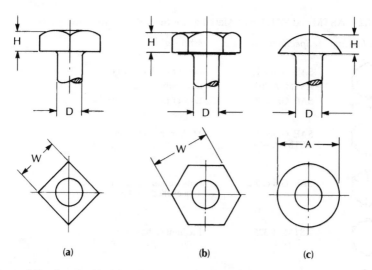

Figure 3.5 Standard bolt head types. (a) Square head; (b) hex head; (c) round head.

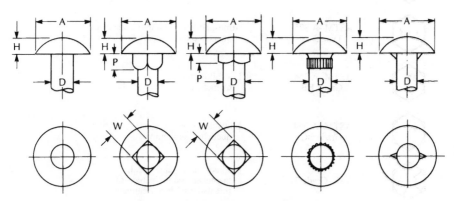

Figure 3.6 Carriage bolt neck shapes.

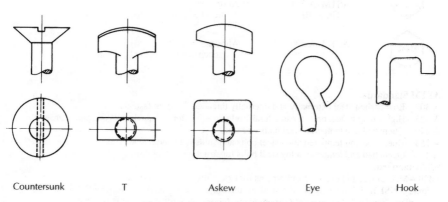

| Countersunk | T | Askew | Eye | Hook |

Figure 3.7 Other bolt head types.

Table 3.2   ASTM and SAE Grade Markings for Steel Bolts and Screws

Grade Marking	Specification	Material
NO MARK	SAE–Grade 1 ASTM–A 307 SAE–Grade 2	Low or medium carbon steel Low carbon steel Low of medium carbon steel
	SAE–Grade 5 ASTM–A 449	Medium carbon steel, quenched and tempered
	SAE–Grade 5.2	Low carbon martensite steel, quenched and tempered

A 325	ASTM–A 325
	  Type 1
		Medium carbon steel,
		   quenched and tempered

A 325	ASTM–A 325
	  Type 2
		Low carbon martensite steel,
		   quenched and tempered

A 325	ASTM–A 325
	  Type 3
		Atmospheric corrosion (weathering) steel,
		   quenched and tempered

BB	ASTM–A 354
	  Grade BB
		Low alloy steel,
		   quenched and tempered

BC	ASTM–A 354
	  Grade BC
		Low alloy steel,
		   quenched and tempered

	SAE–Grade 7	Medium carbon alloy steel,
		   quenched and tempered, roll
		   threated after heat treatment

	SAE–Grade 8
	
	ASTM–A 354
	  Grade BD
		Medium carbon alloy steel,
		   quenched and tempered
		Alloy steel,
		   quenched and tempered

A 490	ASTM–A 490	Alloy steel, quenched and tempered

ASTM Standards:
A 307—Low carbon steel externally and internally threaded standard fasteners.
A 325—High strength steel bolts for structural steel joints, including suitable nuts and plain hardened washers.
A 449—Quenched and tempered steel bolts and studs.
A 354—Quenched and tempered alloy steel bolts and studs with suitable nuts.
A 490—Quenched and tempered alloy steel bolts for structural steel joints.
SAE Standard:
J 429—Mechanical and quality requirements for externally threaded fasteners.
Source: ANSI B18.2.1-1972, as published by the American Society of Mechanical Engineers and R. O. Parmley, *Standard Handbook of Fastening and Joining,* 2nd ed., McGraw-Hill, 1989.

3.3.4 Screws

Screws resemble bolts and, like bolts, use nuts or are threaded into internally threaded parts. Screws are generally made to the same material specifications as bolts, so have the same head markings, but tend to be limited to smaller diameters with fine or very fine threads. Screws come in many types, as shown in Figure 3.8, and include machine, cap, set, thumb, socket, lag, miniature, and self-tapping types. The self-tapping screw is what is commonly referred by most people as a screw, while the other types are generally referred to by their specific name (e.g., machine screw, cap screw, set screw). The self-tapping types are discussed separately in Section 3.3.6.

Like bolts, screws are classified by their head type, but many more head types tend to exist, including hexagonal heads, socket heads, slotted-heads, and machine screws with a variety of head designs for specific purposes, including 100° flat and oval countersunk, pan, fillister, truss, and hex-washer. Figure 3.9 shows schematics of major screw head types, while the following list below provides a more comprehensive summary:

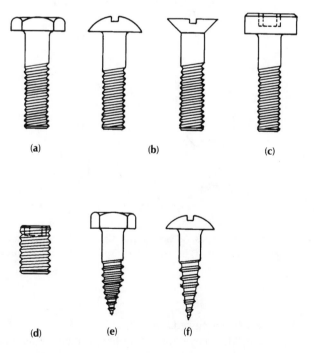

(a)   (b)   (c)

(d)   (e)   (f)

Figure 3.8 Types of screws. (a) Hex or square cap screw; (b) slotted round and countersunk cap screws; (c) socket head cap screw; (d) socket head set screw (e) hex or square head lag bolt; (f) slotted-head self-tapping screw.

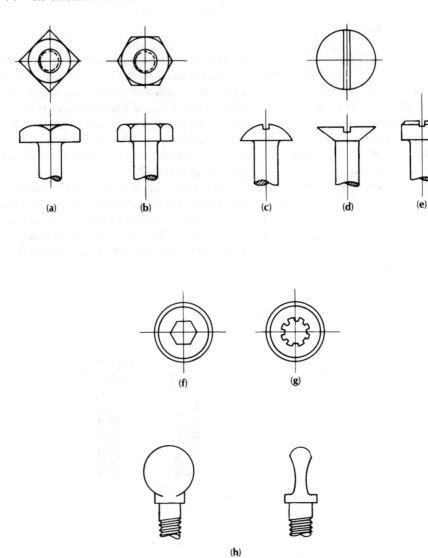

Figure 3.9   Major screw head types. (a) Square cap; (b) hex cap; (c) round, slotted; (d) countersunk, slotted; (e) fillister, slotted; (f) hex-socket; (g) spline socket; (h) thumb.

1. hex-head cap, heavy, and finned screws
2. socket heads (and Allen keys), including hex, forged spline, low, button-head, shoulder, flat-head, pressure plug, and set cap screw varieties
3. slotted-head round, flat-countersunk, and fillister head cap screws

4. machine screws with regular or undercut, slotted or cross-recessed 80° or 100° flat or oval countersunk heads; 100° flat countersunk heads; flat or oval countersunk trim heads; pan, fillister, truss, binding, or hex heads; or hex-washer heads.

Standards for various types of screws are listed in Table 3.1.

3.3.5 Nuts and Locknuts

A nut is a block of material (often a metal), usually of a square of hexagonal shape, which has an internally threaded hole drilled through its center. It is intended to mate with a standard threaded bolt or screw. Nuts are used with bolts or screws to develop clamping action on a joint by moving up the threaded shaft of the fastener to oppose the force applied by the head upon tightening. Usually, the nut is a separate piece used for backing a bolt or screw, but occasionally nuts are attached to (e.g., plate nuts) or integral to the joint element or other part.

Nuts are available in several standardized designs, including square nuts, hexagonal nuts, hexagonal locknuts, slotted or castle nuts, and wing nuts (see Figure 3.10).

Many methods have been proposed and used to prevent the loosening of nuts. The oldest methods involve pinning through the nut with a wire or straight, taper, or Cotter pin (see Section 3.4.4). Other methods include using a second nut to interfere with the regular nut (a so-called jam nut); staking (see Section 3.6.3) or prick punching the exposed thread below the nut (on the exposed face) at one or more points after assembly; adding a drop of adhesive between the threads at the outer surface of the nut; and using off-angle threads in the nut or on the bolt. A special type of nut, called a *locknut* resists loosening through its design. Two major classes of locknut exist: (1) nuts that lock themselves to the thread of the bolt or screw through interference developed between internal and external threads, and (2) nuts that are free turning until they engage a bearing surface on the bolt or screw shank that causes a locking action.

Several common design approaches are used in locknuts. One approach uses an insert made of a soft, resilient material (such as lead, nylon, or resin-impregnated fiber) through which the bolt or screw must form its own thread by plastic deformation. For these nuts, the bolt or screw must be driven to final assembled position, but disassembly and reassembly can be done repeatedly. In another approach, nuts are made with a few threads near their outer surface deformed radially to a slight elliptical shape or with the pitch (or lead-angle) of the threads altered to grip the bolt or screw threads when assembled. A third approach uses any of several designs where the nut is shaped to take advantage of the spring properties of the material from which it is made, including sheet-metal nuts and nuts with slotted segments. Finally, there are nuts that develop a wedging action between an inner and an outer sleeve, tapered so that the inner threaded part is compressed into the threads of the bolt or screw by the outer sleeve.

Locknuts of all of the types described are available under various trade names such as Nylok (with a nylon insert), McClean-Fogg #3 (with an altered pitch), Tri-Lok (with an elliptical thread), Flexlok (with spring action from slotted segments), Tinnerman

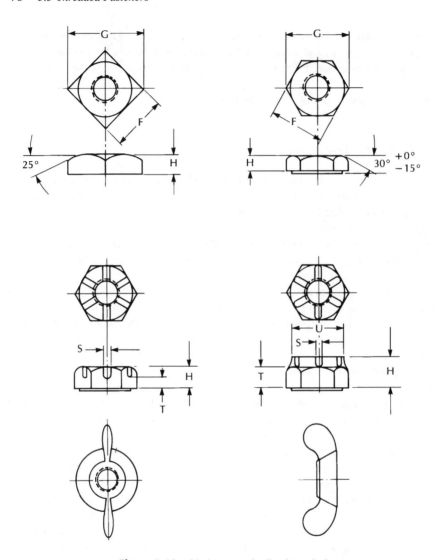

Figure 3.10   Various standardized nut designs.

speed nuts (with spring action from sheet-metal construction), and Klincher (that develop wedging actions between sleeves).

3.3.6 Tapping Screws

Tapping or *self-tapping screws* develop a clamping force in a joint through the friction between the screw and the part as the screw forces its way into the part creating the mating female thread as it goes. No internal thread in the part or supplemental nut is required.

Tapping screws are of three main types, classified by the way in which they produce the threaded path in the material they are being used to fasten: (1) thread-forming, (2) thread-cutting, and (3) thread-rolling. *Thread-forming screws* are used when sufficient joint stresses can be developed to guard against loosening. These screws literally form the mating thread in the part or joint element being fastened by plastically deforming the part or joint material. Various types are shown in Figure 3.11a. *Thread-cutting screws* are used instead of thread-forming screws to lessen the force during tapping and to lessen the internal (residual) stress. Rather than forming threads in the joint material by plastic deformation, these screws form mating threads by a cutting action. Figure 3.11b shows several types. *Thread-rolling screws*, like thread-forming screws, deform threads into the material(s) being fastened but induce significant cold work into the material during tapping, thereby contributing to the joint strength.

3.3.7 Materials and Standards for Major Types of Threaded Fasteners

Bolts, screws, and nuts are generally made from low or medium carbon steels, low alloy steels, medium carbon alloy steels, high alloy steels, or stainless steels, often in quenched and tempered conditions (where applicable). For certain applications, special materials, such as brasses, bronzes, or nickel-copper also have been used. These fasteners are also generally manufactured to standards, such as published by the American National Standards Institute (ANSI). A list of applicable standards is given in Table

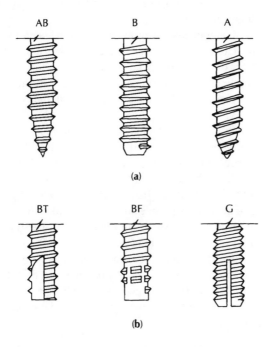

Figure 3.11　(a) Thread-forming screws, and (b) thread-cutting screws.

3.1. Tapping screws are also made to ANSI (and other) standards, and are available in low and medium carbon steels, low alloy steels, stainless steels, and brass, bronze-aluminum, and nickel-copper alloys.

3.3.8 Integral Fasteners and Self-Clinching Fasteners

In addition to the four major types of threaded fasteners described in the previous section, there are two special forms of what are usually threaded fasteners. These are integral fasteners and self-clinching fasteners.

An *integral fastener* is a device, usually threaded, that is installed in a component or unit (such as a chassis, panel, or bracket) to become a permanent part of that component or unit. Such permanently mounted fasteners are used because they facilitate automated assembly, they are dependable in service (i.e., they can't be lost), and they are often extremely cost effective. Permanent mounted fasteners can be mechanically clinched, swaged, riveted, welded, or adhesively bonded in place or may be of a self-clinching or mounting design. Such integral fasteners are available in hundreds of designs and sizes.

One threaded integral fastener of significance is known as a *sems*. Sems are a combination of various standard screws and a captive washer. Screw-washer assemblies or sems are used for their convenience in assembly because they combine two parts into one. This is particularly advantageous for automated fastening or assembly. Figure 3.12 shows a few examples of sems.

Another common example of an integral fastener is the *anchor-type*, or *plate-type nut*, or simply *platenut*. These integral fasteners are usually installed by riveting a plate or housing that holds an internally threaded body or nut. Often, this body has some degree of freedom to move in order to compensate for slight misalignments. A normal threaded bolt or screw is inserted into the part, mating with the permanently anchored nut.

Other types of integral fasteners include clinch nuts, weld nuts, and studs.

Self-clinching fasteners are particularly attractive integral fasteners, providing assembly and service economies as well as cosmetic benefits. Self-clinching fasteners are squeezed into a hole in a sheetmetal part, for example, using a simple press. The sheetmetal, which must always be softer than the fastener, plastically deforms under the installation pressure into an annular groove around the shank and into a knurled step of the fastener. A typical installation is shown in Figure 3.13. A strong, permanent fit is assured by such fasteners, even in very thin (i.e., 0.020 in. or 0.5 mm) sheet metal.

Several fairly standard self-clinching fasteners are nuts with free-floating or self-locking threads; self-locking or non-self-locking floating insert nuts; flush-head, heavy-head, and concealed-head studs; through-hole, blind-hole, and concealed-head standoffs; panel fasteners with captive screws; spring-loaded pins; flush-head pins; and grounding solder terminals.

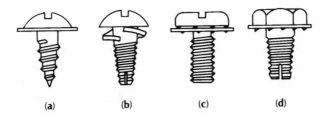

(a)   (b)   (c)   (d)

Figure 3.12 Screw-washer assemblies or sems. (a) Round head type A tapping screw and regular washer; (b) truss head type D tapping screw and helical-spring lock washer; (c) pan head machine screw and internal-tooth lock washer; (d) hex head type T tapping screw and external-tooth lock washer.

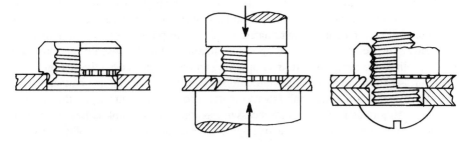

Figure 3.13 Installation of self-clinching fasteners.

3.4 UNTHREADED FASTENERS

3.4.1 General Description

Not all fasteners require threads to achieve interlocking of joint elements. *Unthreaded fasteners* accomplish interlocking through some other means than the clamping produced by a thread. The mechanism may involve simple pinning, relying on varying degrees of bearing and friction between the fastener and the joint element; some elastic spring action in the fastener; or plastic deformation of the fastener. Unthreaded fasteners can be used in shear- or tension-loaded joints, but are usually used in shear, acting principally as pinning points.

Unthreaded fasteners include (1) rivets, (2) blind fasteners, (3) pins, (4) eyelets and grommets, (5) retaining rings or clips, (6) keys, and (7) washers. Most unthreaded fasteners require preprepared holes (or slots, for keys); however, some, like eyelets and grommets, do not require preprepared holes.

3.4.2 Rivets

A *rivet* is an unthreaded fastener typically having a head and a smaller diameter shank and usually made from a plastically deformable material (e.g., metal or thermoplastic). It is used to join several parts by placing the shank through holes in parts that are properly aligned and creating a second head or foot by plastically deforming or upsetting the projecting shank.[4] Depending on the rivet material, upsetting can usually be done cold but may need to be done while the rivet is in its hot-working range. Usually, hot-upsetting rivets are made from steel and are of larger diameters. Such hot-forming rivets were widely used in the erection of steel structure for buildings or bridges. Figure 3.14 shows a schematic of a typical rivet in a joint before and after upsetting.

Rivets offer several advantages, some of which are unique among all mechanical fasteners. These advantages include: (1) low fabrication cost, because they are geometrically simple and can be produced in large quantities by high-speed machines; (2) low installation cost, because insertion and setting is fast and repetitive and requires little operator skill or can be easily automated; (3) ability to produce permanent joints that are readily inspected immediately upon installation simply by visually checking for cracks, splits, or incomplete setting or upsetting and permitting corrective action by the operator; (4) ability to allow relative rotation between parts, acting as pivot shafts; (5) ability to join dissimilar materials (such as metals and polymers) in various thicknesses; (6) ability to join as many parts as necessary, as long as the shank is long enough to project through the stack; (7) attractive appearance compared to many other fasteners and may even be used for decorative purposes; (8) ability to offer aerodynamically smooth contour with the joint element when special countersinking and shaving techniques are used on heads or heads and upset tails; (9) wide diversity of shapes, sizes, and materials; and (10) ability to be installed with a wide selection of methods, tools, and machines.

Despite these numerous advantages, rivets do have limitations. First, they may not be as strong in static tension or fatigue as bolts. Second, high enough tensile loads can pull out the clinch or upset, causing the fastener to pull out of the joint. Third, severe vibration can loosen the joint and retightening can be difficult. Fourth, removal for disassembly is more difficult than for threaded fasteners and many other types of unthreaded fasteners. Fifth, riveted joints may not be either watertight or airtight unless special gaskets, sealing compounds, or caulking are used or very special designs are employed. Another possible disadvantage of most rivets is that installation usually requires access to both sides of the joint. The only exception is the so-called blind fastener (see Section 3.4.3).

Rivets can be upset manually, often by hammering the head while bucking the tail or using automatic, C-frame machines to apply a squeezing force to both head and tail simultaneously. Because of the forged grain structure, most rivets tend to have higher tensile, compressive and shear strengths than many other solid fasteners.

---

[4]There are some two-piece rivets, usually used for higher strength applications such as aerospace, where a malleable collar or retaining piece is formed (by swaging), clinched, or otherwise attached to a high-strength projecting shank with an annular groove. These are called *high-shear rivets*.

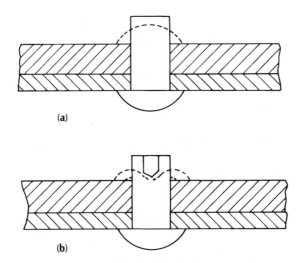

Figure 3.14   Solid (a) and tubular (b) rivets in a joint, before and after upsetting. *Solid line*, before upsetting; *dotted line*, after upsetting.

Rivets can be divided into two major categories: solid and tubular. Tubular rivets can themselves be subclassified as full tubular, semitubular, compression, and, sometimes, self-piercing. A solid and a tubular rivet are compared in Figure 3.14. The most distinguishing feature of rivets is their head shape, of which there are many, as shown in Figure 3.15.

Rivets are available in many types, largely based on their intended application, but also on certain distinguishing geometric features. These types include

1. Standard structural or machine rivet: A solid rivet with a cylindrical shank that is either hot or cold driven. The upsetting force depends on the material, size, and temperature of upsetting. These rivets are standardized and available in many materials and sizes.

2. Slug rivet: A simple cylinder that is inserted into a hole and upset at both ends simultaneously to produce two formed heads or tails.

3. Boiler rivet: A large, solid rivet having a conical head that was formerly widely used for assembling boilers

4. Cooper's rivet: A solid rivet used for barrel hoops or barrel-hoop joints, having a thin, countersunk head with a chamfered crown and shank end

5. Shoulder rivet: A solid rivet with a formed shoulder under the head that is often used for making other attachments

6. Tank rivet: A small, solid rivet with button, countersunk, flat, or truss heads used for sheet-metal work

7. Tinner's rivet: A solid rivet with a large flat head used for joining soft sheet metal.

8. Belt rivet: A special solid rivet with a thin collar below the head used for joining leather or fabrics

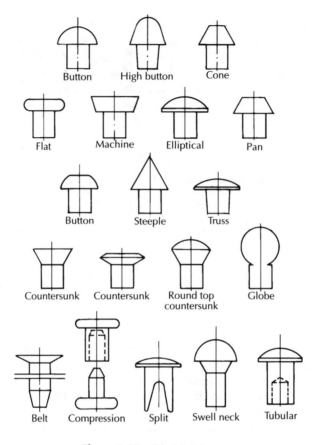

Figure 3.15   Rivet head types.

9. Compression or cutlery rivet: A two-piece rivet, consisting of a tubular portion and a mating solid portion. The hole in the shank of the tubular portion is sized to produce a drive fit when the joint is assembled.

10. Split or bifurcated rivet: A solid, self-piercing rivets having a prong in the shank that cuts its own hole through soft metals or fibrous materials (like wood).

11. Swell-neck rivet: A large, solid rivet with a large bulbous head and a swelled neck below the head. It produces a tight fit when set.

12. Tubular Rivet: A small rivets having a hole in the shank end. The rivet is cold driven with a tool that expands or curls the shank end to set the fastener. Hollow or tubular rivets are the basis for several special types of fasteners, most especially blind rivets.

13. Blind rivet: A special function rivet, with complex design, intended for use where only one side of the joint is accessible. The blind side is not accessible, so a special tool, usually designed and produced by the fastener manufacturer, is needed to install and set the fastener. These special fasteners are described in detail in the next Section 3.4.3.

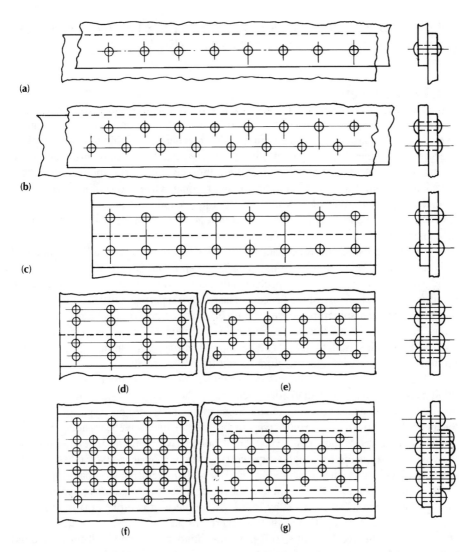

Figure 3.16 Typical designs for riveted joints. (a) Lap joint, single riveted; (b) lap joint, double riveted, staggered; (c) butt joint, single riveted; (d) butt joint, doubled riveted, chain riveted; (e) butt joint, double riveted, stagger riveted; (f) butt joint, triple riveted, chain riveted; (g) butt joint, triple riveted, stagger riveted.

Joint designs for rivets are typically single- or double-lap (or butt) types, and rivets within these joints are arranged (as are bolts in bolted joints) in single, double, and even triple rows of either straight chain or staggered holes as shown in Figure 3.16.

Rivets are made from many materials, but some that are considered standards are: aluminum alloys (such as 1100-S, 3003-S, 2017-S, 2024-S, and 5052-S), brass, and low and medium carbon and high alloy and stainless steels. Most rivets are standardized under ANSI B18.1.1-1977 (covering small rivets) or B18.1.2-1977 (covering large rivets). Tubular rivets are covered by ANSI B18.7-1972(R1980).

3.4.3 Blind Rivets

Blind rivets are a special class of unthreaded fasteners that are one of the most innovative of the entire group in that they often enable mechanical joining that would otherwise be impossible. The designs of blind rivets are intended to allow installation and setting of the fastener from one side, and are thus invaluable for applications where access to the blind side is impossible or impractical. Originally, blind rivets were invented to permit repair of previously riveted joints and were strictly a replacement fastener. Today, blind rivet and other blind fastener designs have proliferated to fill a much needed niche in mechanical fastening.

As distinguished from the standard solid rivet, a blind rivet can be inserted and fully installed (or set) in a joint from one side of a structure or assembly. The back, or blind, side of this type of rivet is mechanically expanded to form a bulb or upset head, using some special feature of the fastener. The result is a permanently installed fastener that duplicates and sometimes exceeds the performance criteria for comparable solid rivets. Figure 3.17 shows how a conventional solid rivet requires access to both sides of the joint to permit installation. Figure 3.18 shows some typical joining applications that prohibit backside access and, therefore benefit from blind rivets.

Figures 3.19 and 3.20 show some examples of some of the principles used in blind rivets. Special tools, usually designed and produced by the fastener manufacturer, are used to install and set these fasteners. Some of the more common types of blind fasteners are the Chobert rivet, Huck rivets or fasteners, Cherry rivets, rivnuts, explosive rivets, Southco rivets, and pop rivets. The Chobert rivet (Figure 3.19a) is upset by drawing a solid mandrel through a tapered hole, leaving what is essentially a tubular rivet. The Huck rivet (manufactured by the Huck Manufacturing Company) is similar to the Chobert in that it is also tubular; however, after the mandrel or pin is pulled through the sleeve to set the rivet by forming a tailpiece, it is broken off to produce a solid, high strength rivet (Figure 3.19b). There are, in fact, several variations of the Huck rivet, including lockbolts, self-plugging blind rivets, and pull-through blind rivets. The Cherry rivet (manufactured by Cherry Rivet Division of Textron) is available in three designs: regular hollow, pull-through, and self-plugging. The regular hollow rivet is used when a high clamping force is desired, where the amount of shank expansion is not a factor, and where the presence of broken stem pieces entrapped in the assembly after fastening is not objectionable. The Cherry pull-through rivet is useful where the broken stem falling into the assembly is objectionable (e.g., aircraft), but where lower clamping or clinching force is tolerable. Where a hollow rivet is unacceptable (because of high loading), the self-plugging Cherry rivet is used. These three types are shown in Figure 3.19c.

Besides these proprietary types of blind fasteners, there are generic rivnuts and explosive rivets. The rivnut is set by pulling in on a threaded pull-up stud that collapses the hollow shank of the fastener to clinch the joint. Figure 3.20a shows the operation of

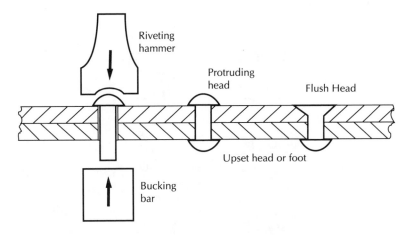

Figure 3.17 Two-side access for installing conventional solid and tubular rivets. (Reprinted with permission after Robert O. Parmley, *Standard Handbook of Fastening & Joining,* 2nd ed., McGraw-Hill, 1989.)

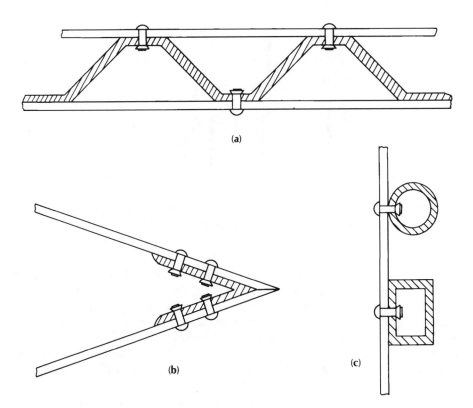

Figure 3.18 Typical applications for blind rivets. (a) Closed structures; (b) acute corners; (c) mounting tubes or pipes.

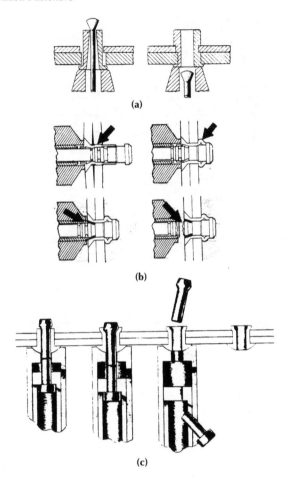

Figure 3.19   Operation of (a) Chobert, (b) Huck, and (c) Cherry blind rivets. (From V. Laughner and A. Hargan, *Handbook of Fastening and Joining of Metal Parts*, McGraw-Hill, 1965.)

rivnuts. A related type is the Southco rivet shown in Figure 3.20b. These rivets form a hollow shank by driving a center pin extending from above the head through the shank. Southco rivets are known generically as drive-pin rivets. While not typical blind fasteners, some rivets can be installed and upset using explosives embedded in their hollow shanks. By detonating the explosive with heat or high-speed impact, the hollow shank expands to set the rivet as shown in Figure 3.20c. There are also some special fasteners, described in Section 3.6.5, that are self-upsetting, employing a metallurgical allotropic transformation.

Pop rivets are a subgroup of blind fasteners. Developed in England, they have been widely used there for many years. Pop rivets are starting to see increased application in the United States, however. The two design types are *break head mandrel* and *break stem mandrel*. Both consist of a hollow rivet and a solid mandrel. A special tool pulls the mandrel into the rivet, upsetting it and creating a clinching force. A recess or

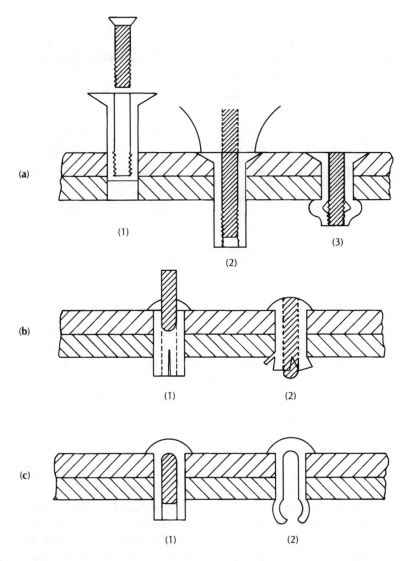

Figure 3.20 Operation of rivnuts, drive-pin rivets, and explosive rivets. (a) Rivnut; (b) Southco blind rivet; (c) DuPont explosive rivet.

undercut causes either the head or the stem to break off with a characteristic pop. Typical designs are shown in Figure 3.21.

Standards for blind fasteners have been established by the Industrial Fasteners Institute (IFI), under IFI Standard Numbers 114, 116, 117, 119, 123, and 126, as well as by the Department of Defense (in Mil-Standards) and by the National Aerospace Standards Committee (under NAS standards).

In order for a blind rivet to work, the pull material or drive pin must be stronger than the rivet body, which must be relatively ductile to permit expansion without

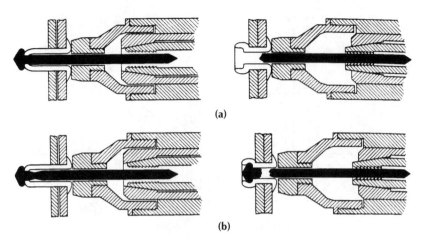

(a)

(b)

Figure 3.21   Break-head mandrel (a) and break-stem mandrel (b) pop rivets. (From V. Laughner and A. Hargan, *Handbook of Fastening and Joining of Metal Parts*, McGraw-Hill, 1965.)

damage. Thus, multiple materials are used in individual systems. Rivet bodies are often made from aluminum alloys, low carbon steels, stainless steels, and some brasses and nickel-copper alloys. The drive pin or pull portion is made from a higher-strength material, selected to be compatible with the material(s) being fastened. Carbon and low alloy steels, stainless steels, and special heat-resistant alloys have been used. For aerospace, titanium has also been used for both rivet bodies and drive pins.

3.4.4 Pins

Undoubtedly, the oldest mechanical fastener is the pin. A *pin* is a machine element or fastening component that secures the position of two or more parts of a structure, an assembly, or a mechanism relative to one another by passing through holes in those parts. Pins usually remain fixed by the friction caused by interference with the part. They often have no heads or feet and, consequently, develop no clamping force. Pins are particularly suited to resisting shear and are cost-effective because of their simplicity of design, ease of installation, and ease of removal for disassembly.

Originally, pins were simple, solid, straight cylinders. Today, there are more elaborate designs for ever-widening applications. Like rivets, pins can and do act as pivot shafts to allow a part to move, usually in rotation.

Pins come in many types, as shown in Figures 3.22 and 3.23, and as described in the following list.

1.   Straight cylindrical pins are headless cylinders, with or without chamfered ends, used for transmitting torque in round shafts.
2.   Dowel pins are often hardened headless cylinders for use in machine and tooling fabrication, including jigs and fixtures, for fixing position.

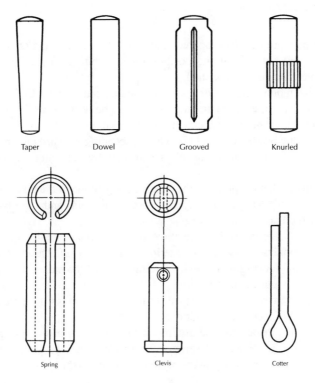

Figure 3.22 Various pin designs.

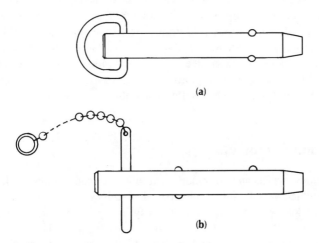

Figure 3.23 Quick-release pin designs (a) Detent-type: push to insert, pull to release; (b) T-handle detent type with chain.

3. Tapered pins are headless tapered rounds used in drilled and/or reamed (often taper reamed) holes for fixing position or transmitting torque. The taper helps assure interference is established.
4. Clevis pins require the use of a second, smaller pin (often a Cotter pin) placed through a hole drilled diametrically through the body at the headless end in order to keep the clevis pin from accidently withdrawing.
5. Cotter pins are headed pins held in place by bending outward the projecting portions of a split body at the headless end, thereby preventing unwanted withdrawal.
6. Spring pins are held in place by the elastic spring action of the body created by an axial slot or a spiral wound design. Slotted tubular pins and spirally-coiled pins are the two predominant types.
7. Grooved pins are typically have three parallel, equal radially spaced grooves impressed longitudinally on the exterior of the pin body. These grooves assure positive radial locking in a hole by forcing some of the material of the part into the grooves by plastic deformation.
8. Knurled pins have a knurled surface for use in die castings or in plastic applications to prevent unwanted withdrawal.
9. Quick-release pins are for use in temporarily fixing the position of parts during assembly or to facilitate disassembly. These pins often use spring-loaded lock-balls located in the pin body in the region that projects through the part being fastened.
10. Barbed pins are usually headed types with projections along the pin body to facilitate locking in soft materials.

Pins are generally manufactured to an ANSI Standard: B18.8.2-1978 for straight, dowel, taper, grooved, and spring types; B18.8.1-1977 for the clevis and Cotter-types. Pins are fabricated from many materials, depending on type and intended use. Most, however, are fabricated from various cold-drawn low or medium carbon or low alloy steels, and, as mentioned, some are hardened for wear resistance. Other materials typically used include various stainless steels (especially 400-series grades), brasses, bronzes, and beryllium copper. Special light-duty pins, for use with plastics, ceramics, or glasses, can be fabricated from other, softer materials.

3.4.5 Eyelets and Grommets

For some applications, eyelets and grommets are trouble-free and economical fasteners. They can be assembled very rapidly using special machines, which punch holes and insert and set the eyelets or grommets simultaneously. Typically, eyelets and grommets are used in relatively soft materials that are prone to tearing and other damage from other fasteners (e.g., cloth, leather, and rubber). Eyelets are used where shearing stresses and pressure-tightness are not important considerations. When this is the case, they can be used in place of rivets, offering savings in weight and cost. An eyelet provides a hole for fastening, with a protected edge. Eyelets can be used with hooks, laces,

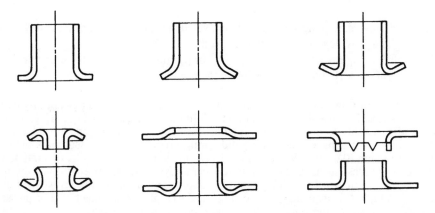

Figure 3.24   Joining with eyelets and grommets.

and ropes, or to provide a passage for wires that prevents wear or abrasive damage to the wire or wire insulation.

The grommet consists of a mating set of male and female units for actually joining two parts by inserting one unit into the other, often with an elastic snapping action. One common form of a grommet is actually called a snap and is the fastener used to close jackets and other garments.

Joining with these fasteners is the result of friction, with or without some elastic deformation, as shown in Figure 3.24.

Eyelets and grommets are usually made from relatively soft, highly formable materials that can be set easily, such as copper, brass, aluminum, zinc, steel, and nickel silver. There are no particular standards, but there are many styles and hundreds of sizes. A few of the most important styles are shown in Figure 3.24.

3.4.6 Retaining Rings

Shoulders (e.g., larger diameter portions) are usually used on shafts or on the interior of bored parts (e.g., recesses) or to accurately position or retain assembled parts to prevent axial motion or play, It is often advantageous to use *retaining rings* or *snap rings* as substitutes for these machined details. Such rings can be used to axially position parts on shafts or in bored housings with great precision and often save a great deal on machining, as well as facilitate disassembly for maintenance or repair. All retaining rings depend on elasticity in both their design and material of construction to function, in that they are either sprung or snapped into position or apply a spring load to the assembly.

Retaining rings come in many varieties but within three basic types based on the purpose to be fulfilled:

1. Axially and radially assembled groove rings are split or crescent-shaped rings, respectively, that require a groove in the part to properly perform their function of locating and retaining position.

2. End-play take-up rings are bowed or beveled in the plane of the ring to allow flexing and provide a spring force to resist play.
3. Self-locking rings have various prongs or protrusions or other features to permit the ring to lock into position without the need for a groove.

Retaining rings are fabricated in either stamped sheet metal or as spiral-wound wires or strips. Because they depend on the ability to deform elastically during assembly or disassembly to perform their intended function, retaining rings are fabricated from materials that have inherently good spring properties. These materials include medium to high (0.6 to 1.0%) carbon (silicon-containing) spring steels, precipitation-hardening stainless steels, and beryllium-copper alloys. Since the strengths and moduli of materials from which retaining rings are fabricated are high, the thrust they can develop is high. Consequently, they can be used to replace nuts, machined or pressed shoulders, collars, Cotter pins, or other positioning or thrust devices.

Specific advantages of retaining rings are (1) they allow quicker assembly and disassembly of parts; (2) they may be accurately positioned without marring; (3) they reduce machining through the use of an auxiliary part; (4) assembly design is simplified; (5) size and weight are reduced compared to heavier nuts; and (6) looseness or end play from accumulated tolerances during manufacture or from wear in service is eliminated.

Retaining rings are generally manufactured to industrial standards, such as those of the Industrial Retaining Ring Company, but are also covered by ANSI standards B27.6- and B27.7-1977. Some typical retaining ring designs and the ways that they are used are shown in Figures 3.25.

3.4.7 Keys

Keys are solid pieces of various geometries used in combination with slots, called *keyways*, to fasten two parts, usually to prevent relative circumferential or rotational motion, and/or to transmit torque. They may or may not prevent relative longitudinal or axial motion, depending on the type of key used. Figure 3.26 shows an example of how a key is used, while Figure 3.27 shows various types of keys. Keys and keyways (or keyseats) are covered by ANSI B17.2 and B17.7-1967(R1978). Materials are usually cold-finished steels, although other materials may be used for compatibility with the parts being fastened.

Another approach to preventing relative circumferential motion is to use raised ridges regularly spaced radially around a shaft, running longitudinally along the shaft. Commonly called splines, they are an integral design feature rather than a fastener (see Section 3.6.2).

3.4.8 Washers

Washers are simple, usually flat, circular secondary fasteners used with bolts, screws, and nuts usually in tension-loaded joints. They serve one or more of several purposes, for example, (1) to spread the load applied by the fastener head and/or nut, especially with soft, deformable materials (e.g., soft metal or polymers) or damage-prone materials

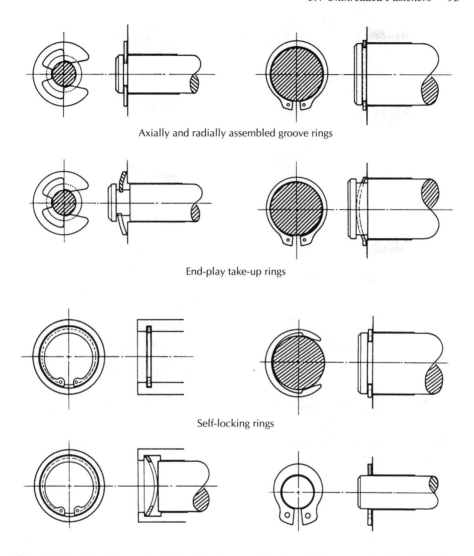

Axially and radially assembled groove rings

End-play take-up rings

Self-locking rings

Figure 3.25   Use of typical retaining ring designs. (Reprinted with permission from J. Shigley and C. Mischke, Standard Handbook of Machine Design, McGraw-Hill, 1977.)

(e.g., brittle ceramics or glasses or reinforced composites); (2) to obtain additional bolt tension in the joint, by acting as a spacer or shim; (3) to take up for relaxing bolt tension or preload or looseness in the joint (i.e., spring washers); or to help prevent loosening. Washers can be applied under the head of a fastener, under the nut, or under both.

Among the variety of washer designs are plain, cylindrically curved, conical or Belleville, slotted, spring, or spring-locking (see Figure 3.28). One important class of washers is the *lockwasher*. As the name implies, these washers have design features that help lock the washer into place, apply a spring action (or force) against the

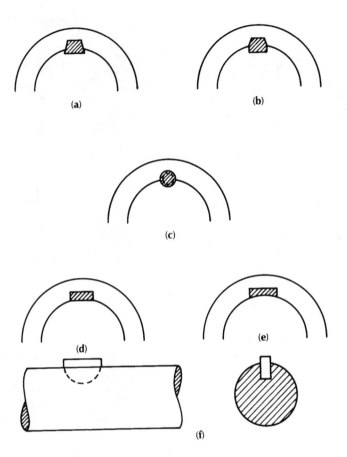

Figure 3.26 Function of keys in keyways. (a) Dovetail key; (b) beveled key; (c) round tapered key; (d) flat saddle key; (e) hollow saddle key; (f) woodruff key.

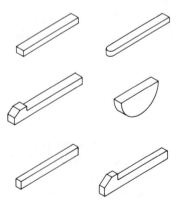

Figure 3.27 Various types of keys.

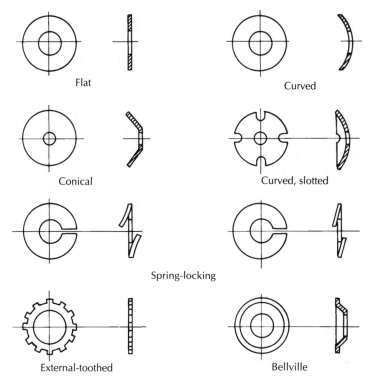

Figure 3.28   Various Washer Designs.

structure or the nut, and take up mechanical "slop" and/or prevent loosening. There are helical-spring, internal-tooth, external-tooth, internal-external-tooth, beveled, and Belleville varieties of lockwashers.

Washers can be made from almost any material. For most applications carbon steel, corrosion-resistant steel (302 or 305 types), aluminum-zinc, phosphor bronze, silicon bronze, and K-Monel are used, especially for spring-type lockwashers. Soft, fibrous materials or electrically insulative materials may be used for special purposes such as for vibration damping, fluid sealing, or electrically insulating the fastener from the joint element(s). Lockwashers are covered by ANSI B18.21.1-1972, while bevel washers (for use on inclined surfaces) are covered by B27.4-1967.

3.4.9 Self-Upsetting Fasteners

There is a special group of unthreaded fasteners that are akin to rivets, except that the foot of the fastener is formed without requiring upsetting from an external source. These *self-upsetting fasteners* create the foot or other locking feature with the part through a reversible phase transformation in the fastener material.

Certain metal alloys undergo a diffusionless, athermal (or martensitic), phase transformation. Often, this transformation results in a fairly significant volume change, and, further, the material exhibits some "memory" of its shape before transformation. One such alloy is Nitinol™, an alloy of 55Ni-45Ti, developed by the Naval Ordnance Laboratory.

When a fastener, say a tubular rivet, is fabricated from Nitinol™, it can have the upset or locking feature fabricated (i.e., formed) in at the same time, while in its higher temperature (i.e., austenitic) phase. By deforming the fastener shank to remove the upset while the fastener is in its lower temperature phase and storing it below the transformation temperature, one may install it into a prepared hole in the structure to be joined. Then, when the fastener warms to above the transformation temperature, it returns to its austenitic shape and produces the needed locking feature by self-upsetting. Figure 3.29 shows some designs developed at Rensselaer Polytechnic Institute in an automated fastening technology research program.

Although these fasteners seem to have potential, they have seen only limited application because of other problems related to corrosion resistance and cost.

3.5 SNAP-FIT FASTENERS AND DESIGN FEATURES

Two trends in modern manufacturing have led to the rapid growth of a relatively new type of fastener or fastening feature, i.e., the snap-fit: the increased use of engineered polymers and organic-matrix composites, and automated assembly. For fastening to polymers, organic-matrix composites, and even thin sheet metals, insertion forces should be low to avoid deforming these easily deformed materials. For automated assembly, especially with the use of robots, insertion forces should be kept low, and the actions or motions needed to accomplish installation should be simple. Snap-fit fasteners and snap-fit design features fill all these requirements. Snap-fits are now widely used for both temporary and permanent assembly.

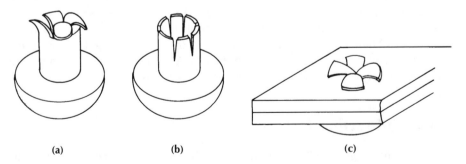

(a)   (b)   (c)

Figure 3.29   Self-upsetting fastener concepts. (a) Preformed in austenitic range; (b) straightened in martensitic range; (c) installed and allowed to self-upset to the memory shape established in (a).

Snap-fits employ some means of interlocking pieces together using a spring action in either the design of the component parts or the fastener or both. Snap-fits offer several attractive features, including (1) they are suitable with many different materials (e.g., sheet-gauge metals, polymers, polymeric composites, ceramics, and glass); (2) they require low insertion force with high pull-out resistance (typically, ten to a hundred times as great as the insertion force); (3) they resist loosening as a result of stress relaxation, since the snap-fit fastener or feature is not under constant stress from the assembly process once it is fully engaged; and (4) they resist to loosening by vibration, since there is less potential (or stored) energy in the fastener. In fact, snap-fits enable far easier assembly than any other mechanical fastening method.

The strength of a snap-fit comes from a combination of significant macroscopic mechanical interlocking and some friction. In the assembly process, the snap fit undergoes an energy exchange with a click sound or a snap (hence the name). During assembly, some feature of the part or the fastener (or both) is elastically deformed. Once fully engaged, the feature springs back to its normal, undeflected position, locking the parts together. In the assembled state, the components are not under load from any stored spring energy. Consequently, the strength of a snap-fit will not decrease with time as a result of any relaxation mechanism.

A snap-fit can be characterized by the geometry of its spring component or feature, with three types predominating:

1. cantilever type
2. hollow cylinder type, including noncircular section tubes
3. distortion type.

Figure 3.30 shows the basic types. The cantilever type includes designs with genuine cantilevers as well as designs with what are more correctly termed *leaf spring components*. Cantilever beams, button-groove combinations, split rings, clevises, and other design features can be and have been used successfully. The hollow cylinder is typified by the lid of many modern, child-proof pill bottles. The hollow cylinder snap design tends to be the strongest and has the most stable geometry, while the cantilever design tends to be the least stable, especially in torsion or impact. Distortion-type snaps include any shape that is deformed or deflected to pass over an interference. While the shapes of the mating parts in hollow cylinder snaps are the same, the shapes of the mating parts in distortion snaps are different. The design of snap-fits, like most fasteners, is highly sophisticated and requires thorough analysis of the deflection forces and actions.

Because snap-fits must deform elastically to install and lock, they tend to be made from polymers. For cantilever and distortion types, hard varieties are preferred, such as polystyrene, ABS, nylon, acetal, polycarbonate, or rigid vinyl. These polymers offer better dimensional stability and consistent shrinkage to permit precise control over the designed interference. Soft polymers, such as polyethylene, EVA, and soft vinyls, are suitable for cylinder type snaps. These polymers offer good sealing qualities.

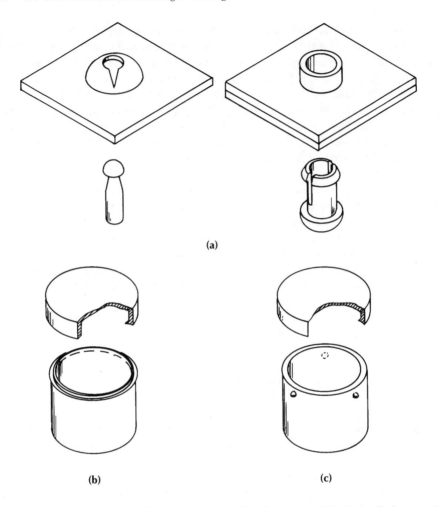

Figure 3.30   Various snap-fit fastener designs. (a) Cantilever type; (b) hollow cylinder type; (c) distortion type.

3.6 OTHER MECHANICAL FASTENING OR JOINING METHODS

3.6.1 General Description

The previous three sections described various fasteners that can be used for mechanically joining structures or assemblies. At the beginning of this chapter, it was stated that mechanical fastening required the mechanical interlocking of two or more parts or structural elements and might or might not require a supplemental device called a fas-

tener. It was pointed out that it is also possible to mechanically join parts using interlocking features of those parts. Those features can be inherent in the parts' designs or
may be introduced into the parts by some process that alters the part's form elastically
or plastically. The following sections will look at some options for mechanically joining or fastening using integral design features, part alterations, and special methods
(including stapling and stitching or sewing).

3.6.2 Integral Design Features for Interlocking

Sometimes designers accomplish mechanical joining or attachment through the use of
features of the mating parts themselves. These *integral design features* are particularly
useful for establishing position or orientation between parts and/or for preventing relative motion, as between parts in shear or between rotating parts (e.g., bearings or bushings in housings). Such features accomplish joining through physically interlocking the
adjoining parts on a macroscopic scale. Besides joining the parts of an assembly to
allow it to function, integral design features often facilitate the process of assembly by
providing self-alignment or keying of details, or, even, serving as self-tooling.

Some examples of integral design features are shown in Figure 3.31 and include (1)
press fits; (2) integral studs or projections or keys; (3) shoulders or flanges; (4) gripping
(mating) surfaces (e.g., using knurling); (5) sliding V-blocks or so-called Morse tapers;
(6) T-slots, dovetails, and tongues-and-grooves; (7) threaded fittings; and (8) bayonette
fittings. All of these rely on physical interference between macroscopic geometric features, with or without relying on some degree of friction or microscopic interference.

One interesting opportunity offered by integral design features or interlocks that
has not been widely exploited is that, instead of replacing fasteners, it might be possible and useful simply to reduce the magnitude of the load or control the direction of the
load on any fasteners that are used. Such an approach could allow less expensive,
easier-to-install fasteners to be used where more expensive, higher strength fasteners
were previously required. One example is the use of integral design features to produce
interlocking to prevent shear, allowing the replacement of high-shear rivets that are difficult to install with robots (given the requirement for both-side access) with blind or
pop rivets.

With increasing use of automation, and especially robots, in assembly, the use of
integral design features to provide self-alignment, self-fixturing, and even joining will
almost certainly increase dramatically.

3.6.3 Part Alteration to Accomplish Joining

Sometimes it is possible and desirable to lock two or more parts together by altering
the features of those parts somehow to prevent further motion or movement during or
after assembly. The most readily apparent method of altering the parts is by plastic

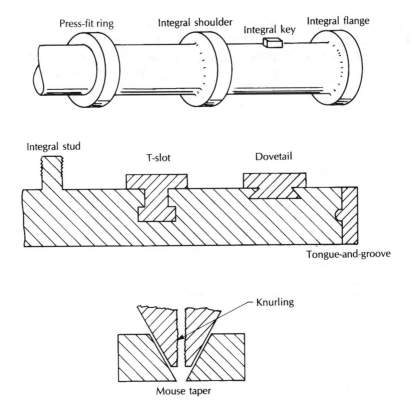

Figure 3.31   Some integral design features for fastening.

deformation, and several methods of joining materials or parts rely on such deformation. These methods include (1) formed tabs or tab fasteners or formed or coformed structures; (2) crimping; (3) hemming; and (4) staking or setting. Naturally, for such methods to work, the materials being joined must be plastically deformable, without degradation. Elastic recovery often contributes to the gripping action of plastically deformed features. For this reason, the methods mentioned are usually used with metals but can also be used with some polymers.

The forming of interlocking tabs, or *tab fasteners*, in sheet metal parts is a simple and inexpensive way to achieve joining. The formed tab is largely useful for holding parts in place only as long as the applied stresses are low. Closely related to formed tabs are *formed* or *coformed features*. A few examples are shown in Figure 3.32.

While with tab fasteners and formed features the detail parts are sometimes deformed during their fabrication before assembly, a new technique called stitch folding joins up to three plies of sheet metal to a total thickness of 2 mm (0.080 in.). The resulting folds resemble the dog-ears schoolchildren form in stacks of paper by folding and tearing at one corner to accomplish simple joining. Shear strengths of stitch folded joints in sheet metal can approach 80 to 90% of that of comparable riveted joints, and

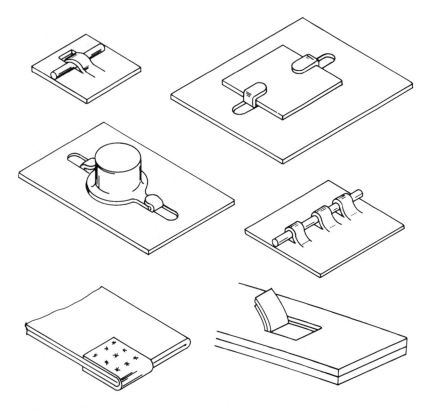

Figure 3.32 Examples of formed tabs or structures for joining sheetmetal. The lower right shows a stitch-folded assembly.

each stitch fold can be made in less than a second with a special tool marketed under the name Tagger™.

Crimping and hemming are similar processes, both of which join metal parts by plastically deforming the two pieces while they are in contact to cause interlocking. In *crimping,* one piece is crushed, squeezed, or otherwise deformed around the other to prevent relative movement. Occasionally, a soft, malleable metal is sandwiched between the folded or crimped features to facilitate clamping. In *hemming,* a linear joint is formed by plastically folding one piece of sheet metal over another to create an immobile seam. Sheet metal parts, such as the outer and inner skin panels of automobiles (e.g., doors, hoods and trunk lids) are often plastically deformed by hemming unless they are spot welded. Metal wires are commonly plastically deformed by crimping into terminal or connector bodies in electrical assemblies.

Staking and *setting* are methods of preventing axial sliding or rotation by creating a deformed impression at the mating surfaces of the two parts using a punch. This results in an interlocking bump and depression that prevents relative movement.

Figure 3.33 shows the methods of crimping, hemming, and staking or setting.

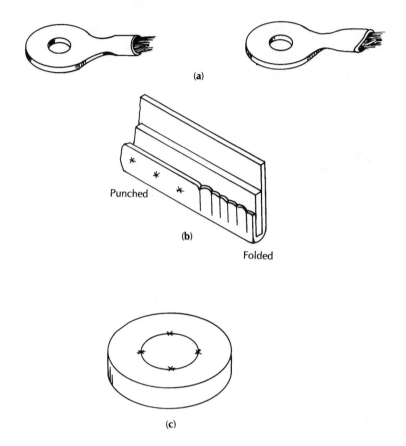

Figure 3.33   Crimping, hemming, and staking or setting. (a) Crimped electrical connection; (b) hemmed sheet metal joint; (c) staked joint.

3.6.4 Stapling and Stitching, or Sewing

It is often possible to join two or more materials or parts together by tying them together with a fine wire, fiber, or filament. *Stapling* and *stitching,* or *sewing* are common examples.

In *stapling,* fine wires are formed into U-shaped fasteners and driven through thin sheets of the materials to be joined. Often, the staple locks into the pieces by having the protruding ends of the wire deform orthogonally inward or outward to prevent withdrawal. In other cases, the staple is held in place just by the friction of the elastically deformed hole it produced. In *stitching* or *sewing,* a continuous fiber or filament (e.g., thread or wire) is passed through holes formed in the mating pieces of material by a needle, locking the pieces together with the continuous fiber. The fiber or stitch behaves like a fastener, preventing relative part movement by bearing and tension.

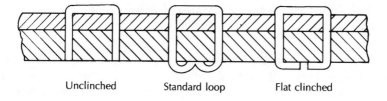

Unclinched   Standard loop   Flat clinched

(a)

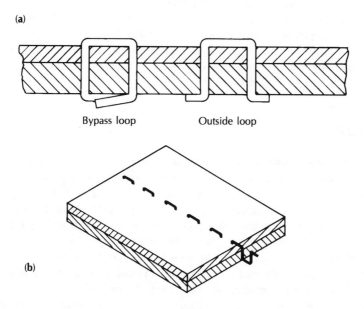

Bypass loop   Outside loop

(b)

Figure 3.34 (a) Metal stapling or stitching; (b) sewing or stitching composites.

Stapling and stitching have long been used with cloth, leather, and even flesh. Later, methods were developed for stapling or stitching thin sheet metals (e.g., tin, aluminum, copper, brass, and steel). Most recently stapling and stitching are finding application with laminated, organic-matrix composites. In these materials, the staples or stitches literally tie the layers together. By selecting the proper staple or fiber material (say, the same as the reinforcing fiber in the composite), the through-the-thickness strength can be increased significantly over the strength provided by the resin matrix or a bonding adhesive. Figure 3.34 illustrates stapling and stitching (or sewing).

Related to these methods of fastening are wire-wrapping, cabling, tying (or roping), and (rope or cable) splicing. Although these methods are important and can produce structural joints, they are not appropriate to advanced materials in general.

3.6.5 Other Methods for Mechanical Joining or Coupling

A host of other devices are used for mechanically joining or coupling the components of an assembly. They will not be described as they are often either more properly machine elements or they have little or no applicability to the joining of advanced materials. Examples include collars (used instead of machined shoulders or flanges); sleeves (for coupling shafts); rigid and flexible couplings for connecting aligned and misaligned shafts, respectively; clutches (which are couplings that can be disengaged or separated during operation); cast-in or molded-in attachments or inserts (which are integral fasteners such as bushings, heli-coil threads, or studs); clamps and clamp fasteners (for applying temporary compression forces for holding parts in place); and snap slides, hinges, hasps, hooks and latches.

SUMMARY

Joining by mechanical fastening requires component parts to physically interlock on a macroscopic scale, with some widely varying contribution from microscopic interlocking in the form of friction at the faying surfaces. This interlocking is often accomplished by inserting a supplemental device called a mechanical fastener through or into holes in the abutting parts to pin or clamp the joint but can also be accomplished using integral design features of the mating parts, which create physical interference either by their inherent geometry or with some alteration (e.g., by plastic deformation). Mechanical fasteners are broadly divided into two types: (1) threaded fasteners that produce a clamping action between the abutting parts using the principle of a helical inclined plane or screw to cause opposing translations of the parts, preventing motion at the joint by friction and bearing, and (2) unthreaded fasteners that rely primarily on bearing or pinning action, with little or no clamping-induced friction. Threaded fasteners include bolts and screws, which require prepared (and, sometimes, internally threaded) holes; nuts for use as backing pieces with bolts or screws when internal threading of the part is not used; and tapping screws, which produce their own hole and opposing threads as they are driven into a part. Unthreaded fasteners include rivets, which require plastic upsetting of a projecting shank to develop some clamping action; blind rivets, which produce the needed upsetting in the projecting shank through their design, without requiring access to the blind side; pins that resist motion by simple bearing; eyelets and grommets, which facilitate fastening with ties or with mating fastener pieces, respectively; retaining rings, which help prevent axial motion and/or take up mechanical slack by a spring action; keys, which transmit torque between or prevent relative motion in mating parts by locking between aligned slots; and washers, which serve to spread the load of fastener heads and/or nuts. Integral fasteners, which permanently mount to structures to facilitate joining using other fasteners; self-clinching fasteners, which lock themselves into the structure; and snap-fit fasteners, which employ elastic deflection or distortion to enable low-force insertion yet high pull-out strength are also available.

Integral design features inherent to the geometric design of the mating structural elements can be used in mechanical joining and include flanges and shoulders; interlocking projections and slots, and press-fits and tapers. It is also possible to join mating parts through part alteration to produce interlocking features, say by plastic deformation as in tab or structure forming, staking, crimping, or hemming. Finally, mechanical fastening can be accomplished using special methods, such stapling with formed wires or stitching, or sewing, with continuous fibers.

PRACTICE QUESTIONS AND PROBLEMS

1. The basis for all mechanical fastening or joining is physical interference or interlocking rather than chemical bond formation. Such interference or interlocking can occur at the macroscopic level (in the actual components or component design features) or at the microscopic level between material features (i.e., surface asperities). Describe several ways that macroscopic interlocking is used to accomplish mechanical fastening or joining. Specifically, consider the role of such interlocking with fasteners, with integral design features, and with part alteration by plastic deformation.
2. Related to the statement made in Problem 1, explain what role microscopic interlocking plays, including in mechanical fastening with threaded fasteners, with unthreaded fasteners and with integral design features.
3. Other than the obvious fact that one is threaded and the other is not, what differentiates the way in which threaded fasteners accomplish joining versus the way unthreaded fasteners do?
4. Explain how the thread on a threaded fastener works to enable joining with these devices.
5. Completely describe the bolt designated by each of the following: (a) 5/16-18 UNC, Class 2; (b) 5/16-24 UNF, Class 3; (c) MJC 70 × 1.5; (d) MJF 10 × 0.75.
6. Explain what is meant by "grade marking" on the head of steel bolts. What is meant by each of the grade markings illustrated?

(a)   (b)

7. What is the principal difference between screws and bolts? What is the principal difference between screws and tapping screws?
8. Describe at least four techniques that are used to prevent the loosening of nuts in service.
9. What is meant by the term *locknut*? Describe three major design approaches used in these fasteners.
10. What is meant by *integral fasteners*? Why are they used? Where are they used? Give some examples of integral fasteners.

11. What are the similarities and the differences among integral fasteners, self-clinching fasteners, and cast-in or molded-in fasteners?

12. Differentiate between the way in which rivets and bolts accomplish joining. For which type(s) of joint loading is each preferred? *Hint:* See Chapter 2.

13. Explain what is meant by *blind rivets*. Give two examples of generic application situations where such fasteners would be useful or required. What sacrifices, if any, must be made in using blind rivets as opposed to conventional rivets or bolts?

14. Describe a couple of designs for blind rivets.

15. Describe *pins* as mechanical fasteners. How do pins differ from rivets and bolts in their primary function?

16. Describe where and how retaining rings are used in mechanical assembly.

17. What are the major functions of washers in mechanical fastening?

18. Describe what is meant by *snap-fit* fastener. Why and where are these fasteners particularly useful? Describe a couple of generic designs for snap-fits.

19. Give some examples of *integral design features* used for accomplishing interlocking. Besides for joining, for what other manufacturing purposes are integral design features useful?

20. Describe several basic ways in which parts can be joined by altering their features during or after assembly.

REFERENCES

1. Chandra, Johan B., "An Investigation of Self-Upsetting Rivets for Fatigue-Life Improvement," Master's thesis, Rensselaer Polytechnic Institute, Troy, New York, August 1988.

2. Chow, W.W., "Snap-Fit Design," *Mechanical Engineering,* July 1977, pp. 35–41.

3. Ernest, Richard B., "Self-Clinching Fasteners," *Design News,* Sept. 27, 1982, pp. 49–51.

4. "Fastening and Joining," *Machine Design,* Reference Issue, Vol. 14, 1967.

5. Faupel, Joseph H., and Fisher, Franklin E., *Engineering Design,* 2nd ed. N.Y., John Wiley, 1981.

6. Haviland, Girard S., "Designing with Threaded Fasteners," *Mechanical Engineering,* Oct. 1983, pp. 17–31.

7. Laughner, Vallory H., and Hargan, Augustus D., *Handbook of Fastening and Joining of Metal Parts,* N.Y., McGraw-Hill, 1956.

8. Parmley, Robert O., *Standard Handbook of Fastening & Joining,* 2nd ed. N.Y., McGraw-Hill, 1989.

9. Shigley, Joseph E., and Mischke, Charles R., *Standard Handbook of Machine Design,* N.Y., McGraw-Hill, 1986, pp. 21.1–21.41 and 22.1–22.29.

Adhesive Bonding

4.1 INTRODUCTION

Materials can be joined mechanically, without the formation of atomic- or molecular-level bonds but, rather, totally through macroscopic and microscopic physical interference or interlocking. They can also be joined by actually forming interatomic or intermolecular bonds. One way of forming bonds and, thereby, joining materials chemically is through the process of adhesive bonding.

Until fairly recently, adhesive bonding was a rather unsophisticated but successful method of joining a wide variety of materials, including paper, fabrics, wood, leather, various polymers and rubbers, and some ceramics and glasses, largely for nonstructural or limited structural applications. It is commonly referred to as gluing, pasting, or, especially for ceramics and glasses, cementing. More recently, largely with the development of synthetic polymers, adhesive bonding has emerged as a unique and highly developed joining process, with applicability to a wide variety of materials of engineering interest, including engineered thermosetting and thermoplastic polymers, electronic and structural ceramics, semiconductors, glasses, metals (especially in thin sections), and various composites, most particularly those having organic matrices. Further, adhesive bonding is now often used for primary structural applications, as shown in Figure 4.1, as well as specialized nonstructural applications such as sealing, insulation, and vibration damping.

As a joining process, adhesive bonding offers many advantages, some of which make it unique. Nevertheless, it has its limitations, and these should be carefully considered during the design process, as well as in service. This chapter looks at the process of adhesive bonding, its general description, its advantages and disadvantages, the mechanisms proposed to explain how joint strength is obtained, modes and mechanisms of bond failure, the requirements for producing a good bond, and joint designs and design criteria. Chapter 5 looks at the constituents of adhesives, various ways the process can be classified, the various types of adhesives, the bonding process itself, and bonded joint performance.

4.2 ADHESIVE BONDING AS A JOINING PROCESS

4.2.1 General Description of Adhesive Bonding

Adhesive bonding is the process of joining materials with the aid of a substance capable of holding those materials together by surface attachment. The materials being joined are called the *adherends*. The bonding substance is called the *adhesive*. The forces that enable the surface attachment are a combination of some mechanical locking at the microscopic level between the adhesive and the adherends and usually substantial secondary (and, occasionally, primary) chemical bonding. Thus, adhesive bonding is fundamentally a chemical bonding process.

There are two fundamental types of adhesive bonding: (1) *structural adhesive bonding* and (2) *nonstructural adhesive bonding*. In structural adhesive bonding, the adhesive has sufficient strength that the adherend or substrate is stressed to near the point that it yields. This allows the designer to take full advantage of the adherend's strength and results in high joint efficiencies. To fulfill their function, structural adhesives must be capable of transmitting stresses without losing their own integrity, within the limits of the design[1]. Typical examples of structural adhesive bonding are shown in Figure 4.1, where metal-to-metal bonding, honeycomb bonding, and fiber-reinforced plastic bonding are all shown for a modern aircraft, namely, a C-5 military transport. In nonstructural adhesive bonding, the adhesive is used for some purpose other than for its structural strength. Examples are sealing against fluid loss or intrusion, electrical insulation, vibration damping or sound deadening. Nonstructural adhesive bonding is widely used in modern automobiles and structural bonding is increasing each year.

The focus of this chapter will be structural adhesive bonding, although most of the principles discussed apply to nonstructural adhesive bonds as well.

4.2.2 The Function of Adhesives

The primary function of adhesives is, obviously, to join materials or parts together. They do this by transmitting stresses from one element of a joint (or one adherend) to another in such a way that the stresses are distributed much more uniformly than with most, if not all, mechanical fasteners and many welds. This is so because the adhesive fills the entire joint and creates bonding forces (of whatever origin) over the entire area of the joint, rather than at discrete points of attachment, as with fasteners. Because of this, adhesive bonding can provide structural load-carrying capability that is equivalent to, or greater than, conventional mechanically fastened assemblies but at lower and more uniform stress levels and at lower cost and weight. This, in turn, permits the use of lighter weight materials and the production of lighter weight structures.

---

[1]With proper selection and use of adhesives, shear strengths up to 50 MPa (more than 7,000 psi) can be obtained.

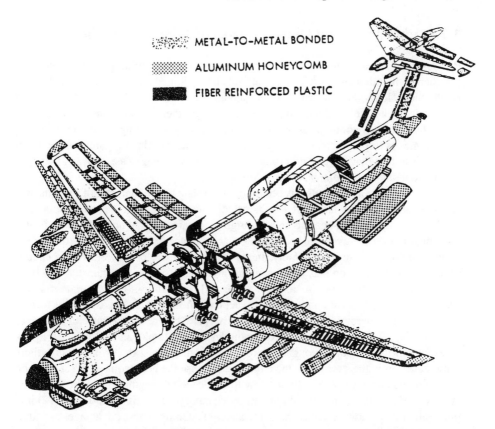

METAL-TO-METAL BONDED

ALUMINUM HONEYCOMB

FIBER REINFORCED PLASTIC

Figure 4.1 Use of adhesives for structural bonding in aircraft. (Reprinted with permission from L. E. Meade, "Adhesives for Aerospace," *Joining Technologies for the 1990's*, J. D. Buckley and B. A. Stein, eds., Noyes Data Corporation, 1986.)

Specially formulated flexible adhesives can accommodate differences in thermal expansion coefficients between adherends and, in this way, prevent damage that might otherwise occur if a stiff or rigid fastening system or joining method were used. This is often an important consideration and function. Flexible adhesives are also useful for imparting mechanical damping to a structure through their high internal friction.

Sealing is another important function of adhesives, particularly for nonstructural adhesives. The continuous nature of the bond that adhesives typically provide, seals out (or in) liquids or gases. Adhesives used for sealing (i.e., adhesive sealants) are often used in place of gaskets made of either solid or cellular materials. Sealing can also be carried out by potting or encapsulation compounds, which are really not true adhesives.

Since adhesives are applied as a thin, continuous film over a large area, they also can be used to improve fatigue resistance and to join thin or fragile parts. In both situations, loading is distributed uniformly, reducing the level of stress and overcoming most severe stress concentrations. The ability of most adhesives to withstand static or cyclic strains and shock loads without cracking usually causes adherends to fail before the adhesives in properly and optimally adhesive bonded structures.

4.2.3 Advantages and Disadvantages of Adhesive Bonding

Besides mechanical fastening, adhesive bonding is the only joining process that does not change the microstructure of the materials being joined and usually causes little or no chemical alteration. This can be important and valuable in itself and leads to another advantage. Adhesive bonding is suitable for joining either similar or dissimilar materials. In fact, adhesive bonding is often the very best way of joining dissimilar materials such as metals to polymers, ceramics to polymers, and metals to organic-matrix composites. The fact that there is no mixing of adherends, in most cases, means there are rarely problems with chemical compatibility.

Other advantages of adhesive bonding that relate to the character of the adhesive and the nature of the large area and, usually, continuous bond produced include (1) there is a larger load-bearing area for high load-carrying potential; (2) there is uniform distribution of stress and reduction of stress concentrations; (3) it is suitable for joining thin or thick materials of any shape; (4) it minimizes or prevents electrochemical or galvanic corrosion between dissimilar materials; (5) it seals against a variety of environments, leading to corrosion resistance and hermeticity; (6) it insulates against heat transfer or electrical conductance (except when the adhesive is specially formulated to provide thermal or electrical conductivity); (7) there is resistance to fatigue or cyclic loads; (8) there is mechanical damping and absorption of shock loads: (9) it provides joints with smooth contours for aerodynamic smoothness or cosmetics; (10) there is no reduction in the strength of the adherend because the heat required to set the joint (i.e., cure the adhesive) is usually too low to reduce the strength of the adherend (e.g., metal or ceramic); (11) it provides an attractive strength-to-weight ratio; and (12) it is frequently faster and/or cheaper than mechanical fastening. These advantages are summarized in Table 4.1.

Although adhesives can produce structures that are often more reliable than those produced by other joining methods, adhesively bonded structures must be carefully designed and used under conditions that do not exceed the known operational limitations of the adhesive. Such limitations include primarily the types or directions and magnitudes of stresses, whether static or dynamic, and environmental factors such as temperature, humidity, or the presence of solvents that can degrade an adhesive's properties. Specific limitations or disadvantages of adhesive bonding include (1) it requires careful surface preparation of the adherends to obtain durable bonds, often with corrosive chemicals; (2) rigid process control is needed, including emphasis on cleanliness, for maximum bond strength; (3) there is a need for holding fixtures, presses, ovens, and autoclaves, not usually required for other joining methods; (4) limited "working" times and shelf lives for adhesives complicate manufacturing logistics; (5) long cure times may be needed, particularly where high cure temperatures are not (or cannot be) used; (6) bonds do not permit direct visual examination of the bond area (unless the adherends are transparent), making inspection difficult; (7) repair of defective or deficient joints is difficult; (8) the upper service temperatures are limited to approximately 180°C (350°F) in most cases, but special adhesives, usually of high cost, are available for limited use to 370°C (700°F); (9) the useful life of the adhesively bonded joint depends critically on the environment to which it is exposed, *and* environment is generally much more of a concern than for other joining methods; (10) exposure to solvents used in cleaning or solvent cementing may present health hazards to workers; and (11) natural (e.g., animal- or vegetable-derived) adhesives are subject to attack by bacteria, mold, rodents, and vermin. These disadvantages are also listed in Table 4.1.

4.3 MECHANISMS OF ADHESION

4.3.1 General Description

Although much is known about particular aspects of adhesion, widely divergent views exist with respect to other aspects. One thing that is clear is that no single mechanism or theory satisfactorily explains all of what takes place in an adhesively bonded joint. The various theories that have been proposed are based on mechanical locking, surface adsorption, electrostatic attraction, interdiffusion, and even weak boundaries. In all likelihood, actual adhesion of materials probably results from a combination of two or more of these various mechanisms, perhaps in different combinations depending on the nature of the particular adhesives and adherends involved.

Regardless of the specific mechanism or mechanisms responsible for adhesion in adhesive bonding, a substantial proportion of the attractive forces between an adhesive and an adherend arise from chemical bonding, and largely secondary bonding at that. Two principal secondary bonding interactions that contribute to adhesion are van der Waals bonds and permanent dipole bonds. Van der Waals secondary bonding arises from the fluctuating-dipole nature of an atom with filled electron shells, while permanent dipole bonding arises when two polar molecules with asymmetrically distributed

Table 4.1 Summary of Advantages and Limitations of Adhesive Bonding

Advantages	*Limitations*
1. Large load-bearing area for high load-carrying capacity	1. Requires careful surface preparation
2. Uniform stress distributions	2. Requires rigid process control
3. Join thin or thick materials of any shape	3. Requires fixtures, presses, autoclaves, etc.
4. Join similar or dissimilar materials	4. Limited working times
5. Seal against many environments	5. Long cure times may be needed
6. Insulate against heat or electricity	6. Direct inspection of joint is not possible
7. Resist fatigue	7. Repair can be extremely difficult
8. Damp vibration or shock	8. Upper service temperature is very limited
9. Provide smooth joint contours	9. Life of joints is sensitive to environment
10. No reduction of strength of adherend by curing heat	10. Solvents can attach adhesive
11. Attractive strength-to weight	11. Natural adhesives are subject to attack by bacteria, mold, rodents, etc.
12. Minimize or prevent galvanic corrosion between dissimilar materials	
13. Often faster and cheaper than mechanical fastening	

Reprinted with permission from Arthur H. Landrock, *Adhesives Technology Handbook,* Noyes Publications, 1985.

charge attract one another's oppositely charged ends to produce a combination with a lower net energy.

The individual mechanisms or theories that have been suggested are described briefly in the following sections, along with the role of secondary bonding.

4.3.2 Mechanical Theory

According to the mechanical theory of bonding, for an adhesive to function properly, it *must* penetrate the microscopic asperities (i.e., hills, valleys, pores, and crevices) on the surface of the substrates and displace any trapped air at the interface. Adhesion is thus believed to be the result of mechanical interlocking or anchoring, with no need for any secondary bonding. The resulting mechanical anchoring between the adhesive and each of the adherends is indeed an important factor in bonding many porous substrates, such as rigid plastic foams or cellular polymers, unglazed ceramics, organic-matrix composites, and even many metals. Adhesives can frequently be made to bond better to abraded nonporous surfaces than to smooth surfaces, such as abraded metals or etched glasses. For this reason, mechanical abrading or chemical etching is an important step in all adhesive bonding for producing strong joints.

The actual effect of abrasion may be a result of one or more of several factors, including (1) enhancing mechanical interlocking, (2) creating a clean surface, (3 forming a highly chemically reactive surface, and (4) increasing the surface area of the bond because of the roughness produced. While the surface unquestionably becomes rougher because of abrasion, it is believed that a change in both physical characteristics and chemical reactivity of the surface layer leads to an increase in adhesive strength.

In any case, some degree of mechanical interlocking or anchoring almost always contributes to the strength of an adhesive bond.

4.3.3 Adsorption Mechanism

A second theory of adhesion attributes the adhesive force to the molecular contact that occurs between two materials (i.e., an adhesive and an adherend) and the resulting surface forces that develop. The process of establishing intimate contact between the adhesive and the adherend is called *wetting*.[2] Wetting is the process in which a liquid spontaneously adheres to and spreads on a solid surface and is controlled by the surface energy of the liquid–solid interface versus the liquid–vapor and the solid–vapor interfaces. A surface is said to be completely wet by a liquid if the angle of contact (θ) between the two is zero and incompletely wet if it is some greater angle.[3] Intermediate angles indicate intermediate degrees of wetting but, practically, any contact angle of less than 90 degrees indicates there is reasonable wetting for bonding. For an adhesive

---

[2]Wetting is an important concept and mechanism for many joining processes, including brazing and soldering, as well as adhesive bonding.
[3]Surfaces are commonly regarded as nonwettable if the contact angle exceeds 90 degrees.

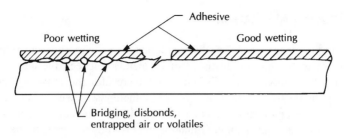

Figure 4.2   Good versus poor wetting by an adhesive.

to wet a solid adherend surface, it should have a lower surface tension (γ) than the adherend.

Good wetting is considered to have occurred when the adhesive flows to fill the microscopic hills and valleys on the adherend surface (as shown in Figure 4.2). Poor wetting is said to have occurred when the adhesive bridges over the valleys, resulting in a reduction of the actual contact area between the adhesive and the adherend, lowering the overall joint strength.

After intimate contact has been achieved between an adhesive and an adherend through wetting, it is believed that permanent adhesion results primarily from the forces of atomic or molecular attraction, i.e., chemical bonding. The chemical bonds involved in adhesion[4] and cohesion[5] can be either primary (e.g., ionic, covalent, or metallic) but are usually secondary (e.g., van der Waals' or permanent dipole). Which type of bonding predominates depends on the chemical nature of the adhesive and the adherend. Secondary bonding from van der Waals forces and permanent dipole bonding undoubtedly contribute to the adsorption mechanism.

4.3.4 Electrostatic Theory

In the electrostatic theory, adhesion is attributed to the development of electrostatic forces of attraction between the adhesive and the adherend at their interface. These forces are assumed to arise from the existence of an electrical double layer of separated charges at the interface and are believed to account for the resistance to separating the adhesive and the adherend. In support of the theory, electrical discharges have been noticed when an adhesive is peeled from an adherend. The reason for charge separation into layers probably arises from the presence of permanent dipoles or polar molecules in either the adhesive or the adherend or both.

---

[4]*Adhesion* is the state in which two surfaces of different materials are held together by physical or chemical valence forces, or both, such that it is necessary to do work to separate them.
[5]*Cohesion* is the state in which the particles of a single substance are held together by primary or secondary chemical valence forces.

4.3.5 Diffusion Mechanism

The fourth predominant theory of adhesion holds the primary mechanism responsible for attachment to be the interdiffusion of molecules between the adhesive and the adherend. The likelihood of such diffusion, and the resulting increase in bond strength, is greater for chemically similar adhesives and adherends, as when both are polymers. In such a case, the long-chain molecules of the adhesive and adherend may be mobile enough to interdiffuse and entangle. Entanglement is known to be extremely important in the joining of polymers, especially thermoplastics. Two types of adhesive bonding where diffusion undoubtedly plays a significant role are solvent cementing and fusion bonding of thermoplastics (see Chapter 13). It is difficult to apply the diffusion mechanism to the adhesive bonding of metals or ceramics.

Secondary bonding does not play a role in the diffusion mechanism.

4.3.6 Weak-Boundary Layer Theory

The weak-boundary layer "theory" tends to explain the failure of adhesively bonded joints more than adhesion but is worth noting. The theory arises from the observation that while many adhesively bonded joints seem to fail at the interface between the adhesive and one of the adherends, they in fact fail immediately adjacent to the interface. This suggests the existence of a weak boundary layer adjacent to the interface. Such layers could arise from some chemical event (e.g., reaction) in the adhesive or in the adherend, possibly from the environment. Another source of weak layers is contamination or impurity concentrations near the bonding surface.

Some examples of sources of weak boundary layers are (1) concentrations of low molecular weight constituents due to separation during bonding; (2) weakly attached oxide, sulfide, or other chemical layers on metals; and (3) entrapped air at the interface.

Figure 4.3 illustrates the five predominant mechanisms or theories of adhesion.

4.3.7 Adhesive Tack and Stefan's Equation

While several theories have been put forth to explain the mechanism or mechanisms by which adhesion is achieved between adherends joined by an adhesive, no single theory is fully satisfactory or accepted. Furthermore, none of the proposed theories of the previous sections explains well how joining is broadly achieved, independent of the specific adherends and adhesive.

An old, somewhat overlooked explanation of adhesion is found in an excellent, classic work on adhesive bonding by J. J. Bickerman entitled *The Science of Adhesive Joints* (1961). Bickerman explains adhesion bonding quite satisfactorily in terms of "tack," and he presents the development of an equation by Stefan that provides an excellent qualitative understanding of how adhesives behave, as well as quantifying joint strength.

When the gap between two adhesives is filled with a liquid adhesive (or, in fact, any liquid), it takes work to separate the adherends again. The resistance to separation

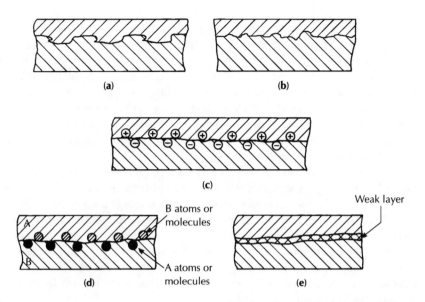

Figure 4.3 Five predominant mechanisms of adhesion. (a) Mechanical theory, locking; (b) adsorption theory, surface wetting; (c) electrostatic theory, double charge layers; (d) diffusion theory, mass interchange or transport; (e) weak-boundary layer theory, impurity layer.

is referred to as *tack*, or *tackiness*, as long as the adhesive remains liquid or at least pliable. In its usual usage, tack or tackiness refers to the situation where the joint can be produced with the application of only a weak external force and the measurement of resistance to separation is made soon after the adhesive is applied. It is interesting to consider the properties necessary for growth of tackiness or development of joint strength in an adhesively bonded system.

One of the factors favoring tack is an initially low viscosity in the adhesive. This is because it is necessary for the adhesive to distribute over the joint area to create full contact by spreading under a squeezing force or with the application of work. However, the resistance to separation clearly increases as the viscosity of the adhesive increases thereafter. In other words, a low viscosity permits rapid establishment of tack, but joint strength is also low. A high viscosity, on the other hand, would increase joint strength but at the cost of great difficulty in establishing tack. Practical adhesive systems seemingly require some compromise value of viscosity or require that the viscosity change from low to high with time.

The gradual establishment of good contact between a viscous liquid adhesive and a solid adherend has been widely studied. In most studies, the measure of good contact is judged from the resistance of the system to subsequent separation. It has been widely observed that this resistance generally increases with the time of contact t as well as with the pressure f used to produce contact and that the product ft during squeezing out of the adhesive is equal to (for Newtonian adhesives) or proportional to (for non-Newtonian adhesives) the product ft during separation. This behavior is exemplified when a piece of essentially any rubber is cut in two and the two new surfaces are immediately pressed back together by a pressure f for a time t. Pulling apart these two pieces is greater the greater f and t. Bickerman cites that with polyisobutylene, the force needed to separate two 1 cm wide strips by peeling was 10^5 dynes/cm after 15 minutes of contact and 2.2×10^5 dynes/cm after 15 hours. Furthermore, after 5 minutes of contact, the peeling force was 5×10^4 dynes/cm when a contact pressure of 5,000 dynes/cm^2 was applied and 18×10^4 dynes/cm when 60,000 dynes/cm^2 was applied. Indeed, after prolonged contact, the stress needed to separate two such pieces could equal the cohesive or breaking strength of the material, with the contact time required for this to occur being dependent on the particular rubber.

The question remains as to the mechanism responsible for the resistance to separation. The most frequently cited example is that of Johannson blocks. These precision ground and polished steel rectangular blocks are commonly used in metrology or gauging. When two clean blocks are pressed together—or, more correctly, are wrung together—in air, the force needed to separate them is considerable. A common, but erroneous, explanation for the origin of this "bonding" force is the force needed to overcome atmospheric pressure due to the vacuum created between the blocks by squeezing out the air. Of course, no such vacuum occurs. Three other reasons for the attraction between blocks are generally considered.

The first reason states that even though seemingly clean, the surfaces of the blocks (or any materials) are actually contaminated with a liquid (e.g., an oil used to protect the blocks from corrosion or from someone's hands or water condensed from the atmosphere). As the blocks are not microscopically smooth nor completely flat, there are ridges and valleys. Thus, contact occurs only at relatively high points rather than everywhere along the interface. This being the case, any traces of liquid between the blocks form droplets around the microscopic points of contact, producing a meniscus at each contact point. If the radius of curvature of the meniscus is R, the capillary pressure in the drop is given by γ/R, where γ is the surface tension of the particular liquid. The attractive force exerted by each such droplet is given by this pressure times the area of the point of contact, $\pi r^2 \gamma/R$, where r is the radius of the point of contact. For a typical Johannson block, R is approximately 10^{-6} cm (i.e., the height of ridges or depth of valleys), γ, for a typical liquid, is 50 g/sec^2, and the radius r of the point of contact is approximately 10^{-4} cm, so the attractive force for each point of contact is 1.5 dynes. For 1,000 contact points along a surface, the total force is 1,500 dynes, and if the volume of liquid is great enough to allow the formation of a continuous film on the order of 10^{-6} cm thick, the attractive force would be 5×10^7 dynes per cm^2 of block surface. Quite a force!

The second reason for attraction assumes no such liquid is present on the block surface. Here, the blocks are said to be attracted because of electrostatic forces that exist in all materials. These electrostatic forces arise from the work function of a solid, which, in turn, arises from the chemical composition of the block, surface contaminants, the crystallographic orientations of the surface layers or surface grains, and internal stresses. This is related to the electrostatic theory presented in Section 4.3.4.

The third reason for attraction is the predominant reason, and it relates to the tackiness of adhesives. Specifically, the product of the time and pressure needed to press together two plates separated by a viscous adhesive leads to an equal product of time and pressure to separate them again by reversing the motion or work. This has been expressed by Stefan as

$$ft = \tfrac{3}{4}\,(\pi\eta a^4)(1/h_1^2 - 1/h_2^2) \tag{4.1}$$

where f is the force required to separate the surfaces, t is the time required to separate the surfaces, η is the viscosity of the adhesive, a is the diameter (or linear dimension) of the contact, h_1 is the initial clearance before pressing together, and h_2 is the final adhesive thickness.

For the typical case, where the adhesive layer is thicker than the initial gap, the equation becomes:

$$ft = \tfrac{3}{4}\,(\eta)(a^2/h_1^2) \tag{4.2}$$

For Johannson blocks, Stefan's equation (in the simplified form of Eq. 4.2) says that a 100 g block with $a = 2.55$ cm and a separation of $h = 10^{-6}$ cm (comparable to the height of asperities) would take 1.8×10^5 sec (or 50 hr) to separate in air (with a viscosity η of 1.8×10^{-4} g/cm-sec) under the force of gravity (f approximately 10^5 g-cm/sec^2). If a liquid adhesive filled the gap between two plates, the time to separate would be much longer. Water instead of air between Johannson blocks would cause them to adhere together for 10^7 sec (or 116 days), given the viscosity of water (i.e., 0.01 g/cm-sec).

The real significance of Stefan's equation is that it provides a superb qualitative understanding of adhesive bonding. It predicts that adhesive bond strength will increase with the viscosity of the adhesive (η), with the area of bonding (a^2), and with the thinness of the bond layer (h). This is all consistent with observations and is the basis for the formulation of adhesives and the design of adhesive bonding joints for adhesive bonding. Adhesives are formulated to have a low initial viscosity to facilitate establishment of contact and then increase in viscosity with time through the evaporation of a solvent, the diffusion of a solvent or diluent, the cross-linking of a thermosetting polymer, or the hardening upon cooling of a thermoplastic polymer. Obviously, bond area and bondline thickness are also important to joint strength. Interestingly enough, no adherend properties appear anywhere in Stefan's equation and, thus, do not affect bond strength.

4.4 FAILURE IN ADHESIVELY BONDED JOINTS

4.4.1 Mechanisms of Failure

While there seem to be many mechanisms that can operate singularly or in various combinations to produce adhesive bond strength, there are also different mechanisms by which an adhesively bonded joint can fail. The two predominant mechanisms of failure in adhesively bonded joints are adhesive failure or cohesive failure. *Adhesive failure* is interfacial failure between the adhesive and one of the adherends and tends to be indicative of a weak-boundary layer, often due to improper preparation. *Cohesive fail-*

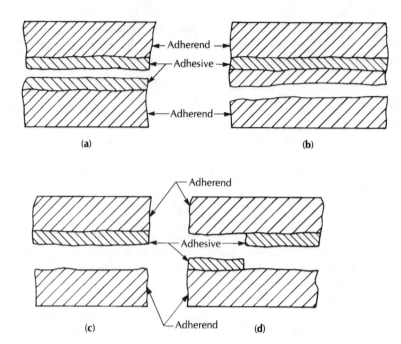

Figure 4.4   Adhesive versus cohesive versus mixed bond failures. (a) Cohesive failure in adhesive; (b) cohesive failure in adherend; (c) adhesive failure at interface; (d) partial (40%) cohesive failure. (Reprinted with permission from Arthur H. Landrock, *Adhesives Technology Handbook,* Noyes Publications, 1985, p. 9.)

ure is when fracture results in a layer of adhesive remaining on both adherend surfaces or, more rarely, when the adherend fails before the adhesive, with fracture totally contained in the adherend. This latter mechanism is known as *cohesive failure of the substrate.* These different modes of failure are shown in Figure 4.4.

The ideal type of failure is when cohesive failure occurs within the adhesive or one of the adherends. With this type of failure, maximum strength of the materials in the joint has been reached, and there is no lingering question about improper preparation before bonding or improper bonding procedures (see Section 4.5).

Joint failure in service or during testing is usually neither purely adhesive or cohesive but rather a mixture of both modes. For this reason, the operative failure mode is often expressed as a percentage of cohesive or adhesive failure, with ideal failure being 100% cohesive!

The operative mode of failure should not be the sole criterion for judging a bonded joint successful, however. Some combinations of adhesives and adherends may fail in adhesion but exhibit greater strength than a similar joint bonded with a weaker adhesive that fails cohesively. In practice, it is the ultimate strength of a joint, regardless of what process is used to make it, that is a more important measure of success than the mode of failure.

4.4.2 Causes of Premature Failure in Adhesively Bonded Joints

Premature failure of adhesively bonded joints, as well as any joint, is always a serious concern. The precise cause of premature failures in adhesively bonded joints is difficult to determine; much more difficult than for joints produced by other processes. For example, if the adhesive fails to wet the surface of the adherend completely during adhesive application, the bond is certain to be less than optimum. Internal stresses arising from adhesive shrinkage during setting, or stresses from different coefficients of thermal expansion (i.e., CTEs[6]) can cause premature failures. The types of stress acting on the completed bonds, their orientation to the adhesive, and the rate of application are also important factors influencing failure. Operating environmental factors, such as temperature, humidity, water, salt, solvents, and radiation, can also seriously degrade the performance of adhesively bonded joints[7].

Whenever possible, candidate adhesively bonded joints, including adhesive composition and form, adherend compositions and forms, joint geometry, and bonding practice, should be evaluated under simulated operating loads in the actual environment the joint is expected to encounter. A good test is worth thousands of words.

4.5 THE KEY REQUIREMENTS FOR QUALITY ADHESIVE BONDING

4.5.1 General Description

The objective of all structural adhesive bonding is to produce a bond that provides the maximum strength and quality possible for the particular combination of adhesive and adherend, usually at the minimum cost. To achieve this objective, several key requirements must be met: (1) cleanliness of the adherend surfaces before bonding; (2) proper wetting of the adherends by the adhesive; (3) proper choice of adhesive for the particular adherends and service conditions; and (4) proper joint design for the types and magnitudes of expected loads. A further requirement in the case where liquid adhesives are used is that the adhesive, once applied, must convert into a solid to produce joint strength.

The following sections look briefly at each of the above key requirements.

---

[6]The effects of differential coefficients of thermal expansion, or CTE, can be offset through the use of an inherently flexible adhesive or through appropriate fillers added to the adhesive.

[7]Combined factors, such as unfavorable stress state and certain environmental conditions can be expected to produce a synergistic effect, reducing joint strength more than might have been expected.

4.5.2 Joint Cleanliness

In order to obtain a sound, strong adhesive bond, it is essential to start with clean surfaces on the adherends. This is true since adhesive bonding is a surface phenomenon. All foreign materials, such as dirt, grease, cutting coolants and lubricants, water or moisture, and weak surface scales (e.g., oxides, sulfides), must be removed. If they are not removed, the adhesive will either not be able to reach and wet the actual adherend surface or will bond to these weak-boundary layers, compromising the final joint strength.

Thorough cleaning with various physical or chemical processes, or a combination of both, removes weak-boundary layers and mechanically and/or chemically conditions the adherend surface. The overall process of cleaning is often called surface preparation or pretreatment and usually involves one, two, or three of the following steps, in sequence, always starting at the first step and progressing to whatever step is felt necessary: (1) solvent cleaning; (2) intermediate chemical and/or mechanical cleaning; and (3) chemical treatment. A process called *priming* may also be carried out (as a fourth step) in some cases to ensure superior, durable bonds under particularly adverse environmental conditions.

Solvent cleaning is the process of removing soil from the surfaces of the adherends using an organic solvent, without physically or chemically altering those adherends. Solvent cleaning can be an end in itself or it can be the preliminary step in a series of progressively more aggressive cleaning and chemical treatment operations. Four basic solvent cleaning procedures, which are progressively more vigorous, are

1. Vapor degreasing for the removal of loose adhering particulate matter or dirt or light soluble soils using hot solvent (e.g., trichloroethylene) vapor that condenses on the adherend and flows away debris
2. Solvent wiping, immersion, or spraying with any of several different solvents (e.g., ethanol, methanol, acetone, or trichloroethylene) for the removal of light or heavy soluble soils (e.g., oils, greases, waxes), dirt, and particulate matter
3. Ultrasonic vapor degreasing for the removal of more tenacious soil and insolubles through the scrubbing action of collapsing bubbles (i.e., cavitation) arising from ultrasonic excitation of a liquid solvent
4. Ultrasonic cleaning in solvent, using the scrubbing action of collapsing bubbles during solvent immersion to break loose tenacious contaminants, followed by a liquid solvent rinse to remove residues. Organic solvents are not necessary. Aqueous solutions with surfactants, detergents, or alkaline or acid cleaners can be used. This process produces high-quality cleaning but is not as efficient as vapor cleaning processes

Following solvent cleaning, intermediate cleaning is often needed to remove especially tenacious contamination or loosely adhering layers of scale. Intermediate cleaning is the process of removing soil or scale from an adherend surface with physical, mechanical, or chemical means, singly or in combinations, without altering the adherend chemically. These cleaning methods are aggressive enough that they may remove some small amounts of the parent material. Some examples of widely used and acceptable intermediate cleaning methods are grit blasting, wire brushing, sanding, abrasive scrubbing, scrapping or filing, and alkaline or detergent cleaning. An intermediate cleaning step should always be preceded by solvent cleaning but may be the last step required for proper surface preparation before bonding.

Chemical treatment is the process of treating a clean adherend surface by chemical means, with the objective of changing the surface chemically to improve its adhesion qualities. The most common chemical treatment is acid or alkaline etching, or pickling, to remove especially tenacious surface films (e.g., oxides) and to roughen the surface on a microscopic scale. Solvent cleaning should always be performed before chemical treatment, and, frequently, intermediate cleaning should be performed as well.

Besides physically roughening the surface of the adherends microscopically to enhance mechanical anchoring, chemical treatment often activates the surface to better accept the adhesive. The mechanism for activation is the removal of adsorbed gases, intervening oxides, or other scales and the exposure of atomically clean material.

Occasionally, after the above cleaning operations, the surface of an adherend is subjected to priming. *Priming* involves applying a dilute solution of the adhesive and an organic solvent on the adherend to a dried film thickness of 0.0015 to 0.05 mm (0.00006 to 0.002 in.). This film protects the adherend from oxidation after cleaning, improves wetting, helps prevent corrosion, helps prevent adhesive peeling, and serves as a barrier layer to prevent undesirable reactions between the adhesive and the adherend. It also tends to help hold the adhesive in place during assembly for bonding.

4.5.3 Assuring Wetting

As stated earlier, wetting is important in adhesive bonding because it increases the contact area between the adhesive and the adherends over which the forces of adhesion may act. For good wetting by the adhesive, the surface of the adherend must be properly and thoroughly cleaned as described above. For proper bonding, the effectiveness of wetting must be assessed by either of two popular tests. The first is the water-break test, while the second is the contact angle test.

In the water-break test, the prepared (i.e., cleaned and pretreated) adherends are immersed in water, lifted from the water, and observed to see that a continuous, unbroken film of water adheres for some time. Break spots or "islanding" indicates poor wetting. A continuous film indicate good wetting. While the test is strictly qualitative, it gives a good indication of suitability for bonding. In the contact angle test, the angle of contact between the liquid adhesive and the solid adherend is actually measured, usually using an optical comparator. Smaller contact angles indicate better wetting. This test is more quantitative but of relatively little consequence.

Figure 4.5 shows how wetting is assessed with each of these tests.

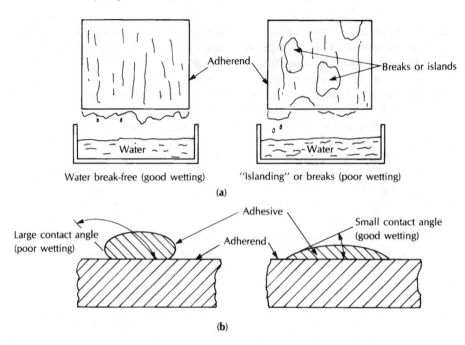

Figure 4.5 Testing for proper wetting for adhesive bonding. (a) Water break test; (b) contact angle test.

4.5.4 Choosing an Adhesive

Assuming joint cleaning and wetting assessment is done properly, successful adhesive bonding is ultimately the result of selecting the proper adhesive. Table 4.2, adopted from Landrock's *Adhesives Technology Handbook,* gives an excellent summary of the factors influencing adhesive selection. These factors include (1) capability for bonding specific adherends or adherend combinations; (2) suitability to service loading requirements and chemical factors (e.g., attack by external chemical agents or undesirable chemical reactions between the adhesive and adherends); (3) suitability to the service environment; (4) meeting specialized functional requirements; (5) meeting production requirements; (6) having acceptable cost; and (7) posing no health or safety hazards. These will be discussed in more detail in Chapter 5, Section 5.4.

One additional essential consideration in selecting an adhesive for any application is that that adhesive be capable of solidifying or stiffening after application in the liquid state. The ways in which liquid adhesives are converted to solids are through a chemical reaction induced by heat, pressure, or curing agent, singly or in combinations; by cooling and solidifying by a phase change; and by drying, after the evaporation of any carrier solvent.

Table 4.2 Factors Influencing Adhesive Selection

Capability of bonding specific adherends

Service requirements
    Stress (tension, shear, impact, peel, fatigue)
    Chemical factors
        external (chemical agents, solvents, etc.)
        internal (reactions, etc.)
Environmentsl factors
    Weathering
    Light
    Oxidation
    Moisture
    Salt spray
    Temperature extremes
    Biological factors
Specialized functional requirements (thermal or electrical
      insulation, etc.)
Production requirements
    Application requirements/method
    Bonding range
    Blocking
    Curing condition (time vs. temperature)
    Storage stability
    Working life
    Coverage
Cost
Health and safety hazards

Reprinted with permission from Arthur H. Landrock, *Adhesives Technology Handbook,* Noyes Publications, 1985.

4.5.5 Proper Joint Design for Bonding

One of the principal causes of failure in adhesive applications is poor engineering design of bonded assemblies and their joints. Joints for adhesive bonding should be designed to take advantage of the desirable properties and characteristics of adhesives while minimizing shortcomings. This means (1) providing the maximum bonding area possible in the design to help spread the applied load and minimize stress in the adhesive, (2) designing the joint in such a way as to force loads to be transmitted to the joint in favorable loading directions (e.g., pure compression, pure shear, or pure tension), (3) orienting joints or designing joint elements or reinforcements in such a way as to minimize unfavorable out-of-plane loading (e.g., peel or cleavage), (4) designing joints to assure uniformity in thickness of the adhesive layer and keeping this layer as thin as possible to maximize tensile and shear strengths, (5) designing joints and their elements in such a way that volatile components of the adhesive can be expelled (or absorbed by the adherend), (6) designing joints from combinations of materials that will minimize stresses arising from differences in thermal coefficients of thermal expansion, and (7) designing joint elements and assemblies in such a way as to facilitate adhesive application, curing, and inspection. Joint design is discussed in detail in the next section.

4.6 JOINT DESIGNS AND DESIGN CRITERIA

4.6.1 Basic Principles

Joints to be adhesively bonded must be designed specifically for the use of adhesives, just as joints for mechanical fastening or welding should be designed for those processes. This may seem obvious but is often forgotten. The aim of good adhesive-bonded joint design is to obtain the maximum strength for a given area of bond. In designing joints specifically for adhesive bonding, the basic characteristics of adhesives dictate the design of the joint, although adherend characteristics also play an important role, as will be shown later (Section 4.6.3). Most importantly, adhesive bonds act over areas rather than at single points, so joints should be designed with the objective of minimizing concentration of stress and maximizing bond area in any way possible.

Besides the obvious performance criteria, the selection of a joint design is also inevitably influenced by limitations in production facilities, production cost constraints, and desired final appearance of the part or assembly.

The strength of an adhesively bonded joint is determined primarily by the following factors: (1) the mechanical properties of the adhesive and the adherends; (2) residual (internal) stresses generated during processing; (3) the degree of true interfacial contact achieved through adhesive application and wetting; (4) the type of loading to which the joint is subjected; and (5) joint geometry.

The properties of specific adhesives and adherends will be discussed in Chapter 5 and in various chapters in Part II, respectively. The effect of residual stresses generated during processing is generally beyond the scope of this treatment, although sources of such stresses will be described. The degree of true interfacial contact was addressed in Section 4.5.3. The following sections address the types of loading to which joints can be exposed, typical joint designs, joint design criteria and analysis difficulties, and methods for improving joint efficiency.

4.6.2 Types of Stress in the Joint

Five types of stress are typically found in adhesively bonded joints, as shown in Figure 4.6, namely, from most favorable to least desirable (1) pure compression, (2) pure shear, (3) pure tension, (4) cleavage, and (5) peel. These types of stress can be found to

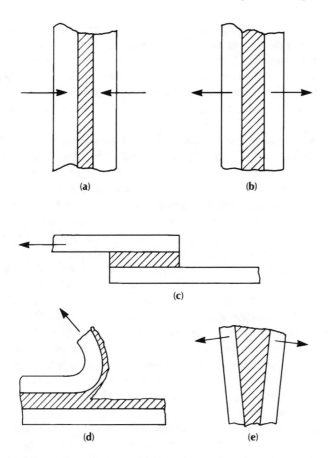

Figure 4.6   Types of stress in adhesively bonded joints. (a) compression; (b) tension; (c) shear; (d) peel; (e) cleavage. (Reprinted with permission from Arthur H. Landrock, *Adhesives Technology Handbook,* Noyes Publications, 1985, p. 32.)

occur singly or in any combination. Even when joints are stressed primarily in compression or tension or shear, these stresses can produce peel and cleavage.[8] Each type is described briefly below.

Pure compression or compressive loading occurs when the applied load is perpendicular to the plane of the joint and tends to squeeze the joint closed. An adhesively bonded joint is least likely to fail under this type of loading. Unfortunately, such loading, especially in the pure state, is rarely found in practice but can occur.

Although not next in the list of five stresses given above, pure tension or tensile loading is most appropriate to describe next. This type of loading is also perpendicular to the plane of the joint but tries to open the joint, pulling the two adherends apart. In

---

[8]In fact, while loading can actually be in shear, compression and tension can also give rise to shear.

theory, under pure tensile loading the stresses on the surfaces of the adherend and in the adhesive are evenly distributed. In reality, offsets in the joint, bending of the adherends, and other complications cause the actual loading, and stress distribution, to be nonuniform, leading to peel or cleavage. The strength of a tension loaded joint should be comparable to a shear loaded joint, barring complications, and can, in a few cases, exceed it. In tension joints, the adherends should be thick to avoid deflection or bending and offset loading.

The most common type of loading found in most adhesively bonded joints is shear. In pure shear, loading is parallel to the plane of the joint and tries to separate the joint elements by sliding them past one another. Pure shear imposes an even stress across the entire bonded area and, thereby, uses the entire joint area to carry applied loads to the best advantage. Whenever possible most of the loading applied to an adhesively bonded structure should be transmitted through the bonded joint in shear. To accomplish this, joints commonly have lap geometries (as described in Section 4.6.3). Lap shear strengths are directly proportional to the width of the overlap, but the unit strength decreases with the length of overlap. The optimum shear strength of a bonded joint is largely dependent on the shear modulus of the adhesive and its optimum thickness. This thickness varies from 0.005 mm (0.002 in.) for high modulus (i.e., stiff) adhesives to 0.015 mm (0.006 in.) for low modulus (i.e., flexible) adhesives.

Again, while not next in the list, peel loading is definitely next in frequency of occurrence. This type of loading is out of the plane of the joint and tries to open the joint nonuniformly, often from an edge. For this type of loading to occur, one or both of the adherends must be flexible and able to deflect. When this is so, a very high stress develops locally at the adhesive–adherend interface. Peel-type loading should be avoided wherever and however possible!

Cleavage is similar to peel, except that it forces one end of a rigidly bonded (as opposed to flexible) assembly to split apart. Cleavage occurs when an offset tensile force or moment is applied, causing stress to be non-uniformly distributed. Like peel, cleavage loading should also be avoided if possible.

4.6.3 Typical Joint Designs or Geometries

The ideal adhesively bonded joint is one in which, under all practical loading conditions, the adhesive is stressed in the direction in which it best resists failure, i.e., in shear. Adhesively bonded joints are designed with and used in many different configurations or geometries to achieve the above objective, but the most common are shown in Figure 4.7. The relative uses of these various joint types depend primarily on the load intensity to be achieved, as shown in Figure 4.8. As a general rule, simpler, less costly joint designs work well with simple, low-level loads, while higher, more complex loading situations demand more elaborate and expensive joint designs.

As a caution, butt joints cannot withstand bending loads because this leads to cleavage forces in the adhesive. Lap joints are the most commonly used in adhesive

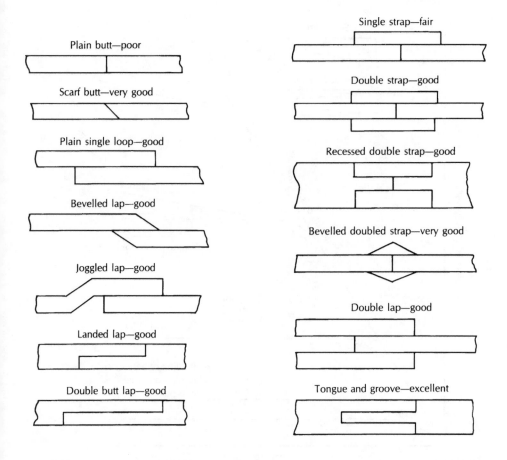

Figure 4.7   Various common joint types. (Reprinted with permission from Arthur H. Landrock, *Adhesives Technology Handbook,* Noyes Publications, 1985, p. 34.)

joints, as they are simple to make (in terms of both joint elements and assembly), can be used with thin adherends, and the stress developed in the adhesive is almost always shear. Unfortunately, bending can easily arise in simple lap joints, leading to cleavage.[9]

As a final consideration, if adherends are too thick to consider lap joints, modified butt joints (i.e., scarf joints or tongue-and-groove joints) can be considered. Obviously, no purpose is served by using unnecessarily complex joint geometries for the lower load intensities, and conversely. As pointed out previously, it is hopeless to expect that simpler and cheaper configurations could ever sustain high load levels.

---

[9]Methods for improving the efficiency of lap joints are presented in subsection 4.6.5.

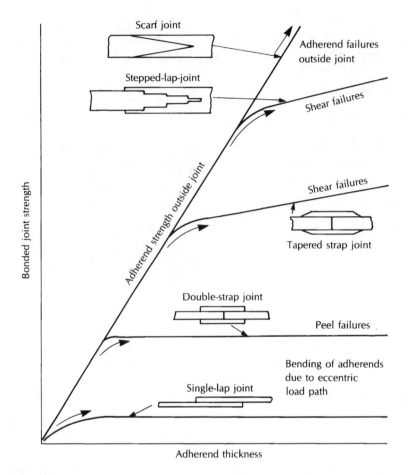

Figure 4.8 Joint type selection versus load intensity. (Reprinted with permission from L. J. Hart-Smith, "Design of Adhesively Bonded Joints" *Joining Technologies for the 1990's*, J. D. Buckley and B. A. Stein, eds., Noyes Data Corporation, 1986, p. 27A.)

4.6.4 Joint Analysis

Aside from the various shortcomings associated with adhesive bonding listed in Table 4.1, some of the major problems that have tended to limit wider application of the process include (1) how the strength of a joint might best be defined and the quality of joints compared; (2) how to predict joint strength; and (3) how a joint's geometry might be altered to optimize its strength and reliability. These problems are mechanical in nature and relate to the ability to perform stress analyses on joints.

The analysis required to properly design joints for adhesive bonding is complex for many reasons. First, adhesively bonded joints rarely see simple or pure loading of any type (e.g., compression, shear, or tension) but, rather, are usually subjected to some bending. Second, none of the components of an adhesively bonded joint is completely rigid. In fact, the adhesive is usually relatively flexible compared to the adherends, and the adherends may flex if thin or act rigidly if thick. This lack of joint rigidity, or relative flexibility, in both components of the joint has a pronounced effect on the distribution of stresses within the joint. Third, the thickness of the adhesive, as well as its modulus, effects distortion under loading and, thus, stress distribution in the joint. Finally, by their very nature, adhesively bonded joints are bimaterial systems, with often widely different properties.

For the above reasons, rigorous analysis of stress distributions and concentrations in adhesively bonded joints is thus rare, even in this the age of sophisticated computers. Two especially important and widely used joints have been analyzed, however, and results have been closely checked by both laboratory tests and actual service experience. These two types are lap and scarf joints. Results from these simple joints have been extended to more complex joints using more sophisticated analytical and numerical techniques.

Figure 4.9 shows the six most common lap and scarf joints for which stress distributions and concentrations have been calculated. These types include colinear scarf joints, butt joints, single flat offset lap joints, double flat lap joints, tubular lap joints, and landed lap joints.

A pure scarf joint is the most efficient of common structural joints and the simplest to analyze since loading is colinear and no bending is introduced. For such joints, the most important stresses are (1) the tensile stress that acts normal to the adherend faces, trying to pull those adherends apart, and (2) the shear stress that acts parallel to these faces, trying to slide them apart. Since the adhesive line in a scarf joint is at an angle to the applied loads, combined stress theory will give the modifying angle functions for calculating these key stresses. Table 4.3 gives modified equations for calculating each stress, assuming adherends are both the same and isotropic. Since both stresses depend upon $\sin \theta$, by designing with a small scarf angle, the joint strength depends only on adherend strength.

Stresses in butt joints can be determined from the equations for scarf joints by making the scarf angle $\theta = 90$ degrees. Butt joints fail to take advantage of the full strength of the adherend because the bond area is minimized, but, as with scarf joints, stress concentration is low.

The more complex joint geometries involve much more complicated analysis. The simplest (and earliest) of these analyses considers the adherends in a single flat offset lap joint to be completely rigid and has the adhesive deform only in pure shear. As shown in Figure 4.10a, if the width of the joint is b, the length l, and the load F, then the shear stress τ is uniform and given by:

$$\tau = F/bl \tag{4.3}$$

If, on the other hand, the adherends are elastic (as shown in Figure 4.10b), the situation is quite different. For the upper adherend, the tensile stress is a maximum at point A and falls to zero at point B. As a result, the tensile strain at A is larger than at B and must progressively be reduced over the length l. For the lower adherend, the converse is true. Thus, assuming that the adhesive–adherend interface remains intact, the uniformly

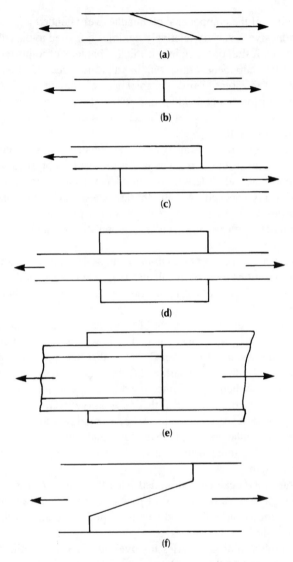

Figure 4.9 The six most common lap and scarf joints. (a) Colinear scarf joint; (b) butt joint; (c) single flat offset lap joint; (d) double flat lap joint; (e) tubular lap joint; (f) landed scarf joint.

sheared parallelograms in Fig. 4.10a become distorted to the shapes shown in Fig. 4.10b, leading to differential shear.

This problem, first analyzed by Volkerson in 1938 in his "shear lag analysis," assumed that the adhesive deformed only in shear while the adherends deformed only in tension. The solution is complicated, and details of the analysis are beyond the scope of this book, so are left to the interested reader by reference to Volkerson's original work (in German) or other sources (Adams, 1984; Anderson, 1977). The solution

Table 4.3 Summary of Equations for Analysis of Various Adhesive Scarf Joints

Type	Loading	Geometry	Shear Stress	Normal Stress
Flat Scarf				
	Tension, compression	F = force / unit width	$\tau = \dfrac{F}{t}\sin\theta\cos\theta$	$\sigma = \dfrac{F}{t}\sin^2\theta$
	Pure bending	M = moment / unit width	$\tau = \dfrac{6M}{t^2}\sin\theta\cos\theta$	$\sigma = \dfrac{6M}{t^2}\sin^2\theta$
Tubular Scarf				
	Tension, compression	P = axial force	$\tau = \dfrac{P}{2\pi r_o t}\sin\theta\cos\theta$	$\sigma = \dfrac{P}{2\pi r_o t}\sin^2\theta$
	Pure bending	M = bending moment	$\tau = \dfrac{2M(r_o + r_i)}{\pi(r_o^4 - r_i^4)}\sin\theta\cos\theta$	$\sigma = \dfrac{2M(r_o + r_i)}{\pi(r_o^4 - r_i^4)}\sin^2\theta$
	Pure torsion	T = torque	$\tau = \dfrac{2T\sin\theta}{\pi(r_o - r_i)^2}$	$\sigma = 0$

Reprinted with permission from H.A. Perry, "How to Calculate Stresses in Adhesive Joints," *Production Engineering*, Vol. 29, No. 27, July 7, 1958.

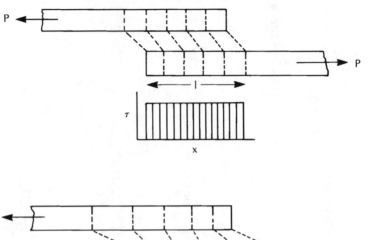

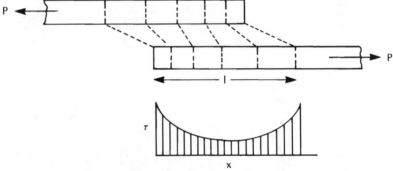

Figure 4.10   Exaggerated deformation in single lap joints. (Reprinted with permission after R.D. Adams, "Theoretical Stress Analysis of Adhesively Bonded Joints," *Joining Technologies for the 1990's,* J. D. Buckley and B. A. Stein, eds., Noyes Data Corporation, 1986, p. 189.)

predicts, however, that the maximum shear stress in the adhesive occurs at the ends of the lap joint overlap and is given by

$$\tau_{max}/\tau_m = (F/2)^{-1/2} \coth (F/2)^{-1/2} \tag{4.4}$$

where τ_m is the average applied shear stress ($= F/bl$), from Eq. 4.1, and F is Gl^2/Et_1t_2, where G is the shear modulus of the adhesive, l is the length of the joint or the overlap, E is Young's modulus for the adherends, t_1 is the thickness of the adherends (assuming they are equal), and t_2 is the thickness of the adhesive.

Example 4.1: Determining Maximum Shear Stress to Average Shear Stress in a Bonded Joint

For the single flat lap joint shown in Figure 4.11, what is the ratio of the maximum shear stress to the average shear stress? Assume the modulus of elasticity for an alu-

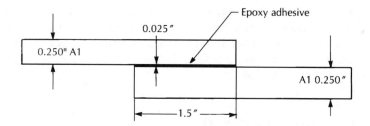

Figure 4.11 Simple single overlap joint.

minum alloy is 72 GPa (10.5×10^6 psi) and the shear modulus for an epoxy adhesive is 2464 MPa (360,000 psi).

Using Eq. 4.4, and substituting the values of shear modulus for the epoxy adhesive ($G = 360,000$ psi), the length of overlap in the joint ($l = 1.5$ in.), the modulus of elasticity for the aluminum alloy adherend ($E = 10,500,000$ psi), and the thicknesses of the adherend ($t_1 = 0.200$ in.) and the adhesive layer ($t_2 = 0.020$ in.), result in

$$\tau_{max}/\tau_m = (F/2)^{-1/2} \coth(F/2)^{-1/2}$$

where

$$F = Gl^2/Et_1 t_2 = (3.6 \times 10^5 \text{ psi})(1.5 \text{ in.})^2/(10.5 \times 10^6 \text{ psi})(0.250 \text{ in.})(0.025 \text{ in.})$$

$$= 12.343$$

so

$$\tau_{max}/\tau_m = (12.343)^{-1/2} \coth(12.343)^{-1/2} = 1.053$$

This is not a very high degree of stress concentration.

Given the form of the equation for τ_{max}/τ_m, as F becomes large, the degree of stress concentration approaches 1.0 (i.e., there is no concentration of stress) at the ends of the overlap as (1) the adhesive gets stiffer relative to the adherend; (2) the modulus of the adherend gets lower relative to the adhesive; (3) the length of overlap increases; and (4) either the adherend or the adhesive or both get thinner. Contrarily, as F becomes small, the degree of stress concentration increases dramatically as (1) the shear modulus of the adhesive decreases relative to the adherend; (2) the adherend gets stiffer relative to the adhesive; (3) the length of overlap decreases; and (4) the thickness of either the adherend or the adhesive or both increases.

The relationship between F and τ_{max}/τ_m is shown in Figure 4.12.

Volkerson's analysis fails to take into account two important factors. First, the two opposing forces applied to the joint are not co-linear, so there will be some bending applied to the joint in addition to the in-plane tension. Second, the adherends are not completely rigid, but bend, allowing the joint to rotate. This rotation further alters the direction of the load line in the region of the overlap. These factors were taken into account by Goland and Reissner in their analysis. This analysis will not be covered here because of its complexity. Suffice it to say that their analysis uses a factor, k, which relates the bending moment on the adherend at the end of the overlap, M_o, to the in-plane loading, F, by the relationship

$$M_o = kF(t/2) \tag{4.5}$$

where t is the thickness of the adherend.

If the load on the joint is very small, no rotation of the overlap occurs, and the load acts along the line shown in Figure 4.13a. It can be seen to pass close to the edge of the adherends at the ends of the overlap, so that $M_o = Ft/2$ and $k = 1.0$. As the load is increased, the overlap rotates. In doing so, the line along which the applied load acts moves closer to the centerline of the adherends, as shown in Fig. 4.13b. This reduces the value of the bending moment.

While the analysis of Goland and Reissner includes some simplifications and assumptions, it is still one of the most rigorous mathematical studies of lap joints. More recent analyses of the more complex designs (e.g., double flat lap tubular lap and landed joints) have been performed by Pantema, Cornell, Lubkin, Lubkin and Reissner, Mylonas, de Bruyne, and McClaren and MacInnes (Patrick, 1967, 1969, 1973). Modern computers have also allowed intricate numerical solutions using finite element methods. Readers interested in the analysis of these more complex joints are referred to specialized sources (e.g., Adams, 1984; Adams, 1987; Anderson *et al*, 1977; Elliott, 1973; Perry, 1958).

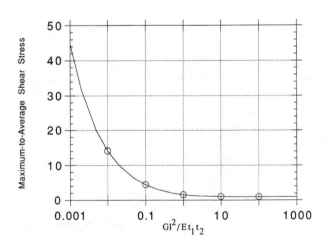

Figure 4.12   Relationship between τ_{max}/τ_m and the value of F, where F is $Gl^2/Et, t_2$.

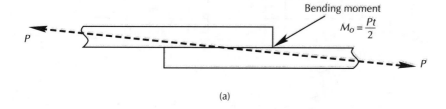

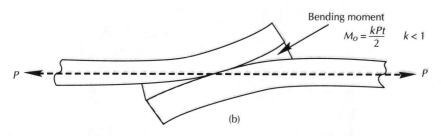

Figure 4.13 Goland and Reissner's bending moment factor. (Reprinted with permission from R. D. Adams, "Theoretical Stress Analysis of Adhesively Bonded Joints," *Joining Technologies for the 1990's*, J. D. Buckley and B. A. Stein, eds., Noyes Data Corporation, 1986, p. 191.)

4.6.5 Joint Design Criteria

Before leaving the subject of joint design, some general comments related to design criteria are worthwhile.

The previous section on joint analysis showed that the stress in the adhesive of a bonded joint is ordinarily a combination of various stresses. Further, the relative flexibility of most adhesives to that of most adherends has a pronounced effect on the resulting stress distribution, leading to often severe stress concentrations. Figure 4.14 shows a typical example of a simple lap joint under tensile loading. It is quickly apparent that most of the stress is concentrated at the ends of the lap, with most of the rest of the lap carrying a comparatively low stress. As a consequence of this distribution, increasing the overlap length does little to increase the load-carrying capacity of the joint. Far more strength could be gained by increasing the joint width. This is shown in Figure 4.15.

A second design criteria is that the bonded area in a joint should be large enough to resist the greatest force that the joint will meet in service, given the allowable stress of the adhesive. Calculation of stress alone is not enough, however, the effects of environmental conditions, age of the adhesive, temperature of cure, composition and size of the adherends, and thickness of the adhesive layer must all be considered.

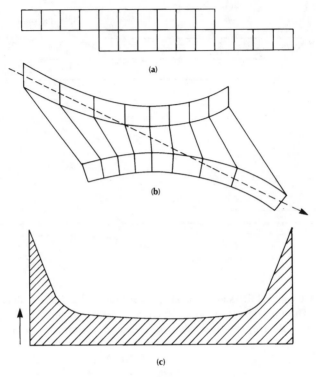

Figure 4.14   A single-lap joint (a) is subjected to a shear stress. Eccentric loading causes the distortion shown in (b), and results in the stress distribution shows in (c). (Reprinted with permission from Arthur H. Landrock, *Adhesives Technology Handbook,* Noyes Publications, 1985, p. 35.)

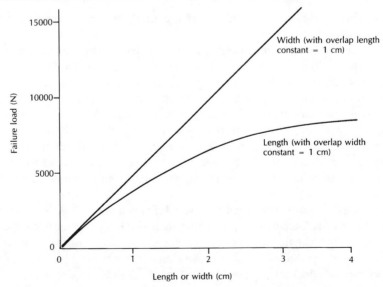

Figure 4.15   Effect of lap joint overlap and width. (Reprinted with permission from Arthur H. Landrock, *Adhesives Technology Handbook,* Noyes Publications, 1985, p. 36.)

As a final criteria, the strength of a lap joint is directly dependent on the yield strength of the adherend.

4.6.6 Methods of Improving Joint Efficiency

There are several ways of improving the effectiveness of an adhesively bonded joint by dealing with the problems of most adhesives' limited strength and flexibility, as well as with adhesive and/or adherend distortion under noncolinear loading. First, the bonded area should be as large as possible, within allowable geometric and weight constraints. Second, the maximum possible percentage of the bonded area should contribute to the strength of the joint through proper wetting. Third, the adhesive should be stressed in the direction of its maximum strength, which is usually in shear. Fourth, stress should be minimized in the direction in which the adhesive is weakest, which is usually peel or cleavage.

It should be obvious that peeling should be avoided wherever and however possible. Peeling can be a particular problem when thin members are bonded to thicker members, in which case the operating loads tend to peel the thin member from the thicker member. Some techniques for resisting peeling include (1) riveting or spot welding (see rivet-bonding or weldbonding in Chapter 10); (2) beading the end of the thin member of the joint; (3) increasing the width of the thin member at the end of the overlap; or (4) increasing the stiffness of the adherend. These techniques are pictured in Figure 4.16. The stiffness of joints composed of thin adherends can be increased by using doublers above or below the primary adherend, strong-back stiffners (e.g., T's, L's), formed beads or other techniques.

In short, adhesively bonded joints function extremely well under pure compressive or pure tensile loads or under shear produced by colinear loading. As soon as shear loading is not colinear, bending occurs, giving rise to a stress distribution that produces the highest stress at the edges of laps. This, in turn, tends to cause peel or cleavage. Out-of-plane loads can also lead to peel or cleavage, which is always undesirable.

SUMMARY

Materials can be joined by creating secondary (or occasionally primary) chemical bonds across an interface by using a substance to facilitate that bonding. This process is called adhesive bonding. The bonding substance is called an adhesive, and the materials being joined are called adherends. The resulting joints can be structural (i.e., intended to transmit stress near the yield point of one or both of the adherends without losing integrity) or nonstructural for the purpose of sealing against fluid leakage or intrusion, to provide electrical or thermal insulation, or to provide mechanical damping. Since the adhesive creates bonding over a large area of the joint in a thin layer, stresses from applied loads are distributed uniformly, resulting in improved fatigue life and allowing thin or fragile structures to be joined. Since the adhesive can be made flexible, differential thermal coefficients of expansion can be accommodated. The process is also ideal for joining dissimilar materials because the adhesive causes no change in the

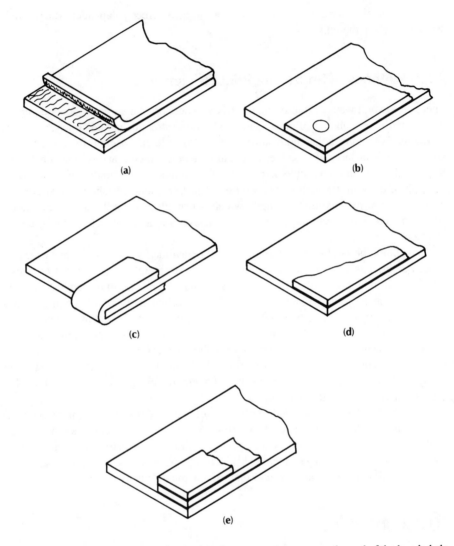

Figure 4.16 Normally, shear loading results in some peel stress near the end of the bonded element, leading to lifting by peel as in (a). Various methods for increasing the resistance to such peeling or for preventing such peeling are shown in (b) through (e). (a) Peel at end; (b) rivet end; (c) bead end; (d) increase end width; (e) increase stiffness. (Reprinted with permission after Arthur H. Landrock, *Adhesives Technology Handbook,* Noyes Publications, 1985, p. 39.)

microstructure and little or no change in the chemical nature of the adherends. Despite these obvious advantages, adhesive bonds have shortcomings, the greatest being the difficulty of inspection and the difficulty of repair of the bonded joint and the susceptibility of the adhesive to environmental degradation in service.

Bond strength is attributed to one or more of several mechanisms, including microscopic mechanical locking, surface adsorption due to wetting, electrostatic attraction between oppositely charged double layers at the interface, interdiffusion across the interface, or the existence of a weak-boundary layer adjacent to the interface. Van der Waals (fluctuating dipole) and/or permanent dipole secondary bonds are involved in many of these mechanisms. Whatever the mechanism, obtaining a good bond demands thorough cleaning, proper wetting, proper choice of the adhesive, and proper joint design. Joints are judged good if they fail predominantly cohesively in the adhesive or in one of the adherends, rather than adhesively at the adhesive-to-adherend interface. The stresses in joints can be compressive or tensile, giving rise to shear, peel, or cleavage under certain conditions. An adhesive tends to perform best in shear, as this maximizes the load carrying area, and worst in peel or cleavage, as these tend to load the adhesive nonuniformly in an opening mode. Joint designs attempt to force as much of the applied loading to act in shear as possible and avoid peeling through the incorporation of special design features or stiffeners. The analysis of adhesively bonded joints is complex, due to the nonperfect shear and the deflection of most adherends, leading to nonlinear strains and bending moments. Volkerson's shear-lag method, and Goland and Reissner's improvement accounting for a bending moment factor give reasonable solutions of shear stress distribution in single lap shear joints.

PRACTICE QUESTIONS AND PROBLEMS

1. Differentiate between *structural adhesives* and *nonstructural adhesives* in terms of the role of the adhesive, the level of loading in the adhesive, and the level of loading in the adherends.
2. What are several functions of adhesives besides joining materials together?
3. What are some of the particular advantages of adhesive bonding as a joining process? What are two specific advantages over mechanical fastening?
4. What are some specific limitations of adhesive bonding?
5. There are several theories as to the mechanism responsible for adhesion, including mechanical, adsorption, electrostatic, and diffusion. The relative proportion of any particular mechanism depends, among other things, on the adhesive–adherend combination. For what types of adherends (by material class) are mechanical forces particularly important? For which types of adherends (by material class) do mechanical forces play little role?
6. Explain the role of van der Waals and permanent dipole bonds in the electrostatic theory of adhesion. What is the role of these bonds in the adsorption theory?
7. An adhesive joint between metal and rubber fails. There is a substantial layer of rubber left on the metal after failure. By what mechanism did failure occur? Is this considered to have been a good joint or not? Would changing to an adhesive with a higher shear strength help? What, if anything, would help?
8. What are the four key steps involved in maximizing the likelihood of producing a good adhesive bond? Give some details associated with each step.

9.   What are the five types of stress that can occur in adhesively bonded joints? Rank these types from most desirable to least desirable.

10. For the scarf joint shown below, what is the shear stress and what is the normal stress in the adhesive? What happens to these stresses if the scarf angle is reduced to 15 degrees? What if the angle is further reduced to 15 degrees? Would it be more advantageous to orient the scarf in the opposite direction to that shown?

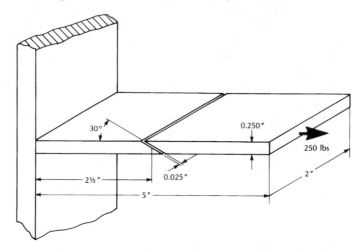

11. Suppose the joint shown in Problem 10 were a butt joint. What would be the shear and normal stresses?

12. Suppose the joint shown in Problem 10 had a scarf angle of 30 degrees and adherend thicknesses of 0.250 in. If a force of 100 lb. were applied downward at the unfixed end (rather than in the plane of the adherends), what would be the shear and normal stresses?

13. What would the shear and normal stresses be if the scarf angle were decreased to 15 degrees? Increased to 90 degrees?

14. For the single flat lap joint shown in Example 4.1 in the text, calculate the ratio of maximum-to-average shear stress if the adhesive is changed to a flexible silicone adhesive with $G = 704$ psi (0.48 MPa). All other materials and dimensions remain the same as in the example. How does this ratio compare to the ratio in Example 4.1 for stiffer or more rigid adhesive?

15. Calculate the maximum-to-average shear stress ratio if the adherends in Problem 14 were changed to a Nylon 6,6 thermoplastic (i.e., polyamide), with $E = 145,400$ psi (993 MPa). How does this ratio compare to the ratio in Problem 14 for a stiffer adherend?

16. Assuming aluminum adherends and an epoxy adhesive as in Example 4.1 in the text, what would the ratio of maximum-to-average shear stress be if the adhesive thickness were reduced to 0.010 in.? What would the ratio be if the adhesive thickness were increased to 0.040 in.? How do these shear stress ratios compare to the ratio obtained in Example 4.1?

17. Assuming aluminum adherends and an epoxy adhesive as in Example 4.1 in the text, what would the ratio of maximum-to-average shear stress be if the thickness of the aluminum were decreased to 0.080 in. for the same adhesive thickness (i.e., 0.020 in.)? What would the ratio be if the thickness of the aluminum were increased to 0.500 in. for the same adhesive thickness (i.e., 0.020 in.)? How do these ratios compare to the ratio obtained in Example 4.1?

18. Suppose the adherend in Example 4.1 were a thermosetting epoxy with a shear modulus $G = 360,000$ (2464 MPa), but everything else was the same as in the example, what would be the ratio of maximum-to-average shear stress? How does this compare to the ratio obtained in Example 4.1?

19. Why is it that increasing the length of overlap in a lap joint offers little benefit after a certain point? Approximately how much overlap is considered optimal?

20. What changes in the design of a simple, single flat lap joint would minimize peel loading near the ends of the overlap?

REFERENCES

1. Adams, Robert D., *Structural Adhesive Joints in Engineering,* New York: Elsevier, 1984.
2. Adams, R.D., "Theoretical Stress Analysis of Adhesively Bonded Joints," *Joining Fibre-Reinforced Plastics* F. L. Matthews, ed., London: Elsevier, 1987, pp. 185–226.
3. "Adhesives and Sealants," *Engineered Materials Handbook,* Vol. 3, Materials Park, Ohio: ASM International, 1990.
4. Anderson, Garron P., Bennett, S. John, and DeVries, K. Lawrence, *Analysis and Testing of Adhesive Bonds,* New York: Academic Press, 1977.
5. Bickerman, J. J. *The Science of Adhesive Joints,* New York: Academic Press, 1961.
6. Cagle, Charles D., *Adhesive Bonding: Techniques and Applications,* New York: McGraw-Hill, 1968.
7. Elliott, S. Y., Techniques for Evaluation of Adhesives, *Handbook for Adhesive Bonding* (C. D. Cagle, ed.), New York: McGraw-Hill, 1973.
8. Landrock, Arthur H., *Adhesives Technology Handbook,* Park Ridge, New Jersey: Noyes Publications, 1985.
9 Meade, L. E., "Adhesives for Aerospace," *Joining Technologies for the 1990's,* John D. Buckley and Bland A. Stein, eds., Park Ridge, N. J.: Noyes Data Corporation, 1986, pp. 340–349.
10. Patrick, Robert L., ed., *Treatise on Adhesion and Adhesives,* New York: Marcel Dekker, Vol. I, 1967, Vol. II, 1969, Vol. III, 1973.
11. Perry, H. A., "How to Calculate Stresses in Adhesive Joints," *Product Engineering,* July 7, 1958, pp. 64–67.
12. Shields, J., *Adhesives Handbook,* 3rd ed., London: Butterworths, 1984.
13. Shigley, Joseph E., *Mechanical Engineering Design,* 3rd ed., New York: McGraw-Hill, 1977, pp. 274–294.
14. Skeist, I., *Handbook of Adhesives,* 3rd ed., New York: Van Nostrand Rheinhold, 1989.

Adhesives and the Bonding Process

5.1 INTRODUCTION

Adhesives are substances that are capable of holding materials together by surface attachment arising from the formation of secondary (and, occasionally, primary) chemical bonds, often with some contribution from mechanical locking of the adhesive into asperities on the substrates or adherends being joined. Other terms used interchangeably with adhesive are glue, paste, cement and mucilage. Adhesives are used to accomplish adhesive bonding, and have applicability to a wide variety of materials, in both similar and dissimilar combinations, including paper, wood, fabric, leather, polymers, ceramics, glasses, metals, and composites of all sorts. Some adhesives are produced from natural substances, but most are synthesized chemically. Adhesives are available in many different chemical types and can be applied in any of several ways in any of several different physical forms for a variety of service conditions. As a result, adhesives can be and are classified by any of several different schemes. The resulting properties of joints can be impressive, whether for structural or nonstructural applications. Degradation from a host of environmental factors can be severe, however, so these factors must be carefully considered in the design of the joint and in the selection and application of the adhesive.

This chapter takes a look at adhesives used in accomplishing structural bonding as well as sealing, vibration or impact damping, and thermal or electrical insulation. First, the general constituents found in adhesives are presented. Then various schemes for classifying adhesives by function, chemical composition, physical form, mode of application or setting, adherends, or use, as well as by other means are presented. Some important types (or compositions) of adhesives are then briefly described. Details of the actual process of adhesive bonding are then presented, from adhesive storage, through joint preparation and adhesive application, to actual joint assembly. A description of the equipment needed is also given. The performance of adhesive bonds is then discussed in terms of expected properties, property testing, and joint quality assurance, with special attention to the effects of the service environment. Finally, a few areas of application are briefly described.

5.2 THE CONSTITUENTS OF ADHESIVES

Most adhesives in use today are actually mixtures of several constituents or components intended to fulfill different functions in creating an adhesive bonded joint, i.e., they are really *adhesive systems*. Basically, an adhesive system can contain various combinations of any or all of the following: (1) the adhesive base or binder; (2) a hardener (for thermosetting types); (3) accelerants, inhibitors, or retarders; (4) diluents; (5) solvents; (6) fillers (for imparting special characteristics or properties); and (7) carriers or reinforcements.

The *adhesive base* or *binder* is the primary component of the adhesive system and is generally the constituent from which the name of the adhesive is derived. For example, an *epoxy adhesive* has an epoxy resin base or binder, possibly with many other constituents such as hardeners, diluents, fillers, and carriers. The base or binder's function is to create the chemical bonds between itself and the adherends to hold those adherends together. A *hardener* is a substance added to certain types of adhesives that require a chemical reaction to set or cure. An example is with thermosetting adhesive binders. Here, a hardener is required to cause these types of polymers to cross-link and, thereby, become rigid. The hardener acts as a catalyst to the cross-linking reaction, although sometimes an additional catalyst is needed in minute proportions. The hardener can be premixed with the base and activated by heat or radiation or it can be mixed in during adhesive use causing the reaction to occur spontaneously. *Accelerants, inhibitors,* and *retarders* are substances that are added to control the rate of curing. Accelerants hasten curing by acting as a catalyst, while retarders slow the rate of curing and inhibitors arrest it.

Diluents are liquid components that are added to an adhesive to reduce the concentration of the solid bonding material or base. They usually have the function of lowering viscosity to facilitate the application of the adhesive and reduce cost. Often diluents evaporate, allowing the adhesive to set. Sometimes, however, they are reactive, participate in the curing process and remain in the cured product. *Solvents* are closely related to diluents, in many respects, in that they are used to thin the adhesive to a spreadable consistency. They do this by dispersing the adhesive base or binder. For many adhesives, however, solvents play an important part in the way the adhesive cures or sets and accomplishes bonding. In so-called evaporative and diffusion types of adhesives, solvents either largely evaporate out or are absorbed away by a porous adherend during curing to allow the adhesive to dry and set, or they soften the adherend to facilitate interdiffusion, respectively. Sometimes some solvent remains behind, either failing to completely evaporate out or becoming entrapped in voids, especially at the adhesive/ adherend interface.

Sometimes it is necessary or desirable to modify (i.e., enhance) the base adhesive's properties by adding a filler. *Fillers* are generally nonadhesive substances themselves but are added to improve the working characteristics of the adhesive (e.g., viscosity or body, working time), reduce the amount of shrinkage, increase the cured strength, enhance the durability, and so on. Sometimes fillers are added to adjust certain functional properties like thermal expansion, electrical or thermal conductivity, or heat resistance. So-called conductive adhesives contain additions of conductive fillers such as metals or conductive carbon to adhesive bases such as epoxies, urethanes, silicones, and polysulfones. Fillers can also be added simply to reduce the cost of the adhesive system by adding inactive bulk.

Carriers or *reinforcements* are sometimes useful for supporting the adhesive during application. Thin fabrics or paper can be used, often as a backing to a semi-cured thermosetting adhesive material or to provide a single- or double-sided tape or supported film. The carrier can also serve as a bond-line spacer, or shim, to help establish the appropriate adhesive thickness. Since, here, the carrier is left in the joint, it is usually designed (or formulated) to reinforce the adhesive both before and after curing.

5.3 CLASSIFICATION SCHEMES FOR ADHESIVES

5.3.1 General Description

There are a number of different schemes for classifying adhesives. While each scheme has its own advantages and serves its own purposes, no one scheme is particularly better than another, and, certainly, no one particular scheme is universally recognized. Classification can be by function, chemical composition, physical form, means of application or setting, specific adherend, primary application, or other means. The Society of Manufacturing Engineers (SME) provides an especially useful, in-depth classification based largely on the chemical nature of the adhesive system. The following sections look briefly at some of the more commonly used classification schemes.

5.3.2 Natural versus Synthetic Adhesives

In perhaps the broadest sense, adhesives can be classified into two categories: natural adhesives and synthetic adhesives. *Natural adhesives* derive their active constituent (i.e., adhesive base or binder) from animal or vegetable sources or, occasionally, from natural inorganic sources. Animal sources include casein (a white, odorless, and tasteless milk or cheese protein), albumin, blood, bone, skin, fish, and shellac (derived from lac, a secretion from certain insects). Vegetable sources include starch, resin, natural rubbers, and asphalt. A natural inorganic adhesive is sodium silicate, known as water glass. Most natural adhesives are water soluble, depend on the evaporation of the water to set, and develop low strengths, so are primarily nonstructural. Natural adhesives are used with paper, wood, leather, and (natural) fabrics for lightly loaded joints. For this reason, natural adhesives are of little interest in the context of this book.

Synthetic adhesives are just that, that is, chemically synthesized, as opposed to naturally occurring. All are polymeric materials, and subclasses include thermosetting polymers, thermoplastic polymers, elastomers, and combinations, or alloys, of these materials. All structural adhesives are synthetic.

5.3.3 Classification by Function

One of the most general and useful ways of classifying adhesives is by their intended function, which may be structural or nonstructural. Structural adhesives consist of adhesive bases or binders, with or without carriers or reinforcements, that provide high strength and reliable performance. Obviously, their primary function is to hold structures together and sustain and/or transmit high loads.[1] Therefore, these are the types of adhesives that are commonly used in load-bearing structures. It is not unusual for structural adhesives to be stressed to a high proportion of their failing load for long periods of time without exhibiting failure. Nonstructural adhesives are also called holding adhesives and they are not intended or required to support substantial loads. These adhesives have primary purposes other than load carrying, including sealing, vibration damping, impact absorption, and thermal or electrical insulation.

5.3.4 Classification by Chemical Composition

Synthetic adhesives can, quite logically, be classified by their chemical composition. As stated previously, synthetic adhesives are polymeric materials and, thus, can be subclassified by the nature of the polymer comprising the base or binder. Subclasses include thermosetting polymers (or thermosets), thermoplastic polymers (or thermoplastics), and elastomeric polymers (or elastomers). In addition, combinations of these basic subclasses, called alloys, are possible.

Thermosetting adhesives are based on polymeric materials that cannot be heated and softened once they have cured. Curing of thermosets takes place by a chemical reaction that leads to the formation of bonds between polymer chains known as crosslinks, shown schematically in Figure 5.1. These reactions may take place at either room or elevated temperature, depending on the type of thermoset, but in either case, strong ionic or covalent bonds result. Solvents are sometimes used to thin the adhesive for easier application, and pressure, ranging from just enough to cause contact between adherends to considerable, is required.

Thermosetting adhesives, while they require a chemical reaction to cure, are available in one-part or two-part systems. In one-part systems, the chemical agent or hardener or catalyst necessary to cause the chemical reaction is premixed with the adhesive base or binder. The reaction is usually activated by heat, so curing takes place at elevated temperature but can be initiated in others ways (e.g., ultraviolet light). Shelf life is usually limited but can be extended by storing the adhesive in cool (possibly refrigerated) dark places. In two-part systems, the hardener or catalyst must be carefully metered, added to, and mixed with the adhesive base to initiate the curing reaction. Curing can usually be accomplished at room temperature but is often accelerated by heating. Once the two components are mixed, working time is limited, although the shelf life of the separate components is usually quite good.

---

[1]The loads carried by structural adhesives can approach levels that cause yielding in one of the adherends.

Thermosetting adhesives are usually structural. The dense cross-linking that occurs during curing results in (1) good shear strength from room temperature to 260°C (500°F); (2) good resistance to heat, with little elastic or creep deformation under loads at elevated temperatures; and (3) good resistance to solvents. Peel strength is only fair, compared to thermoplastic types. Most materials can be bonded with thermosetting adhesives, and major applications are for structural joints that must function at elevated temperatures for long times.

Some important types of thermosetting adhesives include acrylics, anaerobics, cyanoacrylates, epoxies, polybenzimidazole, polyesters, polyimides, polyurethanes, silicones, and urea and melamine formaldehydes.

Thermoplastic adhesives are based on polymers that can be repeatedly softened by heating, since they do not cross-link between chains and form rigid structures during

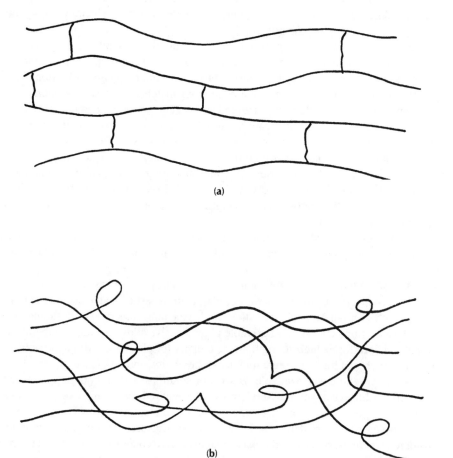

(a)

(b)

Figure 5.1 (a) Cross-linking between chain molecules in thermosetting polymers. (b) Secondary bonding and tangling of chain molecules in thermoplastic polymers.

curing. Thermoplastic adhesives are single-component systems that harden either simply by cooling from a melt, i.e., so called hot melts, or through the evaporation or diffusion of an organic solvent or water. Bonding within the thermoplastic polymer involves the formation of secondary bonds between long-chain molecules and a significant degree of intertwining or tangling of those chains. The secondary bonding and tangling in thermoplastic adhesives is also shown in Figure 5.1. Most thermoplastic adhesives require solvents as vehicles for carrying the active polymer binder.

As a group, thermoplastic adhesives exhibit limited strength, especially as temperature is increased so, until recently, have generally been nonstructural. More recently developed thermoplastic adhesives, and most of the hot-melts, are definitely suited to structural applications, however. Thermoplastic adhesives exhibit the following general properties: (1) service limited to below 65° to 90°C (150° to 200°F); (2) poor creep strength; (3) usually poor solvent resistance; and (4) fair peel strength.

Some important types of thermoplastic adhesives include acrylic, cellulose acetate, cellulose nitrate, ethylene-vinyl-acetate/polyolefin, phenoxy, polyvinyl acetate, polyvinyl alcohol, and polyamide.

Elastomeric adhesives are based on both natural and synthetic polymers that exhibit superior elongation and toughness. Because of these properties, elastomeric adhesives are used exclusively for nonstructural applications such as vibration damping, impact absorption, sealing, accommodating mismatched thermal expansion coefficients, and for joining elastomeric adherends to one another or to another material.

Elastomeric adhesives are available as solvent solutions, dispersions, or cements (i.e., pastes), and pressure-sensitive tapes. Curing methods vary depending on the specific type and form but include evaporation, diffusion, and heat. Usually nonstructural, elastomeric adhesives offer high flexibility and superior peel strength but limited temperature service (up to 70° to 200°C (150° to 400°F) and low bond strength. Elastomeric adhesives offer particular advantages where joint flexure is a key functional requirement.

Some common types of elastomeric adhesives include butyl rubber, natural rubber, neoprene, polyisobutylene, polysulfide, polyurethane, reclaimed rubber, and silicone.

Adhesive alloys are literally combinations, or alloys, of resins from two or more chemical groups from among thermosetting, thermoplastic, and elastomeric types. When properly formulated, adhesive alloys utilize the most important or desirable properties of each component. A good example is an adhesive alloy containing a thermosetting resin chosen for its high strength, made tougher (or plasticized) through the use of an elastomer chosen for its inherent toughness, flexibility, and resistance to impact.

Adhesive alloys are available as solvent solutions and as both supported and unsupported films. Heat and pressure are usually required for curing, except for some epoxy types.

Adhesive alloys are definitely intended to be structural adhesives and should be considered where the highest and most stringent service conditions must be met. Costs can be high but are often secondary to needed performance characteristics. As might be expected, adhesive alloys are excellent for joining dissimilar materials, such as metals, ceramics, glasses, and thermosetting and thermoplastic polymers, to one another.

Some common types of adhesive alloys include epoxy-phenolic (a thermosetting alloy), epoxy-polysulfone (a thermosetting alloy), epoxy-nylon (also a thermoset-thermoplastic alloy), neoprene-phenolic (an elastomeric-thermoset alloy), and vinyl-phenolic (a thermoplastic-thermoset alloy).

The Society for Manufacturing Engineers (SME) classifies adhesives into a number of divisions based largely on chemical composition as described in the following paragraphs. *Chemically reactive adhesives* require something to trigger the curing reaction in a thermosetting base or binder. These adhesives include one-component systems that are cured by moisture from the air or from an adsorbed layer on the adherend, by heat, or by the exclusion of oxygen (i.e., anaerobic); and two-component systems where the hardener or catalyst is premixed (called no-mix types) and activated by heat or where the hardener or catalyst is metered and mixed to start the curing reaction spontaneously (called mix-in types). The most commonly used chemically reactive structural adhesives include epoxies and modified epoxies, modified acrylics, anaerobics, cyanoacrylates, silicones, polyurethanes, phenolics, and the high-temperature polyimides and polybenzimidazoles.

Evaporative or *diffusion adhesives* are organic solvent-based or water-based systems that rely on the evaporation or absorption or diffusion of solvent into the air or into a porous adherend, respectively, to cause the adhesive to cure. Popular solvent-based adhesives include acrylics, phenolics, polyurethanes, vinyls, and a variety of natural and synthetic rubbers. Water-based adhesives include those that are totally soluble and those that are dispersive in water. Popular systems include certain acrylics, certain vinyls, and certain natural and synthetic rubbers.

Hot-melt adhesives are 100% thermoplastic systems that remain solid to a certain temperature, at which they melt abruptly. Upon melting, these adhesives are applied to the adherends and allowed to cool to set. Specific hot-melts are discussed in Section 5.4.8.

Delayed-tack adhesives use solid plasticizers in their formulation which are activated and made tacky by heating. The tackiness persists after heat activation and can last from minutes to days over a wide temperature range. Some common solid plasticizers include dicyclohexyl phthalate, diphenyl phthalate, and *ortho-* or *para-*toluenesulfonamide. Popular bases for delayed-tack adhesives include styrene-butadiene copolymers, polyvinyl acetates, polystyrene, and polyamides.

Pressure-sensitive adhesives are based on rubbers (such as natural rubber, styrene-butadiene rubber, reclaimed rubber, butyl rubber, and nitrile rubber, as well as polyacrylates and polyvinylalkylethers) compounded with tackifiers. The application of pressure causes the adhesive and adherend to adhere. Often, no curing occurs. Adhesion is strictly the result of the adhesive and adherend being in contact.

Film and tape adhesives use high-molecular-weight backbone polymers as carriers to provide toughness, elongation, and peel strength. A low-molecular-weight cross-linking resin and curing agent is used as the adhesive.

Conductive adhesives include both electrically and thermally conductive systems based on epoxies, polyurethanes, silicones, and polyimides. Conductive or insulative properties are obtained with the addition of fillers. For electrical conductivity, as well as thermal conductivity, silver flakes or powder are used (up to 85%), as well as aluminum and copper fillers. For thermal conductivity, beryllia, boron nitride, and silica are also used, while for thermal insulation alumina is used.

Table 5.1 summarizes the SME chemical composition classification scheme.

While by no means comprehensive, Table 5.2 shows which adhesive types work best with which adherends. More complete information on recommendations of specific adhesives for specific adherends is available in various handbooks (ASM International, 1990; Cagle, 1973; Landrock, 1985; Shields, 1984; and Skeist, 1989).

5.3.5 Classification by Physical Form

Adhesives can also be readily classified by the way in which they are applied, which is based on their physical form. These forms include liquid adhesives, paste adhesives, tape and film adhesives, and powder or granular adhesives.

Table 5.1   Summary of Major Structural Adhesives under SME Classification Schemes

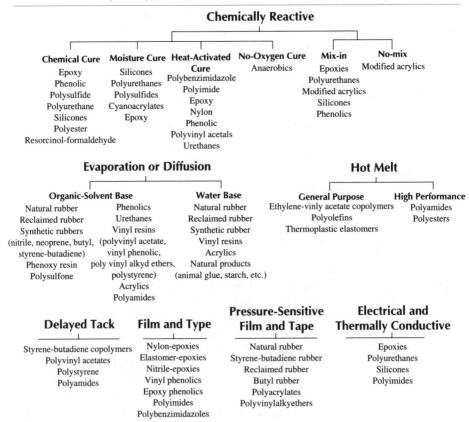

Table 5.2 Summary of Adhesive Types versus Adherends

							Metal										
Adhesive	Al	Ag	Au	Be	Brass\Bronze	Cd	Cu	Mg	Ni	Pb	Sn	Steel	Stainless Steel	Ti	U	W	Zn
Acrylics	•						•										
2nd generation acrylics												•	•				
Anaerobics	•	•	•		•		•										
Cyanoacrylates			•	•		•	•	•		•							•
Epoxies	•		•	•			•	•	•	•		•	•	•	•	•	•
Epoxy-phenolics	•							•		•		•	•	•			
Modified epoxies	•																
Modified phenolics	•																
Neoprene-phenolics	•				•		•	•									
Nitrile-epoxies	•		•	•								•	•	•			•
Nitrile-phenolics			•	•		•	•		•			•		•			
Nitrile-rubbers					•				•							•	
Nylon-epoxy																	
Polyacrylates										•	•						
Polyamides									•								
Polybenzimidazoles			•	•					•			•	•	•			
Polymides									•			•	•				
Polyurethanes					•		•			•							
Polyvinyl acetates			•					•			•						
Polyvinyl alkyl ether		•			•		•			•	•						•
Silicones	•							•									
Styrene-butadiene																	
Vinyl-plasticols	•							•	•								
Vinyl-phenolics																	

Table 5.2 (*continued*)

Adhesives	Polymers																				
	T/P's														T/S's						
	ABS	Cellulosics	Fluoroplastics	Nylons (polyamides)	Polycarbonates	Polyester	PEEK	Polyethersulfane	Polyethylene	Polymethylmethacrylate	Polypropylene	Polystyrene	Polysulfane	Polyvinylchloride	Epoxies	Melamines	Phenolics	Polyester	Polyurethane	Silicone resin	Ureaformaldehyde
Acrylics												●									
2nd generation acrylics	●									●											
Anaerobics		●													●	●	●		●		
Butadiene-nitrile																●					
Cyanoacrylates	●	●		●	●	●	●			●	●				●	●	●	●			●
Epoxies	●	●	●	●	●	●	●	●	●	●			●	●	●	●	●		●	●	●
Epoxy-phenolics															●	●	●				
Melamine-formaldehyde																●					
Modified acrylic				●											●		●		●		
Modified epoxy				●					●						●		●		●		
Neoprene			●											●			●	●	●	●	●
Nitrile-phenolic	●	●													●	●	●	●	●		●
Phenol-formaldehyde				●													●				●
Phenolic																		●			
Phenolic-polyvinylbutyral															●		●		●		●
Polyacrylates		●															●				
Polyamide-epoxy										●											
Polyester															●			●	●		●
Polyurethanes	●	●	●	●	●	●				●	●	●	●						●	●	
Resorcinol-formaldehyde	●	●													●					●	●
Rubber-phenolic																	●				
Silicones						●	●		●											●	
Solvent cementing		●		●	●				●	●		●	●								
Urea-formaldehyde											●					●	●				●
Vinyls	●																				

Reinforced Plastics/Composites
   Solvent cementing
   Adhesive based on matrix

Rigid Plastic Foams
   Solvent cementing
   Water-based adhesives
   Epoxies

Rubbers
   Neoprene
   Nitrile rubber

Glass
   Transparent heat-setting resins
      (e.g., polyvinyl butyral, phe-
      nolic butyral, nitrile-phenolic,
      neoprene, polysulfide, sili-
      cone, vinyl acetate, and epox-
      ies)

Ceramics
   Epoxies
   2nd generation acrylics
   Polysulfides

Liquid adhesives offer easy application on adherends of any shape because of their low viscosity. They can be applied manually or automatically, using brushes, mechanical spreaders, or liquid or mist (including aerosol) sprays. Frequently, liquid adhesive systems include diluents or solvents in their formulation. Paste adhesives have higher viscosities than liquid adhesives to give the adhesive more body. Body permits application in controlled thicknesses and allows application on vertical surfaces without dripping or sagging. Paste adhesives are good for filling gaps between two poorly fitting mated surfaces. Body is usually achieved through the use of diluents or fillers or both. Tape and film adhesives provide a uniformly thick layer of adhesive or bond line without requiring metering, elaborate applicators, or shims. They are limited to geometries with less severe curvature and cannot be used to take up gaps. These adhesives are supported on carriers of paper or cloth or thin, peelable plastic and are thus easily dispensed and handled. Finally, powder or granule adhesives are dry solids in particulate form. They must usually be heat or solvent softened to be made fluid enough to apply but they can be used as powders and melted in place. Hot-melts are a good example.

5.3.6 Classification by Mode of Application or by Setting Mechanism

It is often useful, especially for purposes of planning in manufacturing, to classify adhesives by the way in which they can or must be applied, and/or activated and cured or set. Depending on viscosity, adhesives can be considered sprayable, brushable, trowelable, or extrudable (using pressure pumps, caulking guns, or syringes).

Another useful way of classifying adhesives for manufacturing purposes is by the way in which they flow or solidify or are activated and cured. Some solidify or harden by losing solvent by evaporation or diffusion; others require heat or a catalyst to harden; others require pressure to flow but become stable when pressure is removed; and still others simply need to cool and become less viscous. Thus, there are adhesives that require the following: (1) heat to flow or to cure; (2) pressure to flow during application; (3) time to cure at room temperature (say by evaporation); (4) a catalyst to cure, with or without heat and/or pressure; and (5) reactivation to cure with the aid of heat or solvent.

5.3.7 Classification by Specific Adherend or by Application

It is possible and sometimes useful to classify adhesives according to their end use, either by specific adherend or by intended environment. Examples of adhesives for specific adherends include wood adhesives, metal adhesives, ceramic cements or mortars, and vinyl adhesives. Examples of adhesives intended for specific environments include

---

[2]For example, thermally or electrically conductive adhesives are an important classification for electronic package manufacturing.

heat-resistant adhesives, cryogenic adhesives, acid-resistant adhesives, weatherable adhesives, and general purpose adhesives for noncritical applications.

There are other classification schemes,[2] but the ones presented are fairly comprehensive and are the most common.

5.4 SOME IMPORTANT STRUCTURAL ADHESIVE TYPES

5.4.1 General Description

The most widely used structural adhesives today are the epoxies, modified epoxies, acrylics, modified acrylics, cyanoacrylates, anaerobics, urethanes, silicones, phenolics, and hot melts.

5.4.2 Epoxies and Modified Epoxies

Epoxy resins are the reaction product of acetone and phenol and are the most popular of all structural adhesives used today. Epoxies, or more properly epoxide adhesives, are thermosetting polymer resins usually based on diglycidyl ether of bisphenol-A (DGEBA), so they cure upon heating (in one-part systems) or through the addition of a hardener or catalyst (in two-part systems) or both, with little or no need for pressure, becoming rigid through the formation of dense cross-linking (see Figure 5.1). There are more than 25 different types. Typical hardeners, of which there are more than 60 known, include aliphatic polyamines, fatty polyamides, aromatic polyamines, and boron trifluoride monoethylamine. The cross-linking reaction is exothermic, and this must be taken into account in use to avoid overheating and formation of pores. Aside from hardeners (whether premixed or mixed in), other additives such as accelerators, reactive diluents, plasticizers, fillers, and resin modifiers are often made to modify behavior or properties.

Epoxy adhesives can be used to join most materials, including metals, ceramics, and polymers, with good results. The highest strengths (up to 77.2 MPa or 10,000 psi in shear) and best heat resistance are obtained with heat-cured, two-part types. As a group, shrinkage is low, and bond flexibility tends to be low, so impact strength is poor. Peel strength also tends to be low. Like most thermosetting polymers, resistance to solvents is good.

Because thermosetting adhesives exhibit brittle behavior when cured, an important group of adhesive alloys has emerged called modified epoxies. *Modified epoxies* incorporate various thermoplastics or elastomers to impart flexibility and toughness and better resistance to peel. Examples include epoxy-nylon, epoxy-polysulfide, epoxy-phenolic, and epoxy-nitrile (rubber). These adhesives, made up entirely of solids, contain no solvent. Thus, shrinkage is low, and bonding to impervious surfaces is possible. They also exhibit excellent wetting on metals, glass, and ceramics, so are excellent for bonding similar and dissimilar combinations of these materials.

5.4.3 Acrylics and Modified Acrylics

The acrylic family of adhesives is based on acrylic monomers of ethyl acrylate, methyl acrylate, methacrylic acid, acrylic acid, acrylamide, and acrylonitrile. They are all two-part systems but are not mixed for application. Rather, the resin is applied to one adherend, and the accelerant is applied to the other. When these two are mated, the bonding reaction occurs quickly (in minutes) even at room temperature. Pretreated parts can be stored separately for some time before bonding. Shear strengths can reach 28 MPa (4,000 psi).

New formulations of acrylics, called second-generation or reactive acrylic adhesives, have additives that penetrate hydrocarbon contamination on adherends, incorporate that hydrocarbon into the structure of the cured adhesive, and enhance bonding significantly. These second-generation acrylics make metal-to-metal structural bonding practical and are generally referred to as *modified acrylics*. Modified acrylics also offer good peel strength, impact, and tensile lap strength over the temperature range from $-110°$ to $120°C$ ($-160°$ to $250°F$). High temperature strength, on the other hand, is low, and, as a group, modified acrylics are quite flammable.

Examples of modified acrylics include acrylic-latex, acrylated silicones, acrylated urethanes, and acrylated silicone-urethanes. Silicone additions improve thermal stability, tensile strength, and resistance to solvents, abrasion, and ultraviolet light. Urethane additions increase toughness, thermal stability, and resistance to solvents, abrasion and ultraviolet radiation. Both types offer improved adhesion to ceramics and metals over traditional acrylics.

5.4.4 Cyanoacrylates

Cyanoacrylates, or super glues, are composed of low-viscosity liquid acrylic monomers, which polymerize easily in the presence of adsorbed water, especially where the adherend surface is slightly alkaline. Polymerization is ionic, and strong thermosetting bonds can be created with many materials (especially metals to nonmetals) with no added heat or catalyst, since most surfaces have adsorbed water present. Shear strengths are good, up to 38.6 MPa (5,000 psi), although peel strength and impact resistance are low, and tolerance of moisture is poor. The principal advantage of cyanoacrylates is that they bond in seconds. One important use is for holding components in place, that is, for jigging, until some other, more permanent joining process (such as welding) is accomplished.

Cyanoacrylates will bond with almost any substrate. Methyl cyanoacrylate produces stronger and more impact resistant joints when bonding metals and other rigid materials than the ethyl cyanoacrylate. Ethyl cyanoacrylate, however, produces stronger and more durable joints with elastomeric, thermosetting, or thermoplastic polymer substrates. The peel strength of cyanoacrylates is poor, they tend to be brittle, and they have limited temperature resistance. As a group, these adhesives are also poor at filling gaps.

5.4.5 Anaerobics

Anaerobic adhesives or, more properly, *anaerobic sealants,* are single-component monomeric liquids that only harden satisfactorily in the absence of gaseous oxygen, that is, without air. The mechanism is free radical polymerization. For this reason, they must be stored in the presence of air, usually in permeable containers. As a group, these adhesives have exceptional fluidity and can readily flow into the smallest crevices, achieving thorough sealing when they cure with little or no shrinkage. One important application is in locking threaded fasteners against loosening by vibration (e.g., Locktite).

Second generation anaerobics often contain some urethane to overcome brittleness and improve peel and impact strength. Shear strengths to 77.2 MPa (10,000 psi) and temperature serviceability to 200°C (400°F) are possible.

5.4.6 Urethanes

Urethanes or *polyurethanes* are basically thermoplastic polymers but with the ability to develop cross-links under certain conditions, thereby making them fairly rigid thermosetting adhesives. One- and two-part systems, usually dissolved in solvents, are available. They are distinguished from epoxies by their inherent flexibility and high peel strengths.

Urethanes are generally applied to both adherends and are brought into contact when the proper tackiness is reached. Curing is usually done at room temperature, and, while handling is possible almost immediately, full curing takes many hours or days. Heat can be used to soften the adhesive if it becomes too dry before bonding.

5.4.7 Silicones

Silicones are available as one- and two-component systems that cure to thermosetting solids. The one-component systems undergo acidic or nonacidic cure at room temperature when exposed to atmospheric moisture. (The acidic cure is what gives many silicone sealants or caulking compounds a vinegar smell.) Two-component systems cure by condensation polymerization, and are prone to polymerization reversion or decomposition.

Silicones have good peel strength over a wide temperature range from –60° to 250°C (–75° to 480°F), and some can survive limited exposure to 370°C (700°F). Flexibility, impact resistance, and resistance to moisture, hot water, oxidation, and weathering are very good. Lap-shear strengths tend to be low, however.

As a group, silicone adhesives are expensive but versatile. They are able to bond metals, glass, paper, thermosetting and thermoplastic polymers, and a wide variety of rubbers.

5.4.8 Hot Melts

Hot-melt adhesives are copolymers of polyethylene with polyvinyl acetate, polyolefins, polyamides, polypropylene, nylon, polyester, and thermoplastic elastomers. They are

100% thermoplastic, however, but are not entirely composed of polymers. A portion of most hot-melts is diluent and/or filler. As a group, hot melts tend to soften noticeably at about 80°C (175°F) but are applied much hotter (150° to 390°C or 300° to 550° F). Hot melts can be formulated to make either flexible or rigid bonds and reach 80% of their ultimate bond strength seconds after application. They are excellent for bonding materials with either permeable and impermeable surfaces, and they are quite resistant to solvents and moisture, but they do soften upon exposure to temperatures of 100° to 150°C (210° to 300°F).

High-performance hot-melts, such as polyamides and polyesters, can withstand limited loads at elevated temperatures without undergoing significant creep.

5.4.9 Phenolic Adhesives

Phenolic adhesives, or, more properly, *phenol-formaldehyde adhesives,* are widely used in bonding wood (e.g., plywood) and so constitute one of the largest usages of the synthetic adhesives. These adhesives rely on penetration of the pore or cell structure of the adherend to develop bonding forces. Phenolics can also be used as primers for bonding metals. As a group, phenolics cure to thermosets.

In general, phenolics are very low cost adhesives with good strength and resistance to biodegradation, hot water, and weathering. Elevated temperature resistance is also fairly good. Shortcomings include brittleness, low impact strength, and the development of high shrinkage stresses.

Phenolics are frequently combined with other polymers to produce alloys with one or more enhanced properties such as higher bond strength, higher fatigue strength, higher service temperature, and better resistance to water, humidity, salt, or weather. Examples include nitrile-phenolics, vinyl-phenolics, and epoxy-phenolic. Epoxy-phenolic offers excellent long-term service between 150°C and 260°C (300° and 500°F), as well as good performance to –260°C (–440°F) if specially formulated.

5.4.10 High-Temperature Structural Adhesives

Perhaps the greatest shortcoming of many structural adhesives is their limited tolerance of elevated temperature. This shortcoming has been addressed by developing several special high-temperature structural adhesives. Most are based on synthetic organics having aromatic (i.e., benzene) and/or heterocyclic rings in their main molecule structure. Groups of polymers with such structures include imidazoles and substituted imidazoles. Both have open ring structures that close upon exposure to heat. Thus, they become stronger, as opposed to softening or decomposing.

Two specific high-temperature structural adhesives are the polyimides and polybenzimidazoles. These are both expensive and difficult to handle in that they have long cure times and emit considerable volatiles. Polyimides offer superior long-term strength retention in air to 260°C (500°F), while polybenzimidazoles are stable to 288°C (550°F) only for short times. Both are prone to degradation by moisture. The principal use of such adhesives is in aerospace structures composed of metals bonded to metals or to composites.

Table 5.3 provides a summary of the principal characteristics of the major types of structural adhesives described above.

5.5 THE ADHESIVE BONDING PROCESS: STEPS AND EQUIPMENT

5.5.1 General Description

Many factors must be considered in selecting a particular adhesive bonding method. The size and shape of the parts to be bonded, the specific area to which the adhesive is to be applied, the number of assemblies to be produced and the production speed, the viscosity or other handling characteristics of the adhesive, and the form of the adhesive are all important.

The following sections take a brief look at each of the key steps involved in adhesive bonding, including (1) adhesive storage; (2) adhesive preparation; (3) joint preparation; (4) adhesive application; and (5) joint assembly. In addition, the equipment needed for bonding is described.

5.5.2 Adhesive Storage

The susceptibility of most adhesives to degradation by one or more environmental factors is well known. Extremes of temperature, humidity, water, solvents, and light are but a few of the most important. As a result of this susceptibility, adhesives must be stored properly before their use.

Many adhesives must be stored in the dark, in dark or opaque containers, while others should be stored at low temperatures (e.g., 5°C, or 40°F) to prolong shelf life. Anaerobics need to be stored in air, as the deprivation of oxygen will promote cross-linking. Solvents and solvent-based adhesives, on the other hand, should be stored in impermeable containers that can be sealed quickly after use to prevent loss and the escape of toxic or flammable fumes. As a matter of common sense, resins and curing agents for thermosetting adhesives should be kept apart in storage to prevent accidental contamination should container breakage occur.

The special requirements for storage, while important and not unreasonable, can pose difficulties in many production environments, at least in terms of logistics. Nevertheless, manufacturers' directions on storage should be carefully read and complied with.

5.5.3 Adhesive Preparation

Whenever adhesives are brought out of storage, they should be used as soon as is practical. The reasons for this, depending on the specific adhesive, are to avoid loss of

Table 5.3 Summary of Principal Characteristics of Major Structural Adhesives

Adhesive	Type	Cure	Shear Strength MPA(ks)	Peel Strength N/m(lb/in)	Impact Resistance	Solvent Resistance	Moisture Resistance	Relative Price	Substrates Bonded
Epoxies	One component Two components Film & tape	Heat RT/heat Heat + pressure	15.4 (2.2)	<525 (3)	Poor	Excellent	Excellent	Low	Most
Polyurethanes	One component Two components	Heat Moisture RT/heat	15.4 (2.2)	14,000 (80)	Excellent	Good	Fair	Moderate	Most, smooth & nonporous
Modified Acrylics	One component	RT/heat	25.9 (3.7)	5,250 (30)	Good	Good	Good	Moderate	Most, smooth & nonporous
Cyanoacrylates	One component	Moisture	18.9 (2.7)	<525 (3)	Poor	Good	Poor	Moderate to high	Nonporous, metals or plastics
Anaerobics	One component	No oxygen	17.5 (2.5)	1,750 (10)	Fair	Excellent	Good	Moderate	Metals, glass & thermosets
Silicones	One component	Moisture	1.7–3.4 (0.25)	612 (3.5)	Good	Good	Excellent	Moderate to high	Most
Phenolics	One component Two components Film & tape	Heat Heat Heat + pressure	N/A	N/A	Fair	Good	Good	Low	Most
High-temperature	One component Film & tape	Heat Heat + pressure	8–14 (1–2)	N/A	Fair	Fair	Poor	High	Thermoplastics & metals
Hot-melts	One component	Heat	4.3 (0.6)	N/A	Fair	Fair	Good	Moderate to high	Thermoplastics & metals

necessary but volatile solvents, absorption of undesirable moisture, exposure to light, warming to room temperature, and so on.

For adhesives that are refrigerated, proper preparation means letting them reach application temperature and using them quickly. When separate components are to be mixed (e.g., the adhesive base and the hardener of a two-part system), it is important to measure the proportions correctly to obtain optimum properties. Too little catalyst prevents proper polymerization of the resin, while too much catalyst can cause brittleness. Excess unreacted catalyst may also cause corrosion of the adherend.

Mixing, whenever required, must be complete to assure homogeneity, and only enough adhesive should be mixed to allow working without curing (i.e., within the working life or pot life for that adhesive). While it must be thorough, mixing must also be done carefully to avoid foaming or excessive gas entrapment that can cause incomplete bonding. It may sometimes be necessary to degas a mixed adhesive under vacuum.

Again, proper adhesive preparation can pose some complications in manufacturing but should be done in accordance with manufacturer's recommendations.

5.5.4 Joint Preparation

As described in Chapter 4, it is essential that the surfaces of adherends to be bonded are cleaned completely and are kept clean until the adhesive is applied and the joint assembled. To review what was described in Section 4.5.2, cleaning can involve one or more of the following steps, in sequence to whatever level is required: (1) solvent cleaning to remove oily or greasy contaminants and loose particulates; (2) intermediate cleaning with some combination of chemical and mechanical means to remove more tenacious contaminants or potential weak-boundary layers (e.g., oxide or sulfide scales); and (3) chemical treatment to activate the surface, chemically and physically, using more aggressive agents (e.g., acid etching).

After whatever cleaning steps are necessary, the cleaned surface must be protected from recontamination, including oxidation. Sometimes this is accomplished simply by covering the surface (or bagging) until actual assembly or through the use of a primer. Recall that the primer is usually a dilute mixture of the adhesive to be used with some solvent. It is applied to the adherend surfaces to protect them from recontamination after cleaning and, often, to precondition them for bonding with the adhesive when it is applied.

5.5.5 Methods of Adhesive Application

Selection of the method of adhesive application depends on the physical form of the adhesive being used, which, in turn, depends on the size and shape of the parts to be bonded, the areas where the adhesive is to be applied, and the production rate and volume. Methods can be manual, semi-automated, or fully-automated.

For liquid adhesives, methods of application include brushing, flowing, spraying, roll-coating, knife-coating, silk-screening, squeeze application, and hand dipping. Brushing is best for complex shapes or selected area bonding without masks (or areas covered to prevent the application of adhesive), where limited thickness control and slow application rates can be tolerated. Flowing is best for flat surfaces. It employs a nozzle or flow gun or hollow brush and is capable of providing a uniform film thickness. Spraying is good for large areas with gradual contours, so "runs" are not produced by the thin consistency of this form of adhesive. Application rates can be high, and film thickness control reasonably good, but there can be health hazards from mists and over-spray. Roll coating uses a pick-up roll with fluid adhesives, while knife coating uses a spreader (or "doctor") blade with thicker adhesives. Both methods are best for large flat surfaces. Knife coating enables excellent thickness control, through knife standoff. Silk-screening involves applying the adhesive by filling pores with the desired pattern in an overlay screen. Areas on the adherend not to be coated are blocked by the non-porous portions of the silk-screen. A squeegee or sponge applicator is usually used with the screen. Spot applications of adhesive can be made with a squeeze bottle, while very large surfaces can be covered by dipping, with or without masks, as appropriate.

More viscous paste adhesives can be applied by spreading with spatulas, knives (including doctor blades), and trowels.

Dry powdered adhesives can be applied by sifting onto preheated substrates or by dipping the preheated substrate. Both methods assure uniform coverage since loose powder falls off and no spot is missed. It is also possible to apply dry thermoplastic adhesives by melting into a paste or liquid and applying as appropriate for those forms.

Film adhesives are, in many ways, the easiest to apply. The film form offers excellent repeatability (since no mixing is required); excellent thickness control; clean, easy, hazard-free handling; very little waste (when cut-to-size and shape preforms are used); and excellent properties. As mentioned earlier, these films may be supported (with a carrier) or unsupported (adhesive only). Films are easiest to apply to flat surfaces or parts with gradual, simple curvatures. Complex or severely contoured surfaces may make it difficult to get the film to lie flat without wrinkling.

Finally, hot melts tend to be applied from melt-reservoir systems or progressive-feed systems, using various dispensing devices such as applicator nozzles and spreader blades.

5.5.6 Joint Assembly Methods

Often, assembly of the components of a structure to be bonded is a complicated process. All components must be cleaned in the areas where adhesive is to be applied and kept clean; adhesive must be applied where required, within the restrictions of working time; components must be brought into proper location and orientation and kept there throughout curing; and pressure and/or heat or drying must take place.

A number of methods have been developed for assembling joints for adhesive bonding, including (1) wet assembly using liquid adhesives, especially on porous adherends; (2) pressure-sensitive and contact bonding for adhesives that retain their tack when dry; (3) solvent activation, really postponed wet-assembly, except when unsuitable for nonporous adherends; (4) heat activation for thermoplastic adhesives; and (5) curing for thermosetting adhesives.

All of these methods have some common characteristics. The adhesive coating must become liquid at some point during assembly to promote contact with and assure wetting of the adherends. Excess solvents or diluents must be able to escape as vapor or be absorbed somehow in the assembly or into surrounding materials (often called scrim cloths). Volatile materials, such as solvents and moisture, must be expelled from the joint to prevent the formation of voids, vapor locks, and faults in the glue line. This is often accomplished by providing venting features in the assembly tooling or the assembly itself. Pressure must be applied to the joint until the adhesive sets sufficiently to hold the assembly together without accidental misalignment of the details.

5.5.7 Bonding Equipment

After proper application of the adhesive, which often means to a controlled thickness, the assembly must be mated as quickly as possible to prevent contamination of the adhesive and adherend surfaces and then the substrates held together under pressure and heated, if necessary, until the proper cure is achieved. This requires the use of equipment to control the adhesive thickness, apply pressure, and, when required, apply heat.

Applying an adhesive layer that is uniform and of a controlled thickness is almost always required for sound structural joints. It is almost always desirable to have a uniformly thin (e.g., 0.05 to 0.5 mm, or 2 to 20 mil) adhesive bond line. Three basic methods are used to control adhesive thickness. First, mechanical shims or spacers or stops may be used. They can usually be removed after curing but may have to be left in the cured assembly. If so, this must be considered in the design. Shims are usually made of metal and are removed. Spacers or stops often are or could be an integral feature (e.g., a raised boss) of the components being bonded. Second, a film adhesive can be used. Available in different starting thicknesses, these films become quite viscous during the cure cycle to prevent adhesive flow-out. Another possibility is to use reinforcements in the adhesive, which, because they are always solid, maintain the required spacing. Finally, trial and error can determine the correct pressure-adhesive viscosity factors that will yield the desired bond thickness. Trial and error, while widely practiced, should be avoided.

Pressure is always required to accomplish adhesive bonding, if for no other reason than to hold the components being bonded together until the adhesive has set or cured. Pressure is also necessary to bring all areas of the faying surfaces into contact with the adhesive, to squeeze out air, and, for many adhesives (e.g., pressure-sensitive types) to activate the actual bonding process. Pressure devices should be designed to maintain uniform, constant, controlled pressure in the bond throughout the entire cure cycle. This usually means compensating for thickness reduction from adhesive flow-out and thermal expansion of the assembled parts. It also means accommodating the size and shape of the assembly. Popular types of pressurization systems include (1) pneumatic or hydraulic presses versus screw-activated, mechanical presses, often with heated platens; (2) autoclaves, which apply isostatic pressure from air or inert gas or steam; and (3) vacuum bags, which apply up to atmospheric pressure and withdraw unwanted volatiles generated during curing.

Figures 5.2, 5.3, and 5.4 show a press, an autoclave, and a vacuum bagging set-up for bonding adhesives. Pressure distribution is usually accomplished in all of these systems using special bonding tooling.

For those adhesive systems that are activated and set or cure with heat, a means for applying controlled heat is required. The heat needed to cure adhesives can be applied directly or indirectly to the assembly. Direct methods include ovens (for up to 450°C ±1°C or 842°F ±1.8°F), liquid baths (to 300°C or 572°F in silicone oils), or hot presses or platens. Indirect methods include electric resistance heating, radiation curing (e.g., using infrared), high-frequency dielectric (or radio-frequency) heating, induction heating, and ultrasonic activation.

Figure 5.2   Typical press used for adhesive bonding. (Courtesy of Grumman Aircraft Systems Division of Grumman Aerospace Corp.)

Figure 5.3 Typical autoclave used for adhesive bonding. (Courtesy of Grumman Aircraft Systems Division of Grumman Aerospace Corp.)

Figure 5.4 The vacuum-bagging system for applying pressure during adhesive bonding. (Courtesy of Grumman Aircraft Systems Division of Grumman Aerospace Corp.)

With all these methods, precise control is critical since curing must be complete to obtain optimum properties. Modern manufacturing often calls for the use of embedded sensors for monitoring the state of the cure, perhaps using a measure of the adhesive's dielectric constant, and adaptively controlling the temperature according to the actual progress of the curing reaction in real time.

5.6 ADHESIVE BONDED JOINT PERFORMANCE

5.6.1 General Description

Ultimately, an adhesively bonded joint can only be judged by how well it performs in service. Attaining maximum performance in adhesively bonded joints requires selection of the appropriate adhesive for the adherend and the intended service conditions, proper bonding procedures, inspection of the bonded joint to assure quality, and, at some point, the testing of samples to verify performance.

The following sections look briefly at the testing of adhesive and adhesively bonded joint properties, the inspection of joint quality, the kinds of performance that can be expected, and the effects of environment on joint durability.

5.6.2 Testing of Adhesives and Bonded Joint Properties

Testing is important in materials science and engineering to characterize fundamental properties, during product and process development to optimize design and processing parameters, and in production to verify property attainment as an indication of process and quality control. Testing is especially important for adhesives, where so many variables are critical to the ultimate performance of the bonded joint.

Tests are used, first and foremost, to evaluate the inherent strength of the adhesive but also to evaluate the appropriateness of the joint geometry, the joint preparation technique (i.e., cleanliness, effectiveness of surface treatments, etching of surfaces), the application and coverage of the adhesive, and the effectiveness of the curing cycle. Specific adhesives tests are used for various purposes. First, many tests are used to compare the properties of alternative adhesives, facilitating selection. Properties of interest include tensile, shear, peel, cleavage, impact, and flexural strength; fatigue resistance; environmental durability; and special properties, such as electrical and thermal conductivity. Second, some tests are used to check the quality of a lot or batch of adhesives to determine whether the adhesive is still up to a standard. These tests are usually mechanical or physical but can be chemical in nature. Third, some tests are designed specifically to check the effectiveness of surface and other (e.g., adhesive) preparations. These are strictly mechanical in nature. Fourth, and finally, some tests are designed to determine parameters important in predicting ultimate joint performance, such as curing or drying conditions and bondline thickness.

Whatever the purpose, testing is almost always done in accordance with established standards, such as the methods and practices given by the American Society for Testing Materials (ASTM) or the Society of Automotive Engineers (SAE) in their Aerospace Recommended Practices (ARP's). This chapter will describe only the most common and important tests and refer the reader to the specific standards (e.g., ASTM or SAE) or to various handbooks (Cagle, 1973, Landrock, 1985, for example) for more details. A summary of the most common standards and procedures used in testing adhesives and adhesively bonded joints is given in Table 5.4.

Tensile Tests

Tensile tests are the most common tests used to evaluate adhesives or any material. This is the case despite the fact that, for most adhesives, it is best to load them in other than a tensile mode, e.g., pure shear. While most structural materials have tensile strengths that are high compared to structural adhesives, the advantage of conducting tensile tests on adhesives is that they yield fundamental and uncomplicated data, in the form of tensile strain, tensile strength, and elastic (or Young's) modulus.

Pure tensile tests are those in which the loading is applied normal to the plane of the bondline and in line with the center (i.e., centroid) of the bond area (as shown in Figure 4.6). The original ASTM test method is D 897, which is shown below.[3]

ASTM D 897-78 (1983)—Standard Test Method for Tensile Properties of Adhesive Bonds, 5 pp (DOD Adopted)

This method covers the determination of the comparative "butt" tensile properties of adhesive bonds tested on standard-shape specimens under defined conditions of pretreatment, temperature, and testing machine speed. The method is known as the "butt" joint adhesion tensile test. The method is not as commonly used as the lap-shear test (ASTM D 1002). Blocks or rods of wood or metal are shaped or machined to specified dimensions of 1-13/16 in. (46 mm) diameter for wood and 1-7/8 in. (47.5 mm) for metal for the contact surface. The wood specimens are made from hard maple, and the metal specimens may be brass, copper, aluminum, steel phosphor bronze, magnesium, or nickel silver. Two of the machined circular contact surface buttons are bonded together with the adhesive under test. A tensile testing machine is used under standardized conditions, and the maximum load at failure is recorded with the force normal to the contact area. The wood specimens must be conditioned at 23°C (73.4°F) and 50% RH, but no preconditioning is required for the metal specimens. Results are reported in pounds per square inch (psi) and kilograms per square millimeter (kg/mm^2). (This method replaces Federal Test Method Standard No. 175a, Method 1011.1)

Shortcomings in the original specimen and grips led to revised designs and methods in ASTM D 2094 and 2095.

Shear Tests

Shear tests are quite common because specimens are simple to fabricate and closely duplicate the geometry and service conditions imposed on many structural adhesives. As

---

[3]This standard is shown as an example and is used with the permission of the ASTM.

Table 5.4 Selected ASTM Test Methods & Practices for Evaluating Adhesives

Purpose	ASTM Standard	Title
Aging	D1183-70 (1981)	Resistance of Adhesives to Cyclic Aging Conditions
	D3632-77 (1982)	Accelerated Aging of Adhesive Joints by the Oxygen-Pressure Method
Biodeterioration	D4299-84	Effect of Bacterial Contamination of Adhesive Preparations and Adhesive Films
	D4300-84	Effect of Mold Contamination on Permanence of Adhesive Preparation and Adhesive Films
Chemical Reagents	D896-84	Resistance of Adhesive Bonds to Chemical Reagents
Cleavage	D1062-78 (1983)	Cleavage Strength of Metal-to-Metal Adhesive Bonds
	D3807-79	Strength Properties of Adhesives in Cleavage Peel by Tension Loading (Engineering Plastics-to-Engineering Plastics)
Corrosivity	D3310-74 (1983)	Determining Corrosivity to Adhesive Materials
Creep	D1780-72 (1983)	Conductivity Creep Tests of Metal-to-Metal Adhesives
	D2293-69 (1980)	Creep Properties of Adhesives in Shear by Compression Loading (Metal-to-Metal)
	D2294-69 (1980)	Creep Properties of Adhesives in Shear by Tension Loading
Cryogenic	D2557-72 (1983)	Strength Properties of Adhesives in Shear by Tension Loading in the Temperature Range From –267.8 to –55°C (–450 to –67°F)
Durability	D1151-84	Effect of Moisture and Temperature on Adhesive Bonds
	D1828-70 (1981)	Atmospheric Exposure of Adhesive-Bonded Joints and Structures
Fatigue	D3166-73 (1979)	Fatigue Properties of Adhesives in Shear by Tension Loading (Metal/Metal)
Flexural Strength	D1184-69 (1980)	Flexural Strength of Adhesive Bonded Laminated Assemblies
High Temperature	D2295-72 (1983)	Strength Properties of Adhesives in Shear by Tension Loading at Elevated Temperatures Effects (Metal/Metal)
Impact Strength	D950-82	Impact Strength of Adhesive Bonds
Low and Cryogenic	D2557-72 (1983)	Strength Properties of Adhesives in Shear by Tension Loading in the Temperature Range from –267.8 to –55°C (–450 to –67°F)
Peel Strength	D903-49 (1983)	Peel or Stripping Strength of Adhesive Bonds
	D1781-76 (1981)	Climbing Drum Peel Test for Adhesives
	D1876-72 (1983)	Peel Resistance of Adhesives (T-Peel Test)
	D3167-76 (1981)	Floating Roller Peel Resistance
Radiation Exposure	D904-57 (1981)	Exposure of Adhesive Specimens to Artificial (Carbon-Arc Type) and Natural Light
	D1879-70 (1981)	Exposure of Adhesive Specimens to High-Energy Radiation
Salt Spray	B117-73 (1979)	Salt Spray (Fog) Testing
	G85-84	Modified Salt Spray (Fog) Testing
Shear Strength	E229-70 (1981)	Shear Strength and Shear Modulus for Structural Adhesives
	D905-49 (1981)	Strength Properties of Adhesive Bonds in Shear by Compression Loading
	D1002-72 (1983)	Strength Properties of Adhesives in Shear by Tension Loading (Metal-to-Metal)
Tensile Strength	D897-78 (1983)	Tensile Properties of Adhesive Bonds
	D2095-72 (1983)	Tensile Strength of Adhesives by Means of Bar and Rod Specimens

Reprinted with permission from Arthur H. Landrock, *Adhesives Technology Handbook*, Noyes Publications, 1986.

with tensile tests, the stress distribution is not uniform (see Figures 4.10 and 4.13), and while it is conventional to give the the failure shear stress as the load divided by the bonding area, the maximum stress at the bondline may be considerably higher than the average stress, and the stress in the adhesive may differ from pure shear. The actual failure of the adhesive shear joint can be dominated by either tension or shear, depending on the adhesive thickness, the stiffness of the adherends, and so on

Pure shear stresses are those that are imposed parallel to the bondline and in its plane. While they do not produce pure shear, plain single-lap shear specimens (such as that shown in Figure 4.7) are frequently used because they are simple to prepare and provide reproducible and usable results. The preparation of the specimen and the method of testing are described in ASTM D 1002. Several variations are also used, as shown in Table 5.4.

Peel Tests

Peel tests are designed to measure the resistance of adhesives, especially highly flexible adhesives, to localized stress (see Figure 4.10b) and failure by progressively opening the joint out of plane. The less rigid the adherend and the higher the modulus of the adhesive, the more nearly uniform the stress distribution in the adhesive. Because the area over which the stress is applied is dependent on the thickness of the adherends and the adhesive, it is very difficult to evaluate exactly. Therefore, the applied stress and the failing stress are reported as linear values, that is, Newtons per mm or pounds per linear inch (pli) . The most widely used peel test, especially for thin-gauge metal adherends, is the T-peel test, which is covered by ASTM D 1876. Other peel tests are also covered by ASTM standards for various other situations (see Table 5.4).

Cleavage Tests

As described in Chapter 4, and as shown in Figure 4.6, cleavage is a variation of peel. In the case of cleavage, however, the two adherends are rigid rather than flexible. The cleavage test, covered by ASTM D 3807, is designed to measure the effect of a load applied normal to the bond area at one end of the specimen.

Creep Tests

When a bonded structure is subjected to a sustained load in service, especially when the service temperature is somewhat elevated and/or the loading is vibrational, it is important to measure the resistance of the adhesive to creep. ASTM D 1780 covers the practice for conducting creep tests on adhesives, with the special cases of compressive loading being covered by ASTM D 2293 and tensile loading by ASTM D 2294.

Fatigue Tests

Intermittently or cyclically applied loads impose an especially rigorous condition on adhesives, just as they do on most materials. For this reason, a special test method, covered by ASTM D 3166, is used. This test employs the plain single-lap shear specimen from ASTM D 1002. The specimen is tested on a standard tensile-testing machine capable of imposing cyclic (e.g., sinusoidal) loads, usually at 1800 cycles per minute. Adhesive performance is evaluated from a plot of the number of cycles necessary to cause failure at various stress levels, using a so-called *S-N* curve.

Impact Tests

If an adhesively bonded joint is expected to experience impact loads, it should be tested under impact loading conditions according to ASTM D 950. Here, the ability of the adhesive to absorb or attenuate forces in a very short time interval is assessed, measuring the rate sensitivity of the adhesive to applied loads. Results are reported as the number of foot-pounds of energy absorbed in failing the bond of a 1 in.^{2} specimen. Some test arrangements use a falling weight, simply using the force of gravity, while others use compressed air to accelerate the weights.

Durability Tests

As will be described in Section 5.6.5, the performance of adhesively bonded joints in service can be significantly degraded by environmental factors. Therefore, it is critically important to evaluate the durability of adhesives under the types of environments expected in service.

While there are a number of different tests for evaluating specific environmental effects, the most important is the wedge test, covered by ASTM D 3762. In this test method, a wedge is forced into the bondline of a flat-bonded aluminum specimen, thereby creating a tensile stress in the region of the resultant crack tip. The stressed specimen is then exposed to the appropriate, representative environment (e.g., water or moisture at elevated temperature), and the resultant rate of crack growth with exposure time, as well as the ultimate failure mode, are evaluated. While the test is primarily qualitative, it is useful for determining preferred adhesive choices and joint preparation methods.

5.6.3 Quality Assurance in Adhesive Bonding

One of the disadvantages of adhesive bonding as a joining method is that the bond area cannot be inspected visually, making an assessment of the quality of the joint difficult. Typically, assuring quality during adhesive bonding and of the bonded joints is carried out by two methods: (1) destructive testing using process-control specimens, and (2) nondestructive evaluation involving joint assessment or inspection. Obviously, the level of quality control applied to a particular bonded assembly depends on its structural requirements. Critical joints are controlled by high sampling levels for destructive testing and by tight acceptance requirements. Less critical bonds are controlled by less stringent procedures.

Destructive inspection is carried out on process-control test specimens prepared from the same adherend and adhesive materials as the production parts. The process-control specimen, as the name implies, accompanies the production parts throughout all of the stages of production—from joint cleaning through assembly to curing. The adhesives and adherends in the process-control specimens and in the production parts are all assembled at the same time and are cured in the same press or autoclave run. As an additional control, expendable tabs or protrusions are often used on the actual production parts for subsequent removal and testing with the process-control specimens. Test results are checked against and compared with the specification requirements, and the part is accepted or rejected based on these results. The rejected parts may be subsequently inspected nondestructively for final acceptance or rejection, under discrepancy or discrepant part procedures.

During processing, quality control procedures should be applied at various stages, first, upon receipt of incoming raw material. Incoming inspection should, as a minimum, include assessing the condition of the container in which the adhesive or adhesive components are received, looking for damage and/or leakage, checking the identity of adhesives or components, and conducting appropriate tests on physical properties (such as percent flow, gel time, and percent volatiles) and mechanical properties. This should be done for each individual batch or lot of raw material. Second, quality control procedures should accompany adherend surface preparation. This involves assuring that the proper solvent cleaning, intermediate cleaning, chemical treatment, and priming steps and procedures have taken place; monitoring the sequence, proper identity, and purity or cleanliness of materials used, solution temperatures, and concentrations; and checking cleanliness using wettability tests, such as the water break-free test. Also, once properly cleaned and prepared, the adherend surfaces should be properly protected to maintain cleanliness. Third, quailty control procedures should be applied during bonding. Here, inspection should assure proper prefit of components to be bonded using tool-proofing films to produce an image or imprint of the joint fit; verify presence of process-control specimens; verify proper adhesive application, including complete coverage and checking of thickness, if important; and verify proper assembly to assure all required parts have been assembled in the correct sequence. Fourth and finally, quality control should accompany curing. Here, inspection should verify that the proper cycle of temperature-pressure-time was employed, ideally, using sensors embedded in the production parts or assembly to permit real-time monitoring and adaptive control.

After part processing (i.e., assembly bonding), it is essential to test process-control specimens to assure compliance to specifications and to inspect the joint nondestructively. The tests and testing procedures were discussed in the previous section. Nondestructive inspection techniques could (and should) include visual inspection and various sonic, ultrasonic, thermographic, and radiographic techniques. A brief description of some of the more important and common techniques are given in the following sections.

Visual Inspection

Even though the adhesive bond area cannot be observed directly, many indicators of poor bonding may be evident. By using a strong light, often at shallow angles of incidence to highlight relief features, one can check surface smoothness as an indication of unbonded areas. Such areas often show as bumps or waves in thin adherends. It is also possible to observe the bondline around the joint periphery, observing its thickness (noting whether it is too thick or too thin); the absence of adhesive (indicating too low a clamping force or insufficient adhesive); adhesive flash that breaks away from the adherend too easily (indicating poor surface preparation), is excessively porous (indicating gas entrapment or outgassing), or too soft or softens with solvent or heat (indicating of improper curing).

Sonic Methods

Tapping, with a coin or a solenoid-operated hammer, can detect large voids or *unbonds*.[4] Properly bonded regions produce a sharp, clear tone, while large areas devoid of adhesive or not bonded produce dull, hollow tones. Sonic resonators sense deflection of the adherend, which is greater in unbonded or weakly bonded regions than in properly bonded regions. These resonators employ a vibrating crystal to excite the structure acoustically at sonic frequencies of 5 to 28 kHz. Various eddy-current methods can induce mechanical forces in the bonded (or unbonded) structure through the induced electric eddy current field. The result is an eddy-sonic test method.

Ultrasonic Methods

Ultrasonic methods are based on the response of the bonded joint to loading by low-power ultrasonic energy. The methods are excellent for detecting unbonds between the face sheet and the adhesive or the adhesive and the core in honeycomb sandwich structures or between the adhesive and the adherends in laminated structures. Both pulse-echo (i.e., send-receive approaches using the same transducer) or through-transmission (i.e., also called pitch-catch) modes can be used. The most well known and highly regarded method is the Fokker-bond method, which employs a sweep-frequency resonance technique. Here, the ultrasonic energy introduced into the structure is varied over a wide frequency range. The resonance set up by the probe, face sheet, adhesive, and the remainder of the structure is monitored. Changes in loading are shown by the combination resonance frequency shift and change in amplitude at the resonant frequency. Together, these give a semi-quantitative estimate of bond strength, as affected by the presence of voids, porosity, and incomplete wetting.

Dynamic Thermal Imaging Methods

With dynamic thermal imaging methods, bond discontinuities are revealed through temperature differences during cooling of the surfaces of heated assemblies. Cholesteric liquid crystals, infrared sensing, and other techniques are used.

---

[4] There are actually two different sources of unbonds. The first is called a *disbond*. Here bonding never occurred. The second is called a *debond*. Here, bonding did occur but, because of something improper in adhesive selection, adhesive or joint preparation or curing, separated before inspection.

Radiographic Methods

Radiographic methods permit direct imaging of defects in the adhesive–adherend interface under certain circumstances. X-rays can be used if the adhesive contains a metal filler that absorbs the x-rays. The techniques work especially well with metal-to-metal bonds. Radioisotopes can be used to determine whether electrolytes marked with a radioactive tracer penetrate a bonded joint. Neutron radiography can be used where the adhesive is not absorptive to x-rays. Hydrogen atoms in the adhesive absorb neutrons, making polymeric adhesives radiopaque and allowing detection of any defects.

In conclusion, proper quality control of the adhesive bonding process involves careful process control, largely through monitoring all stages of the process, as well as using process-control specimens to permit destructive testing and nondestructive inspection of the final assembly.

5.6.4 Typical Adhesive Properties

It is not the intent of this book to provide a comprehensive reference for the properties of adhesives or adhesively bonded joints (or for any other joining method, for that matter). It is useful, however, to provide some of the more important properties of some of the more important adhesives to give an indication of what kind of performance can be expected.

Table 5.5 presents the room temperature strength, as well as some low and high temperature strengths, and other important mechanical and physical properties of some common structural adhesives.

As will be discussed in the next section, the properties of adhesives vary considerably depending on the environment, with temperature and humidity being major factors. Other property data for low temperatures, high temperatures, and after exposure to humidity will be presented in Section 5.6.5, and some additional data are presented in Section 5.4. The interested reader is referred to any one of the several excellent handbooks on adhesives for more specific and comprehensive property data. Several excellent handbooks are listed at the end of this chapter.

5.6.5 Effects of Environment on the Durability of Adhesively Bonded Joints

Adhesive bonds must withstand the mechanical forces that act on them, but they must also resist the service environment. The nature of most synthetic and, for that matter, natural adhesives is that they are prone to environmental degradation from many sources. Common environmental factors that can degrade performance are temperature (either high or low), moisture (in liquid or vapor form), solvents or corrosive agents, and outdoor weathering (which includes the combined effects of moisture, drying, temperature cycling, and light). Applied stresses usually cause adhesives to degrade at a

Table 5.5 Important Properties for Major Structural Adhesives

Adhesive	Service Temperature Range °C (°F)	R.T. Shear MPa (psi)	R.T. Peel N/m (lb/in)	Low T Shear MPa (psi)	Low T Peel N/m (lb/in)	High T Shear MPa (psi)	High T Peel N/m (lb/in)
Acrylics		28.4 (4,000)					
Anaerobics	−54 to 232 (−65 to 450)	77.2 (10,000)					
Cyanoacrylates	−54 to 77 (−65 to 170)	13.7 (2,000)					
Elastomerics	66 to 93 (150 to 200)	0.21 to 1.23 (30 to 180)	98.1 (0.56)				
Epoxies	−150 to 260 (−250 to 500)	15.4 (2,200)	525 (3.0)	9.5 (1,000)		5.1 (750)	
Epoxy-phenolic	149–370 (300 to 700)	14–22 (2,000–3,200)	1,050–2,100 (6–12)	14.0 (2,100)		3.3 (450)	
Ethylenevinyl-acetate	−35 to 80 (−30 to 175)						
Hot-melts	−55 to 82 (−67 to 180)	3.4–4.3 (500–630)	5,250 (30)	28.2 (4,100)	700 (4)	2.8 (400)	3,500 (20)
Modified acrylics	−110 to 177 (−160 to 350)						
Natural rubber	−35 to 70 (−30 to 160)						
Neoprene	−50 to 95 (−60 to 200)	0.2–2.0 (30–290)					
Neoprene-phenolic	−57 to 93 (−70 to 200)	14–35 (2,000–4,750)	1,700–10,000 (10–59)				
Nitrile-epoxy		24.5–44.7 (3,550–6,480)					
Nitrile-phenolic	−57 to 150 (−70 to 300)	21–31 (3,000–4,500)	2,625–10,500 (15–60)	19–22 (2,800–3,000)		10 (1,400)	
Nitrile rubber	−50 to 150 (−60 to 300)	1–14 (150–2,000)					
Nylon-epoxy	−240 to 125 (−400 to 250)	34–49 (5,500–7,200)	14,000–26,000 (80–150)	31.7 (4,600)	700 (4)	4.2 (600)	
Phenolics	to 140 (to 280)	7–28 (1,000–4,000)				to 7 (to 1,000)	
Phenoxy	−62 to 82 (−52 to 180)	17–27.5 (2,465–4,000)	3,152–5,225 (18–20)	21 (4,200)		9 (1,350)	
Polyamides	−40 to 185 (−40 to 365)	7.7 (1,000)		22 (4,000)			
Polybenzimid-azole	to 288 (to 550)	14.5 (2,000)					
Polyimides	−196 to 540 (−330 to 1,000)	19 (2,800)	4,300–5,300 (25–30)	29 (4,200)		6 (850)	
Polysulfanes	−101 to 149 (−150 to 300)	27.5 (4,000)					
Polyurethanes	−240 to 127 (−400 to 260)	12–16.6 (2,000–2,400)		55.2 (8,000)	4,550 (26)	2.2 (318)	
Polyvinyl-acetal	0 to 120 (30 to 250)	14 (2,000)					
Silicones	−73 to 260 (−100 to 500)	1.7–3.4 (250–500)	1,700–3,200 (10–19)	3 (500)		2 (300)	
Styrene-butadiene	−40 to 70 (−40 to 160)						
Vinyl-phenolic	−60 to 100 (−70 to 212)	21–31 (3,000–4,500)	2,635–6,065 (15–35)	20 (2,900)		20 (2,900)	
Epoxy-polyurethane	−73 to 121 (−100 to 250)	15 (2,000)	38,700 (210)	4 (600)	26,000 (150)	2.2 (330)	8,600 (50)
Epoxy-polysulfide	−100 to 50 (−150 to 125)	6–17 (1,000–2,500)	1,400–3,200 (8–19)				

faster rate than when unstressed. Individual environmental factors are discussed briefly in the following sections.

High Temperature

All polymers, and therefore most adhesives, are degraded to some extent by exposure to elevated temperatures. Mechanical and physical properties are lowered at elevated temperatures and after continuous or repeated exposure to elevated temperatures.

For an adhesive to withstand exposure to high temperature, it must have a high melting or softening point and must be resistant to oxidation. Thermoplastic adhesives perform well up to their glass transition temperature (T_g), at which point the cohesive strength between chain molecules degrades rapidly and the polymer softens or becomes less viscous. In general, thermoplastic adhesives have more limited elevated temperature resistance than thermosetting types, typically under 120°C (250°F). The best performing thermoplastic adhesives, in terms of temperature service, are the hot-melt polysulfones. Typical strengths as a function of temperature for a variety of liquid, paste, tape, film, and solvent-based structural adhesives are shown in Figure 5.5.

Adhesives that are resistant to high temperatures usually have rigid, cross-linked structures, high softening temperatures, and stable chemical side groups or radicals. Such adhesives are generally thermosetting types. Thermosetting adhesives have no real melting point, so perform well to relatively high temperatures (for polymers), provided thermal oxidation and pyrolysis are avoided. While few thermosets can withstand long-term service over 177°C (350°F), some recently developed polymeric adhesives can withstand sustained exposure of nearly 320°C (600°F), with limited excursions to 370°C (700°F). Epoxy adhesives are usually limited to applications below 120°C (250°F), but some can tolerate short-term service to 260°C (500°F) and long-term service from 150° to 260°C (300° to 500°F). Silicones are good to 260°C (500°F) long term and 320°C (600°F) short term, while epoxy-phenolic alloys can operate successfully to 370°C (700°F). The polyimide (PI) and polybenzimidazole (PBI) adhesives are the best, operating to over 540°C (1000°F) for short intervals, with good thermal stability (see Figure 5.5) but poor tolerance for moisture.

Low and Cryogenic Temperatures

Cryogenic adhesives have been defined as those capable of retaining shear strengths above 6.89 MPa (1000 psi) at temperatures from room temperature down to 20 K (–423°F). Room temperature vulcanizing silicones, called RTV's, exhibit good cryogenic properties, and polyurethanes are very good. Figure 5.6 shows a comparison of strengths for some cryogenic and low temperature adhesives types.

Humidity and Water Immersion

Moisture can affect the strength of an adhesive in two significant ways. First, it can cause *reversion*. This is where the polymer loses its strength and hardness and can liquify in warm, humid air. Ester-based polyurethanes are an example of such

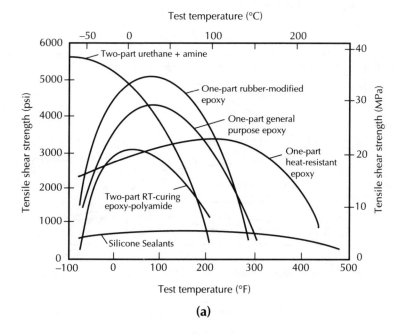

(a)

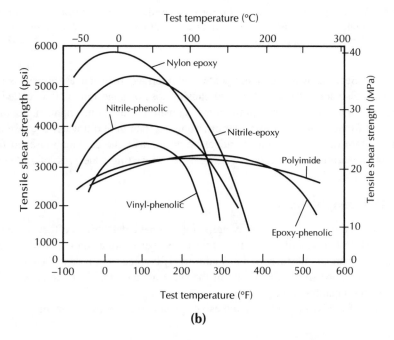

(b)

Figure 5.5   Tensile shear strengths of some structural adhesives as a function of temperature. (a) Paste and liquid adhesives; (b) tape-, film-, and solvent-based adhesives. (Reprinted with permission from Arthur H. Landrock, *Adhesives Technology Handbook,* Noyes Publications, 1986.)

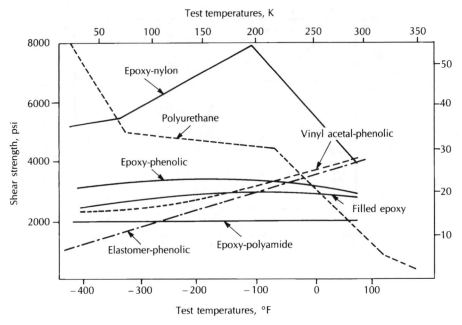

Figure 5.6   Comparison of strengths for some cryogenic and low-temperature structural adhesives. (Reprinted with permission from Arthur H. Landrock, *Adhesives Technology Handbook,* Noyes Publications, 1986, p. 245.)

polymers. Second, water can permeate the interface between an adhesive and an adherend, displacing the adhesive and causing loss of adhesion. This is called *water displacement* and is the more common mechanism. For either mechanism, degradation is usually worse in an aqueous-vapor environment than in liquid water because permeation is more rapid. Stress on the adhesive accelerates the degradation even further. Moisture absorption also causes swelling of thermosetting adhesives, often leading to strength degradation.

Salt Water and Salt Spray

Salt water and especially salt spray are known to have deleterious effects on adhesive joints. Almost certainly, the mechanism is permeation and corrosion at the adhesive–adherend interface. In many adhesively bonded samples exposed to salt water or salt spray, corrosive undercutting of the substrate adjacent to the interface with the adhesive has been observed.

One of the best performing adhesive systems for this environment is the nitrile-phenolics.

Weathering

By far the most detrimental factors affecting adhesives aged outdoors are heat and humidity, but temperature cycling, moisture cycling (i.e., wetting and drying), ultraviolet radiation, oxidation, and low temperatures also have minor degrading effects. In fact, it is the combined action of several of these factors during most outdoor applications that make weathering particularly severe.

When exposed to weather, structural adhesives typically lose strength rapidly during the first 6 months to 1 year. Then, after 2 to 3 years, the rate of degradation usually levels off at 25 to 30% of the initial joint strength. In most cases, corrosion in the adherend or along the adhesive–adherend interface contributes to the degradation of the joint's strength. Other factors, such as the ultraviolet components of light, degrade thermosets by causing too much cross-linking.

Two adhesive systems that tolerate weathering well are epoxy-phenolic and nitrile-phenolic film types.

Chemicals and Solvents

Most organic adhesives tend to be susceptible to corrosive chemicals and solvents, especially at elevated temperatures. Epoxy adhesives are generally more resistant to a wide variety of liquid environments than most other structural adhesives, and thermosetting adhesives are generally more resistant than thermoplastic adhesives. Without question, no one adhesive is optimum for all chemical environments.

Vacuum

Vacuum, such as in outer space, degrades the performance of adhesives based on the degree of adhesive evaporation. Loss of low molecular weight constituents, such as plasticizers or diluents, can result in hardening and porosity. Most structural adhesives consist of predominantly high molecular weight polymers as their base, so they tolerate vacuum well.

Radiation

High-energy particulate and electromagnetic radiation, including neutron, electron, and gamma radiation, have effects similar to vacuum, that is, they cause scission of the polymer molecular chains used in structural adhesives. This, in turn, causes weakening and embrittlement.

Polysulfones, epoxies, polyimides and polyurethanes all perform well in radioactive environments.

Biological

Adhesives in bonded joints may or may not be attacked by biological organisms (e.g., fungi, bacteria, rodents, and insects), depending on how attractive they are to these organisms. Adhesives based on animal or plant materials (natural adhesives) are more likely to be affected than synthetic adhesives, although synthetic adhesives are not immune to such attack.

From the foregoing, it should be clear that the durability of an adhesively bonded joint is critically dependent on the environment in which that joint is to perform, and many environmental factors degrade performance significantly.

5.7 APPLICATIONS OF ADHESIVES

The use of adhesives in both nonstructural and structural joining applications is growing dramatically as the science and engineering of polymeric materials is growing.
Major applications of adhesives have occurred in the automotive, building construction, electrical and electronic, aerospace, marine, packaging, and other industries, and
new and more sophisticated applications are occurring every day. Perhaps nowhere is
the impact of adhesives more obvious than in the aerospace industry, which spear-
headed the development of structural adhesives for joining metals. As an indication of
the importance of adhesives, the Boeing 747 commercial jumbo jet airliner uses 3,700
m^2 (40,000 ft^2) of adhesive film, almost 450 kg (1,000 lb.) of polysulfone sealant, and
23 kg (50 lb.) of silicone rubber sealant. Use of adhesives, both nonstructural and structural, is growing rapidly in the automobile industry. Nonstructural adhesives are used
for sealing, vibration damping and noise deadening, and thermal insulation. Structural
adhesives are being very seriously studied for use in assembling space frames in aluminum intensive vehicles.

The adhesive bonding process provides a much needed and valuable complement
to the other principal joining processes of mechanical fastening, welding, brazing, and
soldering.

SUMMARY

Adhesives are the key ingredient for accomplishing adhesive bonding. Usually, adhesives are mixtures of several constituents, including an adhesive base or binder to provide the chemical bonding agent, a diluent or solvent for controlling the viscosity or
body and for facilitating application and drying, fillers to improve working characteristics or impart certain properties (e.g., electrical or thermal conductivity), and carriers or
reinforcements for supporting the adhesive during application and/or for improving
strength in service. The diversity of adhesives and their use leads to several different
schemes of classification: by function (structural or nonstructural); by chemical composition, whether natural or synthetic; by physical form (liquids, pastes, films); by
mode of application (brushable) or means of setting (heat- or pressure-setting); or by
specific adherend (wood adhesives) or application (weatherable adhesives). Synthetic
adhesives are preferred for structural applications and are, in turn, classified by the type
of polymer making up the base or binder, including: thermosetting, thermoplastic, and
elastomeric types, and adhesive alloys, involving combinations of these. Important
types of structural adhesives include epoxies, modified epoxies (i.e., alloys of epoxies
with other thermosets, thermoplastics, or elastomerics), acrylics, cyanoacrylates (or
super glues), anaerobics, urethanes, phenolics, silicones, hot-melt thermoplastics, and
high temperature types. Another classification scheme based on chemical composition
adapted by the SME considers chemically reactive adhesives, evaporation or diffusion
adhesives, hot-melts, delayed-tack adhesives, pressure-sensitive adhesives, and films
and tapes.

The actual process of adhesive bonding involves the proper storage and preparation of the adhesive, proper preparation of the adherend or joint surfaces, proper application of the adhesive to the joint and assembly of the joint, and proper curing. Necessary equipment includes devices and procedures for controlling adhesive or bondline thickness; presses, autoclaves or vacuum bagging systems for applying pressure uniformly over the entire bond area; and ovens, autoclaves, baths, and heated platens for applying heat. Achieving the desired and optimum bond performance requires pretesting of properties in tension, shear, peel, cleavage, creep, fatigue, and impact using standardized methods such as those of the ASTM, as well as proper quality control during actual bonding using process-control specimens for destructive testing (for comparison to specifications) and nondestructive evaluation techniques for final assembly inspection using vision, sonics, ultrasonics, radiography or other methods. Adhesive properties vary widely with adhesive type, composition, adherend, and environment. Environmental factors, such as temperature, moisture, chemicals, light, vacuum, radiation, and, even, biological agents, can significantly degrade the durability of a particular adhesive.

With proper adhesive selection and proper bonding procedures, the adhesive bonding process provides a much needed and valuable complement to the other principal joining processes of mechanical fastening, welding, brazing, and soldering.

PRACTICE QUESTIONS AND PROBLEMS

1. Adhesives are almost always actually systems comprising of two or more constituents that perform different roles. What are the six different constituents that can be found in adhesive systems? Briefly describe the role or function of each.
2. Some adhesives derive their active constituent from natural sources. What are the three broad categories of natural sources? Give some examples of specific materials from each type of source that can be used as the base for a natural adhesive.
3. What are synthetic adhesives in the most general chemical terms? What are the four major subclasses of synthetic adhesives?
4. Differentiate between a structural and a nonstructural adhesive. What are some of the roles fulfilled by a nonstructural adhesive?
5. Describe how thermosetting adhesives accomplish bonding. Give some examples of important thermosetting adhesives.
6. Describe how thermoplastic adhesives accomplish bonding. Give some examples of important thermoplastic adhesives.
7. What is meant by an elastomeric adhesive? Give some important examples.
8. What is meant by adhesive alloys? Why are these especially important? Give some important examples.
9. Chemically reactive adhesives can consist of one or two components. Give three ways that a one-component adhesive of this type can be activated. Give an example of an important adhesive activated by each mechanism.

10. Some two component adhesives are premixed (i.e., no-mix types) and some must be mixed upon use (i.e., mix-in types). Explain how each type operates. Give an example of an important no-mix and an important mix-in adhesive.

11. Describe how evaporation and diffusion adhesives operate to accomplish bonding. Give an important example of such an adhesive.

12. Differentiate between a delayed-tack and a pressure-sensitive type adhesive. Give an example of an important adhesive of each type.

13. What is meant by the physical form of an adhesive? Why is this an important consideration, worthy of being used for classifying adhesives? What are the popular forms of adhesives?

14. Give the specific means or method by which each of the following adhesives achieve bonding: (a) cyanoacrylates; (b) anaerobics; (c) modified acrylics; (d) polyacrylates; (e) methylacrylates; (f) melamine formaldehydes; (g) polybenzimidazoles; (h) styrene-butadiene copolymers.

15. What are some considerations regarding the storage of an adhesive?

16. What are the major considerations in preparing joints for structural adhesive bonding?

17. What are the three major functions of equipment in bonding? Give some examples of the fundamentally different systems for each of the three purposes.

18. For what four different purposes (not properties) are adhesives tested?

19. What different means are involved in assuring the quality of adhesives and adhesively bonded joints?

20. The performance of an adhesively bonded joint is greatly affected by the environment in which it is meant to function. What are the major environmental factors that can affect adhesive bonded joint performance?

REFERENCES

1. *Adhesives,* Edition 6, D.A.T.A. Business Publishing and International Plastics Selector, Inc., 1990.

2. *Adhesives and Sealants,* Volume 3, Engineered Materials Handbook, Materials Park, Ohio: ASM International: 1990.

3. Buckley, John D., and Stein, Bland A., *Joining Technologies for the 1990's,* Park Ridge, N.J.: Noyes Data Corporation, 1986.

4. Cagle, C. V., editor, *Handbook of Adhesive Bonding,* New York: McGraw-Hill, 1973.

5. Gauthier, Michelle M., Sorting Out Structural Adhesives: Part 1, *Advanced Materials & Processes,* Vol. 138, No. 1, Materials Park, Ohio: ASM International, July 1990, pp. 26–35.

6. Gauthier, Michelle M., Clearing Up Adhesives Confusion: Part 2, *Advanced Materials & Processes,* Vol. 138, No. 2, Materials Park, Ohio: ASM International, Aug. 1990, pp. 41–49.

7. Landrock, Arthur H., *Adhesives Technology Handbook,* Park Ridge, N.J.: Noyes Publications, 1985.

8. Shields, J., *Adhesives Handbook,* 3rd. ed., London: Butterworths, 1984.

9. Skeist, I., *Handbook of Adhesives,* 3rd ed., New York,: Van Nostrand Rheinhold, 1989.

10. Wegman, R.F., *Surface Preparation Techniques for Adhesive Bonding,* Park Ridge, N.J.: Noyes Publications, 1989.

6

Welding as a Joining Process

6.1 INTRODUCTION

Next to mechanical fastening, welding is unquestionably the oldest method of joining materials. While the great majority of welding processes were invented in fairly recent times, some, such as hammer welding or forge welding, have a very long history. Welding first evolved as an important technique for fabricating metals, as part of the iron-making process itself, as well as a means of making finished iron products by joining. *Welding* involves bringing the surfaces of the materials to be joined close enough together for atomic or molecular bonding to occur. In welding, primary (or occasionally secondary) chemical bonds are formed, with the specific type of bonding depending on the materials involved in the joint. In its broadest sense, welding includes any process that causes materials to join through the action of interatomic or intermolecular forces, as opposed to macroscopic mechanical forces. Thus, welding, brazing (Chapter 8), soldering (Chapter 9), and even adhesive bonding (Chapter 4) can all be considered welding processes.

Welding is critically important in modern manufacturing from both a technological and an economic standpoint. It has been estimated that more than half the gross national product of all industrialized nations comes directly or indirectly from welding. Either the process is used to join materials into parts and parts into assemblies and structures or it is used in the machines that make those materials and parts, or it is used for both. One only has to think about the pipelines used to transport natural gas and oil from the Bering Sea across Alaska and Canada to the continental United States or across the Ural Mountains from Russia to Western Europe, or the giant 100,000-ton supertankers that ply the oceans moving oil around the world, or the numerous offshore oil-drilling platforms that tap new reserves of oil and natural gas, or liquid–gas storage tanks, or the numerous pressure vessels in steam and power generation plants, or reaction and containment vessels in the chemical processing industries and the import and impact of welding becomes obvious (see Figure 6.1)

This chapter explores the process of joining by welding. First, the general mechanism by which a joint is formed between materials is described, and the applicability of the process to metals, ceramics, and polymers, as well as their composites, is presented. Then, the relative advantages and disadvantages of welding versus other joining

Figure 6.1   Typical large welded structure. (Courtesy: Chicago Bridge & Iron)

processes are discussed. Next, various schemes for classifying welding processes are presented, including pressure versus nonpressure processes, fusion versus non-fusion processes, classification by heat source, and other schemes. Specific nonpressure and pressure fusion and nonfusion welding processes are then described to give the reader a working knowledge of the many options available. These processes are compared in terms of the efficiency with which energy is transferred from the welding source to the joint to create the desired weld, and the particular advantages and limitations of each are highlighted. Finally, a brief discussion of weld joint designs is presented.

The all important physical metallurgy of the welding process is addressed in the Chapter 7. The subcategories of welding, namely brazing and soldering, are described in Chapters 8 and 9, respectively.

6.2 JOINING MATERIALS BY WELDING

6.2.1 General Description

Welding is a process in which materials of the same type or class are joined together through the formation of primary (and, occasionally, secondary) bonds under the combined action of heat and pressure. These bonds are the natural consequence of bringing two similar materials close enough together to allow interatomic or intermolecular attractive and repulsive forces to balance. The type of bonds formed across the joint are the same as the type of bonds found in the materials being joined.[1] Metals are joined in welding through the formation of metallic bonds. Thermoplastic polymers are joined in welding through the formation of some covalent bonds and substantial secondary bonds, e.g., by van der Waals or permanent dipole forces, as well as by substantial molecular tangling. Glasses can be joined by welding with primary covalent bond formation, while ceramics can be joined by welding through the formation of ionic, covalent, or mixed bonds, depending on the particular ceramic(s) being joined.

The key to all welding is interdiffusion between the materials being joined, whether that diffusion occurs in the liquid, solid, or mixed state.

6.2.2 Creating a Welded Joint

Whenever two or more atoms are brought together from infinite separation, a force of electrostatic or Coulombic attraction arises between the positively charged nuclei and the negatively charged electron clouds, and this force increases with decreasing separation. The potential energy of the separated atoms also decreases as the atoms come together. This is shown in Figure 6.2.

As the distance of separation becomes very small, on the order of a few atom diameters, the outer electron shells of the approaching atoms begin to feel one another's presence, and a repulsive force between the negatively-charged electron shells increases more rapidly than the attractive force.[2] The attractive and repulsive forces combine to create a net force. At some separation distance, known as the *equilibrium distance* or *equilibrium spacing,* the forces of attraction and repulsion just balance, and a net force of zero results. At this equilibrium spacing, the net potential energy[3] is a minimum, the aggregate is stable, and the atoms (or ions) are said to be *bonded.* When atoms are at their equilibrium spacing in a bonded aggregate, all of the atoms achieve

---

[1] By this definition, dissimilar types of materials (e.g., metals and ceramics) cannot be welded to one another, because one material has metallic bonding and the other has ionic, covalent, or mixed ionic-covalent bonding. Dissimilar types of materials can be joined by the subcategories of welding, namely brazing or soldering, or by adhesive bonding.

[2] The attractive force increases inversely to the approximate n power of the atomic separation, while the repulsive force increases as the approximate m power of of the atomic separator, where $n < m,$ m is usually 12, and n is 2–6, depending on bond type.

[3] The potential energy and the forces of attraction and repulsion are related by the relationship, $F = -dU/dx,$ where F is the force of attraction or repulsion, U is the potential energy of attraction or repulsion, and x is the distance of separation. By convention, attraction is negative, and repulsion is positive, based on how work must be done to change the separation.

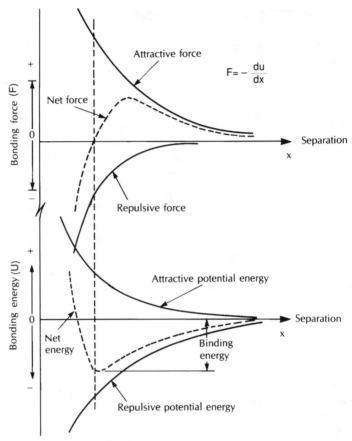

Figure 6.2   The forces and potential energies involved in bond formation leading to welding.

stable outer electron configurations by sharing or by transferring electrons.[4] Obviously, what has just been described for neutral atoms can also occur for molecules having induced or permanent dipoles or for oppositely charged ions, leading to the formation aggregates of molecules in polymers or ions in ionically bonded compounds, called ceramics.

The tendency for atoms to bond is the fundamental basis for welding. The challenge, then, is to bring atoms to their equilibrium spacing to create a weld.

If two perfectly flat surfaces are brought together to the equilibrium spacing for the atomic species involved in the pairs across the interface, bond pairs form and the two pieces are welded together perfectly. In this case, there is no remnant of a physical interface and there is no disruption to the structure of either material involved in the joint.[5] This ideal situation is shown in Figure 6.3a and b. The resulting *weld* has the

---

[4]Electrons are shared in covalent and metallic bonds, while there is a transfer of electrons from one atom to another in a bonded pair in ionic bonds.

[5]This case applies to a crystalline or a noncrystalline (i.e., amorphous) material, with either the crystalline or the amorphous structure being preserved across the interface.

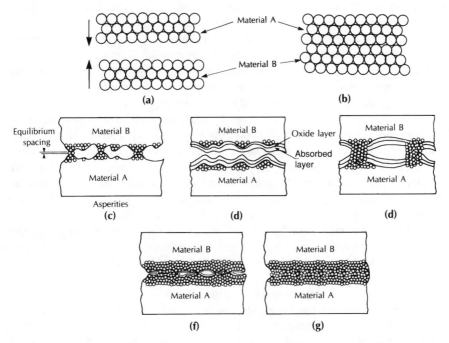

Figure 6.3 Two perfectly smooth and clean versus real materials being brought together to form a weld. (a), (b) Ideal surfaces; (c), (d) real surfaces; (e) heat removes adsorbed layer pressure breaks oxide, (f) increased pressure increases contact; (g) near perfect weld.

strength expected from the binding energy (determined by the depth of well in the net potential curve in Figure 6.2), so the joint efficiency is 100%.

In reality, two materials never have perfectly smooth, planar surfaces, so perfect matching up of all atoms across an interface at equilibrium spacing never occurs, and a perfect joint can never be formed simply by bringing the two materials together. Real materials have surfaces that are highly irregular. Peaks and valleys tens to thousands of atoms high or deep lead to very few points of intimate contact where the equilibrium spacing can be reached. This is shown in Figure 6.3c. Typically, only one out of every billion or so atoms along a real surface[6] can create a bond, so the joint strength is only about one one-billionth of the theoretical strength that can be achieved.

The situation is made even worse by the presence of oxide layers and adsorbed moisture layers usually found on real materials. As shown in Figure 6.3d, bonding, and thus welding, can only be achieved in such cases by removing or breaking these two layers *and* bringing the clean base material atoms within the equilibrium spacing[7] for the materials involved.

There are two ways of improving the situation, that is, to clean the surface of real materials and bring most, if not all, of the atoms of those surfaces into intimate contact

---

[6] The so-called real surface is actually fairly representative of a well machined and polished material, with, say, a 4 rms finish.

[7] Obviously, any other form of surface contamination, say from grease, only further exacerbates the situation.

over large areas. The first is to apply heat; the second is to apply pressure. Heating helps welding to occur in several ways. In the solid state, heating helps by (1) driving off the adsorbed layers of gases or moisture; (2) breaking down the oxide layers through differential thermal expansion or, occasionally, by thermal decomposition; and (3) by lowering the yield strength of the base materials and allowing plastic deformation under pressure to bring more atoms into intimate contact across the interface. This is shown in Figure 6.3e. Alternatively, heating could help by causing melting of the substrate materials to occur, allowing atoms to rearrange by fluid flow and come together to equilibrium spacings, *or* by melting a filler material to provide extra atoms of the same or different but compatible types as the base material This situatio is shown in Figure 6.3f. Pressure helps welding to occur by (1) disrupting the adsorbed layer of gases or moisture by deformation; (2) fracturing the brittle oxide layer to expose clean base material; and (3) plastically deforming the asperities to increase the number of atoms (and the area) in intimate contact (see Figure 6.3e). Figure 6.3g shows a near perfect weld.

The relative amounts of heat and pressure necessary to create welds vary from one extreme to the other. Very high heat and little or no pressure can produce welds by relying on the high rate of diffusion in the solid state at elevated temperatures or in the liquid state produced by melting or fusion. Little or no heat with very high pressure can produce welds by forcing atoms together by plastic deformation, relying on diffusion in the solid state to cause intermixing. Most welding processes involve a fair amount of heat and only enough pressure to hold the joint elements together during the welding process, but there are processes that predominantly employ pressure.[8] In fact, welding processes are frequently classified by the amount of pressure or the amount of heat involved in producing welds. These are called *pressure* and *nonpressure* or *fusion* and *nonfusion* welding processes, respectively.

6.2.3 Welding of Metals versus Ceramics versus Polymers

As was described in Section 6.2.1, any material can be welded if the atoms, molecules or ions composing those materials can be brought into intimate contact in sufficiently large numbers to produce significant bonding. Most people are familiar with the fact that metals can be welded, but fewer realize that ceramics, glasses, polymers, and even composites can also be joined by welding. The requirements are the same in each material; the only difference is the means for bringing atoms into contact and the type of primary (or secondary) bonding that results.

In metals, metallic bonds are formed across the interface. In ceramics and glasses either ionic, covalent or mixed ionic-covalent bonds are formed, depending on the specific nature of the ceramic or glass. In polymers, covalent bonds are formed, provided the polymers are suitably mixed by the action of the applied heat and/or pressure; otherwise secondary bonding by van der Waals induced dipoles or permanent dipoles pre-

---

[8]See Sections 6.3.1 and 6.3.2 on pressure versus nonpressure and fusion versus nonfusion welding processes, respectively.

dominantly contributes to the joint strength. In addition, significant intertwining or tangling of the long-chain molecules of many polymers (e.g., thermoplastics) gives rise to aggregation without the need for substantial bonding. In composites, welding actually occurs within the matrix materials; with metallic bonds formed between metal matrices, ionic or covalent or mixed bonds occurring between ceramic matrices, and secondary or covalent bonds or chain tangling occurring between polymeric matrices.

While dissimilar metals can often be welded to one another (because even dissimilar alloys bond metallically), it is impossible to weld dissimilar material types, for example, metals to polymers, to one another. Likewise, dissimilar ceramics can often be welded to one another, provided both have similar bonding. For polymers, welding, in its true sense, is usually restricted to thermoplastic types. Thermosetting polymers are not normally welded.[9]

The welding of various materials, by type, is discussed in detail in Chapter 11 for metals and alloys, Chapter 12 for polymers, Chapter 13 for glasses and ceramics, and Chapter 14 for various composites.

6.2.4 The Importance of Cleaning for Welding

A critical part of any welding process is cleaning the surfaces to be joined and keeping those surfaces clean throughout the welding process in order to allow the atoms of the materials to be joined to come together to form bonds. Polymers and ceramics and glasses generally only need to have dirt or hydrocarbon (e.g., oil, grease, or wax) removed from their surfaces to be made suitably clean for welding. Metals, on the other hand, usually present additional cleaning requirements, since they are generally reactive, especially when heated to near or above melting. Oxides, sulfides, and other surface scales must usually be removed.

For all materials, solvents can remove oil, grease, and particulate contamination. For some materials, notably metals, mechanical abrasion or chemical cleaning with acids or strong bases may be necessary to remove tenacious scales or oxides. Some processes use a chemical agent called a *flux* to clean and activate the surface of a material to promote bonding. Fluxes chemically reduce oxides and other surface contaminants on metals, producing a perfectly (i.e., atomically) clean surface. To work properly, these fluxes must remain on the metal surface throughout the welding process to prevent recontamination by oxidation, for example. In the molten state, fluxes are often called *slags*. The role of fluxes and slags in welding will be described with the description of specific processes.

Because they are usually reactive, metals must be protected or shielded from oxidation during welding, especially when heating is involved. *Shielding* can be provided by covering or surrounding the pieces being welded, that is, the *weldment*, with an inert atmosphere. Inert atmospheres consist of a gas or mixture of gases or a vacuum that

---

[9]Thermosetting polymers can be readily joined chemically by adhesive bonding. Bonding between the adhesive and the substrate (or adherend) is usually of covalent and secondary, van der Waals.

excludes oxygen. Examples of shielding gases are nitrogen, helium, argon, carbon dioxide, and mixtures thereof. Molten slag can also provide the needed shielding or protection.

6.2.5 Advantages and Disadvantages of Welding

Like other joining processes, welding offers several advantages but has some disadvantages as well. The most significant advantage is undoubtedly that welding provides exceptional structural integrity, producing joints with very high efficiencies. Joint strengths, when continuously welded, approach or exceed the strength of the base material(s). Another advantage is the wide variety of approaches and processes that can be used and the wide variety of materials that can be welded. Almost all metals and many polymers, glasses, and ceramics, as well as the matrices of many composites, can be welded using a variety of pressure or nonpressure fusion or nonfusion processes, with or without auxiliary filler.

Other advantages of welding are that it can be done manually, semiautomatically, or completely automatically and it can be made highly portable for implementation in the field for erection of structure on site or for maintenance and repair. Continuous welds provide fluid-tight joints, so welding is the process of choice for fabricating pressure vessels for gaseous materials and containers for liquid materials. For most applications, welding can be reasonable in cost, although highly critical welds, with stringent quality requirements or involving specialized applications (e.g., very thick section welding) can be expensive.

The single greatest disadvantage of welding is that it prevents disassembly. Often, however, this permanency is the reason for its selection for joining. As mentioned, the process can be expensive if quality requirements are high or for special joining situations. The need for considerable operator skill and the occasional need for capital-intensive equipment (e.g., laser welding systems) contribute to cost. From a material's standpoint, welding can degrade the properties of the base material because of the effects of heat, especially for processes dependent on fusion. Unbalanced heat also leads to shrinkage, residual stresses, and distortion.

All in all, welding is an extremely versatile process, offering exceptional joint integrity to the designer and flexibility to the process engineer. Table 6.1 summarizes the advantages and disadvantages or limitations of welding.

6.3 CLASSIFICATION SCHEMES FOR WELDING PROCESSES

6.3.1 General Description

When one recalls that welding is defined as a process in which materials of the same type or class are joined together through the formation of primary (or, occasionally, secondary) bonds under the combined action of heat and pressure, it should come as no surprise that an extremely wide variety of welding processes exist as well as several

Table 6.1 Summary of Advantages and Disadvantages of Welding

Advantages	*Disadvantages*
1. Joints of exceptional structural integrity and efficiency	1. Prevents disassembly
2. Wide variety of processes	2. Heat of welding degrades base properties
3. Applicable to many materials within a class	3. Unbalanced heat input leads to distortion or residual stresses
4. Manual or automated operation	4. Requires considerable operator skill
5. Can be highly portable	5. Can be expensive (e.g., thick sections)
6. Leak-tight joints for continuous welds	6. Capital equipment can be expensive (e.g., lasers)
7. Reasonable cost, usually	
8. Precludes joint loosening	

schemes by which they can be classified. Processes can be classified by whether pressure is required or not (pressure versus nonpressure processes), whether or not melting or fusion is required (fusion versus nonfusion processes), or by the source of heat, when heat is used. Another less important scheme considers whether a filler material is used or not and, if so, whether that filler is the same as or different from the base material(s). Finally, within one large group of welding processes using an electric arc (i.e., arc welding processes), processes can be subclassified by the nature of the electrode employed. In these processes, electrodes may or may not be intended to be consumed (consumable versus nonconsumable electrode arc welding processes) and, if consumable, may be used in continuous or discontinuous forms (continuous versus discontinuous consumable electrode arc welding processes).

The following sections describe the most common schemes for classifying welding processes and list various processes within these schemes.

6.3.2 Pressure versus Nonpressure Welding Processes

Some welding processes depend upon or are facilitated by the application of pressure[10] to bring the atoms of the material substrates to be joined close enough together to achieve bond formation across the interface. These processes are called *pressure welding* or *pressure bonding processes.*

The effect of pressure in accomplishing welding is multifold. First, it increases the number and area of contacts between the mating substrates through the plastic deformation of the highest asperities, bringing lower height asperities into contact. Second, it enhances diffusion of atoms across the interface by increasing the temperature locally as the result of the mechanical work done in the deformation process and, to a lesser extent, through stress-enhanced diffusion. Obviously, both of these effects require that the materials being welded exhibit reasonable plastic behavior. Finally, pressure holds the joint elements together while the bond formation process occurs.

---

[10]The pressure that is required or used is usually considerably greater than the pressure required with essentially all joining processes to hold the pieces to be joined together during the process of joining.

Fixturing or holding pressures are normally only on the order of a few pounds per square inch (or a fraction of a megapascal), while the pressures used to cause plastic deformation are normally thousands or tens-of-thousands of pounds per square inch (or tens to hundreds of megapascals).

There are very few cold pressure welding processes, although cold welding is possible. Most pressure welding is facilitated by heating, from various sources, to reduce the yield strength of the materials to ease plastic deformation, and to speed diffusion. In true pressure welding processes, however, pressure is far more important than heat, and certainly melting or fusion is not necessary to produce the weld, although some melting may occur.

Nonpressure welding processes rely on heat only, with little or no pressure except to hold the joint elements together.

Various pressure welding processes are listed in Table 6.2, while pressure versus nonpressure processes in an overall taxonomy of welding processes is shown in Figure 6.4. Individual pressure welding processes are described in Section 6.5.

6.3.3 Fusion versus Nonfusion Welding Processes

Many more welding processes rely on heat than on pressure to accomplish joining by creating atomic bonding across the joint interface. This heat may cause melting or fusion or may only serve to soften the material in the solid state to facilitate plastic deformation. When significant melting is involved *and* necessary for welding to take place, the processes are called *fusion welding processes.* Bond formation is aided by the melting process providing a supply of highly mobile atoms throughout the interface, whether or not auxiliary filler material is used. If melting does not occur *or* is not principally responsible for causing welding (i.e., bond formation), the processes are called *nonfusion welding processes.* Heat may still be involved, however.

In fusion welding, the source of heat can be chemical, electrical (arc or resistance), high-energy beams, or others (such as microwave and induction).[11]

Table 6.2   Pressure Welding Processes by Energy Source

	Energy source	
Mechanical	**Chemical**	**Electrical**
Cold welding	Oxyacetylene pressure welding	Spot welding
Friction welding	Exothermic pressure welding	Seam welding
Vibration welding	Forge welding	Projection welding
Ultrasonic welding		Upset butt welding
Forge welding/Roll welding		Flash welding
Diffusion bonding		Percussion welding
Explosion welding		

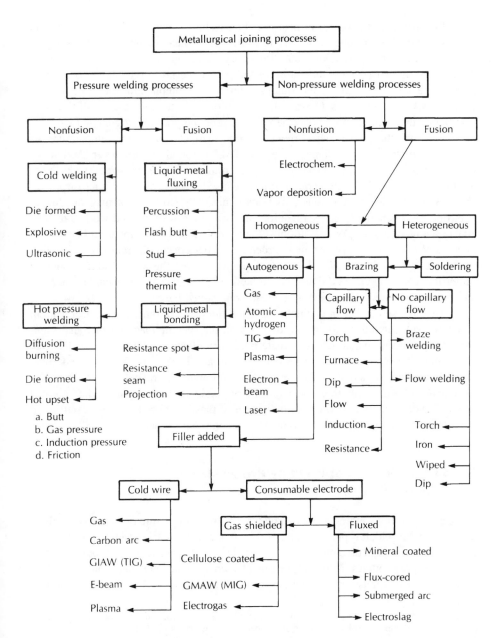

Figure 6.4 Taxonomy of welding processes.

Regardless of the source of heat in fusion welding, fusion welds exhibit distinct microstructural regions due to the various effects of the heat. Figure 6.5a shows a typical fusion weld in a pure and in an alloyed crystalline metallic or ceramic material. Where heating caused the temperature of the pure material to rise above its melting point or the alloy to rise above its liquidus temperature, complete fusion occurred, producing a *fusion zone,* or *FZ.* Outside this region or zone, the temperature of the alloy was below the liquidus but above the solidus, producing a *partially melted zone,* or *PMZ.* Because the pure material melted at one distinct and unique temperature, as opposed to over a range of temperatures, no PMZ is found in pure materials. Farther from the centerline of the heat source and the resulting weld, the temperature was lower but may still have been high enough to have caused some observable microstructural changes due to solid-state transformations. These transformations may be due to allotropic phase changes; recrystallization and/or grain growth (in cold-worked materials); or aging, overaging, or resolutionization (or reversion) in precipitation-hardenable systems. When such observable microstructural changes occur, the region is referred to as the *heat-affected zone, HAZ.*[12] At some point far from the weld (and heat source) centerline, the temperature did not rise high enough to cause any noticeable change in the microstructure.[13] This region is called the *unaffected base material.*

The precise number of distinct zones, and the width of those zones, occurring in a material subjected to welding heat depends on the specific reactions or transformations that can occur in that particular material. In a nonfusion, solid-phase or solid-state weld, there is usually essentially no fusion zone, and little or no heat-affected zone, depending on how much of a role heat played in allowing plastic deformation to occur. As shown in Figure 6.5b, this is shown for a nonfusion weld in a pure and an alloyed material. Note that there is often an indication of plastic deformation, or upsetting, at the original faying surfaces of the joint. This is commonly called *flash.*

6.3.4 Classification of Welding Processes by Energy Source

Besides classifying welding processes by whether or not pressure or fusion is involved in producing a weld, it is possible, and fairly common, to classify welding processes by the specific type of source employed to provide the energy needed to produce welds. Types of energy sources include (1) chemical sources, which generate heat from either (a) exothermic combustion of a fuel-gas using air or oxygen or (b) from an exothermic

---

[11]Some may consider some fusion processes to use mechanical means for generating heat, typically from friction, but usually the little bit of melting that occurs is not necessary for bonding.

[12]It is possible for there to be more than one distinct heat-affected zone, if several different processes or reactions can occur in the material system at different temperatures. One example is a cold-worked alloy that exhibits a phase transformation. Here, there is a region in which recrystallization and grain growth occurred, at a relatively low temperature, and a region of phase change in a region at a higher temperature. These different regions are often referred to as high temperature and low temperature heat-affected zones.

[13]In fact, unnoticeable changes may occur, such as recovery in a cold-worked material, but this is not usually considered part of the heat-affected zone.

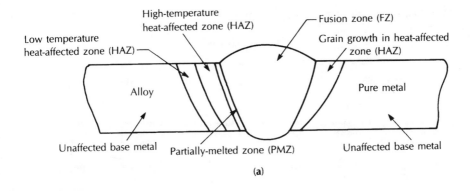

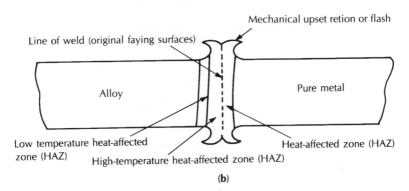

Figure 6.5 The various microstructural zones found in (a) fusion and (b) nonfusion welds.

chemical reaction between a metal and an oxide or oxygen; (2) electrical sources, which generate heat from (a) an electric arc or plasma between an electrode and the weldment, (b) resistance (I^2R) heating of the work as part of an electrical circuit, (c) induction of an electric current, or (d) excitation by microwave or radio-frequency radiation; (3) high-energy beam sources, which generate heat by converting the kinetic energy of a particles in a beam upon collision with the work; and (4) mechanical sources, which generate heat by the conversion of work through friction or plastic deformation under pressure.

Figure 6.4 shows the classification of pressure welding processes by heat source, while Table 6.3 lists fusion processes by energy source. Table 6.2 lists nonfusion processes by energy source.

Table 6.3 Fusion Welding Processes by Energy Source

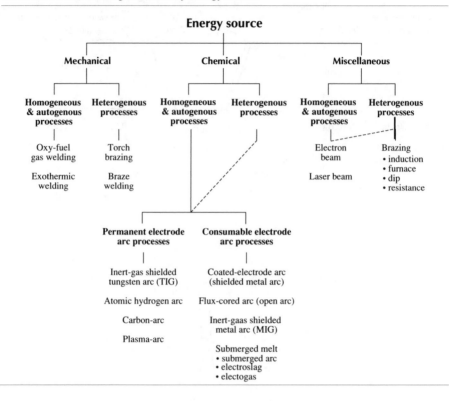

6.3.5 Autogenous versus Homogeneous versus Heterogeneous Fusion Welding

Within fusion welding, the processes can be further subclassified by whether or not an auxiliary filler material is needed and, if so, whether that filler has the same or a different composition as the base material. When no filler is required or used, the process is called *autogenous*. Here, the source of atoms (or molecules or ions) required to help fill gaps at the interface due to microscopic asperities come from the melted base materials themselves. For autogenous welding to produce structurally sound and attractive welds, the base materials making up the joint must be the same or highly compatible to allow mixing without problems *and* the fit of the joint elements must be good (i.e., with little gap) to preclude underfill of the finished joint.

If filler is required or used, the process is called *homogeneous* if the filler's composition is the same as the base material, and *heterogeneous* if it is different. For a process to be homogeneous, all components making up the joint must have the same composition. If the components making up the joint are of different compositions, the filler must be compatible with all of the materials and the process is heterogeneous.

Filler is often needed in fusion welding to make up volume in a joint, first, to compensate for the shrinkage that always accompanies the solidification of a molten material and, second, to add material where there was none because of poor fit. As mentioned earlier, filler also facilitates joining by bridging gaps at the joint faying surfaces by adding atoms, and since it is in the molten state, it facilitates diffusion. For joints that are composed of two materials of the same type but of different compositions, the filler often is selected to make the two material compositions compatible. Fillers are available in several different forms, including electrodes that are consumed by the heat of the arc to provide filler, wires that are either preplaced in the joint or fed into the joint during welding, shims that are preplaced in the joint, or powders that are usually added to the joint during welding.

Autogenous, homogeneous. and heterogeneous processes are shown in the taxonomy in Figure 6.4.

6.3.6 Nonconsumable (or Permanent) versus Consumable Electrode Arc Welding Processes

In arc welding processes, the electrode used to strike an arc with the workpiece can serve only as the means for carrying current to the arc, or it may be consumed in the arc contributing filler as well as heat to the weld. In the first case, the process is referred to as a *nonconsumable* or *permanent electrode process,* while in the second case, the process is referred to as a *consumable electrode process.* Figure 6.4 shows the position of nonconsumable and consumable electrode arc welding processes in the taxonomy of welding processes, while Table 6.3 distinguishes between these two important groups of arc welding processes.

6.3.7 Continuous versus Discontinuous Consumable Electrode Arc Welding Processes

Consumable electrodes used in fusion arc welding processes can be *continuous* in form, consisting of long wires fed into the arc by a mechanized device, or *discontinuous,* consisting of discrete lengths of rods or wire. This distinction is of significance primarily in the difference in productivity between the two broad types. Continuous consumable electrode processes are far easier to automate and, whether operated manually or automated, result in less downtime to change electrodes and, thus, in higher deposition rates. Figure 6.4 and Table 6.3 indicate whether a particular arc welding process uses a continuous or discontinuous consumable electrode.

6.3.8 American Welding Society Classification of Welding Processes

The American Welding Society (AWS) has developed its own classification of welding processes, including brazing and soldering processes. This classification uses many of the classification schemes described in the preceding sections. More than 40 welding processes are recognized, as well as nearly a dozen each of brazing and soldering processes. The AWS Classification of Welding Processes is shown in Figure 6.6. Other prominent welding societies (such as the British Welding Institute and the Canadian Welding Research Institute) use similar classifications.

6.4 FUSION WELDING PROCESSES

6.4.1 General Description

In fusion welding, the edges or surfaces to be joined are heated to above the melting point for a pure material or above the liquidus for alloys and atoms are brought together in the liquid state to create large numbers of primary bonds after solidification. Sometimes, additional filler material must also be melted and added to completely fill the joint gap. Fusion welding processes include all of those processes in which the melting or fusion of portions of the substrates, with or without added filler, plays a significant role in the formation of bonds, producing a joint.[14] As shown in Figure 6.5, fusion welds contain a distinct fusion zone, as well as heat-affected zones and unaffected base material.

The following sections describe some of the main fusion welding processes, in order to highlight the principal characteristics of such processes rather than attempt to be complete. The processes to be described include (1) gas welding using a combustible fuel as the source of heat; (2) arc welding using an electric arc from either a nonconsumable electrode or a consumable electrode as the source of heat; (3) high-energy beam welding using the conversion of the kinetic energy of fast moving particles as the source of heat; and (4) resistance welding using I^2R losses in the workpiece as the source of heat.

6.4.2 Gas Welding

In general, *gas welding* includes any welding process in which the source of heat for welding is the exothermic chemical combustion of a fuel gas with oxygen. While natural gas, propane, butane, and other hydrocarbon gases, or even hydrogen, can be used, oxyacetylene welding, which uses acetylene gas as the fuel, is the most commonly used gas welding process because of its high flame temperature (i.e., source energy).

Oxyacetylene welding (OAW) derives the heat needed to cause melting of the substrates and filler from two stages of combustion. In the first stage, known as *primary combustion*, the acetylene fuel gas partially reacts with oxygen provided from a pressurized gas cylinder to form carbon monoxide and hydrogen, as shown below:

$$2C_2H_2 + O_2 \text{ (cylinder)} = 4CO + 2H_2 \tag{6.1}$$

---

[14] There are some processes in which some melting or fusion occurs, but the principal mechanism for bringing atoms together to form bonds is plastic deformation under pressure. An example is flash welding.

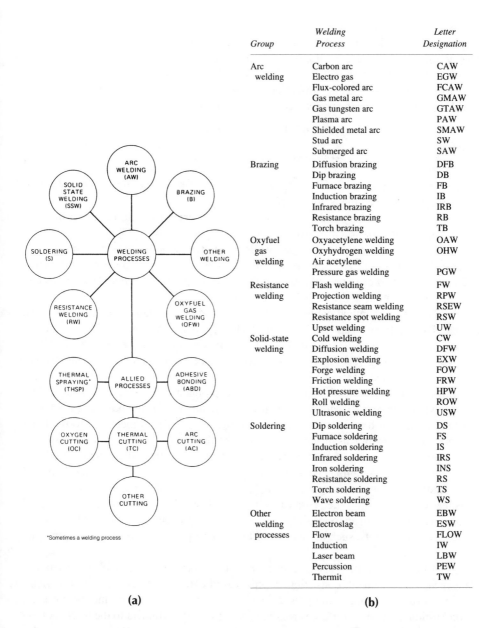

Group	Welding Process	Letter Designation
Arc welding	Carbon arc	CAW
	Electro gas	EGW
	Flux-colored arc	FCAW
	Gas metal arc	GMAW
	Gas tungsten arc	GTAW
	Plasma arc	PAW
	Shielded metal arc	SMAW
	Stud arc	SW
	Submerged arc	SAW
Brazing	Diffusion brazing	DFB
	Dip brazing	DB
	Furnace brazing	FB
	Induction brazing	IB
	Infrared brazing	IRB
	Resistance brazing	RB
	Torch brazing	TB
Oxyfuel gas welding	Oxyacetylene welding	OAW
	Oxyhydrogen welding	OHW
	Air acetylene	
	Pressure gas welding	PGW
Resistance welding	Flash welding	FW
	Projection welding	RPW
	Resistance seam welding	RSEW
	Resistance spot welding	RSW
	Upset welding	UW
Solid-state welding	Cold welding	CW
	Diffusion welding	DFW
	Explosion welding	EXW
	Forge welding	FOW
	Friction welding	FRW
	Hot pressure welding	HPW
	Roll welding	ROW
	Ultrasonic welding	USW
Soldering	Dip soldering	DS
	Furnace soldering	FS
	Induction soldering	IS
	Infrared soldering	IRS
	Iron soldering	INS
	Resistance soldering	RS
	Torch soldering	TS
	Wave soldering	WS
Other welding processes	Electron beam	EBW
	Electroslag	ESW
	Flow	FLOW
	Induction	IW
	Laser beam	LBW
	Percussion	PEW
	Thermit	TW

*Sometimes a welding process

(a)   (b)

Figure 6.6 The AWS classification of welding, brazing, and soldering processes. (a) AWS master chart of welding and allied processes; (b) popular welding processes and AWS letter designations. (Reprinted with permissin from *Welding Handbook,* 8th Ed., Vol. 2, American Welding Society, 1978.)

This reaction is exothermic and is responsible for two-thirds of the total heat generated by burning acetylene completely in oxygen. In the second stage, known as *secondary combustion*, which occurs immediately after the primary combustion, the carbon monoxide resulting from partial combustion of the acetylene reacts with oxygen in the air to form carbon dioxide, while the hydrogen from the primary combustion reacts with oxygen in the air to form water, as shown below:

$$4CO + 2O_2 \text{ (air)} = 4CO_2 \tag{6.2}$$

and

$$2H_2 + O_2 \text{ (air)} = 2H_2O \tag{6.3}$$

These reactions are also exothermic and are responsible for one-third of the total heat generated by burning the acetylene completely.

The actual primary and secondary combustion reactions occur in the gas flame of an oxygen-acetylene torch in two distinct regions, as shown in Figure 6.7a. Primary combustion occurs in the inner cone, while secondary combustion occurs in the outer flame. The inner cone tends to provide the heat needed to accomplish welding, while the outer flame tends to provide the shielding atmosphere (i.e., carbon dioxide) needed to prevent oxidation of the heated and molten metal.

The exact chemical nature, or reactivity, of the flame in gas welding processes, such as oxyacetylene, can be adjusted to be chemically neutral, chemically reducing, or chemically oxidizing. The neutral flame occurs when the molar ratio of acetylene (C_2H_2) to oxygen (O_2) is 1:1. By supplying excess acetylene to the flame, primary combustion becomes less complete, leaving some unburned acetylene. This partially burns during secondary combustion in the outer flame, producing a tell-tale acetylene feather, as shown in Figure 6.7b, and rendering the flame reducing. The neutral flame is attained when the flow of oxygen from the pressurized gas cylinder is increased to the point where the feather just disappears. Increasing the flow of oxygen still further results in an oxidizing flame, where there is actually some oxygen left in the reaction products.

The reducing flame is good for removing oxides from metals, such as aluminum or magnesium, that are being welded and prevents oxidation reactions during welding, such as decarburization (i.e., C to CO_2) in steels. The oxidizing flame causes the metal being welded to form an oxide. This can be useful for preventing the loss of high vapor-pressure components, such as zinc out of brass, through the formation of an impermeable oxide layer.

The oxyacetylene gas welding process is simple, highly portable, and requires inexpensive equipment, consisting of pressurized cylinders of oxygen and acetylene, gas regulators for controlling pressure and flow rate, a torch for mixing the gases for combustion, and hoses for delivering the gases from the cylinders to the torch. A typical torch is shown in Figure 6.8.

The process suffers from limited source energy, so welding is slow and total heat input per linear length of weld can be high.[15] Also, the nature of the process limits the

---

[15]The heat input per linear length of weld is an important measure of how much thermal distortion or metallurgical transformation (i.e., heat-affected zone) can be expected. The analogy is running one's finger through a candle flame. Moving quickly causes no burning sensation, but going slowly does; even though the temperature of the candle's flame is the same in both instances. Moving faster reduces the heat input. Likewise, welding at high speeds results in a lower heat input per unit length of weld than welding slowly.

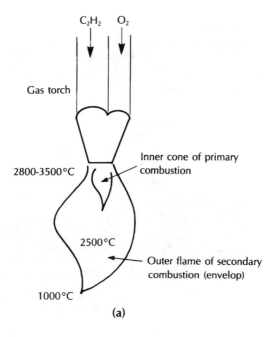

C_2H_2 O_2

Gas torch

Inner cone of primary
combustion

2800-3500 °C

2500 °C

Outer flame of secondary
combustion (envelop)

1000 °C

(a)

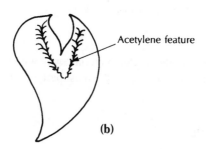

Acetylene feature

(b)

Figure 6.7   The oxyacetylene flame. (a) Neutral flame; (b) reducing flame.

amount of protective shielding provided to the weld, so welding of the more reactive metals (e.g., titanium) is generally impossible. To offset this particular shortcoming, oxyacetylene welding may employ a flux or fluxing agent to provide additional protection to the weld, to prevent oxidation during welding, and/or to clean the workpiece of oxide to promote flow and wetting[16] by any filler metal.

Oxy-fuel processes can be used for cutting, gouging (i.e., grooving) or piercing (i.e., producing holes) as well as for welding. Here, the process involves melting the

---

[16] *Wetting* is the phenomenon of a liquid attaching to a solid at a low angle of contact, tending to form a film rather than beads. The more complete the wetting, the smaller the angle of contact, and the more the liquid spreads as a film. The basis for wetting is surface energy reduction.

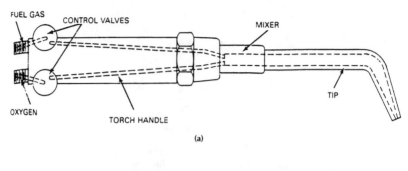

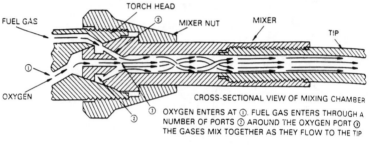

Figure 6.8   A typical gas welding torch. (a) Basic elements of an oxy fuel gas torch; (b) typical detail design of the gas mixer for a positive-pressure type torch. (Reprinted with permission from *Welding Handbook,* 8th Ed., Vol. 2, American Welding Society,1990, p. 362.)

base material and blowing the molten material away with a jet of air or oxygen from a compressed source. Torch designs, for the most part, are the same.

6.4.3 Arc Welding

Fusion welding processes that use an electric arc as a heat source are called *arc welding processes.* The arc consists of thermally emitted electrons and positive ions from both the welding electrode and the workpiece. These electrons and positive ions are accelerated by the potential field (i.e., voltage) between the source (i.e., one electrode) and the work (i.e., the oppositely charged electrode) and produce heat when they convert their kinetic energy by collision with the oppositely charged element. Arc welding includes a large and diverse group of processes, as shown in Table 6.3.

The arc in arc welding is created between an electrode and a workpiece or a weldment at different polarities. The electrode can be intended to be permanent, serving solely as a source of energy from electrons or positive ions, or consumed, in which case it serves as both a source of energy for welding and of filler. If the electrode is intended to be permanent, the processes are called nonconsumable electrode arc welding processes. If the electrode is intended to be consumed, the processes are called consumable electrode arc welding processes. For nonconsumable electrode processes, if filler metal is required, it must be added from a supplemental source (e.g., filler wire). Nonconsumable electrodes are usually composed of tungsten or carbon, because of their very high melting temperatures, but must be protected from oxidation by an inert shielding gas. Consumable electrodes are composed of the metal or alloy needed in the filler and come in the form of rods or sticks (i.e., discontinuous electrodes) or wires (i.e., continuous electrodes).

Whether the arc welding process uses a nonconsumable or consumable electrode, shielding must be provided to the weld by a chemically inert or nonoxidizing gas generated by decomposing the coating on, or flux-core in, a consumable electrode or from an external inert gas source (e.g., pressurized gas cylinder). This shielding is to prevent oxidation of the highly reactive molten weld metal and, also, to help stabilize the arc.

Several of the more common arc welding processes are described in the following paragraphs in an effort to provide the reader with an understanding of key characteristics rather than a comprehensive knowledge. Nonconsumable electrode processes are described first, then consumable electrode processes.

Nonconsumable Electrode Arc Welding Processes

Six predominant arc welding processes use nonconsumable electrodes: (1) gas-tungsten arc welding (GTAW); (2) plasma arc welding (PAW); (3) carbon arc welding (CAW); (4) stud arc welding; (5) atomic hydrogen welding (AHW); and (6) magnetically impelled arc butt welding (MIAB). Of these, the carbon arc and atomic hydrogen processes are rarely used anymore; the magnetically impelled arc butt process is practiced little outside of eastern Europe and Russia, although it has potential; and the stud arc process has a highly specialized role for attaching threaded or unthreaded studs to structures using the heat generated by an arc between the stud and the workpiece and applying a light pressure. The two predominant processes, gas-tungsten arc and plasma arc welding, will be described in some detail.

Gas-Tungsten Arc Welding. *Gas-tungsten arc welding* (GTAW) uses a permanent, nonconsumable tungsten electrode to create an arc to the workpiece. This electrode is shielded by an inert gas, such as argon or helium, to prevent electrode degradation; hence the older, common names tungsten-inert gas (TIG) and heli-arc welding. As shown in Figure 6.9, current from the power supply is passed to the tungsten electrode of a torch (shown in Figure 6.10) through a contact tube. This tube is usually water-cooled to prevent overheating. The gas-tungsten arc welding process can be performed with or without filler (i.e., autogenously). When no filler is used, joints must be thin and have close-fitting square butts (described in section 6.6).

The GTAW process, as well as several other arc welding processes, for example, shielded-metal arc welding (SMAW), gas-metal arc welding (GMAW), and flux-cored arc welding (FCAW), can be operated in several different current modes, including direct current (DC), with the electrode negative (EN) or positive (EP), or alternating current (AC). These different current or power modes result in distinctly different arc and weld characteristics.

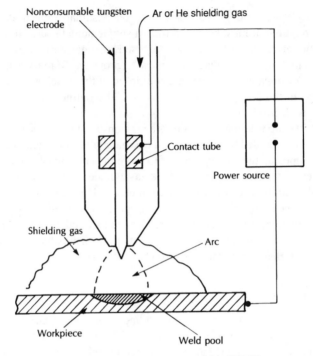

Figure 6.9   The electrical hook-up of the GTAW process.

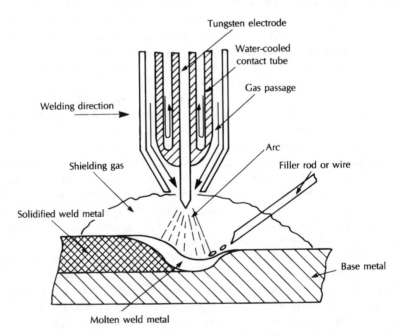

Figure 6.10   A gas-tungsten arc welding torch.

When the workpiece (or weldment) is connected to the positive terminal of a direct current power supply, the operating mode is referred to as direct current straight polarity (DCSP) or direct current electrode negative (DC– or DCEN). When the workpiece is connected to the negative terminal of a direct current power supply, the operating mode is referred to as direct current reverse polarity (DCRP) or direct current electrode positive (DC+ or DCEP).

In DCSP, electrons are emitted from the tungsten electrode and are accelerated to very high speeds and kinetic energies while traveling through the arc. These high energy electrons collide with the workpiece, give up their kinetic energy, and generate considerable heat in the work. Consequently, DCSP results in deep penetrating, narrow welds but with higher workpiece heat input. About two-thirds of the net heat available from the arc (after losses from various sources) enters the work. High heat input to the workpiece may or may not be desirable, depending upon such factors as required weld penetration, required weld width, workpiece mass, workpiece thermal conductivity (high), susceptibility to heat-induced defects, and concern for distortion or residual stress. In DCRP, on the other hand, the heating effect of the electrons is on the tungsten electrode rather than on the workpiece. So larger, water-cooled electrode holders are required, shallow welds are produced, and workpiece heat input can be kept low. This operating mode is good for welding thin sections or heat-sensitive metals and alloys. This mode also results in a scrubbing action on the workpiece by the large positive ions that strike its surface, removing oxide and cleaning the surface. This mode is thus preferred for welding metals and alloys that oxidize easily, such as aluminum or magnesium.

The DCSP mode is much more common than the DCRP mode. There is, however, a third mode, using alternating current, or AC. This mode tends to give some of the characteristics of both of the DC modes, during the corresponding half cycles, but with some bias toward the straight polarity half-cycle. During this half-cycle the current tends to be higher because of the extra emission of electrons from the smaller, hotter electrode (versus larger, cooler workpiece). In the AC mode, reasonably good penetration is obtained, along with some oxide cleaning action. Figure 6.11 summarizes the characteristics of the various current or operating modes of the GTAW process.

The electron emission of tungsten electrodes is occasionally enhanced by adding 1 to 2% of thorium oxide (or cerium oxide) to the tungsten. This addition improves the current carrying capacity of the electrode, results in less chance of contamination of the weld by expulsion of tungsten (as a result of localized electrode melting), and allows easier arc initiation.

While both argon and helium are used for shielding with the GTAW process, argon offers better shielding since it is heavier and stays on the work. Arc initiation is also easier as the required ionization potential is lower than for He. The advantage of helium is a hotter arc.

In summary, the GTAW process is good for welding thin sections because of its inherently low heat input, it offers better control of weld filler dilution by the substrate than many other processes (again because of low heat input), and it is a very clean process. Its greatest limitation is its slow deposition rate, although this can be overcome by using a hot wire variation in which the filler wire is heated resistively by being included in the circuit at a lower potential than the electrode.

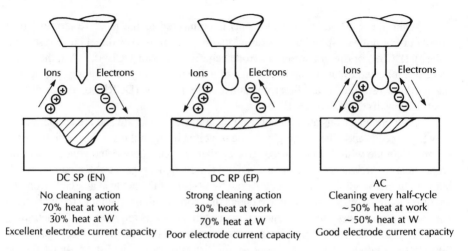

DC SP (EN)	DC RP (EP)	AC
No cleaning action	Strong cleaning action	Cleaning every half-cycle
70% heat at work	30% heat at work	~50% heat at work
30% heat at W	70% heat at W	~50% heat at W
Excellent electrode current capacity	Poor electrode current capacity	Good electrode current capacity

Figure 6.11   Summary of the characteristics of the various operating modes of GTAW.

Plasma Arc Welding.   *Plasma arc welding* (PAW) is similar to gas-tungsten arc welding in that it too uses a nonconsumable tungsten electrode to produce an arc to a workpiece. The difference is that in plasma arc welding the converging action of inert gas at an orifice in the nozzle of the welding torch (Figure 6.12) constricts the arc, resulting in several advantages over the GTAW process. These advantages include greater energy concentration (i.e., higher energy density), higher heat content, improved arc stability, deeper penetration capability, higher welding speeds, and usually cleaner welds since the tip of the tungsten electrode cannot accidentally be touched to the workpiece, causing contamination. Figure 6.13 shows a comparison of the GTAW and PAW processes.

The plasma in PAW is created by the low-volume flow of argon through the inner orifice of the PAW torch. A high-frequency pilot arc established between the tungsten electrode and the inner nozzle ionizes the orifice gas and ignites the primary arc to the workpiece. When the workpiece is connected electrically to the welding torch such that it is of opposite polarity to the permanent electrode, the plasma is drawn to the workpiece electrically, and the plasma generation is referred to as operating in the transferred arc mode. When the workpiece is not connected electrically to the torch, and the plasma is simply forced to the workpiece by the force of the inert gas, the plasma generation is referred to as operating in the nontransferred mode (see Figure 6.14). The transferred arc mode is usually used for welding, while the non-transferred arc mode is usually used for thermal spraying (see Chapter 10, Section 10.3). Concentric flow of inert gas from an outer gas nozzle provides shielding to the arc and the weld in PAW. This shielding gas can be argon, helium, or argon mixed with helium or hydrogen.

Two distinctly different welding modes are possible with the plasma arc welding process: the melt-in mode and the keyhole mode. In the melt-in mode, heating of the workpiece occurs by conduction of heat from the plasma's contact with the workpiece surface inward. This mode is good for joining thin sections (e.g., 0.025 to 1.5 mm or 0.001 to 0.060 in.) and for making fine welds at low currents and for joining thicker

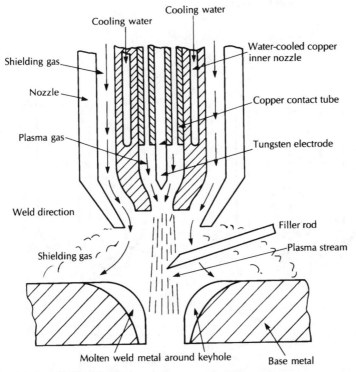

Figure 6.12   A plasma arc welding torch.

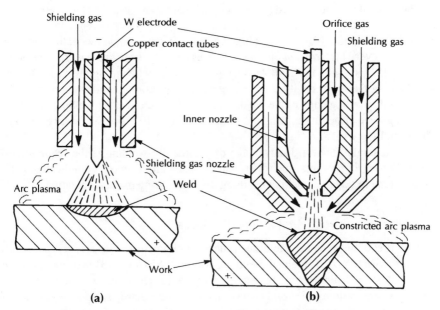

Figure 6.13   Comparison of the (a) GTAW and (b) PAW processes.

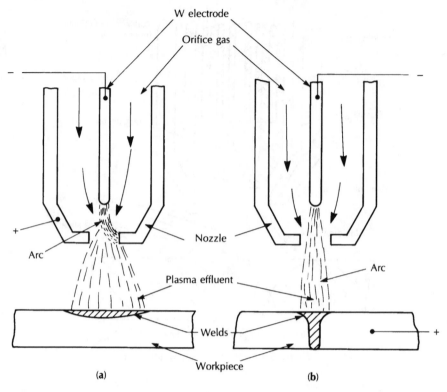

Figure 6.14   Nontransferred (a) versus transferred (b) arc modes of plasma generation.

sections, up to 3 mm (0.125 in.), at high currents. In the keyhole mode, the high energy density of a very high current plasma vaporizes a cavity through the workpiece and creates a weld by moving the keyhole, analogous to a hot wire through wax. Molten metal surrounding the vapor cavity is drawn by surface tension or capillary forces to fill the cavity at the trailing edge of the weld. This mode is excellent for welding applications requiring deep penetration, to approximately 20 mm (0.8 in.). These two modes are shown in Figure 6.15.

The single greatest disadvantage of plasma arc welding is the required equipment. Power sources, gas controllers, and torches are all more complicated, and the torches tend to be large, making handling difficult during manual operation.

Consumable Electrode Arc Welding Processes

The six predominant consumable electrode arc welding processes are: (1) gas-metal arc welding (GMAW); (2) shielded-metal arc welding (SMAW); (3) flux-cored arc welding (FCAW); (4) submerged-arc welding (SAW); (5) electrogas welding (EGW); and (6) electroslag welding (ESW). The gas-metal arc and electrogas welding processes use an inert gas shield provided from an external source, while the shielded-metal and flux-cored arc welding processes achieve shielding from gases generated from within the

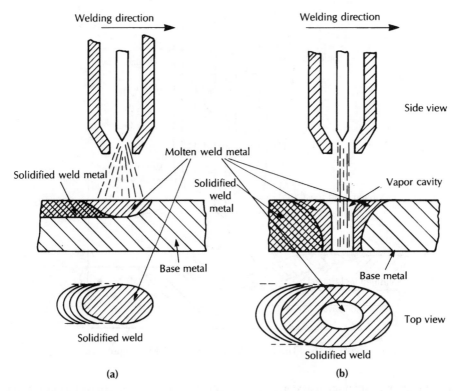

Figure 6.15   Comparison of the melt-in (a) versus the keyhole mode (b) in PAW.

consumable electrode during melting. The submerged arc and electroslag welding processes achieve shielding of the molten weld metal with a molten slag cover. Each of these processes is described in the following sections.

Gas-Metal Arc Welding.   The *gas-metal arc welding* (GMAW) process (or metal-inert gas, MIG) process uses a continuous consumable solid wire electrode and an externally supplied inert shielding gas. A schematic of the process is shown in Figure 6.16. The consumable solid wire electrode provides all of the filler to the weld joint. The externally-supplied shielding gas plays dual roles in GMAW (as it does in the gas-shielded form of the FCAW and in the EGW processes): first, it protects the arc and the molten or hot, cooling weld metal from air and, second, it provides desired arc characteristics through its effect on ionization. A variety of gases (argon, helium, carbon dioxide, and hydrogen, occasionally with oxygen) can be used, depending on the reactivity of the metal being welded, the design of the joint, and the specific arc characteristics that are desired. A variety of DC welding power sources can also be used, hooked up as shown in Figure 6.16. Either DCSP (DCEN) or DCRP (DCEP) may be used, depending on the particular wire and desired mode of molten metal transfer.

A distinct advantage of GMAW is that the mode of molten metal transfer from the consumable wire electrode can be intentionally changed and controlled through a

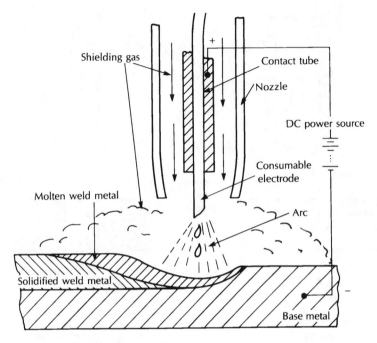

Shielding gas

Contact tube

Nozzle

DC power source

Consumable electrode

Molten weld metal

Arc

Solidified weld metal

Base metal

Figure 6.16   The gas-metal arc welding (GMAW) process.

combination of shielding gas composition, power source type, electrode type and form, and arc current and voltage. There are three predominant metal transfer modes: spray, globular, and short circuiting. There is also a pulsed current or pulsed arc mode that is not specifically related to the molten metal transfer mode.

The spray transfer mode is characterized by an axial transfer of fine, discrete molten particles or droplets from the consumable electrode to the work at rates of several hundred per second. The metal transfer is very stable, directional, and essentially free of spatter.[17] Spray transfer is produced by welding in the direct current electrode positive (i.e., DC+) mode at high voltages and amperages (above some critical value related to the electrode diameter). Argon or argon–helium mixtures are usually used when welding reactive metals like aluminum, titanium, and magnesium, while small amounts of carbon dioxide (e.g., 20%) or oxygen (e.g., 2%) are usually added when welding ferrous alloys to stabilize the arc and give the weld a better and more regular contour.[18]

The high arc energy and heat associated with the spray transfer mode limits the effectiveness for joining sheet-gage metals, but the strong directional spray can be useful for welding out-of-position (i.e., vertically up or down or overhead).

---

[17] *Spatter* refers to residual molten filler metal that is expelled from the consumable electrode onto the workpiece, but not into the weld pool. Spatter is a source of loss of useful energy and mass, so is undesirable. Also, it is often necessary to remove spatter by chipping or machining for either cosmetics or fit or fatigue resistance.

[18] The contour of a weld refers to the shape of its solidified crown or top bead.

Globular transfer is characterized by large globules of molten metal being formed at the tip of the consumable electrode and then being released and carried to the workpiece by gravity and/or arc forces. Spatter is usually considerable compared to spray transfer. When argon or argon–helium is used as for shielding, welding currents must be kept low to achieve this mode. Carbon dioxide–rich gases are usually used when this mode is desired, however, since the spray transfer mode cannot be achieved regardless of the current level and high deposition rates, and welding speeds can be achieved.

Unlike the spray and globular modes, which are known as *freeflight* modes, in the short-circuiting mode, welding currents and voltages are kept low and the slow forming molten globules at the end of the consumable electrode are periodically touched to the weld puddle to cause their release through surface tension forces. This short-circuiting occurs at rates in excess of 50 per second but requires special power sources. The low currents required for this mode enable the welding of thin sections without melt-through or overwelding. Out-of-position welding is facilitated by the direct transfer of the molten metal through contact. Spatter is minimized with this transfer mode.

Rather than using constant currents during welding, as is usually the case, it is possible to superimpose intermittent, high amplitude pulses on a low level steady current that maintains the arc. This is known as the *pulsed current* or *pulsed arc* mode. Here, *mode* refers to the mode of current and not the mode of molten metal transfer. This technique allows spray transfer to be obtained at appreciably reduced current levels during the high amplitude pulses. Argon-rich gases are essential, and programmable power sources are required, but several advantages are obtained. Relatively large diameter electrodes can be used to weld either thin or thick sections, in- or out of position.

The globular, short-circuit ,and pulsed-arc transfer modes usually use the direct current electrode negative (DC–) operating mode, while the spray transfer mode usually uses the electrode positive (DC+) operating mode. The various metal transfer modes are shown in Figure 6.17.

In summary, the GMAW process offers flexibility and versatility, it is readily automatable, it requires less manipulative skill than SMAW, and it enables high deposition rates (i.e., 5 to 20 kg or 10 to 40 lb. per hour) and efficiencies[19] (i.e., 80 to 90%). The greatest shortcoming of the process is that the power supplies[20] typically required are expensive.

Shielded Metal-Arc Welding. The *shielded metal-arc welding* (SMAW) process is also known as the *stick welding process*. As shown in Figure 6.18, metal coalescence is produced by the heat from an electric arc that is maintained between the tip of flux-coated, or coated, or covered, discontinuous consumable (or stick) electrode and the surface of the base metal being welded. A core wire conducts the electric current from a constant current power supply to the arc and provides most of the filler metal to the

---

[19]*Efficiency* in this case refers to the efficiency with which energy is transferred from the heat source (torch) to the workpiece for use in making a weld.

[20]Constant voltage (CV) power supplies are used. These and constant current (CC) types are described in various references (such as References 1 and 3) but not here.

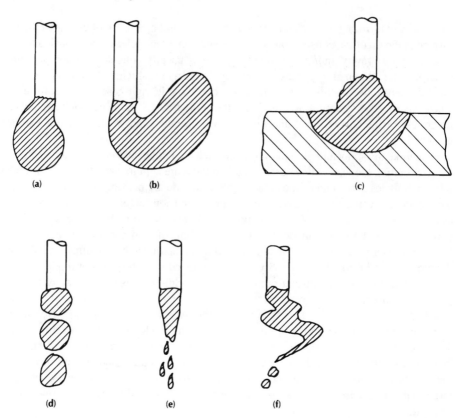

(a)   (b)   (c)

(d)   (e)   (f)

Figure 6.17   Various molten metal transfer modes. (a) Drop globular transfer; (b) repelled globular transfer; (c) short-circuiting transfer; (d) projected spray transfer; (e) streaming spray transfer; (f) rotating spray transfer.

joint. Some portion of the arc heat is lost to the electrode by conduction, and some source power is lost as I^2R heat.

The covering, coating or flux on an SMAW electrode (or, as described later, in the core of an FCAW wire) performs many functions. First, it provides a gaseous shield to protect the molten metal of the weld from the air. This shielding gas is generated by the decomposition of the coating, which may be of several types: cellulosic, which generates H_2, CO, H_2O, and CO_2; rutile (TiO_2), which generates up to 40% H_2; or limestone ($CaCO_3$), which generates CO_2 and CaO slag and little or no H_2, so are known as low-hydrogen types. The different types are selected for different applications, where hydrogen can or cannot be tolerated.[21] Second, the coating provides deoxidizers and fluxing or reducing agents as molten metal compounds to deoxidize or denitrify and cleanse the molten weld metal, as in refining. Once solidified, the slag that is formed from the flux protects the already solidified, but still hot and reactive, weld metal from

---

[21] Hydrogen-generating types of coatings should be avoided when hardenable (i.e., martensite forming) steels are being welded in order to avoid hydrogen embrittlement.

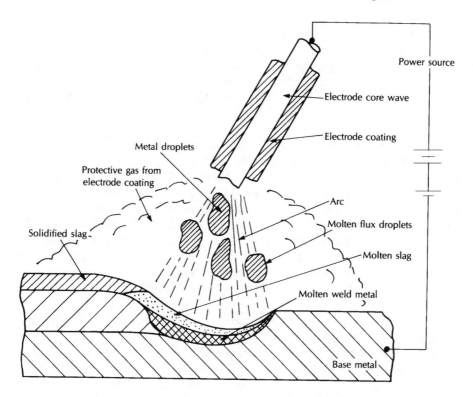

Figure 6.18   The shielded-metal arc welding (SMAW) process.

oxidation. It also aids out-of-position[22] welding by providing a shell, or mold, in which molten weld metal can solidify. Third, the coating provides arc stabilizers in the form of readily ionizable compounds (e.g., potassium oxalate or lithium carbonate) to help initiate the arc and keep the arc steady and stable by helping conduct current by providing a source of ions and electrons. Finally, the coating can provide alloying elements or grain refiners and/or metal fillers to the weld. The former help achieve and control the composition and/or microstructure of the weld, while the latter increase the rate of deposition of filler metal.

SMAW can operate with both direct current (DC) power sources, with the electrode positive or negative, or alternating current (AC) power sources, depending on coating design. Typically, currents range from 50 to 300 amperes, largely based on electrode diameter, at 10 to 30 volts, resulting in 1 to 10 kg (2 to 20 lb.) per hour deposition rates.

Advantages of SMAW are that it is simple, portable, and requires inexpensive equipment (i.e., power supply, electrode holder, and cables). The process is versatile, enabling joining or coating for restoring dimensions or enhancing wear resistance (i.e.,

---

[22] *Out-of-position* welding refers to welding in a position other than with the work in a horizontal plane and the weld deposit being made from above and using the force of gravity to assist in molten metal transfer.

hardfacing or wear facing) for fabrication, assembly, maintenance, or repair, in the plant or in the field. Shortcomings of the process are that it offers only limited shielding protection and limited deposition rates (compared to many other arc welding processes), and it is usually performed manually, rather than automatically. Like all manual processes, but even more than most, SMAW requires fair operator skill for best results.

Flux-Cored Arc Welding.   *Flux-cored arc welding* (FCAW) or *open-arc welding* is similar to SMAW in that it is self-shielding; however, the gas- and flux-generating flux is contained in the core of a roll-formed or drawn tubular wire, rather than on the outside of a core wire as a coating. The cored wire serves as a continuous consumable electrode, with the filler in the core fulfilling the same functions as the coating in SMAW: namely, providing self-shielding gases, slagging ingredients, arc stabilizers, and alloy additions and deposition rate enhancers. The self-shielding provided by the generation of gases from the core through the arc is more effective than when gas is generated from an external coating. By the time gas that is generated reaches the air, to be swept away, it has fulfilled its shielding function. For this reason FCAW is an excellent choice for welding in the field, and it is here that it got the name *open arc welding.*

The FCAW process can also be operated in a gas-shielded mode, in which case it is closely related to the gas-metal arc welding (GMAW) process. Both use a continuous consumable electrode, both provide filler, and both use an externally provided gas to shield the arc and the weld metal. In either mode, FCAW can be operated with DC power supplies, with the electrode positive or negative, depending on the particular wire type and formulation.

Figure 6.19 shows the self-shielded and gas-shielded forms of FCAW. Process advantages include high deposition rates (from 2 to 15 kg or 10 to 30 lb. per hour), with actual rates being high due to the continuous operation at higher currents than SMAW; larger, better contoured welds than SMAW; portability; and excellent suitability for use in the field.

Submerged Arc Welding.   In the *submerged arc welding* (SAW) process, shown in Figure 6.20, the arc and the molten weld metal are shielded by an envelope of molten flux and a layer of unfused granular flux particles. Since the arc is literally buried or *submerged* in the flux, it is not visible. As a result, the process is relatively free of the intense radiation of heat and light and of the fumes typical of most open arc welding processes, and the resulting welds are very clean. The SAW process uses a continuous solid wire electrode that is consumed to produce filler. The efficiency of transfer of energy from the electrode source to the workpiece is very high (usually greater than 90%), since losses from radiation, convection, and spatter are minimal.

The "sub-arc" process is almost always mechanized, as currents are very high (500 to more than 2000 amperes), deposition rate is very high (27 to 45 kg or 60 to 100 lb. per hour), and reliability is high. Thin sections can be welded at very high velocities (up to 500 cm or 200 in. per minute), while very thick sections can be welded at lower velocities, even in the direct current electrode positive (or DCRP) operating mode. At very high currents (over 1000 amperes), AC is often used to avoid problems with arc

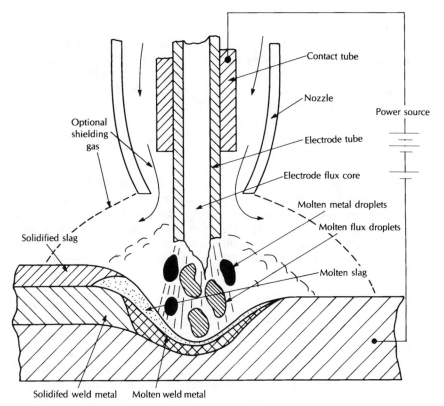

Figure 6.19 Self-shielded and gas-shielded forms of the flux-cored arc welding (FCAW) process.

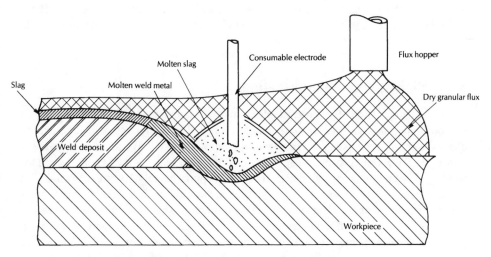

Figure 6.20 The submerged arc welding (SAW) process.

blow.[23] The process can be operated with multiple wires or with strip electrodes to further increase deposition rate. Welding is restricted to flat and horizontal positions, however, because of the effects of gravity on the large molten puddles that typify the process, and the ability to observe the puddle directly can hinder control.

The granular flux employed in the SAW process is specially formulated, often containing additives to compensate for the loss of volatile alloying elements (such as chromium) from the filler. Approximately 1 kg of flux is consumed for every kilogram of filler deposited.

Electrogas Welding.   The *electrogas welding* (EGW) process is a heavy deposition rate arc welding process. The process operates under an inert gas shield provided to a joint enclosed with water-cooled dams or shoes or backing plates, as shown in Figure 6.21. Deposition rate can be as high or higher than SAW, and the quality of weld deposit is excellent because of the extremely effective shielding. A drawback is that the process can only be used for welding vertically up but, in this mode, requires little joint preparation for fit.

Electroslag Welding.   In actual fact, the *electroslag welding* (ESW) process is not a true arc welding process. The energy for melting the base metal and filler is provided by a molten bath of slag that is resistance heated by the welding current. An arc is only used to melt the flux initially, after being struck at the bottom of the joint. Welds are produced in the vertical up direction (and, occasionally, in horizontal fillets), with the joint edges being melted and fused by molten weld filler metal contained in the joint by water-cooled dams or shoes, as shown in Figure 6.22. The molten flux or slag provides excellent protection to the weld. Deposition rates are typically 7 to 13 kg per hour or 15 to 30 lb. per hour) per electrode, and multiple electrodes can be used. In the so-called "guide tube" mode of this process, a consumable, thick-walled tube is used to provide additional filler, as well as to guide the continuous wire to the bottom of the joint. Here, deposition rate can easily reach 15 to 25 kg (35 to 55 lb.) per hour per electrode/guide tube.

6.4.4 High-Energy Beam Welding

Often, the density of the energy available from a heat source for welding (or for cutting) is more important than the absolute source energy. Two major types of high-energy density welding processes are (1) *electron-beam welding* (EBW) and (2) *laser-beam welding* (LBW). Both processes use a very high intensity beam as the heating source for welding. The energy density in these processes is approximately 10^{10} to 10^{13} watts (W) per m^2 versus 5×10^6 to 5×10^8 W/m^2 for typical arc welding processes. Conversion of the kinetic energy of the electrons (in EBW) or photons (in LBW) into heat occurs as these particles strike the workpiece, leading to melting *and* vaporization. Both processes usually operate in the keyhole mode, so penetration can be high, producing deep, narrow, parallel-sided fusion zones with narrow heat-affected zones and minimal

---

[23]*Arc blow* refers to the deflection of an arc by the induced electromagnetic fields in conductive materials.

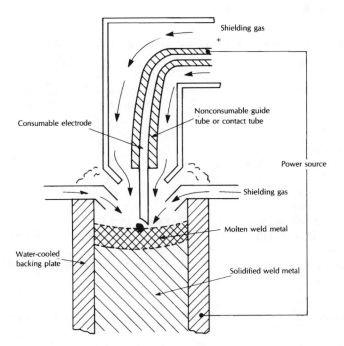

Figure 6.21   The electrogas welding (EGW) process.

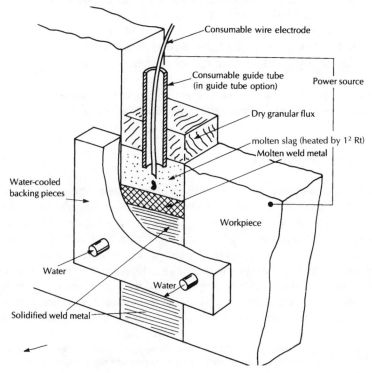

Figure 6.22   The electroslag welding (ESW) process.

Table 6.4   Comparative Advantages and Disadvantages of the EBW vs. LBW Process

EBW	LBW
• Deep penetration in all materials	• Deep penetration in many materials, but *not* in metals that reflect laser light because they are specular or because their vapors are reflective.
• Very narrow welds	• Same
• Low linear heat input (very high energy density)	• Same
• Needs vacuum to operate, to permit electrons to move unimpeded	• Can operate in air or vacuum
• Requires tight fitting joints, usually	• Same
• Almost impossible to add filler	• Filler can be added, if not too deep
• Equipemt expensive	• Same
• Very efficient electrically (99%)	• Very inefficient electrically ($\sim$12%)
• Generates x-rays	• No x-rays

angular distortion due to nonuniform weld metal shrinkage or thermal expansion and contraction.

The electron-beam welding process is almost always performed autogenously,[24] so joint fit must be excellent.[25] Laser-beam welding is usually done autogenously also but can use fillers. Shielding for the EBW process is provided by the vacuum (typically, 10^{-3} to 10^{-5} atmospheres) required to allow the beam of electrons to flow to the workpiece unimpeded. Shielding for the LBW process is accomplished with inert gases, either in so-called dry boxes or from special shrouds over the vicinity of the weld puddle.

These two processes are shown in Figure 6.23a and b. Comparative advantages and disadvantages of the two processes are given in Table 6.4.

6.4.5 Resistance Welding

As a group, *resistance welding* (RW) processes generate heat through the resistance to the flow of electric current through the parts being joined at the point of welding. The parts are an integral part of the electrical circuit. As shown in Figure 6.24, contact resistance, especially at faying surfaces, heats the area locally by I^2R heating, resulting in melting and the formation of a nugget. For the process to work properly, the contact resistance must be higher at the point to be welded than anywhere else. Pairs of water-cooled electrodes, made of copper or copper alloyed with refractory metals to improve erosion resistance, conduct current to the joint, apply pressure, by clamping, to improve contact (i.e., reduce the contact resistance) at the electrode-to-workpiece interface, and

[24]It is possible to provide filler by preplacing shim material in the joint or by laying lengths of wire or shims in an underfilled weld and rewelding with reduced energy, achieving partial penetration to improve the cosmetics of the weld crown.
[25]Usually, joints for EBW or LBW are machined square butts.

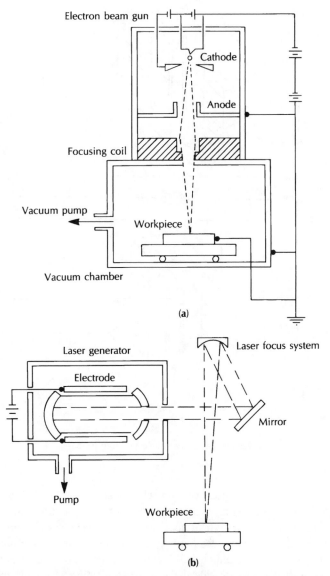

Figure 6.23   The electron-beam (EB) (a) and Laser-Beam (LB) (b) welding processes.

help contain the molten metal in the nugget. The principal process variables are welding current (usually several thousands to tens of thousands of amperes), welding time (on the order of 1/4 sec), electrode force, and electrode shape. Usually, the process is used to join overlapping sheets or plates, which may have different thicknesses.

At least six major types of processes rely on resistance heating to produce welds, with several variations within certain types. The major resistance welding processes include (1) resistance spot welding (RSW); (2) resistance seam welding (RSEW), using high frequency (RSEW-HF) or induction (RSEW-I); (3) projection welding (PW);

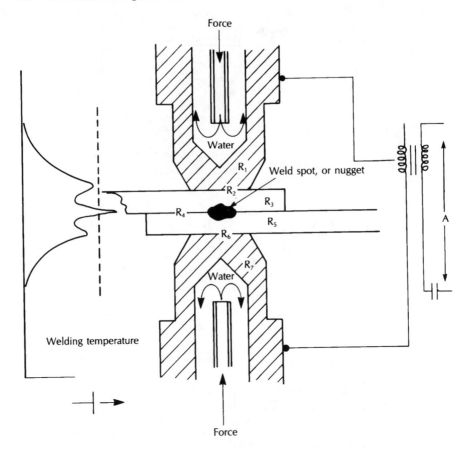

Figure 6.24   Resistance welding process.

(4) flash welding (FW); (5) upset welding (UW), using high frequency (UW-HF) or induction (UW-I); and (6) percussion welding (PEW). Spot welding consists of a series of discrete nuggets produced by resistance heating. Nuggets or welds are usually produced directly under the electrodes but may not be if there is another more favorable path for the current. Spot welding usually requires access to both sides of the work but can be accomplished from one side using a series welding technique where current passes from one electrode on the face of the work, through the work to produce a weld at the interface of the workpieces, through the workpiece farthest from the electrode, back through the workpieces to produce a second weld, and into a second electrode contacting the front face of the work.

Seam welding consists of a series of overlapping spots to produce an apparently continuous seam, or resistance heating, along the edges of two pieces being forced into contact along a seam. In projection welding, projections or dimples in overlapping joint elements are used to concentrate the current during welding, focusing the weld energy

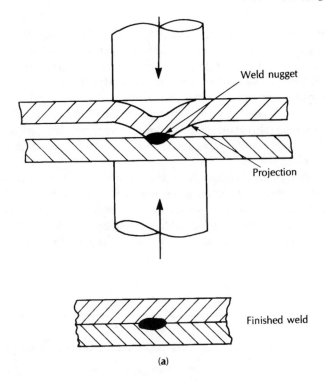

Weld nugget

Projection

Finished weld

(a)

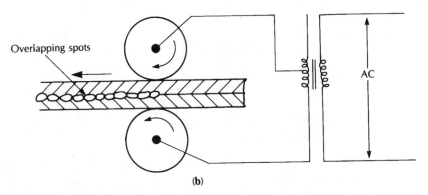

Overlapping spots

AC

(b)

Figure 6.25 Resistance seam (RSEW) and projection welding (PW). (a) Projection welding; (b) seam welding.

and helping locate the weld more precisely. Seam and projection welding arrangements are shown in Figure 6.25.

Flash welding is classified as a resistance welding process but is unique. Heating at the faying surface is by combined resistance and arcing. When the faying surfaces are heated to welding temperature under the action of an applied current, force is applied immediately to consummate a weld. Molten metal is expelled, the hot metal is plastically upset, a weld is produced, and a *flash* of frozen expelled metal is formed. Closely related to flash welding is *upset welding*, with the major difference being the amount of gross plastic deformation,or upsetting, that is used to produce the weld. Figure 6.26 shows flash welding schematically, along with a schematic of a typical flash weld.

Percussion welding (also known as capacitor-discharge welding) produces welds through resistance heating by the rapid release of electrical energy from a storage device (e.g., capacitor).

In all resistance welding processes, the rate of heating is extremely rapid, the time for which the weld is molten is extremely short, and the rate of cooling is usually rapid. This allows these processes to be used where heat input must be limited. On the other hand, resistance welding processes are capable of welding even the most refractory metals and alloys, because of the intense heating that can be made to occur.

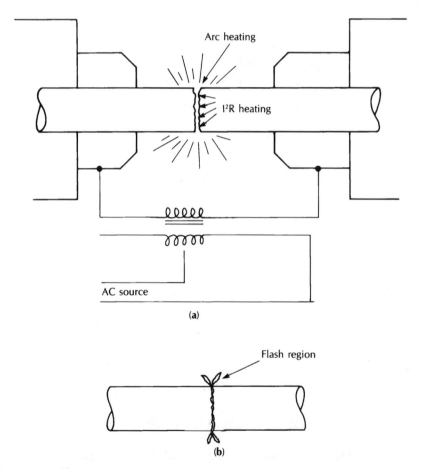

Figure 6.26   (a) Flash welding (FW); (b) a typical flash weld.

6.4.6 Transfer Efficiency in Fusion Welding

There are many, many opportunities for energy to be lost between the welding source and the workpiece in fusion welding processes. The sum of all losses determines the *energy transfer efficiency* of the process. Figure 6.27 shows some of these losses. Anything that prevents or reduces a loss, increases the energy transfer efficiency, η, to a maximum of 1.00 or 100%. Some typical transfer efficiencies for various fusion welding processes are shown in Table 6.5. As an example, in the sub-arc process, transfer efficiency is very high because losses from radiation of light and heat, convection of heat to the air, and through spatter are eliminated. Transfer efficiency is important because it strongly effects the input of heat to the workpiece and subsequent distribution of heat within the workpiece.

6.5 NONFUSION WELDING PROCESSES

6.5.1 General Description

Nonfusion welding processes accomplish welding by bringing the atoms (or molecules) of the materials to be joined to equilibrium spacings principally through plastic deformation due to the application of pressure at temperatures below the melting point of the base material and without the addition of filler that melts. Often some heat is generated by or supplied to the process to allow plastic deformation to occur at lower stresses and to accelerate interdiffusion without causing or, at least, depending upon melting. While

Table 6.5 Typical Energy Transfer Efficiencies

Process	Transfer Efficiency
Gas	0.25–0.80
Gas-tungsten arc	
Low current DCSP	0.40–0.60
High current DCSP	0.60–0.80
DCRP	0.20–0.40
AC	0.20–0.60
Gas-metal arc	
Globular or short-arc	0.70–0.85
Spray	0.65–0.75
Shielded metal arc	0.65–0.85
Submerged arc	0.85–0.99
Electroslag	0.55–0.82
Electron beam	
Melt-in	0.70–0.85
Keyhole	0.80–0.95
Laser beam	
Reflective	0.005$\rightarrow$
Keyhole	0.50–0.70

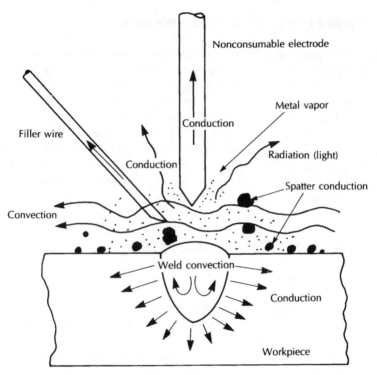

Figure 6.27   Sources of energy losses affecting the transfer efficiency of fusion welding processes.

other sources of heat are possible, mechanical heating is the most common and includes friction sources[26] and pressure sources. Both actually produce heat as the result of the work done in deforming material, but on a microscopic scale for friction processes and on a macroscopic scale for the pressure processes.

As shown in Tables 6.2 and 6.3, the eight major nonfusion welding processes are: (1) cold welding; (2) forge welding; (3) hot pressure welding; (4) roll welding; (5) explosion welding; (6) friction welding; (7) ultrasonic welding; and (8) diffusion welding. Cold welding, forge welding, hot pressure welding, roll welding, and explosion welding all rely on substantial pressure to cause gross (macroscopic) plastic deformation to produce a weld. Friction and ultrasonic welding rely on friction to cause heating and bring atoms or molecules together by microscopic plastic deformation to produce a weld. Diffusion bonding relies on heating to accelerate diffusion to produce welds through mass transport in the solid state. Pressure can play a minor role.

---

[26]Naturally, friction processes also involve pressure, since the force of friction arises from the product of the applied normal force and the coefficient of friction, which is determined by the state of the surface, i.e., $F = \mu N$.

Nonfusion welding processes, as a group, offer several advantages over fusion processes. The general absence of melting and, typically, the low heat involved minimally disrupt the microstructure of the materials being joined. As shown in Figure 6.5b, there is no fusion zone and, usually, a minimal heat-affected zone. By precluding the need for melting, intermixing of the materials involved in the joint is minimal on a macroscopic scale, so materials of dissimilar compositions can often be joined. The joint resulting from nonfusion welding typically has quite high efficiency. Process disadvantages relate to the surface preparation and tooling required to produce acceptable joints and difficulties associated with inspecting and repairing defective joints.

6.5.2 Cold Pressure Welding

There are really only two pressure welding processes that are accomplished cold: cold welding and, possibly, cold roll welding. *Cold welding* (CW), as the name implies, uses (usually substantial) pressure at room temperature to produce coalescence of materials through substantial plastic deformation at the weld. Besides high pressures, extremely clean surfaces are required (as described in Section 6.2.4. The process works best for ductile metals such as aluminum, copper, many of the brasses and bronzes, nickel, lead, and the precious metals. Welding of dissimilar metals, such as aluminum to copper, is possible. Cold welding is normally difficult to accomplish consistently in production but is a viable option for joining in space.

In *roll welding* (ROW), pressure is applied to the elements of the joint to be welded through rollers, and such pressure can be applied while the base materials are at room temperature, i.e., cold. Roll welding is rarely performed without heat, however, as the forces needed to cause sufficient plastic deformation can be very high. Roll cladding of ductile metals, for example, aluminum onto copper, is done this way, however.

6.5.3 Hot Pressure Welding

To reduce the forces needed to cause the gross plastic deformation required to produce nonfusion pressure welds, heat is usually used. Heating base metals or alloys also allows dynamic recrystallization and grain boundary migration to take place and enhances diffusion, thereby facilitating weld formation.

Hot pressure welding processes include hot pressure welding, forge welding, and hot roll welding. In *hot pressure welding* (HPW), coalescence of metals is accomplished with the application of heat and pressure sufficient to cause plastic deformation. Often, a vacuum or inert shielding media is used to prevent oxidation that would impede bonding. A close relative of hot pressure welding is *forge welding* (FOW). In this process, heating takes place in air in a forge, and pressure is applied progressively or in repeated blows. Plastic deformation at the interface between the pieces being welded is extensive. This is an extremely old method of welding, being the basis for early iron making and blacksmithing. To facilitate bonding, a flux is often used at the interface to remove oxides. *Hot roll welding* (ROW) is performed as described for cold roll welding, except that the work is heated to reduce the material flow stress and facilitate deformation.

Explosion welding (EXW) is a pressure welding process that represents a special case. In explosion welding, the workpieces usually start out cold but heat significantly and extremely rapidly very locally at their faying surfaces during the production of the actual weld. As shown in Figure 6.28a, the controlled detonation of a properly placed and shaped explosive charge causes the workpieces to come together extremely rapidly at a low contact angle. When this occurs, air between the workpieces is squeezed out at supersonic velocities. The resulting jet cleans the surfaces of oxides and causes localized but rapid heating to high temperatures. The result is a metallurgical bond. The weld bondline of explosion welds is typically very distorted locally, as shown in Figure 6.28b.

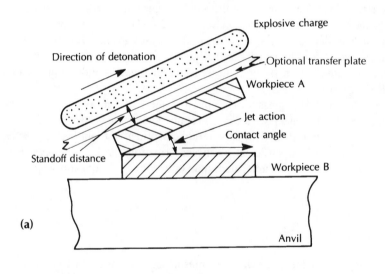

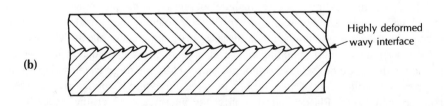

Figure 6.28   (a) Explosion welding (EXW); (b) the interface of an explosion weld

6.5.4 Friction Welding

Friction welding (FRW) processes use machines that are designed to convert mechanical energy into heat at the joint to be welded. Coalescence of materials occurs under the compressive force contact of workpieces moving relative to one another linearly or in rotation. Figure 6.29 shows one version of the friction welding process schematically. Because rotation is frequently used, at least one of the parts involved in the joint has an approximate circular cross section. Frictional heating occurs at the interfaces as they are brought into contact under pressure, raising the temperature of the material to a level suitable for forging. Axial pressure forces the hot metal out of the joint, disrupting and removing oxides and other surface impurities and producing a pronounced upset (as shown in Figure 6.29b).

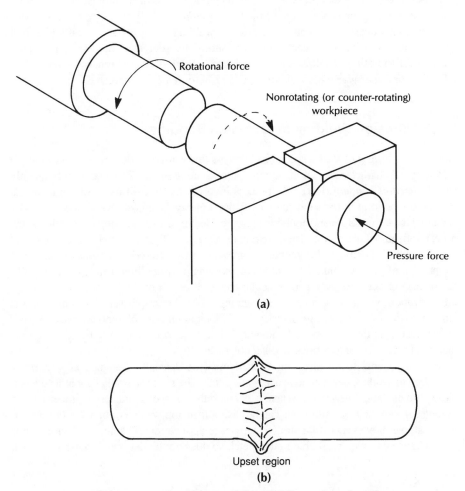

(a)

(b)

Figure 6.29   The friction welding (FRW) process and weld.

Two predominant techniques are used for friction welding. In the first, more conventional technique, known as friction welding, the moving part is held in a motor-driven collet and rotated at a constant speed against a fixed part, while an axial force is applied to both parts. Rotation is continued until the entire joint is suitably heated, and then, simultaneously the rotation is stopped and an upsetting force is applied, producing a weld. Key process variables are rotational speed, axial force, welding time, and upset force or displacement. In the second technique, called *inertia welding,* energy is stored in a flywheel that has been accelerated to the required speed by a drive motor. The flywheel is connected to the motor through a clutch and to one of the workpieces by a collet. A weld is made by applying an axial force through the rotating part to a stationary part while the flywheel decelerates, transforming its kinetic energy into heat at the joint faying surfaces. When done properly, the weld is completed when the flywheel stops. Key process variable are the flywheel moment of inertia, the flywheel rotational speed, the axial force, and the upset force. The actual process for both techniques is usually automated. Some rotational friction welding processes are called *spin welding*.

It is also possible to accomplish friction welding using linear motion while joint components are held in contact under pressure. The source of motion can be pure mechanical vibration or ultrasonically induced vibration. These processes are called *vibration,* or *linear vibration, welding* and *ultrasonic welding,* respectively.

6.5.5 Diffusion Welding or Diffusion Bonding

Diffusion welding (DFW) is a specialized process that is used predominantly when unique metallurgical characteristics of the joining process are required. The process is frequently called *diffusion bonding* (DB). The process involves the interdiffusion of the materials making up the joint in the solid state. Diffusion kinetics is almost always accelerated by elevating the temperature and applying pressure. Heating can be accomplished using a furnace, retort, autoclave, or hot-platen press or by resistance. Pressure can be applied by dead weight loading, a press, or differential gas pressure or by differential thermal expansion of the parts or of tooling. Uniaxial methods of applying pressure limit welding to flat, parallel, planar surfaces, roughly perpendicular to the direction of load application. Isostatic pressurization, using encapsulation, or canning, offers better pressure uniformity and is applicable to more complex geometries. The principal purpose of applying pressure is to obtain contact at the interface to be joined. This contact occurs initially by plastic deformation of microscopic asperities, and later by creep.

Components to be diffusion welded must be specially designed and carefully processed to produce consistently successful joints. The process is only economical when close dimensional tolerances, expensive materials, or special material properties are involved. Even then, not all metals can be easily diffusion welded or bonded. One excellent example, however, is titanium for aerospace applications.Titanium diffusion bonds well because its oxide is soluble in the metal, so does not hinder metal to metal contact.

---

[27] As long as heating is uniform during diffusion welding, the entire structure undergoes the same metallurgical changes, so there is no heat-affected zone, as such, and thus, no metallurgical notch in the structure.

Diffusion welding or bonding has the potential to produce "perfect" welds, with no fusion zone and no heat-affected zone,[27] just unaffected base metal. Resulting joints are 100% efficient.

In summary, nonfusion welding processes offer a valuable alternative to fusion welding processes, enabling the joining of difficult (often dissimilar) materials to fairly precise tolerances, with little or no heat effect in the base material.

6.6 WELD JOINT DESIGNS

6.6.1 General Design Guidelines

Without attempting to cover the subject of weld design, it is worth considering some general design guidelines. The design of a joint for welding, as in other processes, should be selected primarily on the basis of load-carrying requirements. Especially in welding, however, variables in the design and layout of joints can substantially affect the costs associated with welding.

As a first guideline, always select the joint design that requires the least amount of weld metal. This will minimize distortion and residual stresses in the welded structure caused by weld metal shrinkage during solidification and due to thermal expansion and contraction under severe temperature gradients. Where possible, use square grooves (see Section 6.6.3) and welds that only partially penetrate the joint. Partial penetration welds leave unmelted metal in contact throughout welding to maintain dimensions. They also prevent weld metal loss due to drop through under the force of gravity. Use lap and fillet welds instead of groove welds when fatigue is not a design consideration. Use double-V or double-U instead of single-V or single-U groove welds on thick plates to reduce the amount of weld metal and control distortion. For corner joints in thick plates where fillet welds are not adequate, beveling of both members should be considered to reduce the tendency for lamellar tearing.[28]

Finally, always design an assembly and its joints for good accessibility for any welding and inspection that will be required.

6.6.2 Size and Amount of Weld

Overdesign is a common error in welded structures, as is overwelding in production. Control of weld size and the amount of weld begins in design but must be maintained during assembly and welding. Overdesign and overwelding both lead to excessive and unnecessary cost.

Adequate, but minimum size and length, welds should be specified for the forces to be transferred. Oversize welds may cause excessive distortion and higher residual stresses, without improving suitability. The size of a fillet weld is especially important because the amount of weld required increases as the square of the weld size increases.

---

[28] *Lamellar tearing* is tearing in the plane of a plate caused by tensile stresses acting perpendicular to the plane of that plate and aggravated by the presence of nonmetallic inclusions in the plate.

For equivalent strength, a continuous fillet weld of a given size is usually less costly than a larger sized intermittent fillet weld. Also, there are fewer weld starts and terminations that are potential sites for flaws. An intermittent fillet weld can be used in place of a continuous fillet weld of minimum size when a static loading condition exists. An intermittent fillet weld should not be used under cyclic fatigue loading. For automatic welding, continuous welding is preferred.

Welds should always be placed in the section of least thickness, and the weld size should be based on the load or other requirements of that section. The amount of welding should be kept to a minimum to limit distortion and internal stresses and, consequently, the need for stress relieving and straightening. Welding of stiffeners or diaphragms should be limited to that required to carry the load.

Obviously, there are many more considerations important in the design of welds, but these give some appreciation of the problems.

6.6.3 Types of Joints

The loads in a welded structure are transferred from one member to another through welds placed in the joints. The type of joint, or the joint geometry, is predominantly determined by the geometric requirements or restrictions of the structure and the type of loading. Accessibility for welding, process selection, accessibility for inspection, and cost constraints are other factors affecting the choice of joint type.

The fundamental types of joints used in welding are shown in Figures 6.30–6.32 and include butt, corner, tee, lap, and edge geometry groove and fillet types; full and partial penetration types; and continuous and intermittent types. Within these types are several variations, primarily based on specific geometries and preparation methods,

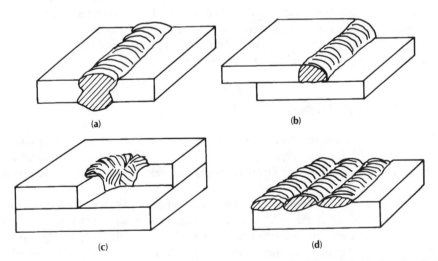

(a)

(b)

(c)

(d)

Figure 6.30 Fundamental types of welds. (a) Groove weld; (b) fillet weld; (c) plug (or slot) weld; (d) surfacing weld.

including single and double grooves; single and double fillets; plug and slot; and squares, V's, U's, and J's.

Briefly, the square groove is simple to prepare, economical to use, and provides satisfactory strength but is limited by joint thickness. For thick joints, the edge of each member of the joint must be prepared to a particular geometry to provide accessibility for welding and ensure the desired weld soundness and strength. For economy, the opening or gap at the root of the joint and the included angle of the groove should be selected to require the least weld metal necessary to give access and meet strength

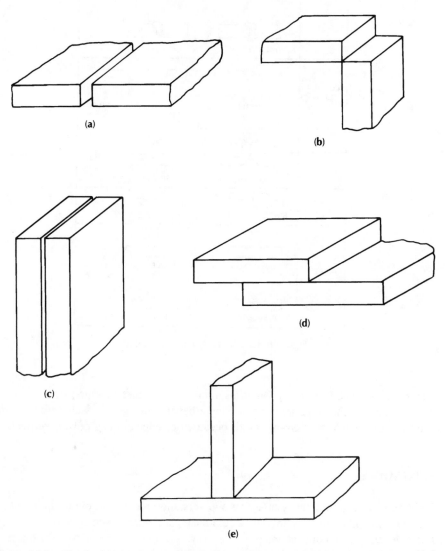

Figure 6.31 Five basic joint designs. (a) Butt joint; (b) corner joint; (c) edge joint; (d) lap joint; (e) tee joint.

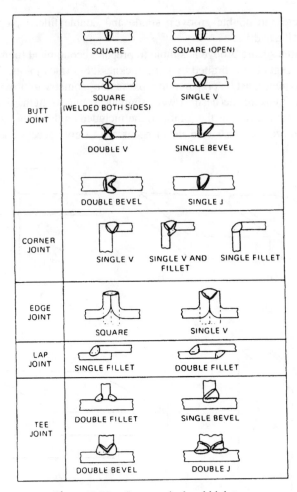

Figure 6.32   Some typical weld joints.

requirements. J and U groove geometries minimize weld metal requirements compared to V's but add to the preparation costs. Single-bevel- and J-groove welds are more difficult to weld than V or U groove welds because one edge of the groove is vertical.

SUMMARY

Welding is an extraordinarily effective and versatile process for joining like types of materials, including metals, polymers, ceramics, glasses, and even certain composites. It uses a combination of heat and pressure to bring atoms or molecules in the materials making up the joint to equilibrium spacing to allow primary bonds, like those in the parent materials, to form and produce joint strength. Welding of real materials is complicated by the existence of adsorbed layers of gases, moisture, oxides, and other contaminants on surfaces, as well as by the fact that roughness on the atomic scale limits the number and area of effective contacts needed to allow bonds to form. The challenge in practical welding is to overcome these contaminants by cleaning and these asperities by causing melting or plastic deformation in the solid state. In fact, welding processes are broadly classified as fusion or non-fusion types, depending on whether melting or plastic deformation is the principal means for bringing atoms or molecules together to effect a weld. Other classification schemes consider whether or not the process requires pressure (pressure versus nonpressure) and the source of heating (chemical, electrical, mechanical, or other). Subclassification of fusion welding processes considers whether or not filler is required (i.e., is autogenous) and, if it is, whether that filler is homogeneous or heterogeneous with respect to the base materials, or, if the process is an arc welding process, whether the electrodes used are consumable or nonconsumable, and discontinuous or continuous.

The most common gas welding process (oxyacetylene), arc welding processes (gas-tungsten arc, plasma arc, gas-metal arc, shielded metal-arc, flux-cored arc, submerged arc, electrogas, and electroslag), high-energy beam welding processes (electron-beam and laser-beam), and resistance welding processes (resistance spot, resistance seam, projection, flash, upset, and percussion) were described, along with nonfusion pressure, friction, and diffusion processes. For fusion welding processes, several potential sources for loss of energy between the source and the workpiece result in a transfer efficiency for these processes that varies widely based on the process but is always less than 1.00 or 100%.

Finally, the design of a weld joint (or the joint type) is dependent on the geometric requirements and restraints of the structure being welded, as well as on the expected loading. Specific preparations of joint types depend on the process, loading, and cost.

PRACTICE QUESTIONS AND PROBLEMS

1. Define *welding* in the broadest sense. Based on this definition, which of the following joints could, at least in theory, be welded? (a) pure copper to pure copper; (b) AISI 1030 steel to 304 stainless steel; (c) 1100 Al alloy to 304 stainless steel; (d) polyvinyl chloride (PVC) thermoplastic to PVC; (e) PVC to pure aluminum; (f) two different composition glasses with matching coefficients of thermal expansion; (g) silicon nitride ceramic to silicon nitride ceramic; (h) silicon nitride ceramic to pure molybdenum; (i) graphite-fiber reinforced polyetheretherketone (PEEK) composite to unreinforced PEEK; (j) SiC-particulate reinforced 6061 Al alloy to 6061 Al alloy.

2. Fundamentally, what is required to achieve chemical bonding to produce a weld? What are the practical problems encountered in real materials? How are these problems overcome by various welding processes?

3. If two pieces of pure aluminum are brought together such that every atom across the interface forms a bond, the resulting joint strength is equal to the theoretical cohesive strength of the aluminum. What would be the strength of the joint (per square millimeter) if 1 out of every 10^8 atoms along the faying surface of one piece bonded to an atom along the faying surface of the other piece? Assume the faying surfaces are both close-packed planes.

    Assume the following in arriving at a solution: The binding energy, E_o, at equilibrium spacing r_o is:

    $$E_o = -A / r_o^6 + B / r_o^{12}$$

    $$F = -\frac{dE}{dr}$$

and E_o for aluminum is 3.4 eV/atom or 324 kJ/mole, while r_o is 2.86 Angstroms (where 1 Angstrom = 10^{-7} mm). *Hint:* Find the force expression from E_o, where the attractive and repulsive terms are equal or their sum is zero, and solve the resulting two simultaneous equations to find values of A and B. Then, differentiate the force equation to obtain the expression for the maximum force and solve using A, B and r_o.

4. Cleaning is critical to the production of a sound weld. By what different general means are the surfaces of metals cleaned and/or kept clean by various processes? Cite an example of a process that employs each means you name.
5. What are some of the particular advantages of welding versus(a) mechanical fastening; (b) adhesive bonding; (c) brazing or soldering? What is the single greatest disadvantage for some situations?
6. Describe several schemes by which welding processes can be classified.
7. Sketch and label all of the specific microstructural regions or zones you would expect to find for a fusion weld made between a piece of annealed pure copper and a piece of cold worked 70Cu–30Zn yellow brass. Show the relative sizes of the various regions in each piece.

    Assume the two pieces are of equal thickness and that the welding source was centered directly over the faying surface of two square blocks.
8. Explain how an oxyacetylene welding flame can be adjusted for welding (a) an aluminum alloy; (b) a high carbon steel; (c) a 60Cu-40Zn alloy; (d) a piece of austenitic stainless steel.
9. If instead of acetylene (C_2H_2), propane (C_3H_8) were used as the fuel gas in oxy-fuel welding, what would the primary and secondary combustion reactions would be?
10. Which operating mode of gas-tungsten arc welding (GTAW) would you choose for welding each of the following? (a) 0.1 mm thick stainless steel; (b) 3 mm thick commercially pure titanium; (c) 1.5 mm thick magnesium alloy casting; (d) 15 mm thick Inconel 625.
11. Compare the GTAW and the PAW processes. How are they similar? How do they differ? Which process would you choose for welding critical aircraft gas tur-

bine components where any contamination could be disastrous? Which would you choose for repair welding tool steel dies?

12. Compare the GMAW and the FCAW processes. How are they similar? How do they differ? Which process would you choose for welding a reactive metal? Which would you choose for repair welding on offshore drilling platform pipes?

13. Give at least five functions of the coating of an SMAW electrode. Given these functions, and the similarity between the coating on SMAW electrodes and the core in FCAW wires, are there any limitations on the formulation of fluxes for use in FCAW versus SMAW? If so, what limitations? If not, why not?

14. Compare the SAW and the ESW processes. How are the processes similar? How do they differ? Which process would you choose for overlaying a wear protective alloy on large diameter rolls? Which process might you choose for welding thick hull plates on ships?

15. Differentiate between the melt-in and the keyhole modes of welding. Which processes are capable of welding in either mode? For what types of situations would you use each?

16. Compare the EBW and the LBW processes. How are the processes similar? How do they differ? Which process would you choose for welding highly reactive zirconium alloys for nuclear applications? Which process might you potentially use for welding thick section structures in shipbuilding?

17. Describe how a resistance spot weld is made. Is resistance spot welding a good choice for welding dissimilar metals? Can it be used for welding polymeric materials? What about ceramics?

18. Rationalize the differences in transfer efficiency in each of the following: (a) $\eta = 0.9$ for SAW versus 0.7 for GMAW; (b) $\eta = 0.5$ for GTAW versus 0.7 for GMAW; (c) $\eta = 0.5$ for melt-in PAW versus 0.85 for keyhole PAW; (d) $\eta = 0.1$ for LBW of aluminum versus 0.8 for LBW of tungsten.

19. Would you expect a difference in transfer efficiency when the DC polarity is changed from EN to EP in GMAW? If so, why? If not, why not?

20. Give several major advantages of nonfusion welding over fusion welding. For which of the material combinations in Problem 1 that can be welded would a nonfusion process be preferable to a fusion process?

21. Can nonfusion welding processes be applied to polymeric materials? What about application with ceramics?

22. What are some of the manufacturing considerations in selecting one weld joint design or geometry over another?

REFERENCES

1. American Welding Society, *Welding Handbook, Vol. 1, Welding Technology,* 8th ed., Miami: American Welding Society, 1987.
2. American Welding Society, *Welding Handbook, Vol. 2, Welding Processes,* 8th ed., Miami: American Welding Society, 1990.

3.   American Welding Society, *Welding Handbook, Vol. 2, Welding Processes—Arc and Gas Welding and Cutting, Brazing and Soldering,* 7th ed., Miami: American Welding Society, 1978.

4.   American Welding Society, *Welding Handbook, Vol. 3, Welding Processes—Resistance and Solid-State Welding and Other Joining Processes,* 7th ed., Miami: American Welding Society, 1980.

5.   Arata, Yoshiaki, *Plasma, Electron & Laser Beam Technology,* Metals Park, Ohio: American Society for Metals, 1986.

6.   Cary, Howard B., *Modern Welding Technology,* 2nd ed., Englewood Cliffs, N.J.: Prentice Hall, 1989.

The Metallurgy of Welding

7.1 INTRODUCTION

The process of joining materials by welding, including the processes of brazing and soldering, almost always involves the application of significant heat to aid in the formation of the bonds that create the joint. As described in Chapter 6, this heat may or may not cause melting of the parent or base materials and/or a compatible filler material. When it does cause melting of the base material and/or filler, the process is referred to as a *fusion process.* When the base material is melted, the process is called *fusion welding.* When only the filler is melted, the process is called *brazing* or *soldering,* depending on the temperature at which the filler melts. When no melting occurs and bonds are created by bringing atoms, ions, or molecules together in the solid state by pressure and/or diffusion, regardless of the temperature, the process is referred to as *nonfusion welding.*

Whatever the level of heating during welding (or for any process), above a certain temperature heat causes materials to undergo changes in state or structure or both. For crystalline materials such as metals and ceramics, these changes can be significant and can include melting, gas–liquid chemical reactions (e.g., gas adsorption or oxidation), solidification, liquid–solid reactions (e.g., eutectic or peritectic formation), grain growth, recrystallization, allotropic transformations, solid-phase reactions (e.g., eutectoid formation), phase precipitation, or phase dissolution. Which changes actually occur depend on the level of heating, the composition of the material(s) being heated, the environment, previous thermal-mechanical processing of the material(s) being heated, time at temperature, and cooling rates. During welding, metallurgical changes occur under conditions that deviate from equilibrium, often significantly, because of the high rates of heating and cooling involved. As a result, many of the reactions or transformations are quite different than expected from and predicted by equilibrium phase diagrams, and, thus, the resulting structure and properties are often much different than expected and predicted.

This chapter looks at the effects of heating and subsequent cooling on the final structure of the newly created joint and its surrounding regions during the general process of welding, brazing, or soldering. First, the distribution of heat in the parts being joined and its general effect on the materials making up those parts (i.e., the creation of various microstructural zones) is considered. Then, the reactions that can occur in the fusion zone of a weld, or in the molten filler of a brazed or soldered joint, are described. Next, the process of solidification of the melt is described. Similarities and differences between the solidification of pure crystalline materials (e.g., metals or ceramics) and crystalline alloys are presented, with special emphasis on the effects of nonequilibrium (e.g., creation of solidification substructure and microsegregation). Then, reactions and transformations in the region of welded, brazed, or soldered joints outside any fusion zone, that is, in the heat-affected zone, are described. Effects of heat in work-hardened, age-hardened, transformation hardenable, or sensitization-prone materials are considered. Finally, sources of defects in the fusion, partially melted, and heat-affected zones are presented, along with methods for testing susceptibility to defect formation (i.e., weldability).

It must be reemphasized that the nonequilibrium physical metallurgy of solidification and of heat-affected zone transformations is the same irrespective of whether the specific process involved is fusion welding, brazing, soldering, or nonfusion welding. For this reason, the concepts presented in this chapter are vital to an understanding of any or all of these important joining processes.

7.2 HEAT DISTRIBUTION IN FUSION WELDING

7.2.1 General Description

In fusion welding, an intense source of energy is necessary to cause the required melting of the materials to be joined. The source is usually chemical or electrical in nature, but can be high energy beams as well, with losses occurring between the source and the workpiece (see Section 6.4.6). Even after the net energy from the source is transferred to the workpiece as heat, not all of that heat contributes to producing the weld. Some is conducted away from the weld fusion zone, raising the temperature of the surrounding material and causing unwanted metallurgical and geometric changes. This surrounding region is called the *heat-affected zone*. Some heat is lost in other ways.

In considering the effects of heat on melting and solidification in the fusion zone and transformations in the surrounding heat-affected zone of a fusion weld, it is important to first consider how that heat is distributed. How that heat is distributed directly influences the extent of melting, the extent and nature of peripheral heating, and the rate of cooling. The extent of melting, in turn, directly affects the weld shape, homogeneity through convection, the degree of shrinkage and attendant weldment distortion, and, often, susceptibility to defects. The extent of peripheral heating, in turn, affects the development of thermally induced stresses acting on the fusion zone (which contribute to fusion zone defect formation), the rate of cooling in the fusion zone (which controls solidification mechanics and determines final fusion zone structure), the level of heating in the heat-affected zone (which can cause degradation of properties), the rate of cooling in the heat-affected zone (which determines the final structure in this zone), and the degree and nature of distortion and/or residual stresses in the weldment.

For these reasons, the heat distribution in fusion welds[1] will be addressed in the following subsections. What is described for fusion welds applies equally to other joining processes using heat, such as certain nonfusion welding processes and brazing or soldering.

7.2.2 Heat Flow Analysis

The thermal conditions in and near the fusion zone of a fusion weld must be maintained within specific limits to control metallurgical structure, residual stresses and distortions, and chemical reactions (e.g., oxidation) that result from a welding operation. Of specific interest are (1) the distribution of heat between the fusion zone and the heat-affected zone; (2) the solidification rate of the weld metal; (3) the distribution of maximum or peak temperature in the weld heat-affected zone; and (4) the cooling rates in the fusion and heat-affected zones.

The transfer of heat in a weldment is governed primarily by the time-dependent conduction of heat, which is expressed by the following equation:

$$\frac{\rho C \delta T}{\delta t} = \frac{\delta}{\delta x} \frac{k(T) \delta T}{\delta x} + \frac{\delta}{\delta y} \frac{k(T) \delta T}{\delta y} + \frac{\delta}{\delta z} \frac{k(T) \delta T}{\delta z} - \rho C \frac{V_x \delta T}{\delta x} + \frac{V_y \delta T}{\delta y} + \frac{V_z \delta T}{\delta z} + Q \quad (7.1)$$

where x is the coordinate in the direction of welding (meters), y is the coordinate transverse to the welding direction (meters), z is the coordinate normal to the weldment surface (meters), T is the temperature of the weldment (Kelvin), $k(T)$ is the thermal conductivity of the metal (or ceramic) (joules per meter second Kelvin) as a function of temperature, $\rho(T)$ is the density of the metal (or ceramic) (grams per cubic meter) as a function of temperature, $C(T)$ is the specific heat of the metal (or ceramic) (joules per kilogram Kelvin) as a function of temperature, V_x, V_y, and V_z are components of velocity (meters per second), and Q is the rate of any internal heat generation (watts per cubic meter).

Whether this general equation needs to be solved for one, two, or three dimensions depends on the weldment and weld geometry, including whether the weld is penetrated fully or partially, is parallel sided or tapered, and the relative plate thickness (as shown in Figure 7.1). A one-dimensional solution can be used for welding a thin plate or sheet with a stationary source or for welding under steady state (i.e., at constant speed and in uniform cross sections remote from edges) in very thin weldments; a two-dimensional solution is most useful in relatively thin weldments or in thicker weldments where the weld is full penetration and parallel-sided to assess both longitudinal

---

[1]When the heating needed to produce a joint is uniform (i.e., the entire workpiece temperature is raised uniformly), there are no temperature gradients in the workpiece, so there are no zones as such. Whatever metallurgical changes take place take place uniformly and based solely on the peak temperature reached, the time at that peak temperature, and the rate of cooling from that temperature. This is the situation in most brazements, the subject of Chapter 8. In soldered joints, the heat input may be nonuniform, resulting in heat distribution and temperature gradients in the workpiece, but the peak temperatures are usually too low to cause metallurgical changes or produce a heat-affected zone. In nonfusion welding, heating is nonuniform, and only heat-affected zone (as opposed to fusion zone) changes need to be considered.

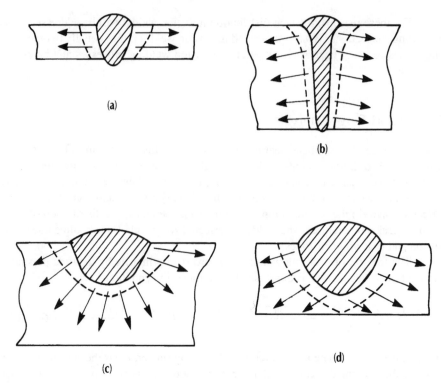

Figure 7.1 Effect of weldment and weld geometry on heat flow characteristics. (a) Two-dimensional heat flow, full-penetration weld, thin plate; (b) two-dimensional heat flow, full-penetration parallel-sided weld, thick plate; (c) three-dimensional heat flow, partial-penetration weld, thick plate; (d) intermediate condition, near-full-penetration weld.

and transverse[2] heat flow; and a full, three-dimensional solution is required for a thick weldment in which the weld is partial penetration or nonparallel-sided.

Theoretical analyses of the weld thermal cycle, starting from the general heat flow equation, Eq. 7.1, have long been attempted, although before the advent of computers (and even with computers) these usually require a number of simplifying assumptions. Without going into detail, it is worth looking briefly at one of the earliest and most well known simplified solutions, that of Rosenthal. The key to Rosenthal's solution is the assumption of *quasi-* or *pseudo-steady state.*

Rosenthal's first critical assumption is that the energy input from the heat source used to make the weld is uniform and moves with a constant velocity v along the x-axis of a fixed rectangular coordinate system, as shown in Figure 7.2. This is not a bad assumption for many situations. The net heat input to the weld is given by

$$H_{\text{net}} = q/v \text{ (in joules per meter)} \tag{7.2}$$

---

[2]*Longitudinal* is in the direction of welding; *transverse* is perpendicular to that direction.

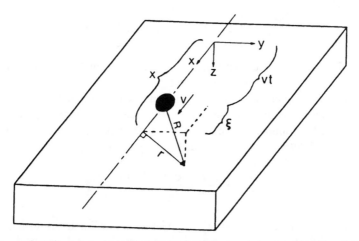

Figure 7.2   Coordinate system for Rosenthal's simplified, steady-state solution of the general heat flow equation. Welding configuration in terms of a point heat source q and constant velocity v. (Reprinted with permission from K. Easterling, *Introduction to the Physical Metallurgy of Welding*, London: Butterworths, 1985, p. 18.)

and the heat or energy q is given by:

$$q = \eta E I \tag{7.3}$$

where η is the transfer efficiency of the process (see Section 6.4.6), E and I are the welding voltage (in volts) and current (in amperes), respectively, and v is the speed or velocity of welding in meters per second. Rosenthal further assumed the heat source to be a point source, with all of the energy being put into the weld at a single point. This avoided complexities with density distribution in the energy from different sources and restricted heat flow analysis to the heat-affected zone, beyond the fusion zone (or weld pool) boundary.

Rosenthal next simplified the general heat flow equation in two ways: first, by assuming that the thermal properties (i.e., thermal conductivity, k, and product of the specific heat and density, $C\rho$) of the material being welded are constants; and, second, by modifying the coordinate system from a fixed system to a moving system. The assumption that thermal properties are constants is seriously flawed, as these properties typically change dramatically with temperature (often by two to five times), especially considering the wide range of temperatures normally involved in fusion welding. The moving coordinate system was a perfectly reasonable simplification, replacing x with ξ, where ξ is the distance of the point heat source from some fixed position along the x axis, depending on the velocity of welding, v (where there is only a velocity component in the x direction) by the relationship

$$\xi = x - vt \text{ where } t \text{ is the time} \tag{7.4}$$

When the general heat flow equation (Eq. 7.1), with constant thermal properties, is differentiated with respect to ξ rather than x, the result is

$$\frac{\delta^2 T}{\delta \xi^2} + \frac{\delta^2 T}{\delta y^2} + \frac{\delta^2 T}{\delta x^2} = -\frac{C\rho}{k} v \frac{\delta T}{\delta \xi} + \frac{C\rho}{k} \frac{\delta T}{\delta t} \tag{7.5}$$

where k is the thermal conductivity, in joules per meter second Kelvin and v is the speed or velocity of welding, in meters per second.

This can be further simplified, in accordance with Rosenthal, if a so-called quasi-stationary temperature distribution exists. This means the temperature distribution around a point heat source moving at constant velocity will settle down to a steady form, such that $dT/dt = 0$, for $q/v = $ a constant. The result is

$$\frac{\delta^2 T}{\delta \xi^2} + \frac{\delta^2 T}{\delta y^2} + \frac{\delta^2 T}{\delta z^2} = -\frac{C\rho}{k} v \frac{\delta T}{\delta \xi} \tag{7.6}$$

This situation is, in fact, achieved in many welds, so the simplified form given in Eq. 7.6 is reasonable except for the assumption of constant thermal properties, which can be dealt with in other ways.

Rosenthal solved the simplified form of the heat flow equation (Eq. 7.6) for both thin and thick plates, in which the heat flow is basically two dimensional and three dimensional, respectively. The solutions are:

$$T - T_0 = \frac{q}{2\pi k} \exp\left(\frac{-v\xi}{2a}\right) Ko\left(\frac{vR}{2a}\right) \tag{7.7}$$

for the thin plate, where q is the heat input from the welding source (in joules per meter), k is the thermal conductivity (in joules per meter second Kelvin) a is the thermal diffusivity, $=k/\rho C$ (square meters per second), K_o is a Bessel function of the first kind, zero order, and $R = (\xi^2 + y^2 + z^2)^{1/2}$, the distance from the heat source to a particular fixed point, in meters.

For the thick plate,

$$T - T_0 = \frac{q}{2\pi kd} \exp\left(\frac{-v\xi}{2a}\right) \frac{\exp\left(-\frac{vR}{2a}\right)}{R} \tag{7.8}$$

where d is the depth of the weld (which for symmetrical welds is half of the weld width, since $w = 2d$). These can each be written in a simpler form, giving the time-temperature distribution around a weld, when the position from the weld centerline is defined by a radial distance, r, where $r^2 = z^2 + y^2$. For the thin plate, the time-temperature distribution is

$$T - T_0 = \frac{q/v}{d(4\pi k\rho Ct)^{1/2}} \exp\left(\frac{-r^2}{4at}\right) \tag{7.9}$$

and for the thick plate is:

$$T - T_0 = \frac{q/v}{2\pi kt} \exp\left(\frac{-r^2}{4at}\right) \tag{7.10}$$

The form of the temperature distribution fields given by these equations is illustrated in Figure 7.3.

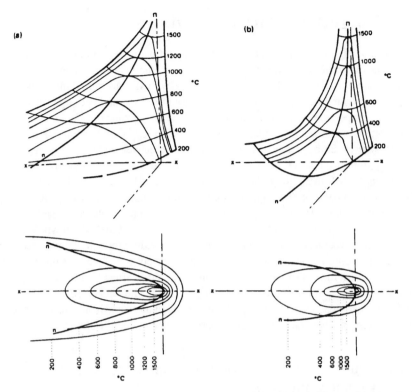

Figure 7.3 The temperature distribution fields in thin (a) and thick (b) plates. (Reprinted with permission from K. Easterling, *Introduction to the Physical Metallurgy of Welding*, London: Butterworths, 1985, p. 21.)

Since Rosenthal's outstanding early contribution in this area, there have been numerous refinements. Adams recognized the existence of the weld pool and other problems and used the fusion line as the boundary condition to obtain expressions for the peak temperature at any distance (*y*) from this boundary. He did this for two- (i.e., thin plate) and three-dimensional (i.e., thick plate) heat flow conditions. One of Adams's solutions is given in the section "Peak Temperatures."

Others, like Adams, tried to modify Rosenthal's solutions for two- and three dimensions. Various modifications of the two-dimensional solution were developed by Grosh, Trabant, and Hawskins; Swift-Hook and Gick; Jhaveri, Moffatt, and Adams; Meyers, Uyehari, and Borman; Ghent, Hermancz, Kerr, and Strong; and Trevedi and Shrinivasan. Various modifications of the three-dimensional solution were developed by Jhaveri, Moffatt, and Adams; Grosh and Trabant; and Malmuth, Hall, Davis, and Rosen. Because of the complexity of the problem, most of these efforts found only limited success. In particular, heat flow and solidification rate in and immediately outside the weld pool (i.e., at the FZ/HAZ boundary) is extremely difficult to handle, and predictions are inaccurate.

7.2.3 Zones in Fusion Welded Materials

As shown in Figure 6.5 and again in Figure 7.4, a fusion weld produces several distinct microstructural zones in a pure crystalline metal or ceramic or in a crystalline alloy of one of these materials. These different zones correlate with various transformations on appropriate phase diagrams, adjusted for non-equilibrium heating and cooling effects on transformation temperatures and microconstituents. The fusion zone (or weld metal in a metal) is the portion of the weld that melted during welding, by being heated to above the temperature where melting just begins for the material being welded (T_{melting} for the pure metal and T_{solidus} for the alloy in Figure 7.4). The fusion zone is often designated as the FZ or the weld metal (WM). In an alloy, there is a partially melted zone (or PMZ) where the temperature rose to between the liquidus and the solidus temperature (T_{solidus} and T_{liquidus} for the alloy in Figure 7.4). Naturally, there is no PMZ in a pure material. The heat-affected zone (or HAZ) is the portion of the base material that was not melted but whose properties (and, usually, structure) were altered by the heat of welding through some phase transformation or reaction. The unaffected base material is the portion of the base material wherein the welding heat did not exceed the minimum required to affect its structure or properties (i.e., below $T_{\text{allotropic}}$ for the pure metal and T_{solvus} for the alloy in Figure 7.4). The so-called weld zone encompasses the fusion, partially melted (if any), and heat-affected zones, since this is the region where base material properties are affected by welding.

7.2.4 Simplified Equations for Approximating Welding Conditions

In the past, results of experiments were typically fit to simplified equations suggested by heat transfer theory, and techniques were developed to calculate important factors like peak temperatures (to cause melting or produce heat affects), fusion and heat-affected zone widths, solidification rates, and cooling rates at various points from the boundary of the weld fusion zone and the heat-affected zone. Later, simplifying assumptions, such as those of Rosenthal, allowed simplified analytical solutions. Computers can now be used to solve the complex differential equations that model heat flow in welding, and the general solution can be solved numerically, yielding remarkable correlation between predicted and measured fusion and heat-affected zone sizes and shapes, weld pool convection, and so on. Simplified equations, supported by engineering data, however, still play an important role in approximating conditions. For this reason, several useful simplified equations are presented in the following sections.

Peak Temperatures

Predicting or interpreting metallurgical transformations (e.g., melting, austenitization, and recrystallization of cold worked material) at a point in the solid material near a weld requires some knowledge of the maximum or peak temperature reached at a specific location. For a single-pass, full-penetration butt weld in a sheet or a plate, the distribution of peak temperatures (T_p) in the base material adjacent to the weld is given by:

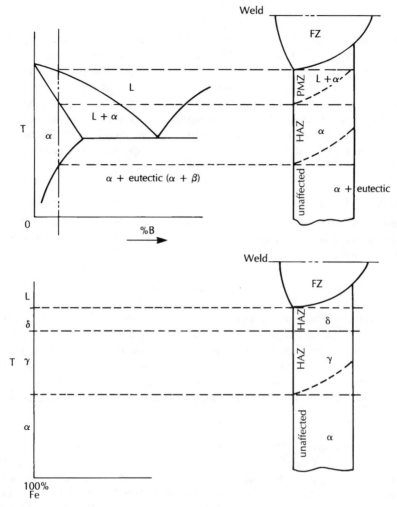

Figure 7.4 Correlation between microstructural zones in a fusion weld in an alloy (a) and a pure metal exhibiting several allotropic transformatins (e.g., Fe) (b) and the corresponding phase diagram.

$$\frac{1}{T_p - T_0} = \frac{(2\pi e)^{0.5} \rho C h y}{H_{\text{net}}} + \frac{1}{T_m - T_0} \tag{7.11}$$

where T_0 is the initial temperature of the weldment (in Kelvin), e is the base of natural logarithms (2.718), ρ is the density of the base material (in kilograms per cubic meter), C is the specific heat of the base material (in joules per kilogram Kelvin), h is the thickness of the base material (in meters), $y = 0$ at the fusion zone boundary and where $T_p = T_m$, T_m is the melting (or liquidus) temperature of the material being welded (in Kelvin), and H_{net} is obtained from Eq. 7.2 with 7.3.

Width of Fusion or Heat-Affected Zone

If one defines a peak temperature as either the melting point or as the temperature at which the structure and properties of the base material are altered by some metallurgical transformation (i.e., creating a heat affect), then the peak temperature equation (Eq. 7.11) can be used to calculate the width of the fusion zone and heat-affected zone, respectively. The width of the fusion zone is the value of *y* at which T_p equals the melting point for a pure metal or the solidus temperature for an alloy. The width of the heat-affected zone is determined by the value of *y* that yields a T_p equal to the pertinent transformation temperature (e.g., recrystallization temperature, austenitizing temperature).

Solidification Rates

The rate at which weld metal (or ceramic) solidifies can have a profound effect on its microstructure, properties, and response to postweld heat treatment. The solidification time, S_t, in seconds, is given by

$$S_t = \frac{LH_{net}}{2\pi k \rho C (T_m - T_0)^2} \tag{7.12}$$

where *L* is the heat of fusion in joules per cubic meter and *k* is the thermal conductivity of the base material in joules per meter second Kelvin. ρ, *C*, T_m and T_0 are as before. S_t is the time elapsed from the beginning to the end of solidification (in seconds).

The solidification rate, which is derived from the solidification time, helps determine the nature of the growth mode (with temperature gradient) and the size of the grains or the coarseness or fineness of the microstructure (e.g., dendrite arm spacing).

Cooling Rates

The final metallurgical structure of a weld zone (i.e., FZ and HAZ) is primarily determined by the cooling rate from the peak temperature attained during the welding cycle. The rate of cooling influences the coarseness or fineness of the resulting solidification structure and the homogeneity, as well as the distribution and form of the phases and constituents in the microstructure, of both the fusion zone and the heat-affected zone for diffusion controlled transformations. When competing diffusionless or athermal transformations can occur (as in steels), cooling rate determines which transformation reaction will occur and, thus, which phases or constituents will result. These, in turn, directly determine the properties of the resulting weld.

If cooling rates are too high in certain steels, for example, hard, untempered martensite can result, embrittling the weld outright and adding to the susceptibility to embrittlement by hydrogen. By calculating the cooling rate, one can decide whether undesirable microstructures are likely to result. If so, preheat[3] can be used to reduce the cooling rate. Cooling rate is primarily used to determine the need for preheat.

---

[3]*Preheat* refers to the intentional elevation of the temperature of the base material of a weldment before welding in order to reduce the temperature gradient and the cooling rate in the fusion and heat-affected zones.

For a single weld pass used to make a butt joint between two plates of equal thickness, where the plates are relatively thick (i.e., requiring more than six passes[4])

$$R = \frac{2\pi k(T_c - T_0)^2}{H_{net}} \qquad (7.13)$$

where R is the cooling rate at the weld centerline (in Kelvin per second), k is the thermal conductivity of the material (joules per meter second Kelvin), T_0 is the initial plate temperature (in Kelvin), T_c is the temperature at which the cooling rate is calculated (in Kelvin), and where H_{net} is given by Eq. 7.2, using Eq. 7.3.

If the plates are relatively thin, requiring fewer than four passes:

$$R = 2\pi k\rho C(h/H_{net})^2(T_c - T_0)^3 \qquad (7.14)$$

where h is the thickness of the base material (in meters), ρ is the density of the base material (kilograms per cubic meter), C is the specific heat of the base material (joules per kilogram Kelvin), ρC is the volumetric specific heat (in joules per cubic meter Kelvin). Increasing the initial temperature, T_0, i.e., applying preheat, decreases the cooling rate.

7.3 CONSIDERATIONS IN THE FUSION ZONE

7.3.1 General Description

The processes of melting and solidification that take place during fusion welding are keys to the structure and properties of the welds that result. The features of fusion welds that have to be taken into consideration in the fusion zone itself are (1) the weld pool contains impurities or contaminants; (2) dilution of filler by base metal occurs; (3) there is considerable turbulence from convection and, therefore, good mixing in the molten zone; (4) the volume of the molten material is small compared to the volume of the overall base material which acts as a mold; (5) the composition of the molten zone and the mold are similar (or at least are compatible); (6) there are large temperature gradients across the melt, giving rise to stresses and nonequilibrium transformations; (7) since the heat source moves, weld solidification is a dynamic process, dependent on the welding speed; and (8) in high energy welds or multipass welds in which the base material is preheated, temperature gradients and, hence, solidification behavior are affected.

This section takes a look at some of the more important considerations in the fusion zone of a weld, including weld pool composition, weld pool size and shape, and solidification for pure materials and alloys, respectively.

---

[4]It is often necessary to deposit weld metal in more than one step or *pass*. When this is the case, welding is said to be *multipass*.

7.3.2 Weld Pool Composition

The composition of the weld pool is affected by many factors, including the composition(s) of the base material(s), the extent of base material melting or dilution, filler composition, cleansing or fluxing effects of molten slags, and degree of shielding and effects of dissolved gases. Figure 7.5 illustrates some of the various contributions to the composition of a fusion weld pool.

High temperatures during fusion welding cause chemical reactions to occur between molten weld metal and the surrounding atmosphere, unless proper protection is provided. Ductility and toughness are degraded by the formation of brittle nonmetallic compounds or inclusions. Fatigue strength, especially, is also degraded by entrapped gas porosity. Various techniques can be used to protect the weld pool during welding through solidification and early cooling, including (1) inert shielding gases in GTAW, PAW, GMAW, and ESW; (2) molten slag in SAW and ESW; (3) generated gas and slag in SMAW and FCAW; (4) vacuum in EBW; and (5) self-protection in resistance welding processes. Different protection systems provide different degrees of weld metal protection.

Despite all attempts to protect the weld pool from undesired reactions, its composition is still subject to alteration by one or more of absorbed gases (even from the

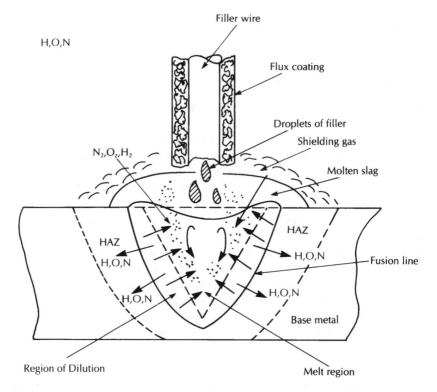

Figure 7.5   The various contributions to the composition of a fusion weld pool.

shielding gases), metallurgical refinement by slag-metal reactions, and dilution by melted base material.

Absorbed Gases

Absorption of gases from surrounding air, from decomposed or dissociated contaminants (e.g., paints, oils, greases, organic solvents, and water), or from the shielding gases themselves are one problem. Absorbed gases alter the composition of the weld metal in the fusion zone, so can affect the structure and properties of the resulting weld. For example, dissolved gases can stabilize certain phases, as dissolved nitrogen stabilizes austenite in steels or oxygen stabilizes the alpha phase in titanium alloys. In excess, absorbed gases can lead to the formation of bubbles in the melt and porosity in the resulting solid. Absorbed gases can also react to form unwanted, and often brittle, compounds.

The three most common gases that can be absorbed by molten welds are nitrogen, oxygen, and hydrogen. All three dissolve approximately in accordance with Sievert's law for diatomic gases:

$$K = [N]/P_{N_2 \text{gas}}^{1/2} \qquad (7.15)$$

where K is the equilibrium constant, $[N]$ is the concentration of nitrogen dissolved in liquid metal (in percentage of weight), and P is the pressure of nitrogen gas in the atmosphere (in atmospheres). In words, this law states that the amount of a diatomic gas that dissolves in a molten metal is proportional to the square root of the gas's pressure in the surrounding atmosphere. In fact, the actual amount of dissolved gas is almost always greater than predicted by Sievert's law as there is some additional absorption of the nacent and ionic forms of the gases.

Nitrogen is usually absorbed from air, of which it constitutes 78%, but can also come from nitrogen-containing shielding gases. From either source, nitrogen either goes into solution or reacts to form nitrides. Nitrogen can reduce toughness and ductility when present as brittle, often needlelike or acicular nitrides. Even when nitrides do not form and the nitrogen simply dissolves, it can act as an potent interstitial solid solution strengthener, can stabilize certain phases (such as austenite in steels), and can increase susceptibility to hot cracking by forming low-melting eutectics. If the concentration of dissolved nitrogen exceeds its solubility limit in the base metal because of temperature and liquid-to-solid state changes, it can produce porosity. Figure 7.6 shows the solubility of nitrogen in iron as a function of temperature. The pronounced drop in solubility with solidification gives rise to the formation of porosity. This behavior is typical of most diatomic gases.

Oxygen can be absorbed from the air (of which it constitutes 20%) under improper shielding, as the result of being present in excess in oxy-fuel gas welding, or from oxygen-containing shielding gas (as in active GMAW or metal-active gas or MAG welding). It can react with the solute or solvent metals to form oxide inclusions and lead to reduced toughness and ductility; it can oxidize carbon in steels and can lead to softening through decarburization; it can cause interstitial solid solution strengthening; it can dissolve and stabilize a particular phase (e.g., alpha in Ti alloys); or it can

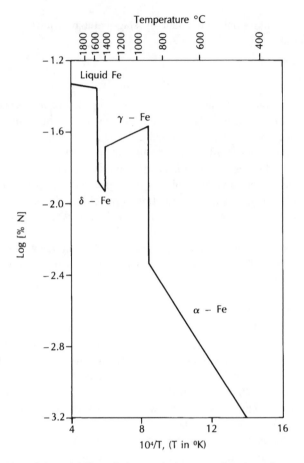

Figure 7.6   Plot of the solubiilty of nitrogen in iron as a function of temperature. (Reprinted with permission after Sindo Kou, *Welding Metallurgy*, John Wiley, 1987, p. 66.)

cause porosity upon solidification as solubility decreases with decreasing temperature and with the phase change.

Finally, hydrogen can be absorbed from partial combustion of fuel gases; decomposition of coatings (e.g., cellulosics); dissociation of hydrated oxides, adsorbed moisture, or hydrocarbon contaminants; or from dissolved hydrogen in the workpiece, from a shielding gas, or from water in the air, in a flux coating or slag or on the workpiece. Once again, hydrogen can react to form embrittling hydrides (as in titanium) or can lead to formation of porosity because of decreasing solubility in the liquid phase with decreasing temperature in many metals and a discontinuous decrease in solubility in the solid phase. Even more insidious than the effects of hydrides and porosity, hydrogen can lead to embrittlement in steels (and certain other metallic alloy phases) in the dissolved state. This phenomenon, known as *hydrogen embrittlement*, becomes a risk in steels when three necessary conditions are present: (1) there is a source of hydrogen; (2) there is a tensile stress (either applied or internal); and (3) there is a susceptible microstructure, such as martensite. The result is cracking that occurs after solidification, in the region of the susceptible microstructure (i.e., fusion, partially melted, or heat-affected zone), when the weldment is cool. For these reasons, hydrogen embrittlement is also referred to as *delayed cracking* and *cold cracking*.

It is especially important to avoid hydrogen embrittlement or cracking at all costs. The best way is to reduce the content of hydrogen by one or more of the following: (1) avoid hydrogen-containing shielding gases or coatings; (2) clean and dry workpieces, electrode coatings, fluxes and wires; (3) postweld heat treat, or bake, to cause hydrogen to diffuse out of the workpiece (sometimes using vacuum assists); or (4) allow hydrogen bubbles to escape from the molten weld pool by stirring. It is also prudent to prevent the formation of untempered martensite in susceptible steels by keeping cooling rates low by preheating.

Metallurgical Refinement

Reactions between the slag formed by melting fluxes from coatings (e.g., in SMAW), cores (e.g., in FCAW) or other sources (e.g., in SAW or ESW) and the molten weld metal helps control the weld composition. Unwanted, dissolved gases like oxygen are removed. The interested reader is referred to other sources (e.g., Reference 2) for a treatment of slag-metal reactions in fusion welding.

Dilution

The composition of the weld pool of a fusion weld that uses a filler can be altered by melting the base material. This phenomenon is called *dilution*, and its source is shown in Figure 7.5. Dilution becomes greater as the heat input of a particular fusion welding process is increased but is also a function of the particular process, the joint design, and the welding procedure or technique.

7.3.3 Weld Pool Size and Shape

The size and, especially, the shape of the melt region in a fusion weld affects the mechanics and kinetics of solidification and, therefore, the structure and properties of the resulting weld. The shape and, especially, the size of the weld pool also affect the thermally induced stresses that act on the weld, leading to the formation of defects or residual stresses or distortion. The shape of the melt, although not necessarily its physical size, is a function of material, welding speed, and welding power (i.e., voltage times current) and can be obtained from the isotherms predicted by the solutions to the heat flow equation. Weld pool convection as affected by process welding parameters and chemistry also has a strong effect on the shape of the melt. Figures 7.7, 7.8 and 7.9 show the effect on weld pool size and shape caused by changes in material (i.e., thermal properties), welding speed, and plate thickness, respectively.

The specific effects of weld pool shape on solidification are described in Section 7.3.6.

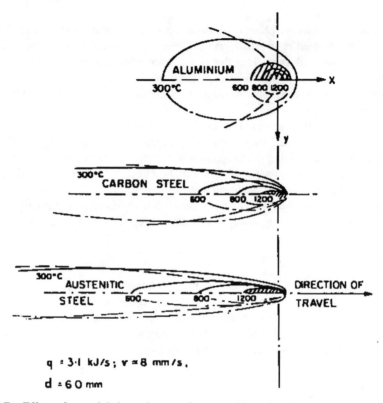

Figure 7.7   Effect of material thermal properties on weld pool and HAZ size and shape. (Reprinted with permission from T. G. Gray *et al.*, *Rational Welding Design*, London: Newnes-Butterworth, 1975.)

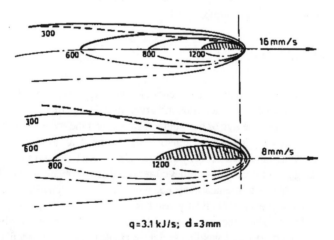

Figure 7.8   Effect of welding speed on weld pool and HAZ size and shape. (Reprinted with permission from T. G. Gray *et al.*, *Rational Welding Design*, London: Newnes-Butterworth, 1975.)

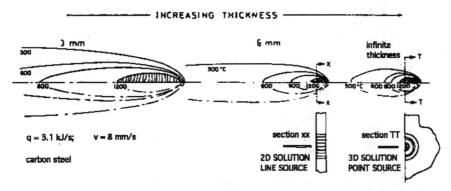

Figure 7.9 Effect of weldment thickness on weld pool and HAZ size and shape. (Reprinted with permission from T. G. Gray *et al.*, *Rational Welding Design*, London: Newnes-Butterworth, 1975.)

7.3.4 Solidification in Pure Metals (or Crystalline Solids)

Many pure materials, including all metals and most ceramics, form a crystalline solid phase (or phases) from a liquid phase when the temperature is lowered sufficiently.[5] The driving force for solidification is the reduction of the volume (or Gibbs') free energy, G_v, with the crystalline aggregate of atoms in the solid state having a lower Gibbs' free energy, G_S, than the same number of atoms in the liquid state, G_L. The volume free energy for the liquid (G_L) and the solid (G_S) phases of a pure metal or crystalline ceramic are shown by the curves in Figure 7.10. The solid phase forms from the chance arrangement of atoms into a regular, crystalline array with long-range order versus the random arrangement typifying the liquid, where only short-range, if any, order is found. These clusters of atoms, arranged in a crystalline state, are called *nuclei*, and the process of their formation is called *homogeneous nucleation*. Solidification occurs homogeneously when

$$\Delta G_v = G_S - G_L \text{ is negative} \tag{7.16}$$

This occurs when there is supercooling below the temperature, T_E, at which the liquid and the solid phases have equal free energies (i.e., $G_L = G_S$) and can co-exist in equilibrium (see Figure 7.10).

Newly forming nuclei become stable and grow in size by adding additional atoms to the solid crystalline aggregate if the size of the nuclei reach or exceed a critical size r_c.[6] At this size, the sum of the volume free energy, G_v (which is decreasing as the size of the nucleus increases further by $4\pi r^3 \Delta G_v / 3$) and the surface free energy,[7] γ (which is

---

[5]Crystalline phases form if solidification occurs at reasonable rates. At extremely high rates of cooling (e.g., over 10^6°C/sec) the regular atomic arrangement found in crystalline materials can be prevented from occurring, resulting in an amorphous, *glassy* state. A glassy state can occur even in metals if cooling occurs rapidly enough.

[6]The nuclei are assumed to be spherical (since a sphere minimizes the surface area to volume ration), so r represents the radius of the sphere.

[7]The surface free energy arises from the formation of an interface between the liquid and the solid phases.

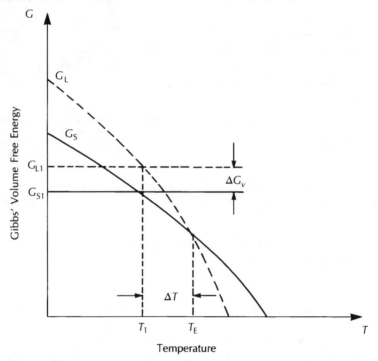

Figure 7.10   Plot of volume free energy curves for liquid and solid phases of a metal showing the net volume free energy change upon melting or solidification.

increasing as the size of the nucleus increases further by $4\pi r^2\gamma$) reduces more and more with further growth. This is shown in Figure 7.11. Under the condition that the nuclei reach the critical size r_c, growth of the nuclei occur dendritically,[8] as equiaxed grains.

The number of atoms required to form a nucleus of a stable size (i.e., with a radius greater than r_c) can be reduced, and growth can thus occur more easily, if a substrate is present on which the nuclei can form. The substrate helps reduce the surface free energy, and the process is called *heterogeneous nucleation*.

When a substrate, B, is present, atoms can begin to form a cluster with the structure and volume free energy of the solid, creating new interfaces between the solid and the substrate and between the solid and the liquid. As shown in Figure 7.12, the sum of the energies of these new interfaces must equal or be less than the energy of the liquid-substrate interface:

$$\gamma_{BL} = \gamma_{BS} + \gamma_{SL} \cos \theta \tag{7.17}$$

where θ is the angle of wetting or contact angle of the solid on the substrate. This relationship, known as Young's equation, implies that if there is *any* wetting, so that $\theta < 180°$, the number of atoms required to form a nucleus with a radius r_c is reduced from

---

[8]*Dendritic growth* refers to growth that proceeds in certain preferred crystallographic directions, forming treelike primary and secondary branches or arms.

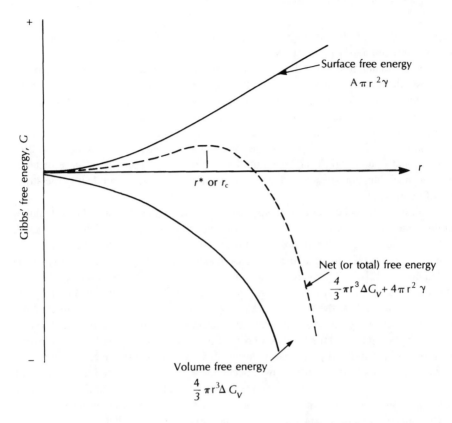

Figure 7.11 Total free energy during homogeneous nucleation.

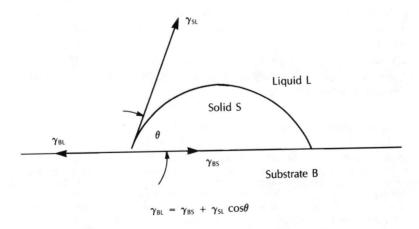

$$\gamma_{BL} = \gamma_{BS} + \gamma_{SL} \cos\theta$$

Figure 7.12 Schematic of wetting leading to heterogeneous nucleation.

the number required for homogeneous nucleation (where $\theta = 180°$), and nucleation occurs heterogeneously more easily.[9]

In real material systems, heterogeneous nucleation almost always tends to occur during solidification, with a mold wall, the unfused portion of the weld adjoining the fusion zone, or, even, unmelted impurities (i.e., inclusions) acting as the substrate. When the substrate is similar in composition and crystalline structure to the solid that is trying to form (as in welding), nucleation occurs heterogeneously and growth takes place *epitaxially*. In *epitaxial growth*, the solidifying material reproduces the crystalline structure and orientation of the substrate. During welding, grains in the base material serve as the nucleation substrates for growth in the fusion zone, reproducing the structure of the base material.

For a pure metal or ceramic to solidify, two criteria must be met: (1) the metal or ceramic must supercool to below T_E, the equilibrium melting point; and (2) the latent heat of fusion released during solidification must be dissipated either by conduction through (a) the solid to the much cooler base material, or (b) to the surrounding supercooled liquid until $T = T_E$, where T is the instantaneous temperature of the solid–liquid interface. The two ways in which these criteria can be met are shown in Figure 7.13a and b. In Figure 7.13a, the temperature at the solid–liquid interface is infinitesimally below T_E, decreasing with distance from the interface back into the solid, and increasing into the liquid toward the weld centerline. For such a situation, with a positive temperature gradient in the liquid, solidification occurs by nucleating heterogeneously on the base material substrate, with growth occurring epitaxially into the liquid as a plane at the rate at which weld heat and latent heat are withdrawn through the solid. Such *planar growth* results in the presence of a positive temperature gradient into the liquid because any protuberance on the solid–liquid (S–L) interface that extends past the region of supercooling into the liquid above T_E remelts.

When there is a negative temperature gradient into the liquid, as in Figure 7.13b, any protuberance that forms on the solid–liquid interface extends into the volume of supercooled liquid and grows preferentially until T at the S–L interface rises to T_E or above because of the release of latent heat. In this instance, *dendritic growth* occurs, with dendritic spikes and side-branches[10] forming heterogeneously and growing epitaxially. Latent heat is dissipated into both the liquid and the solid. Only one or the other of these two solidification growth modes, i.e., planar or dendritic, can occur in pure crystalline solids.

---

[9]A portion of a sphere (i.e., a spherical cap) always has a smaller volume, and requires fewer atoms, than a full sphere. Thus, heterogeneous nucleation always occurs more easily than homogeneous nucleation. The extreme cases are where there is absolutely no wetting, i.e., $\theta = 180$ degrees, or where there is complete (or total) wetting, i.e., $\theta = 0$ degrees. In the first case, the nucleus is completely spherical and to grow must have a radius r_c, so nucleation is homogeneous. In the second case, the volume of the spherical cap approaches zero (being a flat film of zero thickness), with $\theta = 0$ degrees. Here, nucleation occurs heterogeneously, and very easily, with very few atoms needed to form a critical cluster.

[10]Side branches form because the primary spike establishes a new solid–liquid interface and a similar temperature gradient situation.

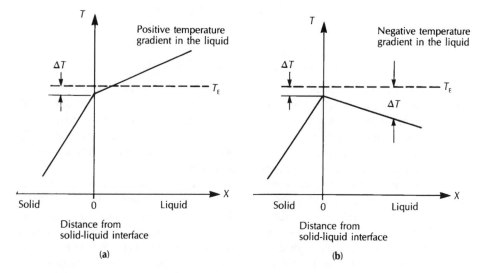

Figure 7.13   Two ways of satisfying the criteria for solidification of a pure crystalline solid: (a) planar growth, latent heat flows into cooler solid; (b) dendritic growth, latent heat can flow into cooler liquid.

7.3.5 Solidification in Crystalline Alloys

Generally, most of the metals and many of the ceramics of engineering interest are actually alloys since they consist of two or more components, including those added intentionally to develop properties and those that are present as impurities that cannot be removed economically. Solidification of alloys is more complex than for pure materials because of the requirement for *solute redistribution.* This redistribution leads to solute buildup in the liquid at the solid–liquid interface and, under nonequilibrium solidification, microsegregation in the resulting solid. As a result of the buildup of solute in the liquid at the S–L interface, solidification modes are altered in alloys versus pure materials. Furthermore, microsegregation in the solid arising from nonequilibrium solidification may lead to increased susceptibility to solidification defects.

Consider a hypothetical binary alloy of solute B in solvent A, having the equilibrium phase diagram shown in Figure 7.14. For an alloy of nominal composition C_0, upon cooling from the melt, solid of composition C_1 begins to form at T_L, with solute being rejected into the liquid because of the alloy's distribution coefficient, k, where $k = C_L/C_S$ at any temperature. As the temperature continues to fall (at an infinitesimally slow rate to allow equilibrium to be established), the compositions of both the remaining liquid and the solid formed at any lower temperature increase and adjust by diffusion to preserve equilibrium and homogeneity. According to equilibrium, at any intermediate temperature, T_i, the entire volume of solid present would have a composition C_2, determined by the intercept of the T_i isotherm or horizontal tie line with the solidus. The amount of each phase present at any time (say, the time the temperature reaches T_i) can be calculated from the *lever law* at T_i:

$$\% \text{ Solid} = [(C_3 - C_o)/(C_3 - C_2)] \times 100 \tag{7.18}$$

and

$$\% \text{ Liquid} = [(C_o - C_2)/(C_3 - C_2)] \times 100 \tag{7.19}$$

At just above T_s, the composition of the solid would be slightly to the left of C_o and an infinitesimal volume of liquid with a composition slightly to the left of C_4 would be present in equilibrium. If cooled to T_s at an infinitesimally slow rate, these two phases would interact and by diffusion convert the entire alloy to a solid of composition C_o, i.e., the starting or nominal composition. Obviously, equilibrium of this sort never occurs, and nonequilibrium leads to certain deviations.

For an alloy to solidify, the same two criteria required for pure materials to solidify must be met, plus one additional criterion: The metal or ceramic alloy must supercool to below the liquidus temperature for the particular alloy composition, *and*, for solidification to proceed, the latent heat of fusion released during solidification must be dissipated by conduction through the solid to a heat sink or by conduction to the surrounding supercooled liquid until $T = T_{\text{liquidus}}$, where T is the instantaneous temperature of the solid–liquid interface. In addition, however, the solute(s) present must be redistributed between the coexisting solid and liquid phases at the solid–liquid interface. Satisfying the first criterion is a thermodynamic problem; satisfying the second criterion is a heat transfer problem; and satisfying the third criterion is a mass transport problem.

How the required redistribution of solute occurs depends on how much mixing occurs in the liquid weld pool, which is usually a lot, and how fast diffusion occurs in the newly formed solid, which is usually limited.

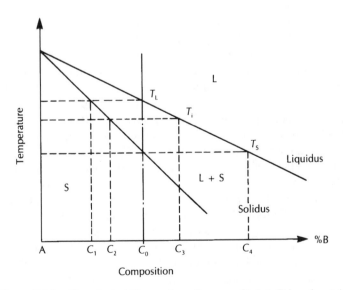

Figure 7.14   Hypothetical binary phase diagram of solute B in solvent A.

There are four boundary conditions that can define solidification of alloys, of which the second and third best represent welding: In Case 1, equilibrium is maintained at all times, so there is no composition gradient (or $dc/dx = 0$) in both the liquid and the solid. In Case 2, complete mixing takes place in the liquid, so $dc/dx = 0$, and no diffusion takes place in the solid, so there is no change in composition with time (or $dc/dt = 0$). In Case 3, there is no mixing in the liquid, so there is a composition gradient, dc/dx; changes in the liquid composition can occur as the result of diffusion, however, but no diffusion takes place in the solid, so $dc/dt = 0$. In Case 4, there is no mixing or diffusion in the liquid and no diffusion in the solid. The liquid is supercooled to a glassy state.

Case 1 is the case under equilibrium, which was described at the beginning of this section. Solidification under equilibrium conditions essentially never occurs and certainly does not occur after welding, so this case will not be considered further. Case 4 represents the situation in *splat cooling*, in which a molten metal, for example, is cooled so rapidly (often, by dropping, or splatting it onto a cold spinning plate) that atoms are unable to arrange themselves into the desired crystalline structure but, rather, remain in an amorphous or glassy state. This case is not pertinent here. Cases 2 and 3 border the situation in real welding, where there is some stirring or mixing of the weld pool because of various convection sources and little or no diffusion in the solid state because of the rapid cooling rate. These two cases will, thus, be considered in detail.

During the nonequilibrium solidification of alloys, solute redistribution occurs somewhere between Cases 2 and 3, with two major consequences. First, the absence or low degree of diffusion in the solid leads to microsegregation, or *coring*. The solid that forms first is least rich in solute, and the solute concentration increases in later and later formed solid. The effect is to create a cored microstructure, having layers of increasing solute concentration as solidification and growth proceed. As shown in Figure 7.15, the first solid forms at T_1 and is of composition C_1, as under equilibrium. At some lower

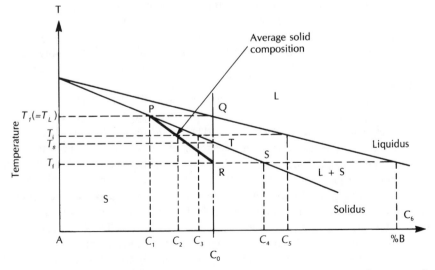

Figure 7.15   Nonequilibrium solidification of alloys under case 2, leading to microsegregation.

temperature, T_i, however, the composition of the solid that is forming is C_3, while the average composition of all of the solid that has formed is somewhere between C_1 and C_3 (as shown by the heavy line). Rather than ceasing at T_s (as under equilibrium), solidification continues to T_f, where the composition of the solid that is being formed last is C_4, but the average composition (given by the intersection of the heavy line with the alloy's nominal composition C_o) is C_o. At this point, the need for material balance is satisfied.

Obviously, the effect of solidification in which there is no diffusion in the solid is to cause solidification (and subsequent melting upon reheating) to occur well below the temperature expected under equilibrium. The resulting microsegregation significantly increases the susceptibility to the formation of defects during solidification, or hot cracks, and can lead to unexpected, localized melting (i.e., liquation) at grain boundaries upon heating in service or as a result of heat effects upon deposition of subsequent welding passes.

The second effect of nonequilibrium solidification in alloys arises from the incomplete mixing that takes place in the liquid weld pool, just ahead of the solid–liquid interface, where solute is being rejected from the newly forming solid into the liquid. Since mixing is incomplete, a solute concentration profile tends to build up in front of the advancing interface, leading to the phenomenon of *constitutional supercooling*. Constitutional supercooling is supercooling relative to the gradient that exists in the liquid as the result of changes in the effective liquidus temperature of the liquid caused by its constitution or composition. As shown in Figure 7.16, the buildup of solute concentration immediately adjacent to the advancing solid–liquid interface causes the effective liquidus of the liquid here to be low compared to the effective liquidus of liquid further removed from the S–L interface and less enriched by rejected, but unmixed, solute. If the actual temperature gradient that exists in the liquid causes the actual temperature to be below this effective liquidus, solidification will occur.

Just as in the pure material, if the temperature gradient in the liquid is negative, dendritic growth mode will persist. If the gradient in the liquid is positive, however, unlike the case of the pure material, there can be some supercooling caused by the constitution of the liquid just in front of the S–L interface. So, rather than simply a planar growth mode, there can be other growth modes, such as *cellular, cellular dendritic* and *columnar dendritic* .

As a consequence of solute redistribution during the solidification of an alloy, composition gradients can and do build up in the incompletely mixed liquid ahead of the advancing solid–liquid interface in a weld and in the resultant solid, where cooling is too rapid for much diffusion to occur. As a result of the buildup of solute in the liquid, the liquid immediately adjacent to the S–L interface can be supercooled relative to the actual temperature gradient in the liquid, even if that gradient is positive. This constitutional supercooling gives rise to additional growth modes besides planar, including cellular, cellular dendritic, and columnar dendritic. For a negative temperature gradient in the liquid, dendritic growth occurs, just as in the pure material. The significance of the existence of different growth modes is that their structure, and their effect on solute distribution in the resulting solid (i.e., microsegregation), can be intentionally varied and

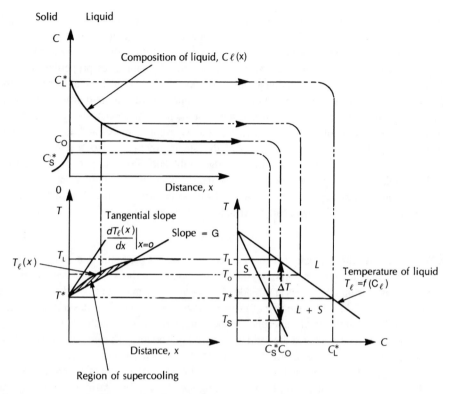

Figure 7.16 Nonequilibrium solidification of alloys under case 3, leading to constitutional supercooling in the liquid.

controlled by properly selecting welding parameters to effect cooling rates (see Section 7.2). These various structures, in turn, affect the weld properties.

7.3.6 Competitive Growth in Welds

Besides the effects that welding parameters can have on the resulting fundamental growth modes that can occur in pure materials and alloys, welding parameters, especially welding speed, can also affect resulting structure through effects on growth direction. All crystalline materials solidify or grow along certain preferred (close-packed) crystallographic directions. These directions are along cube edges for face-centered or body-centered cubic metals and along diagonals of the base hexagon for hexagonal close-packed metals. Similar preferred growth directions exist for the various crystal structures found in ceramics.

At any instant of time during solidification, the operative growth direction will be that which is most closely aligned with the steepest temperature gradient in the weld pool. This results from the necessity for quickly dissipating the latent heat of fusion liberated at the advancing solid–liquid interface and leads to the phenomenon of *compet-*

itive growth. Competitive growth occurs for pure crystalline materials or their alloys. It also occurs whether nucleation occurred homogeneously or heterogeneously.

Because welding velocity has the most pronounced effect on weld pool shape, this parameter strongly affects competitive growth and thereby influences properties. When a fusion welding heat source is stationary, the weld pool is circular in shape at its surface and hemispherical in three dimensions. For this pool shape, growth occurs uniformly from the periphery of the weld, radially toward the center of the pool. As soon as the welding source is moved, the weld pool becomes elongated, first becoming elliptical and, then, becoming teardrop shaped.[11] For an elliptical weld pool, the rate of growth of the solid at the S–L interface varies with location around the trailing edge of the pool. For constant velocity v, if the welding voltage and current are held constant, the shape of the weld pool remains constant.

Because, on average, growth of the solid will occur parallel to the maximum temperature gradient, the rate of growth can be written

$$R = v \cos \phi \qquad (7.20)$$

where v is the welding velocity (meters per second or inches per minute), and ϕ is the angle between the maximum temperature gradient in the liquid and the welding direction. For an ellipse, ϕ varies from 90 degrees at the side of the pool to 0 degrees at the weld centerline (as shown in Figure 7.17. Thus, the rate of growth varies from zero at the sides of the pool to v (i.e., $v \cos 0° = v$) at the weld centerline. So the maximum rate of growth and the maximum evolution of latent heat occur at the weld centerline. For a teardrop shaped pool, along the tail the angle ϕ is constant and a minimum, so the growth rate is constant and minimum.

The typical effects of weld pool shape on the direction of growth of columnar grains in an alloy is shown in Figure 7.18. Higher welding speeds cause increasingly more elliptical and then teardrop shaped puddles, increasing the tendency toward centerline microsegregation and cracking, due to grain impingement at the centerline for these higher welding speeds.

7.4 CONSIDERATIONS IN THE PARTIALLY MELTED ZONE

The *partially melted zone,* or PMZ, is the area immediately adjacent to the fusion zone (FZ) in welds made in alloys where the peak temperatures reached during welding were between the liquidus and the solidus for the alloy or above its effective solidus if nonequilibrium has resulted in pronounced previous microsegregation of solute. This zone is subjected to localized melting principally along grain boundaries, because of microsegregation and, to a lesser extent, in the grain interiors. Pure metals and ceramics do *not* exhibit partially melted zones.

For a typical eutectic alloy, the PMZ is located in a region surrounding the fusion zone, where the temperature rises to between the equilibrium eutectic temperature, T_E, and the equilibrium liquidus temperature, T_L. Under nonequilibrium, the values of T_E

---

[11]The *teardrop shape* occurs at some critical velocity where the rate of latent heat evolution at the weld centerline exceeds the ability of the system to dissipate it. At this point, the rate of growth parallel to the welding direction decreases and the weld pool assumes a teardrop shape.

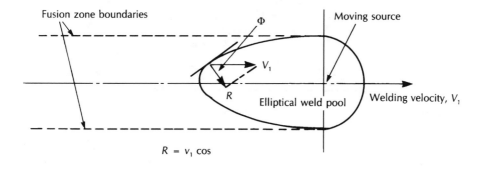

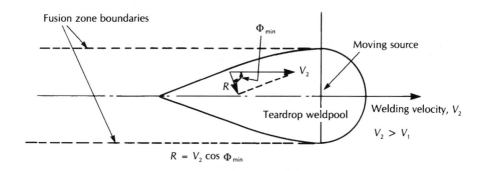

Figure 7.17 The effect of weld pool shape on solidification growth rate.

and T_L are higher during heating (i.e., at the leading edge of the weld pool) and are lower during cooling (i.e., at the trailing portion of the weld pool). These effects are shown in Figure 7.19.

Partially melted zones in welds can suffer from hot cracking due to microsegregation, loss of ductility due to localized melting, and hydrogen cracking due the presence of untempered martensite resulting from rapid cooling of segregation-enriched (and more hardenable) regions. These adverse effects can be minimized in any of several ways: first, by keeping the welding heat input as low as possible to minimize the extent of this region; second, by minimizing restraint on the weldment to reduce tensile stresses on the weld that could cause cracking; third, by selecting base materials carefully to avoid residual elements; and fourth, by choosing a filler with a melting point higher than the base material so that solidification occurs in the PMZ before the FZ.

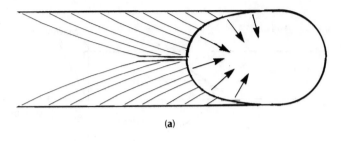

(a)

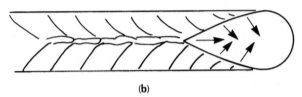

(b)

Figure 7.18   Typical growth directions as a function of weld pool shape. (a) Low v; (b) high v.

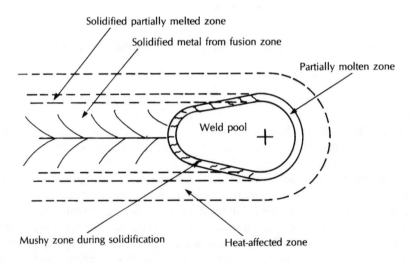

Solidified partially melted zone

Solidified metal from fusion zone

Partially molten zone

Weld pool

Mushy zone during solidification

Heat-affected zone

Figure 7.19   The location of the PMZ around a moving weld pool during welding.

7.5 CONSIDERATIONS IN THE HEAT-AFFECTED ZONE OF WELDS

7.5.1 General Description

Metals and ceramics and alloys of engineering interest are generally strengthened by one or more of the following basic mechanisms: (1) work hardening or cold working; (2) solid solution strengthening or alloying; (3) precipitation hardening or aging; (4) transformation hardening; and (5) dispersion strengthening.

The heat of welding (whether sufficient to cause fusion or simply to lower the yield strength to facilitate nonfusion welding) often significantly alters the properties of a material in the heat-affected zone surrounding the fusion zone (or partially melted zone, if there is one) of such materials. Especially affected are work-hardened, precipitation-hardened, and transformation-hardened materials. Relatively unaffected are solution-hardened and dispersion-strengthened materials, except for some grain growth effects. In ceramic materials, thermal stresses from temperature gradients or from rapid temperature excursions (i.e., shock) can lead to formation and/or growth of microflaws, such as cracks.

A complete understanding of welding demands an understanding of the effects of welding on the heat-affected zone, since there is so much heat involved with most processes.

7.5.2 Work-Hardened Materials—Recovery, Recrystallization, and Grain Growth

When a metal is plastically deformed below a certain temperature (approximately 0.4 to 0.5 of the absolute melting point of the metal or alloy), existing one-dimensional, line crystal imperfections known as dislocations are set in motion, new dislocations are generated, and these dislocations interact to form tangles that hinder the further motion of these and other dislocations. This, in turn, restricts further plastic deformation. In this way, the metal or alloy is strengthened by cold working and the mechanism is known as *work hardening*. Most of the energy expended in plastically deforming and work hardening appears as heat, but some fraction is stored in the material as strain energy. This creates an energetically unfavorable condition that is metastable at best.

When a work-hardened material is exposed to elevated temperature (i.e., greater than approximately $0.4T_{MP}$, absolute), the deformed grains tend to nucleate[12] new, strain-free grains, a process generally called annealing and specifically known as *recrystallization*. The driving force for recrystallization is the reduction of stored strain

---

[12]This process of nucleation is essentially identical to the process of nucleation that occurs during solidification. A new phase tries to form since it has a lower volume free energy, resulting in a net negative volume free energy change. The creation of a new phase results in the creation of an interface with the parent phase and gives rise to a positive surface free energy. At some critical size, the total of these opposing energies reaches a maximum, and further growth results in a stable structure.

energy, which is released as the new, strain-free grains are formed. With the onset of recrystallization, mechanical properties change drastically: strength decreases, hardness decreases, ductility increases, and toughness increases.

Even before recrystallization takes place, certain changes occur. Specifically, zero-dimensional, point crystal imperfections are reduced and certain properties change. Electrical resistivity decreases as scattering sites to charge carriers are reduced. This process of *recovery* is usually of little concern in welding but is important generally as a predecessor of recrystallization.

The extent of recrystallization increases with increasing annealing temperature and time, and the rate at which recrystallization occurs is greater (and the temperature is lower) for more severely cold-worked materials and for purer materials. During welding, this means that the extent of recrystallization increases as the peak temperature increases as the distance from the fusion zone boundary is reduced (i.e., closer to the FZ).

Once recrystallization is complete, the newly formed, strain-free grains grow in order to reduce the surface energy associated with grain boundaries, which are two-dimensional, planar crystal imperfections. Grain growth also increases with increasing temperature and time at temperature, so is greatest very close to the fusion zone (where the peak temperature is the highest in the HAZ) and as net heat input increases. All materials exhibit grain growth on heating, regardless of whether or not they have been work hardened.

Any effect of work hardening that existed in a material before welding is completely lost in the fusion zone owing to the processes of melting and solidification that occur and is partially lost in the heat-affected zone owing to the processes of recovery, recrystallization and grain growth that occur. A typical effect of weld heat in the HAZ of a work-hardened material is shown in Figure 7.20.

The only way to recover the mechanical properties lost in the heat-affected and fusion zone by welding a work-hardened material is to re-work the weldment. This is often impractical but is not impossible, with rolling, forging, shot-peening, burnishing, and other methods being options.

7.5.3 Precipitation-Hardened Materials—Overaging

Certain alloys can be strengthened and hardened by the presence of a second phase formed by precipitation from a supersaturated solid solution. Most aluminum alloys (e.g., Al-Cu, using Al_2Cu)) and nickel-base superalloys (e.g., Udimet 700, using $Ni_3[Al,Ti]$) are prime examples. The three requirements for an alloy to exhibit *precipitation hardening* are (1) limited solid solubility of the solute in the solvent (e.g., Cu in the α phase in Al-Cu alloys); (2) decreasing solid solubility with decreasing temperature; and (3) an alloy composition that contains less solute (e.g., 4% Cu) than the maximum amount soluble in the terminal solid solution (e.g., 5.6% in α). These requirements are shown being satisfied in Figure 7.21.

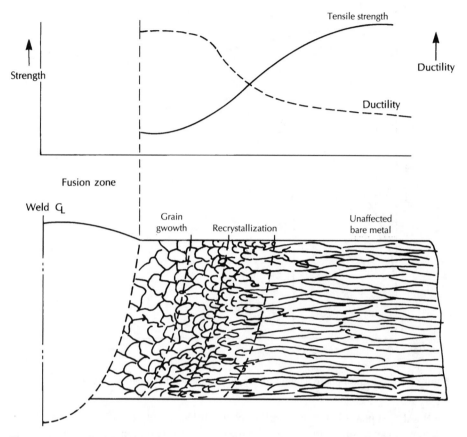

Figure 7.20 The effect of welding heat on the structure and properties of a work-hardened material.

The first step in precipitation hardening an alloy, Al–4% Cu for example (shown by the dashed vertical line in Figure 7.21), is to heat the alloy into the single-phase (α) solid solution region and hold or *soak* for sufficient time to allow diffusion to produce a homogeneous single phase. This process step is called *solution heat treatment.* The second step is to *quench* this solutionized alloy to room temperature or below so that it becomes *supersaturated* with solute (e.g., with Cu). The third and final step is to heat this metastable, supersaturated solid solution to an elevated temperature (e.g., 190°C for Al–4% Cu) to accelerate diffusion kinetics. This treatment—called *aging*—must continue for an optimum period of time so that the strengthening second phase can precipitate uniformly and with the proper and particle shape or morphology and size.

In every case, the precipitate develops progressively, resulting in different degrees of strengthening, with maximum strengthening at some point before *overaging* occurs. For Al–Cu alloys, the progression is

$$\text{supersaturated SS} > \text{GP} > \theta" > \theta` > \theta \ (\text{Al}_2\text{Cu})$$

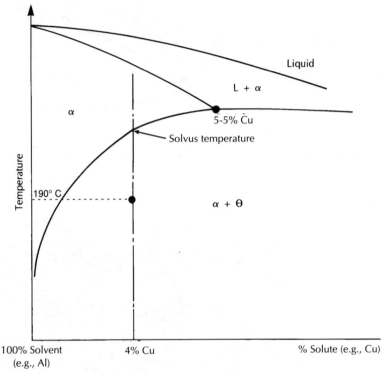

Figure 7.21   Typical phase dagram for a precipitation hardening alloy system (e.g., Al-Cu).

The θ (Al$_2$Cu) phase is the equilibrium phase. It has a body-centered tetragonal (bct) structure and an incoherent[13] lattice with the matrix (α). The strengthening effect of this phase is less than and past optimum. The GP (Guinier–Preston) zones (or GP1's), the θ'' phase (called GP2), and the θ' phase are all metastable, representing progressively more stoichiometric forms of the newly forming phase. Given sufficient time, these phases will progress to the equilibrium θ phase but not necessarily passing through each step.

GP1 and GP2 (θ'') are coherent with the matrix lattice. This results in greater lattice strain (θ'' more than GP1 because of its greater physical size) and produces a larger strengthening effect by more effectively impeding dislocation motion (i.e., the strain fields of the dislocation and the coherent precipitate particles interact). The θ' phase is semicoherent and nucleates heterogeneously on dislocations. As θ' grows in size and increases in amount, coherent strains decrease, and overaging occurs.

The strengthening effect of each of the above precipitation stages is shown in Figure 7.22. Strengthening is greatest for θ'' (GP2) at a maximum concentration. For this phase, coherency and lattice accommodation strain are greatest.

---

[13]To be *coherent*, the atomic lattice of a second phase precipitate (i.e., a daughter phase) must have nearly the same lattice parameters as the parent phase, so that the lattice of the precipitate can distort to match the parent in spacing and orientation. When this is not the case, i.e., there is no matching, the precipitate is said to be *incoherent* with the matrix or parent phase.

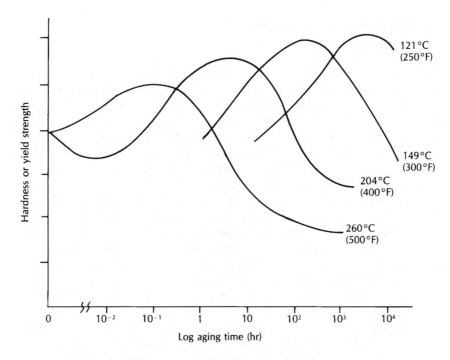

Figure 7.22   Hardness versus aging time for precipitation-hardenable aluminum-copper alloys.

Heating aged alloys to temperatures higher than the solvus[14] temperature for the composition, for example, in the heat-affected zone of a weld, results in resolutionizing, or *reversion,* of all precipitate, regardless of its particular form (i.e., θ, θ′, θ″, or GP1). Heating to above the optimum level for age hardening results in overaging. The GP zones redissolve, or revert, more highly developed precipitates increase in size and progressively lose coherency, and softening occurs throughout the HAZ. Cooling rates after welding are usually high enough to prevent reprecipitation after reversion in and near the FZ.

During postweld artificial aging,[15] different areas of the HAZ respond differently, depending on whether reversion, coarsening, or overaging has occurred. Full aged strength can only be obtained after welding by resolutionizing, quenching, and artificially aging. Since this is often impractical, because of the size of the weldment or adverse effects of high solutionizing temperatures on adjoining structure or on overal structural integrity (e.g., shape distortion), some compromise is needed. Usually, it is best to weld precipitation hardening alloys in the solutionized condition and, then, postweld age.

---

[14]The *solvus temperature* is the temperature at which the last trace of one phase in a two-phase region just disappears upon heating into the single phase region. That is, the solvus line separates one-phase and two-phase regions of solids.
[15]*Artificial aging* is aging that is achieved by intentionally heating the supersaturated solid solution to accelerate precipitation. Natural aging occurs at the temperature to which the alloy is quenched and held (i.e., usually room temperature).

7.5.4 Transformation-Hardened Materials Hardenability

For materials (such as steels) that undergo cooling-rate dependent transformations to hard, strong phases (martensite), welding can alter properties dramatically in both the fusion zone and the heat-affected zone (as well as in the PMZ). Welding exposes steels to more severe conditions than heat treating, so effects on properties can be dramatic. Peak temperatures during welding are higher in the HAZ (e.g., >1400°C near the FZ) than even the most severe heat treatments (e.g., austenitizing at about 900°to 1050°C). The heating rate is very high, and the retention or dwell time is very brief versus heat treatment, thereby accentuating nonequilibrium effects.

The higher peak temperatures of welding versus heat treatment can result in partial melting from previous microsegregation or constitutional liquation[16] of certain second phases; grain coarsening, especially near the FZ, from exposure on both heating and cooling; and reaustenitizing or solutionizing, with attendant phase transformations (e.g., formation of untempered martensite). The high heating rates associated with welding make diffusional transformations (e.g., ferrite + pearlite > austenite) more difficult, increasing effective transformation temperatures (such as lower critical temperatures, A_{C1}, and upper critical temperatures, A_{C3}, in the Fe-Fe$_3$C phase diagram). The effect of heating rates is more pronounced in steels with greater amounts of carbide-forming elements (e.g., V, W, Cr, Ti, and Mo) because these elements have low diffusion rates themselves and hinder the diffusion of carbon as well. Thus, they retard diffusional transformations.

Higher heating rates *with* short high-temperature retention time also can result in the formation of nonhomogeneous austenite, which upon subsequent rapid cooling can cause the formation of localized high hardness (i.e., high carbon) martensite areas. The result is widely scattered microhardness in the HAZ and stress concentrations.

Steels are classified by both their carbon and alloy content, and include mild steels (with less than 0.3% carbon and no intentional alloy additions other than Si and Mn), higher-carbon steels (with >0.30% carbon and no alloy additions), and alloy steels (with specific alloy additions of nickel, chromium, molybdenum, tungsten, or vanadium to enhance certain properties such as corrosion resistance, hardness, or hot or red hardness. The amount of carbon and alloy additions in a steel greatly influences the hardenablility of the steel, i.e., the susceptibility of that steel to form martensite during cooling. Hardenability increases most dramatically with increasing carbon and, to a lesser extent, with most alloy additions (except Co).

Martensite formation in the HAZ, PMZ, or FZ, especially when uncontrolled, increases strength and hardness but at the expense of ductility and toughness. Also, the inherent brittleness of martensite can lead to cracking under cooling stresses and can render the structure prone to hydrogen embrittlement (see Section 7.3.2). The tendency to form undesired martensite in the weld can be offset by preheating the weldment to reduce cooling rates to below the critical cooling rate (see Section 7.2.4).

---

[16]*Constitutional liquation* is melting that is caused during nonequilibrium heating as a result of the diffusion of one component from a second phase particle into the matrix, producing a low-melting eutectic composition at some point.

7.5.5 Sensitization in Corrosion-Resistant Stainless Steels—Weld Decay

Stainless steels are a class of iron-base alloys with high corrosion and oxidation resistance achieved through extensive solid solution alloying. Typically, 12–27% Cr, 1–2% Mn, often 6–22% Ni, and, sometimes, Mo, Nb, Zr, or Ti are added. A small amount of carbon (0.03 to 0.25%) is also present, either deliberately added for strength or as an unavoidable impurity. Stainless steels can be classified into three main categories, based on their primary microstructure, i.e., ferritic, austenitic, and martensitic.

Austenitic stainless steels containing about 0.1% C or more are usually susceptible to intergranular corrosion in the weld heat-affected zone. The phenomenon is known as *sensitization* or *weld decay*. Sensitization is caused by the precipitation of complex $Cr(X)_{23}C_6$ and $Cr(X)C_6$-carbides preferentially at grain boundaries. The diffusion of chromium to the carbon to form the chromium-rich carbides depletes or *denudes* the matrix of chromium in the vicinity of the carbide. If the level of chromium drops below the level required to provide corrosion resistance (i.e., about 14%), this area is rendered anodic compared to the bulk of the alloy, and intergranular corrosive attack of the denuded region can occur. This type of carbide formation occurs most rapidly in the range of temperatures between 600° and 850°C, known as the *sensitization range*.

Sensitization or weld decay is more severe when the alloy contains higher amounts of carbon, the weld heat input is high, the alloy is cold worked before welding, or there are no alloying additions with greater affinity for the carbon than chromium. Sensitization tends to occur only in a narrow band or temperature range because, at cooling rates greater than a certain level, precipitation cannot occur because the time for diffusion is too short in the sensitization range, and if the peak temperature of a weld thermal cycle is too low, precipitation cannot occur because nucleation does not occur.

Weld decay can be prevented in welded stainless steels in one of several ways. In the most common approach, strong carbide formers, such as Ti (in 321 SS) or Nb (in 347 SS), are added to promote the formation of Ti- or Nb-carbides preferentially to chromium carbides, precluding chromium depletion. This approach is known as *stabilization,* and the resulting stainless steels are called *stabilized grades.* In a second approach, the carbon content of the stainless steel is lowered to less than 0.035%, resulting in a so-called low-carbon grade (e.g., 304L or 316L). In these low-carbon grades, there can never be enough carbide precipitation to cause chromium depletion below the threshold level. A third approach involves heat treating after welding. Here, the weldment is heated to 1000° to 1100°C and soaked at temperature sufficiently long to dissolve any Cr-carbides, restoring the chromium to solution in the matrix. This *solutionizing treatment* is often impractical, however, as either structures are too large to place in heat treating furnaces or the high temperatures required would cause severe distortion.

While most common in austenitic stainless steels, sensitization can also occur in ferritic grades.

7.5.6 Effect of Welding Heat on Solid-Solution- or Dispersion-Strengthened Alloys

The heat of welding has little or no effect on the structure and properties of materials that derive their strength purely from solid-solution strengthening or dispersion strengthening. For solid-solution strengthened alloys, other than the effects of welding on grain substructure and possible microsegregation in the fusion zone, and grain growth in the high-temperature heat-affected zone, there is little adverse effect on properties. There can be some degradation of fracture toughness or fatigue strength, however. In most dispersion-strengthened alloys, the dispersed phase is highly thermally stable, so there is little degradation by peak temperatures in the HAZ, and there may be little effect in the FZ as well, if the dispersoids are sufficiently refractory to survive the melting of the matrix. For less thermally stable dispersoids, degradation can occur from dispersoid coarsening, decomposition, constitution liquidation cracking, or dissolution.

Figure 7.23 shows the patterns of typical hardness traverses across single-pass welds in the basic types of materials, including solid-solution-strengthened, work-hardened, precipitation-hardened, transformation-hardened, and dispersion-strengthened types.

7.6 DEFECT FORMATION IN WELDS

7.6.1 General Description

The mechanical integrity of a weld can be severely compromised by the presence of defects, as such defects almost always act as points of stress concentration, often reduce the effective cross-sectional load-bearing area, and sometimes also degrade the material's properties (especially toughness and/or ductility). Defects in welds can arise from one or more of several sources, including (1) improper joint design or preparation or fit-up (i.e., alignment); (2) inherent base or filler material characteristics; (3) process characteristics; and (4) environmental factors. Defects from these sources can occur anywhere; in the fusion zone, in the partially melted zone, in the heat-affected zone, or in combinations of these.

7.6.2 Joint-Induced Defects

Improper or inappropriate joint design or preparation or fit-up (i.e., alignment) can lead to the following types of defects: (1) lack of complete penetration (from improper design or inappropriate process selection or parameters); (2) mismatch or surface offset (from misalignment); (3) severe distortion (from unbalanced masses or excessive heat input); (4) porosity (from entrapped air or contaminants); (5) shrinkage voids or cracks (from poor fit or restraint); (6) underfill (from poor fit); and (7) excessive dilution (from improper design or process selection).

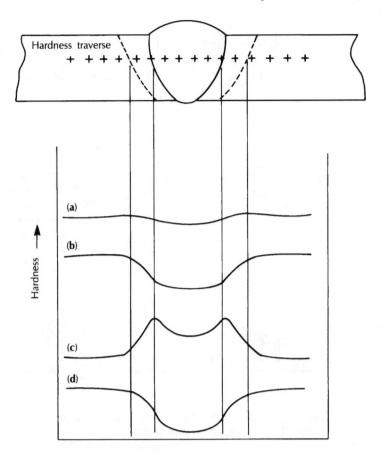

Figure 7.23 Typical hardnesses across single-pass welds in materials strengthened by different mechanisms. (a) Solid solution-strengthened or dispersion-strengthened alloy; (b) work-hardened alloy; (c) transformation-hardenable steel; (d) precipitation-hardened alloy.

Figure 7.24 shows some of the more common joint-induced defects found in welds.

7.6.3 Fusion Zone Defects

One potential defect exclusive to the fusion zone is porosity. As discussed in Section 7.3.2, porosity typically arises from the evolution of trapped or dissolved gases during solidification. These gases can come from improper joint preparation (e.g., poor cleaning and the decomposition of oxides or dissociation of water or hydrocarbon contaminants), the environment (e.g., oxygen, nitrogen, hydrogen, or water in the atmosphere or shielding gas or water dissolved in the coatings or flux-cores of certain consumables), or the base material (e.g., tramp or residual elements).

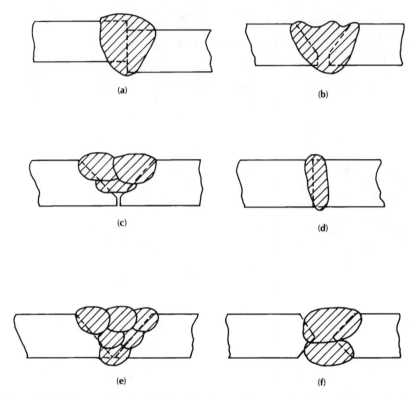

Figure 7.24 Common joint-induced weld defects. (a) Mismatch; (b) underfill; (c) lack of penetration; (d) missed seam; (e) lack of fusion; (f) undercut.

The other major source of defects in the fusion zone is *solidification cracking* also called *hot cracking*. Such cracking is intergranular, occurring along grain or substructure boundaries as the result of microsegregation or exceeding the equicohesive temperature.[17] This form of cracking occurs at the terminal stages of solidification, when the stresses arising from volumetric shrinkage and thermal contraction and acting across the newly formed solid exceed the strength of the almost completely solidified material. The severity of hot cracking increases with both increasing degree of constraint in the weldment (due to fixturing or surrounding structure and joint thickness) and increasing total heat input.

The predominant theory of solidification cracking is that a coherent interlocking solid network separated by essentially thin liquid films is ruptured by contraction stresses. Factors that influence solidification cracking include the freezing temperature range of the alloy, the presence of low melting segregates, the structure found in the fusion zone, the nature of any grain boundary liquids (i.e., their surface tension), the magnitude of contraction stresses, and the degree of restraint imposed on the weld.

---

[17]The *equicohesive temperature* is that temperature above which the grain boundaries become weaker than the bulk structure because of their inherent imperfection.

The wider the difference between the liquidus and the solidus of an alloy, i.e., the wider the freezing temperature range, the greater the cracking tendency. This is because a wider range allows thermal contraction stresses to build to higher levels than narrower ranges, causing tearing of incompletely solidified networks. Alloying elements or impurities like copper, magnesium, or zinc in aluminum or sulfur or phosphorus in steels or nickel aggravate the situation. Related to the width of the freezing range is the final temperature of solidification. Certain residual elements segregate by solute redistribution and form low-melting compounds (e.g., FeS in steels) in grain and substructure boundaries that can tear under contraction stresses.

As always in materials, the structure present in the fusion zone strongly affects the tendency to hot cracking. Coarse structures (such as columnar grains) are more prone to cracking than fine structures (such as cellular grains), because in finer structures the effects of segregates are more spread out, possibly lowering the concentration to below critical levels. Related to this structural effect is the distribution of any liquid present in grain or subgrain boundaries. Liquids that wet the grains, i.e., have lower surface tensions, promote cracking through the formation of a permeating film, as opposed to localized globules or pockets.

Obviously, hot tearing or cracking cannot occur unless a tensile stress is present. Tensile stresses arise during contraction upon cooling and are aggravated by high thermal expansion coefficients, large volume shrinkages upon solidification, and excessive heat input. The more restrained the weld, the higher the stresses that develop during solidification and the greater the likelihood of cracking.

7.6.4 Partially Melted Zone Defects

Hot cracking is the most severe problem associated with PMZs, and, like solidification cracking in FZs, is intergranular. The cause of hot cracking in the PMZ is the combination of grain boundary liquation or melting (caused by diffusion gradient effects from carbides or other second phase compounds during nonequilibrium heating, for example) and the stresses induced by both solidification shrinkage and thermal contraction during welding. Cold cracking from hydrogen embrittlement of martensite formed in the PMZ can also occur, especially as a result of back-filling of incipient cracks by solute-rich, and thus highly hardenable, liquid.

The location of cracks in a weld fusion or partially melted zone tells a lot about their source.

7.6.5 Heat-Affected Zone Defects

Defects in the heat-affected zone must develop in the solid state (i.e., without melting) and usually involve thermally induced stresses, often combined with embrittling phases. Three predominant types of defects are found in HAZs: (1) hydrogen or cold cracks, (2) reheat cracks, and (3) lamellar tears.

Hydrogen cracking, or *cold cracking,* or *delayed cracking,* requires four factors to be present simultaneously: (1) hydrogen must be introduced into or present in the weld metal or base metal to diffuse into the HAZ, (2) high tensile stresses must be present, (3) a susceptible microstructure (such as martensite in steels) must be present, and (4) relatively low temperatures (between $-100°$ and $200°C$) must usually prevail. Sources of hydrogen were described in Section 7.3.2, but the most prevalent sources are water or moisture on parts, in shielding gases, or in fluxes; hydrated oxides; and hydrocarbon contaminants (e.g., oil, grease, paint). High stresses can be induced by high heat input and steep gradients and/or constraints on the weld. These constraints can come from tooling or fixturing that is purposely used to prevent movement or from the inherent rigidity of the surrounding structure. A susceptible microstructure, such as martensite in steels, arises from rapid cooling after welding.

An incubation time is required for crack development while hydrogen diffuses up prevailing stress gradients. For this reason, hydrogen cracking is often called delayed cracking. The exact mechanism of hydrogen cracking is uncertain, but two prominent theories are (1) reduced cohesive strength of the lattice of certain materials (i.e., the Troiano theory) and (2) reduced surface energy of any crack that forms (i.e., the Petch theory). The high hydrostatic pressure developed in the crack is often considered a major contributor as well.

Hydrogen cracking can be prevented by (1) preheating to prevent martensite formation by lowering cooling rates and reduce stresses by lowering the flow stress of the material; (2) postheating to relieve stresses, temper the martensite, and drive out hydrogen (i.e., by baking); (3) using low hydrogen processes and consumables to avoid introducing hydrogen into the weld; (4) using lower strength fillers to help reduce stresses in the HAZ through yielding in the FZ; and (5) using austenitic fillers when welding steels to absorb and hold any hydrogen from the HAZ, innocuously.

Reheat cracking is typically a problem in corrosion- and heat-resisting alloys, especially ferritic steels with chromium, molybdenum, or vanadium and nickel-based age-hardenable alloys. The problem is caused by the formation of complex, brittle carbides at grain boundaries in the HAZ within certain temperature ranges. Cracking occurs under the action of shrinkage and cooling contraction stresses. There are heat-treating procedures that can offset the tendency toward reheat cracking.

Lamellar tearing is the result of a combination of high localized stresses due to weld contraction and low ductility of the base metal in its through-the-thickness direction due to the presence of elongated nonmetallic inclusions or stringers. Tearing is triggered by the fracture of these inclusions. This is a material problem that should be addressed by proper alloy composition control and processing but can sometimes be overcome by proper joint design to orient aggravating stresses in the direction of the stringers, as opposed to across the stringers.

7.7 TESTS OF WELDABILITY

7.7.1 General Description

The suitability of a material to welding can often be assessed using tests that evaluate the susceptibility of that material to defect formation. To be most meaningful and useful, these tests should be representative of the process and the joint design being considered. They principally assess potential problems with the materials being welded or the fillers being considered. Some assess solidification (FZ) cracking susceptibility, some assess PMZ cracking, and some assess HAZ cracking. There are many, many testing devices and methods for assessing weldability.

7.7.2 Solidification Crack Susceptibility Tests

Solidification crack susceptibility tests assess the susceptibility of a material and/or a process to hot crack formation in the fusion zone owning to the existence of unfavorable freezing ranges, low-melting eutectics, grain boundary films, and excessive thermally induced stresses. Three popular tests are the Houldcroft test, the Varestraint test, and the circular patch test. In addition, weld simulators are used, such as the Gleeble.

The Houldcroft test is used predominantly to assess sheet materials. The specimen (shown in Figure 7.25) is free of external constraints, but a progression of slots of varying length across the width of the specimen allows the dissipation of stresses internally. If there is a tendency toward cracking, cracking starts immediately and progresses to where sufficient stress is dissipated by the slots that propagation is arrested. The crack length from the starting edge to the point of arrestment is an index of the solidification cracking tendency.

The Varestraint test applies an augmented strain to a test specimen by bending it over a controlled radius at an appropriate time during welding. Both the amount of strain applied and the crack length (i.e., either total length of all cracks or the maximum crack length) serve as an index of cracking susceptibility. This test is shown in Figure 7.26.

The circular patch test welds progressively smaller diameter plugs, or patches, back into a plate from which they were removed. Cracking, if it is to occur, occurs when the stress from shrinkage exceeds the ability of the material to sustain it.

A much more sophisticated approach for assessing weldability is to use weld simulators and measure hot ductility. These simulators typically program the thermal cycle produced by a particular welding process and measure the hot ductility of specimens of the material to be welded in appropriate temperature ranges. One of the most popular weld simulators is the Gleeble. The Gleeble applies a controlled tensile load (versus strain) to a specimen that is resistance heated to the temperature range of interest to assess hot ductility. The ability to program the Gleeble time-temperature-load cycle allows regions of welds to be expanded to cover the gauge length of the test specimen, reproducing the desired, and questionable, microstructure.

7.7.3 PMZ Crack Susceptibility Tests

The Varestraint, circular patch, and Gleeble tests can also be used to assess crack susceptibility in the PMZ of base materials. When the Varestraint test is used, the length of cracks in the PMZ only is assessed. In the circular patch test, cracking usually occurs 360 degrees at the outer (versus the inner) edge of the weld. In hot ductility tests, such

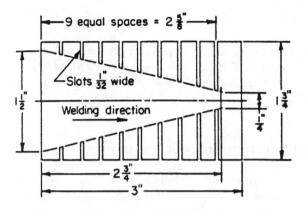

Figure 7.25 Houldcroft cracking test. (Reprinted from J. J. Vagi, *et. al., Weldment Evaluation Methods*, Defense Metals Informatin Center Report 165, 1961.)

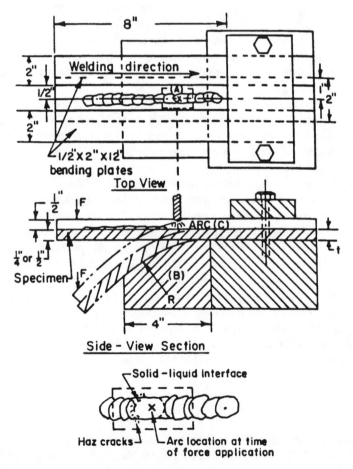

Figure 7.26 Varestraint cracking test.(Reprinted from J. J. Vagi, *et. al., Weldment Evaluation Methods*, Defense Metals Informatin Center Report 165, 1961.)

as those of the Gleeble, cracking is assessed in the temperature range between the liquidus and the solidus.

7.7.4 HAZ Crack Susceptibility Tests

Hydrogen cracking susceptibility is assessed using such tests as the Lehigh restraint test, the RPI Varestraint test, the Lehigh slot welding test, and the implant test, as well as the Gleeble. Lamellar tearing can be assessed using the Lehigh cantilever lamellar tearing test, the Cranefield lamellar tearing test, and the tensile lamellar tearing test.

The interested reader is referred to excellent source books on weldability testing, such as those available from the Defense Material Information Center.

SUMMARY

Welding frequently involves the application of heat to affect the formation of chemical bonds to create a sound joint between materials. In fusion welding, brazing, and soldering, this heat is sufficient to cause melting of either the base material and, possibly, a filler material for fusion welding or simply the filler for brazing and soldering. Understanding the distribution of this heat, its effects on the microstructure of the fully melted material, partially melted material, and of the solid heat-affected zone surrounding these regions is critical to fully understanding the versatile and diverse joining process of welding. It is also essential to understand the physical metallurgy of the highly non-equilibrium processes of melting, solidification, and solid-state phase transformations involved in welding to fully understand the process, its capabilities, and its limitations.

Not all of the energy from a fusion welding source contributes to the formation of the weld fusion zone proper. Some of the energy is lost in transfer to the workpiece (reflecting the transfer efficiency of the particular process), while that which does arrive at the workpiece is distributed between the fully melted fusion zone (FZ), the partially melted zone (PMZ) in alloys, and the surrounding heat-affected zone (HAZ). Solutions to the general equation of heat flow using simplifying assumptions (such as Rosenthal's steady state moving heat source), employing advanced computers and numerical solutions using finite element and finite difference methods, and empirically by fitting experimental data to equations allow prediction of peak temperatures, fusion zone and heat-affected zone widths, fusion zone solidification rates, and heat-affected zone cooling rates. These, in turn, enable assessment and prediction of weld structure, substructure, and properties.

Within the fusion zone, weld composition is affected by absorbed gases from the environment, leading to the formation of porosity, embrittling inclusions, or embrittlement of susceptible microstructures; by metallurgical refinement by fluxes and their slags; and by dilution of filler by melted base material. Weld size and shape are affected by the thermal properties of the base material, the process heat source characteristics, process operating mode, welding parameters (especially welding speed), and thickness of the weldment.

The process of solidification in pure crystalline materials and alloys is similar in that both require the temperature to fall below the equilibrium melting point (or liquidus), both require the dissipation of latent heat evolved during solidification, and both depend on nucleation and growth, whether that nucleation occurs without the aid of a substrate (i.e., is homogeneous) or with the aid of a substrate (i.e., is heterogeneous). Growth always occurs competitively by trying to take place in those preferred crystallographic directions that are aligned most closely with the steepest temperature gradients. In pure materials, a *planar mode* is observed when the temperature gradient in the liquid is positive and a *dendritic mode* is observed when the gradient is negative. Solidification of alloys has the added complexity of needing to deal with solute redistribution. Nonequilibrium leads to solute buildup in the liquid ahead of the advancing solid–liquid interface, and this leads to constitutional supercooling. The result is additional growth modes for positive temperature gradients in the liquid, including cellular, cellular dendritic, and columnar. In the solid, nonequilibrium leads to microsegregation that, in turn, leads to hot crack defect formation.

In the HAZ, the heat of welding causes different structural changes in different materials, depending on the predominant method of strengthening in that material. Work-hardened materials soften because of recovery, recrystallization, and grain growth. Aged (precipitation hardening) alloys soften because of overaging and reversion, while alloys that exhibit cooling-dependent transformation hardening can embrittle. Certain stainless alloys become sensitized to corrosive attack or weld decay due to carbide precipitation at grain boundaries. Proper welding requires that these problems be dealt with.

Defects can occur in welds because of improper joint design, improper joint preparation, improper fit-up or alignment, improper process selection or operation, inherent material susceptibility to hot or cold cracking, and other sources. Defects can be avoided by assessing the inherent *weldability* of a material by using suitable tests or weld simulations. Tests are available for assessing FZ, PMZ, and HAZ defect susceptibility.

PRACTICE QUESTIONS AND PROBLEMS

1. Long, wide sheets of 8 mm thick commercially pure titanium are butt welded using the plasma arc process with direct current, electrode negative. Near parallel-sided welds are produced using a current of 200 A, a voltage of 20 V, and at a travel speed of 240 mm/min. The weldment starts out at 25°C. What is the peak temperature at distances of 4.0 mm and 8.0 mm and 16.0 mm from the fusion zone boundary? Assume a transfer efficiency of 80%. The melting temperature of T_i is 1700°C and the value of ρC is 3.0×10^6 J/m^3–°C.

2. For a 2.5 mm thick plain carbon steel (i.e., assume an Fe-C binary alloy) containing 0.5 wt.% carbon, calculate the width of the fusion zone and the width of the heat-affected zone to the point where any austenite would form for a weld made at a current of 200 A, a voltage of 10 V, and a travel speed of 5 mm/sec with

gas-tungsten arc welding ($\eta = 0.6$). Assume the melting temperature of 0.5% C steel is 1500°C and the eutectoid transformation temperature (where austenite just forms) is 727°C. The thermal conductivity (k) of such a steel is 41.0 J/m-sec-°C and the value of ρC is 4.5×10^6 J/m³-°C.

3. For the weld in Problem 2, calculate the cooling rate in this thin plate or sheet at the edge of the fusion zone and at the edge of the heat-affected zone where austenite just formed. If the critical cooling rate for 0.5% C steel is 100°C/sec, will martensite form in the HAZ of this weld?

4. What would be the width of the heat-affected zone (as defined in Problem 2) if a weld were made by an electron beam using 15.0 kV, 100 mA (0.1 A) at a travel speed of 1500 mm/min? Assume the transfer efficiency of electron beam welding to be 0.9.

5. What would be the cooling rate at the fusion zone boundary and at the edge of the HAZ for the EB weld in Problem 4?

6. How would preheating the plate or sheet in Problem 4 above to 300°C change the cooling rate at the fusion zone boundary and at the edge of the HAZ? Would martensite form if the critical cooling rate is 100°C/sec?

7. What are the possible sources of nitrogen and oxygen in a molten weld pool made by the gas-metal arc (GMAW) process? Are there possible sources different if the weld is made by SMAW? If so, what are these differences?

8. What are the possible sources of hydrogen in a molten weld pool made by flux-cored arc welding (FCAW) during maintenance welding outdoors?

9. The composition of the fusion zone of a weld is affected by dilution by the base metal. For a joint shown to scale below, estimate the composition of a weld made in 0.4 wt. % C plain carbon steel if an 18 wt. % Cr–8 wt. % Ni stainless steel filler is used. Assume the plain carbon steel is a simple binary alloy of Fe-C and the stainless steel in a simple ternary of Fe-Cr-Ni. *Hint:* Use some technique for quantitatively estimating the relative area of base metal making up the final weld.

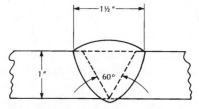

10. How does the relative size of a weld pool change for a given energy input as the weld is made in an austenitic stainless steel versus a plain carbon steel? Explain this change. What if the weld was made in pure copper instead of in pure aluminum? Explain this change.

11. Explain why a weld pool takes on a teardrop shape at some value of travel speed.

12. Explain why heterogeneous nucleation occurs much more easily than homogeneous nucleation in welds.

13. The equation below is the solution of the diffusion equation under steady state that gives the concentration profile in the liquid ahead of the solid–liquid interface during solidification under Case 3 conditions:

$$X_L = X_o \{1 - (1 - k)/k \exp[x/(D/R')]\}$$

for X_o/k at $x = 0$ and X_o at some distance away from the interface, where X_L is the peak concentration in the liquid at the interface, X_o is the alloy's nominal composition, k is the distribution coefficient, D is the diffusion rate of a solute in the molten solvent, and R' is the solidification rate. Using this equation, calculate the approximate distance from the interface at which the composition in the liquid reaches X_o, if $k = 0.5$, $D = 5.0 \times 10^{-5}$ cm²/sec, and $R' = 0.2$ cm/sec.

14. Using the equation and parameters in Problem 13, calculate the concentration in the liquid 10^{-5} cm away from the interface.

15. Explain how the structure in the weld shown in the illustration below developed.

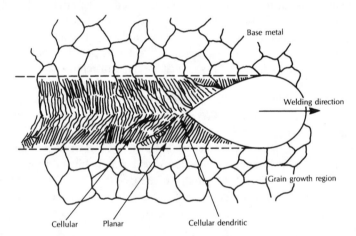

16. The new structure formed in a weld upon solidification grows competitively, attempting to have preferred crystallographic growth directions align with maximum temperature gradients. Sketch the growth pattern you would expect if all the grains in the base metal had their easy growth direction aligned parallel to the welding direction and the weld pool were circular. *Hint:* Do not forget that the weld pool moves as growth progresses from each fusion zone boundary toward the weld centerline, so the direction of the maximum temperature gradient changes.

17. Explain why the partially melted zone appears as it does in Figure 7.19.

18. Sketch the expected structure of a joint made in a heavily cold-worked pure metal from the fusion zone boundary through the HAZ to the unaffected base metal. Superimpose on this sketch a plot of the expected hardness.

19. Explain the differences in expected hardness or (strength) for a fusion weld made in a precipitation-hardened alloy in the fully aged condition and then aged for 4

hours at 300°C versus a weld made in the same alloy in the solution-treated condition and aged for the same time and at the same temperature.

20. A low alloy steel containing 0.4 wt. % C, 0.75 wt. % Mn, 0.30 wt. % Si, 1.83 wt. % Ni, 0.8 wt. % Cr and 0.25 wt. % Mo is to be welded. Would you be concerned about the possibility of cold cracking? If not, why not? If so, what would you do to minimize the chances of such cracking?

REFERENCES

1. Easterling, Kenneth, *Introduction to the Physical Metallurgy of Welding,* 2nd ed., London: Butterworths, 1992.
2. Kou, Sindo, *Welding Metallurgy,* New York: John Wiley, 1987.
3. Lancaster, J.F., *Metallurgy of Welding,* 4th ed., Allen & Unwin, London, 1987.
4. Linnert, G.E., *Welding Metallurgy of Carbon and Alloy Steels, Vol. 1, Fundamentals,* 3rd ed., Miami: American Welding Society, 1965.
5. Randall, M.D., Monroe, R.E., and Rieppel, P.J., *Methods of Evaluating Welded Joints,* Defense Metals Information Center (DMIC) Report 165, Columbus, Ohio: Battelle Memorial Institute, December 1961.
6. Vagi, J.J., Meister, R.P., and Randall, M.D., *Weldment Evaluation Methods,* Defense Metals Information Center (DMIC) Report 244, Columbus, Ohio: Battelle Memorial Institute, August 1968.

Brazing

8.1 INTRODUCTION

It is often desirable to produce a permanent, mechanically acceptable and leak-tight joint without fusing the bulk base material. While non-fusion welding processes are an option, there are practical limitations as to what can be joined, based on shape, size or joint area, material combinations, and fit-up, to name a few (see Section 6.5). Adhesive bonding is another option, but joint strength, particularly in peel, and operating conditions are limited, and environmental degradation can be a serious problem. The processes of brazing and soldering offer two other options.

Unlike in solid-phase joining or nonfusion welding, where the joint is made by bonding the two surfaces directly, without filler and without requiring fusion of the substrates, in brazing and soldering, a liquid is made to flow into and fill the space between solid joint faces and then solidify. The liquid used as a filler has a lower melting point than the materials being joined, so there is no melting of the bulk substrate. Brazing and soldering are distinguished from one another primarily on the basis of the melting temperatures of the fillers, with braze fillers melting over 450°C (840°F) and solder fillers or solders melting below this temperature. Brazing produces a stronger joint than soldering primarily because of the inherently higher melting and stronger fillers employed, the lower homologous temperatures[1] at which brazed (versus soldered) joints operate, and, often, to greater degrees of interdiffusion between filler and substrate. For this reason, brazing has become an extremely important and versatile method for joining both commonplace and high-performance structures, from automobile radiator cores to aerospace honeycomb sandwich cores.

By precluding the need for fusing the base materials, yet producing mechanically strong, environmentally stable joints, brazing offers an extremely attractive joining alternative. This chapter looks at the process of brazing in detail. First, the process is defined, as a subcategory of the broader process of welding and as distinct from soldering. Then, the relative advantages and disadvantages of brazing are identified. Next, a detailed description is given of the principles of operation of the general process and specific brazing methods, for both manual and automated operation. This is followed

---

[1] *Homologous temperature* is the actual temperature versus the melting temperature of a material on an absolute temperature scale.

by a discussion of brazing filler materials, their basic characteristics, their general physical metallurgy, and their most important types. The critically important role of fluxes and/or atmospheres for providing a clean wettable surface for brazing is then described. Finally, a brief description of joint designs and general joint properties is given.

8.2 BRAZING AS A JOINING PROCESS

8.2.1 General Description

Brazing is a subgroup of welding that produces coalescence of materials by heating to a suitable temperature and by using a filler material having a liquidus above 450°C or 840°F (some sources say 425°C or 800°F) and below the solidus of the base materials. The filler material, or *filler,* is of necessity (based on the process definition) of different composition than the base materials and is distributed between close fitting surfaces of the joint by surface wetting, chemical reaction, and spreading. Capillary action plays a major role in filler flow through the joint. Chemical bonds are formed between the filler material and the base material, and substantial diffusion of elements in the filler into the base material usually occurs. Braze fillers can be metallic for joining metals or ceramics to themselves or one another or can be ceramics for joining ceramics. Bonding is often primary and can be metallic, ionic, covalent, or mixed ionic-covalent, depending on the specific natures of the filler and the base material or materials.

A process called *braze welding* is not, in fact, a brazing process, even though base material melting is not involved. In braze welding, a filler metal is deposited in a groove or fillet exactly at the point where it is to be used, and spreading by surface tension or capillary action is not a factor. Limited base metal fusion may also occur. This process is discussed in Chapter 10.

To be a true brazing process, all of the following criteria must be met: The parts must be joined without melting the base materials by using a filler, *and* the filler material must have a liquidus temperature above 450°C (840°F), *and* the filler material must wet the base material surfaces and be drawn into or held in the joint by capillary action, depending on whether the filler is added during brazing or preplaced. Brazing is arbitrarily distinguished from soldering by the filler material melting temperature. In soldering, the filler materials melt below 450°C (840°F).

To achieve a good finished joint using any of the various brazing processes, the following four basic elements must be considered: (1) joint design; (2) filler material; (3) joint heating; and (4) protective or reactive shielding. Briefly, the joint design must afford a suitable capillary for the filler material when properly aligned. The need is to enable filler spreading and assure coverage. Filler material must melt at a lower temperature than the base material(s) (but above 450°C or 840°F) and must be compatible with the base material(s) to allow wetting and interdiffusion. This means at least some components of the filler must be soluble in the substrate solvent and vice versa. To allow brazing, heat can be applied locally at the joint or the entire assembly to be brazed can be heated. In either case, a temperature must be reached at the joint to allow filler melting, wetting, and spreading, and temperature must be uniform, at least within the joint proper. Protective shielding is required during brazing to prevent oxidation of cleaned joint faying surfaces during heating and throughout braze flow. This can be accomplished with either a chemical flux or an inert atmosphere. Sometimes a reactive flux or atmosphere is required to actually clean and chemically activate the surfaces to be brazed.

8.2.2 Advantages and Disadvantages of Brazing

Like all joining processes, brazing offers certain advantages but has certain limitations and shortcomings as well. Perhaps most important, brazing has little or no effect on the composition of the base material(s), thereby preserving special metallurgical characteristics where desired. In brazing, there is no melting of the substrate and, thus, no gross mixing. There may be, and often is, however, some interdiffusion, but this usually does not change the composition significantly.[2] For some brazing processes and base materials, there is very little heat effect on the substrate, since heating is either relatively low compared to the melting temperature of the substrate (as in metal brazing of some refractory metals or ceramics) and/or heating is highly localized. Since brazing temperatures can be kept relative low (compared to the substrates' melting temperatures), the process can be the ideal choice where base material melting cannot or should not occur. For other brazing processes, the entire structure is raised to the brazing temperature, so heat effects can be significant if the brazing temperature is high relative to the melting temperature of the substrates. In any case, any heat effect is uniform.

Related to these process characteristics is the fact that brazing is often an ideal choice for joining dissimilar material types and compositions. Provided a filler can be found that wets all of the substrate materials involved in the joint, sound joints can usually be made. Limited intermixing again contributes to this important processing advantage.

The brazing process has the ability to join cast, wrought or powder-processed (including porous) metals, dissimilar metals, oxide and nonoxide ceramics, metals to ceramics, and fiber- or dispersion-strengthened composites with metallic or ceramic matrices. It provides a simple means for bonding large joint areas or long joint lengths. The surface bonding action of the thin braze layer in lap-type joints results in excellent stress distribution, as in adhesive bonding (see Chapter 4). Like adhesive bonding, the uniform distribution of stress over large areas allows the joining of very thin sections, as well as the joining of thick-to-thin sections. The only drawback in this regard is that there can be a pronounced notch effect due to the thinness of the joint if loading is in tension (i.e., little strain compliance). The process also allows the affordable fabrication of complex assemblies, often consisting of multiple components. Finally, brazing offers the ability to join large structures under relatively stress-free conditions and is capable of producing precision tolerances reproducibly in production.

In terms of service, braze joints often have limited elevated temperature serviceability and stability, because of the low melting point and strength of the filler and the high

---

[2] An exception is *diffusion brazing,* in which diffusion of one of the components from the filler into the base material is essential to the creation of the joint. The filler changes dramatically in composition, as does the immediately surrounding base material.

Table 8.1   Summary of Relative Advantages and Disadvantages of Brazing versus Other Joining Processes

Advantages	*Disadvantages*
1. Little or no effect on composition of base material.	1. Can have limited elevated temperature service and stability.
2. Often little heat affect on microstructure or, for many specific processes, a uniform effect.	2. Can be pronounced notch effect due to thinness of braze layer.
3. No melting of base material.	3. Requires fluxing or atmosphere control and removal of flux residue.
4. Excellent for joining dissimilar material types and compositions within a type.	4. Requires use of joints with controlled gaps or clearance.
5. Allows large area bonding; distributes stresses evenly over area.	
6. Allows joining of widely different thicknesses, especially thin-to-thin or thin-to-thick.	
7. Allows joining of porous materials (PM).	
8. Affordable fabrication of complex multi-component assemblies.	
9. Very low distortion and residual stresses (if C.T.E.'s are matched).	
10. Easy to automate.	

diffusion rate of some filler components. With proper choice of filler and process (e.g., diffusion brazing), joint temperature capability can approach that of the base material.

Table 8.1 summarizes the relative advantages and disadvantages of brazing as a joining process.

8.3 PRINCIPLES OF PROCESS OPERATION

Capillary flow is the dominant physical principle that assures good brazements.[3] Capillary flow is the result of a liquid lowering the surface energy of a solid–vapor interface by wetting that solid in accordance with Young's equation (see Section 7.3.4, Eq. 7.17), but for different phase combinations. Specifically, for capillary flow to enable braze filler flow,

$$\gamma_{vs} = \gamma_{ls} + \gamma_{vl}\cos\theta \qquad (8.1)$$

where γ_{vs}, γ_{ls}, and γ_{vl} are the surface energies of the vapor–solid, liquid–solid, and vapor–liquid interfaces, respectively, and θ is the angle of contact or wetting angle. Capillary flow is facilitated by low angles of θ, i.e., good wetting. In actual practice, brazing filler material flow characteristics are also influenced by dynamic considerations involving the molten metal's fluidity or viscosity and vapor pressure, gravity, and, especially, the effects of any chemical or metallurgical reactions between the filler and the base material(s).

---

[3] A *brazement,* like a *weldment,* refers to the assembly being joined by brazing or welding, respectively.

In general, the brazed joint is one of relatively large area and very small thickness, and the process of brazing is predominantly controlled by surface conditions. In the simplest application of the brazing process, four critical steps are involved. First, the surfaces to be joined are cleaned to remove contaminants and oxide on metal substrates. Then, in brazing metals and alloys, the surfaces are coated with a material, called a flux, capable of dissolving solid metal oxides still present *and* also preventing new oxidation. The joint area is then heated until the flux melts and cleans the base materials through its chemical reactivity and, then, is protected against further oxidation by a layer of liquid flux. Finally, the brazing filler material is melted at some point on the surface of the joint area[4] and any flux is displaced by the filler because of the combination of the lower surface energy that results from wetting and the filler's higher density.

Instead of fluxes, brazing is sometimes performed with an active gaseous atmosphere, such as hydrogen for parts needing cleaning, or an atmosphere of inert gas or vacuum for parts that are already clean. When atmospheres are used, the need for postcleaning the potentially corrosive flux residues is eliminated.

Joints to be brazed are usually made with relatively small clearances, typically 0.025 to 0.25 mm (0.001 to 0.010 in.). High fluidity of the molten filler is thus an important characteristic. Joint assembly for proper positioning, alignment, and fit-up or gapping usually requires tooling or fixturing, which includes some means of assuring the proper gap (e.g., shims). Shrinkage stresses arising from the volume change of the filler after solidification and from thermal contraction are usually low (compared to welding), so brazed joints are made in a relatively stress-free condition.

In summary, brazing is an economically attractive process for the production of high strength metallurgical bonds while preserving desired base material properties and achieving precision tolerances.

8.4 BRAZING PROCESSES

8.4.1 General Description

Brazing processes are customarily designated according to the source or method of heating, not unlike welding processes. Manual processes are possible, but automated processes predominate. Some processes restrict heating to the joint proper, while others heat the entire braze assembly or brazement uniformly. The methods currently of most industrial significance, and described in detail in the following subsections, are torch brazing, furnace brazing, induction brazing, resistance brazing, dip brazing, infrared brazing, and diffusion brazing. Also several specialized brazing processes are also worth noting, including laser brazing, electron-beam brazing, ultrasonic brazing, vapor-phase brazing, and step-brazing. Braze welding, although often considered a brazing process, is actually a hybrid process related to welding (see Chapter 10, Section 10.4.3). All of the processes can be performed by localized heating except the furnace, dip, diffusion, and vapor-phase processes.

---

[4] Filler can be added to some point on the surface of a joint during the actual process of brazing or can be preplaced in the joint just prior to the start of actual brazing. This depends on the form of the filler (see Section 8.5).

8.4.2 Torch Brazing

Torch brazing (TB) is accomplished by heating the joint area locally with one or more gas torches. Depending upon the temperature and amount of heat required, the fuel gas can be, among others, natural gas, propane, or acetylene, burned with air, compressed air, or oxygen. Torches for accomplishing brazing are identical to those used for oxy-fuel welding (see Figure 6.8) and can be used manually or automatically (i.e., machine operation). Single or multiple flame tips are available, with the multiple tips being employed for larger, heavier brazements. During torch brazing, filler may be preplaced at the joint in the form of rings (e.g., for tubular assemblies), washers, strips, slugs, shims, powder, or special shapes (preforms), or may be fed from hand-held wire or rod. Fluxes should always be used.

Manual torch brazing is especially useful on assemblies involving sections of un-equal mass, since heat input can be manipulated. Machine operation is usually only used where the volume and rate of production, or the large size of the assembly, warrants. Manual systems are inexpensive (i.e., approximately $300) and highly portable, while machine-based systems range in cost from $5,000 to well over $50,000, based on complexity.

8.4.3 Furnace Brazing

Furnace brazing (FB) is used exclusively where the parts to be brazed can be assembled with the filler material preplaced in or near the joint. It is particularly applicable for high production. Preplaced filler can be in the form of wire, foil, filings, slugs, shims, powder, paste, tape, or shapes. Fluxing is used except when an atmosphere is specifically introduced into the furnace to perform the same shielding and/or reducing function. Hydrogen and either exothermic or endothermic combusted gas are usually used for reduction, while inert (e.g., argon or helium) gases are used for special cases. Vacuum atmospheres are also widely used, especially in aerospace, and often preclude the need for and use of flux. Vacuum brazing cannot be performed with certain high vapor pressure filler or base materials, however.

Furnaces generally are batch or continuous types, are heated by electrical resistance elements, gas or oil, and should have automatic time and temperature (and, possibly, atmosphere) controls. Cooling chambers or forced atmosphere injection cooling is often used to speed up and control solidification and subsequent cooling of the assembly. Figure 8.1 shows a typical furnace brazing system.

8.4.4 Induction Brazing

The heat necessary for brazing with the *induction brazing* (IB) process is obtained from an electric current induced in the parts to be brazed. Heating can be restricted to the immediate area to be brazed or can be more general. Typically, water-cooled coils carrying high-frequency alternating current are used by being placed in close proximity to the parts to be brazed but not as a part of the electrical circuit. Figure 8.2 shows examples of induction brazing coils.

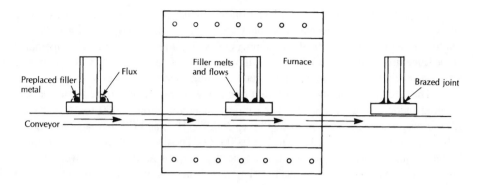

Figure 8.1   A typical furnace brazing system.

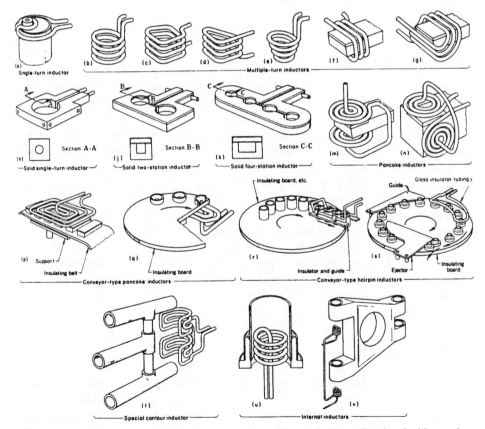

Figure 8.2   Examples of typical induction brazing coils and devices. (Reprinted with permission from *Metals Handbook*, 9th ed., Vol. 6, American Society for Metals, 1983.)

Brazing filler is usually preplaced during induction brazing, with forms being similar to those used with furnace brazing. Careful joint design and coil set-up are essential to assure that all surfaces of all members of the joint reach the brazing temperature at the same time. Otherwise, filler flow will not occur as intended. Flux is used with this process, except when an atmosphere, often a vacuum, is specifically employed.

The three common sources of high frequency alternating electric current for performing induction brazing are: (1) the motor-generator (from 5,000 to 10,000 Hz); (2) the resonant spark gap (from 20,000 to 300,000 Hz); and (3) the vacuum tube or solid-state oscillator (from 20,000 to 5,000,000 Hz). The depth of heating is determined by the frequency of the power source. Higher frequencies produce shallower heating, until, finally, only skin-heating occurs. The rate of heating is always fairly fast, typically 10 seconds to 1 minute. Part thicknesses for induction brazing are generally thin, in the range up to 3 mm or 0.125 in.

A related process is *resistance brazing* (RB) where joint heating is produced by the resistance at the joint contact surfaces for current introduced through electrodes. Conventional resistance welding machines are typically used, with alternating current.

8.4.5 Dip Brazing

There are two methods of dip brazing (DB): (1) chemical bath dip brazing and (2) molten metal bath dip brazing. In *chemical bath dip brazing*, the filler, in suitable form, is preplaced in or near the joint, and the assembly is immersed in a bath of molten salt. Molten salts have high specific heat capacities, so are excellent for furnishing the heat needed for brazing. In addition, the molten salts are usually aggressive enough to provide the necessary cleaning and oxidation protection. If not, a suitable flux may be required. In *molten metal bath dip brazing*, the parts to be brazed are immersed in a bath of molten brazing filler metal. The parts must first be cleaned and fluxed (if necessary), and a cover of molten flux should be maintained over the molten metal bath to protect it from oxidation. Figure 8.3 shows a typical dip brazing set-up.

The dip brazing process is usually restricted to small parts, and, since the entire assembly is heated to the brazing temperature (and may be exposed to aggressive chemicals), the assembly must be tolerant of heating to the required brazing temperatures, corrosive chemicals, and subsequent cleaning.

8.4.6 Infrared Brazing

Infrared brazing (IRB) uses infrared (IR) heating through irradiation with long-wavelength light in the visible spectrum. High wattage (e.g., 5,000 watts) lamps are often used. These lamps should be placed close to the workpiece, since heat input intensity drops off as the square of the distance from the source. It is best if the lamps follow the contour of the part, to provide uniform proximity, but it is not essential. Heating can

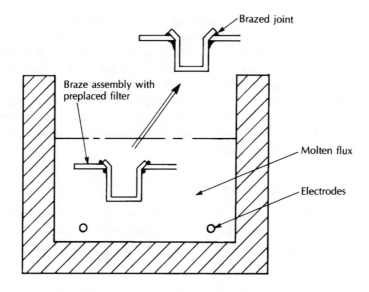

Figure 8.3   A typical dip brazing setup.

take place in air, vacuum, or under inert atmosphere. After heating, the part is moved out of the IR source, often to cooling platens.

The infrared brazing process is particularly suited to the brazing of very thin materials, such as honeycomb and honeycomb sandwiches for aerospace applications. It is rarely used with material thicker than approximately 1.3 mm or 0.05 in.

8.4.7 Diffusion Brazing

Unlike the previous brazing processes, *diffusion brazing* (DFB) is not defined by its heating source but, rather, by the mechanism involved in forming the bond at the joint. A joint is formed by holding the brazement at a suitable temperature for sufficient time to allow mutual diffusion of the base and filler materials. Rather than the filler melting and flowing, the liquid that fills the joint is produced from a reaction between the filler and the base material, producing a transient liquid (eutectic) phase. For the preceding reasons the process is called *reaction brazing* and *transient liquid phase bonding*. The resulting joint ends up with a composition considerably different from either the filler or the base material(s), and no filler should be discernible in the final microstructure of the completed joint.

The typical thicknesses of base materials that are diffusion brazed range from foil to several inches (i.e., more than 50 mm). Relatively heavy parts can be brazed using this process because the process is not sensitive to joint thickness. Many brazements can be made by DFB that are difficult or impossible to make by other brazing methods.

The diffusion brazing process produces much stronger joints than normal brazing processes, frequently approaching the strength of the base material. Butt and lap joints, in particular, exhibit excellent mechanical properties. Mechanical fixturing or tack welding may be required to hold parts together during brazing. Unlike diffusion welding, however, no auxiliary pressure is needed.

DFB requires a relatively long time (1/2 to 24 hours) to complete, but many joints can usually be done at the same time. The DFB joint also has a remelt temperature approaching that of the lower melting base material, so service temperatures can be extended over most other brazing processes.

8.4.8 Other Special Brazing Methods

Several other methods of brazing are used for special applications. These include blanket (resistance) brazing, exothermic brazing, laser brazing, electron-beam brazing, ultrasonic brazing, vapor-phase brazing, and step-brazing. *Blanket brazing* simply uses a resistance-heated blanket to transfer heat to the brazement by a combination of radiation and conduction. The process is good for contoured parts. *Exothermic brazing* obtains the heat for brazing from an exothermic chemical reaction, usually solid-state or nearly solid-state metal–metal oxide reactions (e.g., Fe_3O_4 + Al, thermite). One modern version of this process is self-propagating high-temperature synthesis (SHS) joining for use in joining refractory metals, ceramics or intermetallics. This process, while only rarely used in industry, is receiving attention for its potential for joining structures in space, using the heat of the sun to initiate the exothermic chemical reaction in preplaced reactants. The *laser-* and *electron-beam brazing processes* use beams of photons or electrons, respectively. Heating is highly localized, and the processes are attractive for precision assemblies. The EB process is limited to materials that can be brazed under the vacuum required for the beam generation and transmittal. *Ultrasonic brazing* really is ultrasonic-assisted brazing by other, more conventional methods. The ultrasonic vibration energy helps remove voids and gas pores from the molten filler and improves fill and fill integrity.

Vapor-phase brazing and *step-brazing* are highly specialized processes. In *vapor-phase brazing,* vapors of metals are combined in the vapor state to form a low melting alloy that deposits on the brazement and produces a bond by being drawn into the joint by capillary action. Copper and zinc vapors have been combined to produce brass brazing filler alloys. *Step-brazing* is really a procedure for accomplishing multiple brazes, sequentially. By choosing braze fillers with progressively lower melting temperatures, brazes can be made near previously made joints without causing remelting.

8.5 BRAZING FILLER MATERIALS

8.5.1 Basic Characteristics

For satisfactory use in brazing, filler materials must possess certain basic characteristics. First, filler materials must have the ability to form brazed joints possessing suitable mechanical and physical properties for the intended application. This often means strength but may include ductility, toughness, electrical or thermal conductivity, temperature resistance, and stability. An extremely important physical property for intended fillers is that they have a coefficient of thermal expansion that closely matches the substrates being joined or, where severe temperature gradients persist, bridges the difference in coefficients of thermal expansion between the two joint element materials. This is so that thermal mismatch stresses across the joint do not cause failure by fracture. Second, the melting point or range of an intended filler material must be compatible with the base materials being joined *and* have sufficient fluidity at the brazing temperature to flow and distribute into properly prepared joints by capillary action. "Suitable" melting range means below the solidus of the base materials but as high as necessary to meet service operating temperature requirements. Third, the composition of the intended filler must be sufficiently homogeneous and stable that separation of constituents, known as *liquation*, does not occur under the brazing conditions to be encountered. Obviously, the intended filler material composition must also be chemically compatible with the substrates to avoid adverse reactions during brazing or by subsequent sacrificial (i.e., galvanic) corrosion. Fourth, intended fillers must have the ability to wet the surfaces of the base materials being joined to form a continuous, sound, strong bond. Fifth, depending on requirements, intended fillers must have the ability to produce or avoid reactions with the base materials. Usually it is desirable to avoid such reactions, since brittle intermetallics may result, degrading joint properties. However, for so-called active-metal or reactive brazing, it is necessary for the filler and the substrates to react chemically in a particular way (see Chapter 13).

The characteristics of melting and fluidity, liquation, and wetting and bonding deserve special consideration, so are dealt with in more detail in the following paragraphs.

Melting and Fluidity

Pure crystalline metals and ceramics melt at a constant temperature, and molten metals, in particular, are generally very fluid. Alloys, on the other hand, whether metallic or ceramic, melt over a range of temperatures from the solidus to the liquidus for the particular composition and can have a fluidity that varies widely depending on the relative amounts of liquid and solid present at the brazing temperature. This "mushy" state always reduces the fluidity versus the fully liquid state. The wider the mushy range, the more sluggish the flow of the filler by capillary action. So, special care must be taken in selecting and employing alloy fillers for brazing to assure proper fluidity.

Most braze alloys are designed to be more complex than simple binaries, often for the purpose of altering the liquidus–solidus ranges and phase proportions as much as for any other property. A brazing alloy will always have a liquidus that is below the melting point of the lowest melting component of the alloy, so will always be suitable for brazing that component.

Liquation

Because the compositions of the solid and liquid phases of brazing filler materials differ because of the distribution coefficient for the alloy solute, the proportion and composition of each phase will undergo gradual changes as the temperature increases from the solidus to the liquidus. If the portion that melts first is allowed to flow away from the remainder of the unmelted filler by capillary spreading, the remaining solid has a higher melting point than the original composition, never melts, and remains behind as a solid *skull*. This phenomenon is known as *liquation*. Obviously, such separation is undesirable. The tendency for liquation should be minimized in properly designed brazing filler alloys. This is accomplished by employing filler alloys with narrow melting ranges and heating rapidly through the melting range during brazing. Optimum brazing temperature for a particular filler alloy is usually about 10° to 90°C (50 to 200°F) above the liquidus of the filler alloy. This super heat assures flow without liquation.

Wetting and Bonding

To be effective, a brazing filler material must alloy with the surface of the base material without undesirable degrees of diffusion into the base material, dilution by the base material, base material erosion, or formation of brittle compounds at the interface. These effects are dependent upon the mutual solubility between the filler and the base materials, the amount of brazing filler material present, and the temperature and time profile of the brazing cycle.

8.5.2 Filler Selection Criteria

The following factors should be considered when selecting a brazing filler material, whether it is a metal or a ceramic: First, it should be compatible with the base material and the joint design (see Section 8.7). Compatibility with the base material means properly matching chemical, mechanical, and physical properties. Compatibility with the joint design means proper mechanical properties for the type and magnitude of loading (e.g., static or fatigue; tension, shear, or peel). Second, the filler material must be suitable for the planned service conditions for the brazed assembly, including service temperature, thermal cycling, life expectancy, stress loading, corrosive conditions, radiation stability, and vacuum operation (i.e., outgassing). Third, the filler material must be selected based on the brazing temperatures required and acceptable to the components of the assembly and to the production environment. Low temperatures are usually preferred for economizing on heat energy, to minimize heat effects on the base material (e.g., annealing, grain growth, and warpage), and to minimize interactions (e.g., embrittlement by intermetallics). On the other hand, high temperatures are preferred to take advantage of higher melting alloys for their economy, to combine stress relief or heat treatment of the base material with the process of brazing, to promote interactions that will increase joint remelt temperatures, or to promote the removal of certain refractory oxides in vacuum or certain atmospheres. Finally, filler material selection depends on the method of heating to be used. Alloys with a narrow (i.e., 30°C or 50°F) melting range can be used with any heating method. Fillers with wider melting ranges that are prone to liquation should be brought to brazing temperature quickly, so processes with more intense heating are preferable.

8.5.3 The Metallurgy of a Key Filler Material System (Cu-Ag)

While it is certainly not possible to consider the metallurgy of every possible, or even major, brazing filler alloy to understand how these alloys behave, neither is it necessary. It is possible to understand much of the physical metallurgy of brazing filler alloys by considering copper-silver alloys, a key brazing alloy filler system.

Figure 8.4 presents the constitutional diagram for the copper–silver binary alloy system. This diagram is fairly representative of all major brazing filler alloys, whether metallic or ceramic. The diagram shows the phase or phases that exist under equilibrium conditions as a function of alloy composition and temperature. This type of constitutional diagram is called a eutectic system with partial solid solubility in the

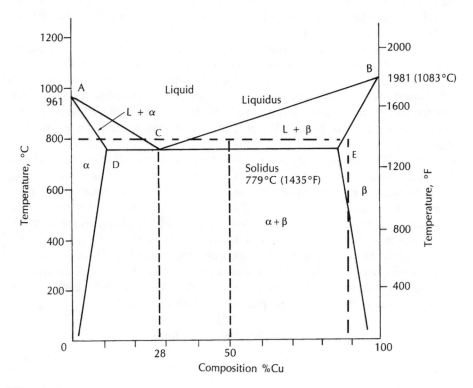

Figure 8.4   Constitutional diagram for the Ag-Cu system to explain the metallurgy of brazing filler materials.

terminal phases. The other major type of constitutional diagram appropriate to brazing filler alloys is the eutectic system with no solid solubility in the components. The only difference is that instead of having terminal phases that are solid solutions, this latter type of diagram has terminal phases that are pure materials.

The highest temperature at which a pure crystalline metal or ceramic or alloy is completely solid is called the *solidus,* while the lowest temperature at which a pure crystalline metal or ceramic or alloy is completely liquid is called the *liquidus.* Between the solidus and the liquidus, an alloy consists of both a liquid and a solid phase in equilibrium. In the process of brazing, the solidus is considered the melting point of the brazing filler material, and the liquidus is considered its flow point.

In Figure 8.4, the solidus temperatures of all alloy compositions from pure silver to pure copper, with all compositions in between, is given by the solidus line ADCEB. This represents the start of melting on heating of any of these alloys. The liquidus temperature of all alloy compositions is given by the liquidus line ACB. This represents the temperatures at which all alloy compositions are completely liquid.

At point A, pure silver melts at a single temperature (i..e, 961°C, or 1762°F), while at point B, pure copper also melts at a single temperature (i.e., 1083°C, or 1981°F). At point C, the two lines (i.e., the liquidus and the solidus) meet, indicating that a particular alloy (i.e., 72 wt. % Ag - 28 wt. % Cu) melts at a constant temperature and at the lowest temperature of any composition. This point is called the *eutectic point,* and the alloy is called the *eutectic alloy.*

The alloy of eutectic composition is as fluid as the pure materials that compose it, since there is no solid present, ever. At all other compositions, melting occurs over a range of temperatures between the liquidus and the solidus, the alloy is made up of both a liquid and a solid phase, and is mushy, like slushy snow that consists of solid ice crystals and water. How fluid is this mixture of two phases depends on the relative proportions of liquid and solid present. The greater the proportion of solid, the less fluid the mixture.

The silver-rich end of the diagram is terminated with a substitutional solid solution of copper in silver called α. The copper-rich end of the diagram is terminated with a substitutional solid solution of silver in copper called β. The central compositions consist of an intimate mixture of α and β solid solutions, actually: α plus α–β containing eutectic below 72 wt. % Ag–28 wt. % Cu; β plus α–β containing eutectic above 72 wt. % Ag–28 wt. % Cu; and just α–β containing eutectic at 72 wt. % Ag–28 wt. % Cu. At any particular composition and temperature, the proportions of liquid and solid phase making up the mixture, and creating a mushy state, can be calculated from the *lever law* (see Section 7.3.5). For example, a 50 wt. % Ag–50 wt. % Cu alloy at 800°C (1472°F) contains the following proportions of liquid and solid β:

$$\% \text{ liquid } L = (88 - 50/88 - 37) \times 100\% = 74.5\%$$

and

$$\% \text{ solid } b = (50 - 37/88 - 37) \times 100\% = 25.5\%.$$

Above the liquidus, the copper and silver atoms are thoroughly interdispersed as a liquid solution.

8.5.4 Filler Metal Types

The American Welding Society (AWS), in its "Specification for Brazing Filler Metal," AWS A5.8, lists eight categories of brazing filler metal types. These are presented in Table 8.2 and include the following categories.

Aluminum–Silicon Alloys (Designated BAlSi). Aluminum–Silicon alloys are used primarily for brazing aluminum and its alloys to themselves or various other metals and alloys. Because of the presence of oxides on aluminum and its alloys, these fillers always require flux.

Magnesium Alloys (Designated BMg). Magnesium alloys are used for brazing magnesium and its alloys to themselves and, because of the presence of oxides on magnesium and its alloys, always require flux.

Copper and Copper–Zinc Alloys (Designated BCu or RBCuZn).   Copper and copper–zinc alloys are widely used with both ferrous and nonferrous metals and alloys. They are, without question, the workhorse braze alloy system. As a group, they have limited corrosion resistance but are highly fluid. They may or may not require fluxes, depending on the base material being brazed.

Copper–Phosphorus Alloys (Designated BCuP). Copper–phosphorus alloys are good for brazing copper and copper alloys to themselves; however, they tend to liquate.

Silver Alloys (Designated BAg). Expensive but popular, silver alloys are used for brazing most ferrous and nonferrous metals and alloys except aluminum and magnesium. They offer excellent flow.

Gold Alloys (Designated BAu). Very expensive, gold alloys are used for brazing parts in electronic assemblies and vacuum tubes where volatile components are undesirable and where electrical conductivity must be high. They offer excellent corrosion resistance with iron, nickel, and cobalt alloys.

Nickel Alloys (Designated BNi). Nickel alloys are used for corrosion and heat resistance (up to 980°C or 1800°F continuous; 1200°C, or 2200°F short term). They are excellent in vacuum systems and they are the usual choice for use with nickel–base alloys.

Cobalt Alloys (Designated BCo). Cobalt alloys are used for service at high temperatures and for compatibility with Cobalt–base alloys.

Table 8.2   AWS Classification of Braze Filler Alloys

AWS Classification	Brazing Temperature Range		Nominal Composition (%)
	°F	°C	
Aluminum-silicon alloys			
BAlSi-2	1110-1150	599-621	92.5 Al, 7.5 Si
BAlSi-3	1060-1120	571-601	86 Al, 10 Si, 4 Cu
BAlSi-4	1080-1120	582-604	88 Al, 12 Si
BAlSi-5	1090-1120	588-604	90 Al, 10 Si
Magnesium alloys			
BMg-1	1120-1160	604-627	89 Mg, 2 Zn, 9 Al
BMg-2a	1080-1130	582-610	83 Mg, 5 Zn, 12 Al
Copper-phosphorus alloys			
BCuP-1	1450-1700	788-927	95 Cu, 5P
BCuP-2	1350-1550	732-843	93 Cu, 7 P
BCuP-3	1300-1500	704-816	89 Cu, 5 Ag, 6 P
BCuP-4	1300-1450	704-788	87 Cu, 6 Ag, 7 P
BCuP-5	1300-1500	704-816	80 Cu, 15 Ag, 5 P
Copper-copper zinc alloys			
BCu-1	2000-2100	1093-1149	99.9 Cu (min)
BCu-1a	2000-2100	1093-1149	99.0 Cu (min)
BCu-2	2000-2100	1093-1149	86.5 Cu (min)
RBCuZn-A	1670-1750	910-954	57 Cu, 42 Zn, 1 Sn
RBCuZn-D	1720-1800	938-982	47 Cu, 11 Ni, 42 Zn
Silver alloys			
BAg-1	1145-1400	618-760	45 Ag, 15 Cu, 16 Zn, 24 Cd
BAg-1a	1175-1400	635-760	45 Ag, 15 Cu, 16 Zn 24 Cd
BAg-2	1295-1550	700-843	45 Ag, 26 Cu, 21 Zn, 18 Cd
BAg-2A	1310-1550	710-843	30 Ag, 27 Cu, 23 Zn, 20 Cd
BAg-3	1270-1500	688-816	52 Ag, 15 Cu, 15 Zn, 15 Cd, 3 Ni
BAg-4	1435-1650	780-899	40 Ag, 30 Cu, 23 Zn, 2 Ni
BAg-5	1370-1550	743-843	45 Ag, 30 Cu, 25 Zn
BAg-6	1425-1600	774-871	53 Ag, 31 Cu, 16 Zn
BAg-7	1205-1400	651-760	56 Ag, 22 Cu, 17 Zn, 5 Sn
BAg-8	1435-1650	780-899	77 Ag, 23 Cu
BAg-8a	1410-1600	766-871	77 Ag, 23 Cu
BAg-13	1575-1775	857-635	54 Ag, 40 Cu, 5 Zn, 1 Ni
BAg-13a	1600-1800	871-982	56 Ag, 42 Cu, 2 Ni
BAg-18	1325-1550	718-843	60 Ag, 40 Cu
BAg-19	1610-1800	877-982	92 Ag, 8 Cu
Precious metals			
BAu-1	1860-2000	1016-1093	37 Au, 63 Cu
BAu-2	1635-1850	890-1010	79.5 Au, 20.5 Cu
BAu-3	1885-1995	1030-1090	34 Au, 62 Cu, 4 Ni
BAu-4	1740-1840	949-1004	82 Au, 18 Ni
Nickel alloys			
BNi-1	1950-2200	1066-1204	14 Cr, 3 Br, 4 Si, 4 Fe, 75 Ni
BNi-2	1850-2150	1010-1177	7 Cr, 3 Br, 4 Si, 3 Fe, 83 Ni
BNi-3	1850-2150	1010-1177	3 Br, 4 Si, 2 Fe, 91 Ni
BNi-4	1850-2150	1010-1177	1 Br, 3 Si, 2 Fe, 94 Ni
BNi-5	2100-2200	1149-1204	19 Cr, 10 Si, 71 Ni
BNi-6	1700-1875	927-1025	11 Br, 89 Ni
BNi-7	1700-1900	927-1038	13 Cr, 10 Br, 77 Ni

Howard B. Cary, *Modern Welding Technology,* 2e, ©1989, pp. 220, 221, 222. Reprinted by permission of Prentice-Hall, Englewood Cliffs, New Jersey.

Table 8.3 Brazing Fillers for Use with Refractory Metals and Alloys

Brazing Filler Metal	Liquidus Temperature		Brazing Filler Metal [a]	Liquidus Temperature	
	°C	°F		°C	°F
Cb	2416	4380	Mn-Ni-Co	1021	1870
Ta	2997	5425			
Ag	960	1760	Co-Cr-Si-Ni	1899	3450
Cu	1082	1980	Co-Cr-W-Ni	1427	2600
Ni	1454	2650	Mo-Ru	1899	3450
Ti	1816	3300	Mo-B	1899	3450
Pd-Mo	1571	2860	Cu-Mn	871	1600
Pt-Mo	1774	3225	Cb-Ni	1190	2175
Pt-30W	2299	4170			
Pt-50Rh	2049	3720	Pd-Ag-Mo	1306	2400
			Pd-Al	1177	2150
Ag-Cu-Zn-Cd-Mo	619-701	1145-1295	Pd-Ni	1205	2200
Ag-Cu-Zn-Mo	718-788	1325-1450	Pd-Cu	1205	2200
Ag-Cu-Mo	780	1435	Pd-Ag	1306	2400
Ag-Mn	971	1780	Pd-Fe	1306	2400
			Au-Cu	885	1625
Ni-Cr-B	1066	1950	Au-Ni	949	1740
Ni-Cr-Fe-Si-C	1066	1950	Au-Ni-Cr	1038	1900
Ni-Cr-Mo-Mn-Si	1149	2100	Ta-Ti-Zr	2094	3800
Ni-Ti	1288	2350			
Ni-Cr-Mo-Fe-W	1305	2380	Ti-V-Cr-Al	1649	3000
Ni-Cu	1349	2460	Ti-Cr	1481	2700
Ni-Cr-Fe	1427	2600	Ti-Si	1427	2600
Ni-Cr-Si	1121	2050	Ti-Zr-Be[b]	999	1830
			Zr-Cb-Be[b]	1049	1920
			Ti-V-Be[b]	1249	2280
			Ta-V-Cb[b]	1816-1927	3300-3500
			Ta-V-Ti[b]	1760-1843	3200-3350

[a]Not all the filler metals listed are commercially available.
[b]Depends on the specific composition.
Reprinted with permission from *Welding Handbook,* Vol. 2, 8th Ed., American Welding Society, 1991, p. 395.

In addition to these major categories, there are specialty braze fillers. One example is fillers for brazing refractory metals and alloys, some examples of which are listed in Table 8.3. Table 8.4 lists various base metal–filler metal combinations.

Metal braze alloy fillers can be and are used for brazing metals to metals, metals to ceramics, or ceramics to ceramics (including graphite to itself and to metals). Braze filler alloys are commonly listed under the base metals (or ceramics) for which they are most suited. Just to show the diversity of alloys available, Table 8.5 lists BAg filler alloys suitable for brazing stainless steels. There are literally thousands of commercially available brazing filler alloys and probably an equal or larger number of proprietary or specialty fillers.

Table 8.4 Base Metal–Braze Filler Metal Combinations

	Al & Al Alloys	Mg & Mg Alloys	Cu & Cu Alloys	Carbon & Low Alloy Steels	Cast Iron	Stainless Steel	Ni & Ni Alloys	Ti & Ti Alloys	Be, Zr, & Alloys (Reactive Metals)	W, Mo, Ta, Cb & Alloys (Refractory Metals)	Tool Steels
Al & Al alloys	BAlSi										
Mg & Mg alloys	X	BMg									
Cu & Cu alloys	X	X	BAg, BAu, BCuP, RBCuZn								
Carbon & low alloy steels	BAlSi	X	BAg, BAu, RBCuZn, BNi	BNi							
Cast iron	X	X	BAg, BAu, RBCuZn, BNi	BAg, BAu, BCu, RBCuZn, BNi	BAg, RBCuZn, BNi						
Stainless steel	BAlSi	X	BAg, BAu	BAg, RBCuZn, BNi	BAg, BAu, BCu, BNi	BAg, BAu, BCu, BNi					
Ni & Ni alloys	X	X	BAg, BAu, RBCuZn, BNi	BAg, BAu, BCu, BNi	BAg, BCu, RBCuZn	BAg, BAu, BCu, BNi	BAg, BAu, BCu, BNi				
Ti & Ti alloys	BAlSi	X	BAg	BAg, BCu, RBCuZn, BNi	BAg, BNi[a]	BAg, BNi[a]	BAg	Y			
Be, Zl & alloys (reactive metals)	X	X	BAg	BAg, BNi[a]	BAg, BCu, BNi[a]	BAg, BCu, BNi[a]	BAg, BNi[a]	Y	Y		
W, Mo, Ta, Cb & alloys (refractory metals)	BAlSi(Be)	X	BAg, BNi	BAg, BCu, BNi[a]	BAg, BAu, BCu, BNi	BAg, BAu, BCu, BNi	BAg, BCu, BNi[a]	Y	Y	Y	
Tool steels	X	X	BAg, BAu, RBCuZn, BNi	BAg, BAu, BCu, BNi	BAg, BAu, RBCuZn, BNi	BAg, BAu, BCu, BNi	BAg, BAu, BCu, RBCuZn, BNi	X	X	X	BAg, BAu, BCu, RBCuZn, BNi

Note: Refer to AWS Specification A5.8 for information on the specific compositions within each classification.
X—Not recommended; however, special techniques may be practicable for certain dissimilar metal combinations.
Y—Generalizations on these combinations cannot be made. Refer to the Brazing Handbook for usable filler metals.
[a]—Special brazing filler metals are available and are used successfully for specific metal combinations.

Filler Metals:

BAlsi—Aluminum	BCuP—Copper phosphorus	BAg—Silver base	RBCuZn—Copper zinc
BAu—Gold base	BMg—Magnesium base	BCu—Copper	BNi—Nickel base

Reprinted with permission from *Welding Handbook*, Vol. 2, 8th Ed, American Welding Society, 1991, p. 393.

Table 8.5 Silver-Based Braze Filler Alloys for Use with Stainless Steels

Filler Metal	Composition, %									Solidus Temperature		Liquidus Temperature		Brazing Temperature Range	
	Ag	Cu	Zn	Cd	Ni	Sn	Li	Mn	Other Elements (total)	°C	°F	°C	°F	°C	°F
BAg-1	44.0-46.0	14.0-16.0	14.0-18.0	23.0-25.0	...	...	...	...	0.15	607	1125	618	1145	618-760	1145-1400
BAg-1a	49.0-51.0	14.5-16.5	14.5-18.5	17.0-19.0	...	...	...	...	0.15	627	1160	635	1175	635-760	1175-1400
BAg-2	34.0-36.0	25.0-27.0	19.0-23.0	17.0-19.0	...	...	...	...	0.15	607	1125	701	1295	701-843	1295-1550
BAg-2a	29.0-31.0	26.0-28.0	21.0-25.0	19.0-21.0	...	...	...	...	0.15	607	1125	710	1310	710-843	1310-1550
BAg-3	49.0-51.0	14.5-16.5	13.5-17.5	15.0-17.0	2.5-3.5	...	...	...	0.15	632	1170	688	1270	688-816	1270-1500
BAg-4	39.0-41.0	29.0-31.0	26.0-30.0	...	1.5-2.5	...	...	...	0.15	671	1240	779	1435	779-899	1435-1650
BAg-5	44.0-46.0	29.0-31.0	23.0-27.0	...	...	...	...	...	0.15	677	1250	743	1370	743-843	1370-1550
BAg-6	49.0-51.0	33.0-35.0	14.0-18.0	...	...	...	...	...	0.15	688	1270	774	1425	774-871	1425-1600
BAg-7	55.0-57.0	21.0-23.0	15.0-19.0	...	...	4.5-5.5	...	...	0.15	618	1145	651	1205	651-760	1205-1400
BAg-8	71.0-73.0	Rem	...	...	...	...	...	...	0.15	779	1435	779	1435	779-899	1435-1650
BAg-8a	71.0-73.0	Rem	...	...	...	...	0.25-0.50	...	0.15	766	1410	766	1410	766-871	1410-1600
BAg-9	64.0-66.0	19.0-21.0	13.0-17.0	...	...	...	...	...	0.15	671	1240	713	1325	713-843	1325-1550
BAg-10	69.0-71.0	19.0-21.0	8.0-12.0	...	...	...	...	...	0.15	690	1275	738	1360	738-843	1360-1550
BAg-13	53.0-55.0	Rem	4.0-6.0	...	0.5-1.5	...	...	...	0.15	713	1325	857	1575	857-969	1575-1775
BAg-13a	55.0-57.0	Rem	...	...	1.5-2.5	...	...	...	0.15	771	1420	893	1640	871-982	1600-1800
BAg-18	59.0-61.0	Rem	...	...	...	9.5-10.5	...	...	0.15	601	1115	713	1325	713-843	1325-1550
BAg-19	92.0-93.0	Rem	...	...	...	...	0.15-0.30	...	0.15	760	1400	885	1635	877-982	1610-1800
BAg-20	29.0-31.0	37.0-39.0	30.0-34.0	...	...	...	...	...	0.15	677	1250	766	1410	766-871	1410-1600
BAg-21	62.0-64.0	27.5-29.5	...	...	2.0-3.0	5.0-7.0	...	...	0.15	690	1275	801	1475	801-899	1475-1650
BAg-22	48.0-50.0	15.0-17.0	21.0-25.0	...	4.0-5.0	...	...	7.0-8.0	0.15	682	1260	699	1290	699-830	1290-1525
BAg-23	84.0-86.0	...	...	...	...	...	...	Rem	0.15	960	1760	971	1780	971-1038	1780-1900
BAg-24	49.0-51.0	19.0-21.0	26.0-30.0	...	1.5-2.5	...	...	...	0.15	660	1220	707	1305	707-843	1305-1550
BAg-25	19.0-21.0	39.0-41.0	33.0-37.0	...	...	...	...	4.5-5.5	0.15	738	1360	790	1455	790-846	1455-1555
BAg-26	24.0-26.0	37.0-39.0	31.0-35.0	...	1.5-2.5	...	...	1.5-2.5	0.15	707	1305	801	1475	801-871	1475-1600
BAg-27	24.0-26.0	34.0-36.0	24.5-28.5	12.5-14.5	...	...	...	...	0.15	607	1125	746	1375	746-857	1375-1575
BAg-28	39.0-41.0	29.0-31.0	26.0-30.0	...	...	1.5-2.5	...	...	0.15	649	1200	710	1310	710-843	1310-1550

Reprinted with permission from Mel M. Schwartz, *Brazing*, 2nd Ed., ASM International, 1990, p. 104.

8.5.5 Ceramic Braze Fillers

Ceramics can be joined to themselves or to metals by brazing with ceramic fillers rather than metal fillers. Glasses are a common ceramic braze (or "solder") material, since glasses soften and flow well at elevated temperatures. However, mixtures of glasses and crystalline ceramics, or even all crystalline phases, often as eutectics, can be used as braze filler materials.

The same criteria apply for selecting ceramic braze filler materials as for metal braze filler materials, i.e., they must have low melting temperatures relative to the substrates for which they are intended, they must wet the substrate, they must be fluid, they should not liquate, and they should have appropriate mechanical, chemical, and physical properties for the application.

Chapter 13 discusses ceramic braze fillers in more detail, where the joining of ceramics and glasses is addressed specifically.

8.5.6 Brazeability

The base material has a prime effect on the strength and durability of a brazed joint. Obvious factors include the strength of the base material, the hardenability of the base material, the coefficient of thermal expansion of the base material compared to the filler, and the reactivity of the base material. High-strength base materials, especially high-strength metals, tend to produce higher strength braze joints. Most important of all, however, is the receptivity of the base material to brazing in the first place, and this depends directly on the ease or difficulty of wetting the base material. Table 8.6 lists the relative ease of brazing various base materials, i.e., their *brazeability* .

Table 8.6   Relative Ease of Brazing Various Base Materials

Degree of Difficulty	Materials
Easy	• Cu, Ni, Co, and their alloys
	• Steels
	• Precious Metals
Fair	• Al, W, Mo, Ta, and alloys with more than 5% of metals that form refractory oxides
	• Cast iron
	• WC
Difficult	• Ti, Zr, Be, and their alloys
	• Ceramics
	• Graphite
	• Glass
Impossible	• None

Modified with permission from Mel M. Schwartz, *Brazing,* 2nd Ed., ASM International, 1990, p. 15.

8.6 FLUXES AND ATMOSPHERES

8.6.1 General Description

Metals and their alloys tend to react with various constituents of the atmosphere to which they are exposed. This tendency increases as the temperature is raised and is especially great once melting occurs. The most common reaction is oxidation, but nitrides, carbides, and hydrides, are sometimes formed, and any of these reaction products hinders the production of consistently sound brazed joints. Fluxes, gas atmospheres, and vacuum are used to exclude reactants and thus prevent undesirable reactions during brazing. Under some conditions, fluxes and atmospheres may also actually reduce oxides that are present, cleaning the metal surface and refining the bulk melt.

8.6.2 Fluxes

A *flux* is a nonmetallic chemical compound that will react with metal oxides to reduce those oxides and clean the metal. Typically, fluxes contain chlorides, fluorides, fluoroborates, borax, or borates of active metals (such as Na, K, Ca, or Ba), boric acid, wetting agents, and water. To effectively protect the surfaces of metals that are to be brazed, a flux must completely cover and protect those surfaces until the brazing temperature is reached. The flux must then remain chemically active and react throughout the brazing cycle.

Fluxes become active and effective in their molten state. The viscosity and surface tension of the flux and the interfacial energy between the flux and the substrate are important since the molten filler metal must be able to displace the molten flux from the joint at the brazing temperature.

There is no single flux that is best for all brazing applications. Rather, fluxes must be matched to the application by considering the base metal and the particular process. Fluxes are classified by the AWS into six groups by base metal and are recommended for use in rather specific temperatures ranges. Table 8.7 lists these classes and use temperatures.

Within a particular AWS type, there are several criteria for selecting a specific flux for maximum efficiency. As guidelines: (1) for dip brazing, water (including water of hydration) must be avoided in fluxes; (2) for resistance brazing, the flux must be electrically conductive (usually, wet and dilute); (3) the ease of removal of flux residue should be considered; and (4) the corrosive action of the flux to the filler or substrate should be minimal to do the job.

Fluxes are available in a variety of physical forms, including powders, pastes, and liquids. The form selected depends on the individual work requirements, the specific brazing process, and the particular procedure.

Flux residue and excess flux must be thoroughly removed from the brazement as soon as possible after brazing is completed; otherwise corrosion can occur. Methods for removing flux and flux residues include washing in hot water, taking advantage of thermal shock by quenching to break up brittle oxides or other non-metallics, and often using agitation or scrubbing. Chemical cleaners (e.g., solvents, acids, or alkalies) can

Table 8.7   List of Fluxes for Brazing Various Base Materials

AWS Brazing Flux Type No.	Base Metals Being Brazed	Recommended Filler Metals	Recommended Useful Temp. Range °F	Recommended Useful Temp. Range °C	Major Flux Ingredients	Forms Available
1	All brazeable aluminum alloys	BAlSi	700–1190	371–643	Chlorides Fluorides	Powder
2	All brazeable magnesium alloys	BMg	900–1200	482–649	Chlorides Fluorides	Powder
3A	Copper and copper-base alloys (except those with aluminum) iron-base alloys; cast iron; carbon and alloy steel; nickel and nickelbase alloys; stainless steels; precious metals	BCuP BAg	1050–1600	566–871	Boric acid Borates Fluorides Fluoborates Wetting agents	Powder Paste Liquid
3B	Copper and copper-base alloys (except those with aluminum); iron-base alloys; cast iron; carbon and alloy steel; nickel and nickel-base alloys; stainless steel; precious metals	BCu BCuP BAg BAu RBCu Zn BNi	1350–2100	732–1149	Boric acid Borates Fluorides Fluoborates Wetting agent	Powder Paste Liquid
4	Aluminum bronze, aluminum brass and iron or nickel-base alloys containing minor amounts of Al and/or Ti	BAg (all) BCuP (copper-base alloys only)	1050–1600	566–871	Chlorides Fluorides Borates Wetting agent	Powder Paste
5	Same as 3A and B above	Same as 3B excluding BAg through -7)	1400–2200	760–1204	Borax Borac acid Borates Wetting agent	Powder Paste Liquid

Howard B. Cary, *Modern Welding Technology*, 2e, ©1989, pp. 220, 221, 222. Reprinted by permission of Prentice-Hall, Englewood Cliffs, New Jersey.

be used, as can some mechanical aids (e.g., ultrasonics, shot or grit blasting). Many of the cleaning procedures are similar to those used for proper preparation of adherends for adhesive bonding (see Chapter 4, Section 4.5.2).

8.6.3 Controlled Atmospheres

Controlled atmospheres or a vacuum can be employed to prevent the formation of oxides during brazing and, in many instances, to remove oxide films to promote wetting and flow. Techniques include (1) the use of gaseous atmospheres alone; (2) the use of gaseous atmospheres together with fluxes; (3) the use of high vacuums; and (4) combinations of vacuum and gas atmospheres.

Gaseous atmospheres include pure forms and mixtures of CO_2, CO, H_2, inert gases (such as Ar), N_2, and CH_4. Of these, CO and H_2 are reducing, while the others are simply inert. Table 8.8 lists the AWS categories of atmospheres for brazing. It should be noted that controlling the moisture content of the atmosphere, i.e., the *dew point*, is extremely important to avoid oxidation by water vapor. Recommended maximum dew points are also given in Table 8.8.

8.7 BRAZE JOINT DESIGN

From a mechanical standpoint, the design of a brazement is no different than the design of any other part. Static loading, dynamic loading, stress concentrations, and environmental factors must all be considered. In addition, some important factors specific to brazements are (1) the composition of the base materials and the filler, (2) the type and design of the joint, and (3) the service requirements, including mechanical performance, electrical or thermal conductivity, pressure tightness, corrosion resistance, and service temperature.

In general, the strength of the filler metal in a brazed joint is lower than the strength of the base materials. The specific joint strength will vary according to joint clearance, degree of filler–substrate interaction, and presence of defects in the joint.

Basically two types of joints are used in brazing designs: the lap joint and the butt joint. A variant of the butt joint, which forces loading more toward shear, is called the *scarf joint*. These joint types are shown in Figure 8.5. In lap joints, overlap of at least three times the thickness of the thinner member will usually yield the maximum joint efficiency. Scarf joints are attractive because they increase the joint area without increasing the thickness of the joint members. Figures 8.6 illustrates some lap and butt joint designs for various static and dynamic loading situations.

Joint clearance, or *gap,* is a key parameter in the design and production of brazed joints for several reasons. First, the clearance in a joint is important purely from the standpoint of the effect of mechanical restraint. The plastic flow of the braze filler is restrained by the higher strength base material only if the filler does not get too thick. On the other hand, for purposes of strain accommodation, the braze filler layer cannot be too thin either. Second, joint clearance must be sufficient to prevent the entrapment

Table 8.8   Atmospheres for Brazing

AWS brazing atmosphere type number	Source	Maximum dew point incoming gas	Approximate composition, %				Application		Remarks
			H_2	N_2	CO	CO_2	Filler metals[e]	Base metals	
1	Combusted fuel gas (low hydrogen)	Room temp.	5-1	87	5-1	11-12	BAg,[f] BCuP, RBCuZn[a]	Copper, brass[a]	
2	Combusted fuel gas (decarburizing)	Room temp.	14-15	70-71	9-10	5-6	BCu, BAg[a], RBCuZn[a], BCuP	Copper[b], brass[a], low-carbon steel, nickel, monel, medium carbon steel[c]	Decarburizes
3	Combusted fuel gas, dried	−40°C (−40°F)	15-16	73-75	10-11		Same as 2	Same as 2 plus medium and high-carbon steels, monel, nickel alloys	
4	Combusted fuel gas, dried (carburizing)	−40°C (−40°F)	38-40	41-45	17-19		Same as 2	Same as 2 plus medium and high-carbon steels	Carburizes
5	Dissociated ammonia	−54°C (−65°F)	75	25			BAg[a], BCuP, RBCuZn[a], BCu, BNi	Same as for 1, 2, 3, 4 plus alloys containing chromium[d]	
6	Cylinder hydrogen	Room temp.	97-100				Same as 2	Same as 2	Decarburizes
7	Deoxygenated and dried hydrogen	−59°C (−75°F)	100				Same as 5	Same as 5 plus cobalt, chromium, tungsten alloys and carbides[d]	Decarburizes
8	Heated volatile materials	Inorganic vapors (i.e., zinc, cadmium, lithium, volatile fluorides)					BAg	Brasses	Special purpose. May be used in conjunction with 1 thru 7 to avoid use of flux

	Purified inert gas	Inert gas (e.g., helium, argon, etc.)			Special purpose. Parts must be *very* clean and atmosphere must be pure
9			Same as 5	Same as 5 plus titanium, zirconium, hafnium	
10	Vacuum	Vaccum above 2 Torr[f]	BCuP, BAg	Cu	
10A	Vacuum	0.5 to 2 Torr	BCu, BAg	Low carbon steel, Cu	
10B	Vacuum	0.001 to 0.5 Torr	BCu, BAg	Carbon and low alloy steels, Cu	
10C	Vacuum	1×10^{-3} Torr and lower	BNi, BAu, BAlSi, Ti alloys	Ht. and corr. resisting steels, Al, Ti, Zr, refractory metals	

Note AWS Types 6, 7, and 9 include reduced pressures down to 2 Torr.
a. Flux is required in addition to atmosphere when alloys containing volatile components are used.
b. Copper should be fully deoxidized or oxygen free.
c. Heating time should be kept to a minimum to avoid objectionable decarburization.
d. Flux must be used in addition if appreciable quantities of aluminum, titanium, silicon, or beryllium are present.
e. See Table 8.2 for explanation of filler metals.
f. 1 Torr = 133 Pa.
Reprinted with permission from *Brazing Manual*, American Welding Society, 1976, p. 62.

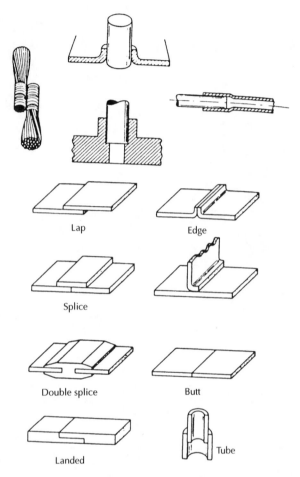

Lap

Edge

Splice

Double splice

Butt

Landed

Tube

Figure 8.5   Various geometries used in braze joints. (Howard B. Cary, *Modern Welding Technology,* 2e, ©1989, pp. 220, 221, 222. Reprinted by permission of Prentice-Hall, Englewood Cliffs, New Jersey.)

of slag from fluxes or gas released by volatiles or absorbed from atmospheres and trapped in the form of voids. Third, the relationship between joint clearance and capillary force controls proper flow and distribution of the filler. Finally, for diffusion brazing, clearance controls the amount of filler metal that must be diffused away. Table 8.9 gives the recommended joint clearances for brazing with various filler classes.

Finally, it is vitally important that the braze filler completely fills the joint and that defects are avoided during brazing. Defects to be avoided include voids (i.e., entrapped gas pockets), unbrazed areas, pores or porosity (i.e., from evolved or absorbed gases), flux inclusions, cracks, and brittle (often, intermetallic) compounds. One of the most effective ways of preventing defects of these types is to properly place filler. Figure 8.7 shows how filler can be preplaced for maximum effectiveness, using various physical forms. Braze filler, whether applied during brazing or preplaced and whether the process is performed manually or automatically, should be applied to the joint in such a way that when it flows through the joint properly, filler will appear at

Table 8.9 Recommended Joint Clearances for Brazing

Filler Metal AWS Classification	*Joint Clearance*		*Brazing Conditions*
	Inches	*Millimeters*	
BAlSi group	0.002–0.008	0.051–0.203	For length of lap less than 1/4 in. (6.4 mm)
	0.008–0.010	0.203–0.254	For length of lap greater than 1/4 in. (6.4 mm)
BCuP group	0.001–0.005	0.025–0.127	No flux or mineral brazing fluxes
BAg group	0.002–0.005	0.051–0.127	Mineral brazing fluxes
	0.000–0.002	0.000–0.051	Gas atmosphere brazing fluxes
BAu group	0.002–0.005	0.051–0.127	Mineral brazing fluxes
	-.000–0.002	0.000–0.051	Gas atmosphere brazing fluxes
BCu group	0.000–0.002	0.000–0.051	Gas atmosphere brazing fluxes
BCuZn group	0.002–0.005	0.051–0.127	Mineral brazing fluxes
BMg	0.004–0.010	0.102–0.254	Mineral brazing fluxes
BNi group	0.002–0.005	0.051–0.127	General applications flux or atmosphere
	0.000–0.002	0.000–0.051	Free flowing types, atmosphere brazing

Howard B. Cary, *Modern Welding Technology,* 2e, ©1989, p. 222. Reprinted with permission of Prentice Hall, Englewood Cliffs, New Jersey.

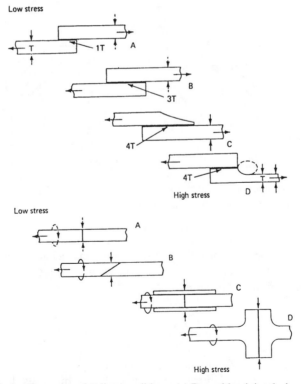

Figure 8.6 Joint design for various loading conditions. (a) Brazed lap joint designs for use at low and high stresses—flexure of right member in C and D will distribute the load through the base metal; (b) brazed butt joint designs to increase capacity of joint for high stress and dynamic loading. (Reprinted with permission from *Welding Handbook,* 8th Ed., Vol. 2, American Welding Society, 1991, p. 407.)

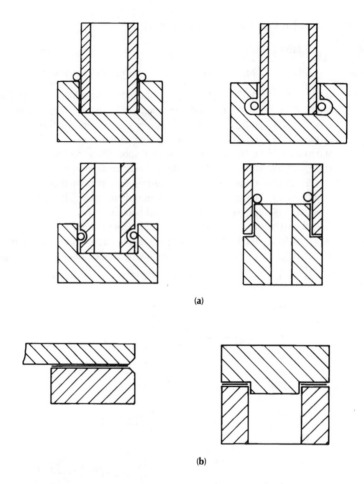

(a)

(b)

Figure 8.7   Suggested methods for preplacing brazing fillers for maximum effectiveness. (a) Preplaced filler-metal wires; (b) preplaced filler-metal shims.

some point remote from the point of application. This technique relies on the sweeping action of the filler under the influence of the capillary forces and is a reasonable sign that proper flow and fill occurred.

SUMMARY

Brazing is a process for producing permanent, mechanically-acceptable, leak-tight joints without necessitating melting of the bulk base materials. It is accomplished by using a filler material that melts above 450°C (840°F) but below the solidus of the substrate materials, and that flows by capillary action into a properly designed joint consisting of a controlled gap, or clearance. Wetting of the substrate surfaces by the filler is critical to the process as it enables capillary flow and proper distribution. Joint strength is achieved through the formation of chemical bonds; metallic bonds between metals; ionic, covalent, or mixed bonds between ceramics; and ionic, covalent, or mixed bonds between metals and ceramics.

Brazing is capable of producing strong, continuous joints over large areas or long lengths with little or no alteration of the microstructure or chemistry of the substrates. This allows dissimilar materials, by type or composition within a type, to be readily joined. The surface bonding mechanism spreads loading, so both thin and thick sections can be joined in any combination, under relatively stress-free conditions, producing close-tolerance structures or brazements. The most serious limitation of brazing can be service temperature because of the relatively low melting point of the filler versus the base materials. This can be overcome, however, with proper filler choice or use of a special diffusion process.

A variety of brazing processes allow either manual or automatic operation, with either localized heating of the joint or general, uniform heating of the entire braze assembly. Classification of processes is normally by the method of heating, such as torch brazing, furnace brazing, induction brazing, resistance brazing, dip brazing, and infrared brazing. Other special processes include diffusion brazing, laser, electron-beam, ultrasonically assisted, vapor-phase, and step-brazing.

Braze fillers can be metallic or ceramic. Metallic fillers are used for joining metals or ceramics to themselves or one another, while ceramic fillers are used exclusively for joining ceramics. Important characteristics of brazing filler materials are chemical, mechanical, and physical compatibility with the base materials; melting temperature and fluidity; the tendency for liquid and solid phases to separate or liquate; and the ability to wet and bond the substrates. Typically, braze filler alloys exhibit a low melting eutectic constituent in their constitutional diagram.

It is critical to avoid oxidation of heated metal substrates and molten metal fillers throughout brazing, so a chemically reducing nonmetallic flux or inert or reducing gaseous atmosphere or vacuum are usually used. Joint design for brazements can be lap, butt, or scarf geometries, and must have controlled clearances during actual brazing to assure proper flow of filler and filler thickness. If the filler layer is too thin, strain across the joint cannot be accommodated without fracture, while if it is too thick, plastic flow occurs and the joint is weakened. Joint clearance and filler placement are two important factors in preventing formation of defects during brazing.

PRACTICE PROBLEMS AND QUESTIONS

1. Differentiate the process of *brazing* from fusion welding, nonfusion welding, and soldering.
2. What are the specific advantages of brazing versus fusion welding? What are the specific advantages of brazing versus mechanical fastening? What are the specific advantages of brazing versus adhesive bonding?
3. What is the single greatest limitation of brazing as a joining process compared to fusion welding? What about compared to adhesive bonding? What about compared to mechanical fastening?

4. Which specific brazing processes require the entire assembly to be heated to the brazing temperature? What are some disadvantages of such processes?
5. What is the particular advantage of the induction brazing process? What is the particular advantage of the infrared brazing process? Are these processes similar or different? Explain.
6. How does diffusion brazing differ from all other brazing processes? What are some particular advantages of the diffusion brazing process?
7. What is meant by step brazing? Cite an example of where step brazing would be useful.
8. What is considered a suitable melting range for a braze filler for a particular substrate? Is a wide melting temperature range for a braze filler alloy considered advantageous or disadvantageous? Explain.
9. Cite an example of when it would be advantageous for a braze filler alloy to react with the substrate. Cite an example of when it would be considered disadvantageous for a braze filler alloy to react with the substrate.
10. What factors, other than base metal melting point, determine what melting temperature range should be used for brazing a particular material joint combination? What are some particular manufacturing concerns related to the temperature required to accomplish brazing?
11. Which AWS class of brazing alloy filler would you select for each of the following? (a) maximum temperature serviceability; (b) maximum general corrosion resistance; (c) optimum suitability for high vacuum seals; (d) optimum electrical conductivity; (e) joining aluminum alloys.
12. Differentiate between the role of all chemical fluxes used for brazing versus most controlled atmospheres. What is one of the most significant factors affecting the oxidizing potential of all gaseous atmospheres?
13. Compare the design of joints that are to be brazed to (a) typical fusion welding joints; (b) typical adhesive bonding joints.
14. What would you expect are the various failure modes observed in brazed joints? Explain.
15. Why is the gap, or clearance, of a braze joint important? How is the gap established and maintained during brazing?

REFERENCES

1. American Society for Material, *Metals Handbook, 9th edition, Vol. 6, Welding, Brazing and Soldering,* Metals Park, Ohio: American Society for Materials, 1983, pp. 929-1066.
2. American Welding Society, *Welding Handbook, Vol. 2, Welding Processes—Arc and Gas Welding and Cutting, Brazing and Soldering,* Miami: American Welding Society, 1978.
3. American Welding Society, *Brazing Manual,* Miami: American Welding Society, 1976.
4. Gilliland, R. G., and Slaughter, G. M., "The Development of Brazing Filler Metals for High Temperature Service," *Welding Journal,* Vol. 48, No. 10, pp. 463-s–469-s, 1969.
5. Lancaster, J. F., *The Metallurgy of Welding, Brazing and Soldering,* London: Allen & Unwin, 1985.
6. Schwartz, Mel M., *Brazing,* 2nd ed., Metals Park, Ohio: ASM International, 1990.
7. Witherell, C. E., and Ramos, T. J., "Laser Brazing," *Welding Journal,* Vol. 59, No. 10, pp. 267-s–277-s, 1980.

9

Soldering

9.1 INTRODUCTION

Metals and ceramics and alloys can be joined by causing a molten material to flow to
fill the space between closely gapped joint faces and then solidify without requiring or
causing melting in the base materials. One process, described in Chapter 8, is brazing;
the other, to be described here, is *soldering*. In both cases, the process of filling occurs
by capillary action due to wetting of the substrate by the liquid filler and spreading by
the lowering of surface energy. Brazing and soldering are distinguished from adhesive
bonding, which also uses a filler, which can be in a liquid form, by the nature of the
filler and filler distribution. Brazes and solders are always inorganic, while adhesives
are usually organic and are never metallic. Brazes can be metals or ceramics, while solders are always metals.[1] Both braze and solder fillers flow into and fill the joint by capillary action. Adhesives are usually applied to the faces of the surfaces to be bonded and
do not depend upon capillary flow for their distribution within the joint. Brazing and
soldering are distinguished from one another primarily on the basis of the melting temperature of the fillers employed. Braze fillers melt above 450°C (840°F), while solder
fillers, or *solders,* melt below this temperature.

The forces responsible for joining in soldering arise from the formation of bonds[2]
between the filler and the substrate or substrates, but there is often a significant contribution from a purely mechanical component as a result of interlocking between the
solder and microscopic asperities on the substrate(s) surfaces. As a result of lower bond
strength and the lower strength of the solder alloys themselves, solder joints are generally less strong than brazed joints. While soldering was used by the ancients for joining
metals as far back as the Bronze Age, this was primarily a matter of simplicity because
of the low melting temperatures of the fillers and the low heat requirement. In modern
times, soldering is used predominantly to provide electrical continuity and conductivity (i.e., connectivity) and/or for hermeticity.[3] Sound, conductive, and hermetically
tight joints can be formed at low temperatures, with little or no thermal degradation of

---

[1] Solders are always metals, although there are so-called solder glasses, which will be described in Chapter 13.
Solder glasses are simply low melting point glasses used to fuse other glasses to one another.

[2] The bonds formed between solders and a substrate depend upon the substrate but are almost always metallic. This
is true when the substrates are metals and, often, when they are ceramics. When the substrate is a glass, bonding between either a metallic or glass solder can be covalent or, more likely, van der Waals or permanent dipole.

[3] *Hermeticity* means the property of being sealed completely against the escape or entry of a gas, liquid, or
vacuum.

the substrate materials. Because of these properties, soldering has been and continues to be an extremely important process for joining electrical and electronic components.

This chapter looks at the process of soldering for joining. First, the process of soldering is defined, and its relative advantages and disadvantages compared to other joining processes are described. Then, the principles of operation of soldering are discussed. Next, the metallurgy of solder alloys is presented, in detail for the representative and predominant tin–lead system of solders and as overviews for other important, but lower usage, systems. The physical forms of solders are then described, and important individual soldering processes are presented. The critical role of fluxes and their composition and physical forms are then described. Next, solder joint design is presented. Finally, methods for evaluating the solderability of materials are addressed.

9.2 SOLDERING AS A JOINING PROCESS

9.2.1 General Description

Soldering is defined as a subgroup of welding processes that produce coalescence of materials by heating them to a suitable temperature and by using a filler material having a liquidus not exceeding 450°C (840°F) and below the solidus of the base materials. The filler material used in soldering, i.e., the *solder*, is a low melting metal (e.g., lead or tin or indium) or metallic alloy, regardless of whether the base materials being joined to themselves or one another are metals, ceramics, or glasses. The molten filler flows into and fills a prepared joint between base materials by capillary action and, thus, requires that the filler wets the substrate(s) and spreads by lowering the surface energy of the substrate(s), often through a chemical reaction. So-called solder glasses are low melting-point glasses that are used for joining one glass to another or to a metal, but these are not truly solders because their melting point can (and frequently does) exceed 450°C (840°F) and they do not flow by capillary action.

The bond between solder and base materials is almost always a combination of chemical bond formation and mechanical interlocking into microscopic surface asperities in the base material. An essential feature of soldered joints is that bonds are produced by a solvent action. The solder dissolves, but does not melt, a small amount of the base material to form a layer of an intermetallic compound. This inherently brittle layer often limits the strength that can be developed but facilitates the spreading and adhesion of the solder. The ease with which this essential solvent action occurs is related directly to the ease of wetting of the substrate by the molten solder and is described by the *solderability* of the base material.

While the strength, or mechanical integrity, of soldered joints is limited compared to other welding and brazing processes, electrical continuity, or *connectivity,* and leak-tightness, or *hermeticity,* are excellent. For these reasons, soldering is most important for the joining or interconnection of electrical components in electrical or electronic devices and assemblies. A standard radio receiver contains about 500 soldered joints; a color television set about 5,000; and a computer or telephone system more than 100,000 (see Figure 9.1). With the correct joint design and application of a good process, reliable joints are obtained, having negligible contact resistance and acceptable

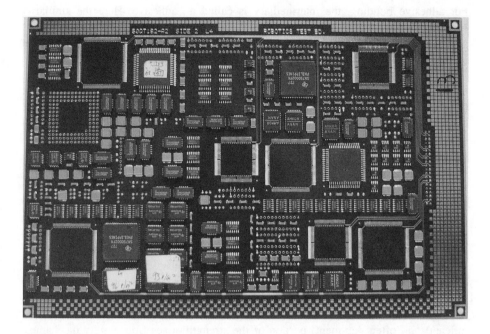

Figure 9.1   A printed circuit board from a modern computer, an example of a typical soldered electrical assembly.(Photo courtesy of Digital Equipment Corporation.)

strength. Reliability implies that the joints not only have the desired properties immediately after production but can be expected to ensure failure-free performance during the life of the equipment.

9.2.2 Soldering Compared to Solid-Phase Welding, Brazing, and Adhesive Bonding

Soldering, brazing, and adhesive bonding are all processes that are capable of forming a joint between base materials by forming chemical bonds without requiring or causing any melting of those materials. While nonfusion welding processes (e.g., friction welding, explosive welding, and diffusion bonding) are also capable of forming such joints, bonding is direct for these processes. Atoms, ions, or molecules of the base materials are brought into contact (i.e., to equilibrium interatomic spacing) under the action of pressure, with or without bulk heating. No filler material is needed or used. In soldering, brazing, and adhesive bonding, bonding is accomplished through an intermediary. The intermediary is added to provide a source of atoms or molecules to allow bonding to the base materials through the intermediary. Pressure, beyond that needed to hold the joint elements in contact, is usually not required and is almost never applied. The intermediary is called a filler material or filler alloy (for brazing) or a solder (for soldering).

For adhesive bonding, the intermediary is called an adhesive. Here the similarity between the processes breaks down.

Both brazing and soldering require that the filler melt, flow into, and fill the joint by capillary action. In adhesive bonding, there is no such requirement of the adhesive. First, the adhesives used for adhesive bonding only melt if they are thermoplastic; otherwise, they start as liquids before the cure (in the case of thermosets) or dry (in the case of solvent-thinned thermoplastics). Also, capillary action is only rarely involved in the distribution of certain liquid adhesives. More often, adhesives are applied to the faces of the parts to be joined so that distribution of the adhesive is accomplished as part of the preparation (.i.e., adhesive application) process. The composition of the intermediaries used with soldering and brazing versus adhesive bonding also differ considerably. Braze and solder fillers are always alloys of inorganic materials. Braze fillers can be alloys of metals or ceramics, while solder alloys are always metal alloys (although there are solder glasses for use in joining glasses, as well). Adhesives, on the other hand, are usually organic materials (e.g., resins) but can be inorganic compounds (e.g., sodium silicate).[4]

The bonding that develops between the intermediary and the base material(s) is often primary for soldering and brazing, while it is usually secondary for adhesive bonding. The source of the strength of joints produced by brazing, however, comes entirely from these primary bonds, while for soldering, as for adhesive bonding, a significant (and often dominant) portion of the strength arises from purely mechanical sources. As with adhesives in adhesive bonding, solders fill the microscopic valleys on real surfaces through wetting and, by so doing, mechanically lock the substrate to the adhesive or solder.

In summary, soldering joins metallic or ceramic or glass base materials together, in similar or dissimilar combinations, by using a filler that melts, flows, and fills a properly fitted joint by surface wetting and spreading by capillary action. No melting of the base materials is required or caused. The strength of the joint arises from a combination of primary bond formation between the solder and the substrate and mechanical interlocking of the solder into microscopic asperities on the substrate surface. As such, soldering is a subgroup of welding and is especially closely related to brazing, with the principal difference being that solder fillers melt below, rather than above, 450°C (840°F).

9.2.3 Advantages and Disadvantages of Soldering

Obviously, the low temperatures needed to accomplish soldering mean that base materials can be joined with little or no thermal damage in most cases. This important general advantage notwithstanding, soldering has some clear specific advantages over competi-

---

[4] The adhesives used in cementing or mortaring ceramics are almost always inorganic materials and include silicates, carbonates, and various mixtures of glass frits and crystalline ceramics (see Chapter 13, Section 13.3.2).

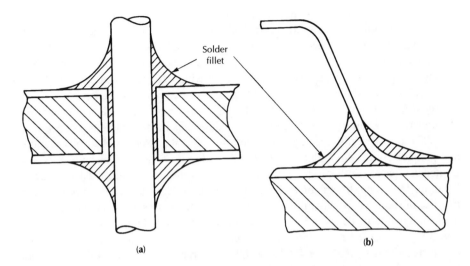

Figure 9.2   A typical solder joint showing the natural formation of a fillet by wetting to reduce stress concentrations. (a) Through-hold lead; (b) preformed lead in surface mount.

tive joining techniques such as welding or adhesive bonding using conductive adhesives. First, the solder joint forms *itself* by the nature of the wetting process, even when the heat and solder are not directed precisely to the places to be soldered. The meniscus that forms upon wetting of the substrate by the molten solder automatically produces a joint that gives rise to minimal stress risers under loading. This is shown in Figure 9.2, where surface tension causes the solder to form a natural fillet or radius, rather than a sharp notch. Second, because solder does not adhere to insulating materials (unless a special effort is made), it may, in many cases, be applied in excess quantities without detrimental consequences (e.g., conductive adhesives). This enables mass soldering of joints, another distinct advantage in achieving high manufacturing productivity. Third, as the soldering temperature is relatively low, there is no need for the heat to be applied locally as for welding; rather, the entire assembly can usually be heated without detriment. In situations where the entire assembly cannot be heated to the soldering temperature, local heating can easily be accomplished with simple heating devices.

A fourth advantage is that soldering allows considerable freedom in the dimensioning of joints. Because of this it is possible to obtain good results even if a large variety of components are used on the same soldered product without having to adjust process parameters. Fifth, the equipment needed for both manual (i.e., hand) soldering and machine soldering is relatively simple, and the process can be easily automated, offering production economics (e.g., in-line processing). A sixth rather unique advantage, shared only completely by mechanical fastening and occasionally by adhesive bonding, is that soldered connections can be disassembled, disconnected, or desoldered if necessary, meaning that repair or upgrade can be accomplished easily.

The adoption of mechanized methods for soldering, in particular, and for manufacturing, in general, is stimulated on the one hand by considerations of efficiency and on the other by the desire to obtain a more closely controlled quality than is realizable by manual techniques. This all means improved productivity. Machine soldering affords a gain in quality because the machine, unlike a human being, never relaxes its attention and never tires. At the same time, however, machine soldering involves losses because the process is generally incapable of adapting to extreme or unexpected situations. Unless special measures are taken (e.g., elaborate sensing and adaptive or intelligent control schemes), no automated method is able to cope with such contingencies as inadequate solderability of certain joints or deficiencies in the soldering process.

The principal disadvantage of soldering is that the strength of a soldered joint, and particularly the strength under thermally induced or mechanical fatigue, is quite limited compared to other joining methods such as welding, brazing, and even adhesive bonding. The creep strength of soldered joints is also inferior to welded or brazed joints because of the lower melting point of the filler and the correspondingly higher homologous temperature. To show how serious this can be, room temperature (i.e, 25°C, or 298 K) is 65% of the absolute melting temperature of the most common, eutectic tin–lead solder (i.e., 183°C or 456 K), well into the range where creep (normally beginning around 50 to 60% of T_{MP} absolute) can be expected. Typical operating temperatures of electronic devices can easily cause tin–lead eutectic solders to operate at near 90% of their homologous temperature.

A practical difference between soldering and welding is that in welding the welds almost always have to be made consecutively, whereas soldered joints can be made simultaneously in one single operation. This factor, in principle, makes soldering a cheap joining method. On the other hand, assuring the quality of *every* joint becomes difficult and demands either expensive, off-line, postprocess inspection or, in most modern operations, embedded inspection.

Table 9.1 summarizes the relative advantages and disadvantages of soldering versus other joining processes.

Table 9.1   Summary of Relative Advantages and Disadvantages of Soldering versus Other Joining Processes

Advantages	*Disadvantages*
1. Solder joint forms itself by nature of wetting process; self-controlling	1. Strength is very limited, as degree of metallurgical bonding can be minimal
2. Amenable to producing many joints at once	2. Low metling temperature of solder means service is usually at high homologous temperature; creep is a problem
3. Heat need not be applied locally but can be applied to entire assembly without detriment	3. Uniform heating can be detrimental to some heat-sensitive components
4. Almost no significant change of composition or microstructure of base materials	4. Production of multiple joints at a time can make inspection of every joint difficult and can make process control difficult
5. Considerable freedom in joint dimensioning	
6. Allows disassembly by desoldering	
7. Manual or automated; highly automatable	

9.3 SOLDERING PROCESS CONSIDERATIONS

9.3.1 General Description

A sound soldered joint is achieved by selecting and using the proper materials and processes, along with the proper joint design. Key factors include (1) base material selection; (2) solder selection; (3) flux selection; (4) joint design; (5) part and joint pre-cleaning; (6) soldering process selection and operation; (7) flux residue removal; and (8) joint inspection.

9.3.2 Base Material Selection

Base materials are usually selected for specific property requirements that are needed for the components' or assembly's design function, such as strength, ductility, electrical conductivity, thermal conductivity, weight, corrosion resistance, and vacuum compatibility. When soldering is required to join components, the solderability of the base material(s) must also be considered in their selection. The *solderability* of a material can vary widely and depends directly on the ease or difficulty of wetting, or *wettability*. Table 9.2 lists metallic materials in terms of their wettability in the uncoated condition when a common tin–lead solder is used. There is a reasonably clear relationship between the wettability of a base metal by a solder and the tendency of that base metal to form a tenacious oxide. Some metals, like chromium and magnesium, form tenacious oxides so readily that they are practically impossible to solder. Others, like gold, have very good wettability because they do not form oxides under most conditions. Still others form oxides, but these oxides can be easily removed.

Solderability or wettability can be enhanced by removing any oxide layer with an appropriate flux or fluxing agent. For materials with inherently good wettability, mild or mildly aggressive fluxes can be used. For materials that are difficult to wet, very aggressive fluxes are required. For materials that are considered practically unwettable, coatings are required to serve as intermediaries to the solder. For many materials,

Table 9.2   Rating of the Wettability of Various Uncoated Materials by Tin–Lead Solder

Good	Fair	Moderate	Difficult	Practically Impossible
Gold	Bronze	Kovar	Aluminum bronze	Chromium
Tin–lead	Brass	Nickel–iron	Alloyed steel	Magnesium
Tin	Monel	Nickel	Aluminum	Molybdenum
Silver	Nickel silver	Steel		Tungsten
Palladium				Beryllium
Copper		Zinc		
Mild flux	⟶		Aggresive flux	

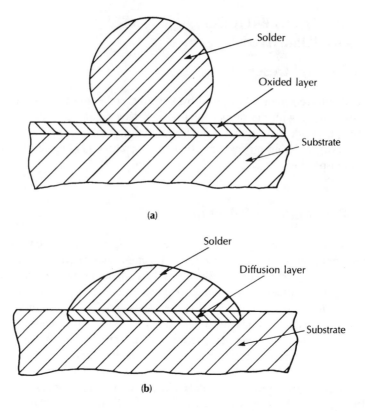

Figure 9.3   The difference in wetting of a clean versus an oxidized base metal by a drop of molten solder. (a) Oxidized substrate; (b) clean substrate.

removing the oxide initially must be followed by keeping the oxide from reforming. This is usually accomplished with coatings.

Figure 9.3 shows the difference in wetting of a clean versus an oxidized base metal by a drop of molten solder. When the base metal is clean, the drop spreads and some diffusion occurs at the interface.[5] To be truly solderable, a material must be wettable and remain wettable even after some storage. If the wettability deteriorates during storage, the surface is said to *age*.[6] Such aging is problematic in a production environment, as precleaning and soldering operations must be carefully timed.

Base materials, whether metallic or not, may be coated for several reasons. These include decorative, protective, and functional purposes, as well as for improving wet-

---

[5] Often the diffusion that occurs at the interface between a clean base metal and molten solder results in the formation of a brittle intermetallic. This intermetallic may limit the strength of the joint but is believed to play an important role in the spreading and adhesion of the solder.

[6] The phenomenon of *aging* as it pertains to a surface prepared for soldering does not involve the same mechanism as the aging responsible for strengthening a precipitation-hardenable alloy. In soldering, aging involves the reoxidation of the surface.

tability for soldering. Coatings can provide protection against the environment or can improve function by increasing wear resistance, providing high reflectivity or specularity, or providing electrical contact or insulation. Coatings for improving solderability can be applied by electrolytic deposition, thermal deposition, mechanical cladding, or chemical deposition. Different coating methods and systems are used for different purposes.

In addition to being used to enable soldering of normally unsolderable base materials, intermediate layers are sometimes used during soldering to provide (1) improved adhesion; (2) a barrier against the formation of brittle intermetallics; (3) a barrier against zinc diffusion in zinc-bearing alloys (to prevent "dezincification"); and (4) improved corrosion resistance.

There are two other considerations in selecting base metals for solder joints. First, the materials making up the joint elements, as well as the solder alloy, must be electrochemically compatible to avoid galvanic corrosion. Large differences in electrochemical potential will lead to more severe galvanic corrosion. Second, the metals composing the soldered joint, as well as the solder alloy, should be compatible in terms of their relative coefficients of expansion. Too great a difference in coefficient of thermal expansion (CTE) can give rise to severe thermally induced stresses, leading to thermal fatigue. As a rule, CTEs should be kept as close as possible, not exceeding about 10% difference between adjacent materials. Large differences in CTE between base metals can often be handled by using several intermediate materials with intermediate values of CTE.

9.3.3 Solder Selection

The solder to be used in a soldered joint is selected to provide good wetting, spreading or flow, and joint penetration in the actual soldering operation *and* the desired joint properties in the finished product. As mentioned earlier, all solders are metal alloys. Major solder alloy systems and their metallurgy are described in some detail in Section 9.4.

9.3.4 Flux Selection

A flux is intended to enhance the wetting of the base material by the solder by removing tarnish films from precleaned surfaces and by preventing reoxidation during the soldering operation. The selection of the type of flux usually depends on the ease with which a material can be soldered, i.e., its solderability or wettability. This, in turn, usually relates to the ease with which any tarnish can be removed by chemical reduction. The need to obtain an atomically clean base metal goes back to the nature of how a bond is formed in the first place.

Chemically nonaggressive or mild rosin fluxes are used with solderable base metals or metals that are precoated with a solderable finish. Rosins are naturally occurring resins derived from certain pine trees. Strong, chemically aggressive inorganic fluxes are often used on metals that are difficult to wet, like stainless steels. These are usually acids of various types. Table 9.3 summarizes the flux requirements for various metals, alloys, and coatings.

Table 9.3   Summary of Flux Requirements for Various Metals, Alloys, and Coatings

Base Metal, Alloy, or Applied Finish	Flux Type			Special Flux and/or Solder	Soldering not Recommended[a]
	Rosin	Organic	Inorganic		
Aluminum	-	-	-	X	-
Aluminum-bronze	-	-	-	X	-
Beryllium	-	-	-	-	X
Beryllium-copper	X	X	X	-	-
Brass	X	X	X	-	-
Cadmium	X	X	X	-	-
Cast iron	-	-	-	X	-
Chromium	-	-	-	-	X
Copper	X	X	X	-	-
Copper-chromium	-	-	X	-	-
Copper-nickel	X	X	X	-	-
Copper-silicon	-	-	X	-	-
Gold	X	X	X	-	-
Inconel	-	-	-	X	-
Lead	X	X	X	-	-
Magnesium	-	-	-	-	X
Manganese-bronze (high tensile)	-	-	-	-	X
Monel	-	X	X	-	-
Nickel	-	X	X	-	-
Nickel-iron	-	X	X	-	-
Nichrome	-	-	-	X	-
Palladium	X	X	X	-	-
Platinum	X	X	X	-	-
Rhodium	-	-	X	-	-
Silver	X	X	X	-	-
Stainless steel	-	X	X	-	-
Steel	-	-	X	-	-
Tin	X	X	X	-	-
Tin-bronze	X	X	X	-	-
Tin-lead	X	X	X	-	-
Tin-nickel	-	X	X	-	-
Tin-zinc	X	X	X	-	-
Titanium	-	-	-	-	X
Zinc	-	X	X	-	-
Zinc die castings	-	-	-	-	X

[a]With proper procedures, such as precoating, most metals can be soldered.
Reprinted with permisson from *Welding Handbook,* Vol. 2, 8th Ed., American Welding Society, 1991, p. 426.

9.3.5 Joint Design

Joints that are to be soldered should be designed to fulfill the requirements of the finished assembly and to permit application of the flux and solder by the soldering process to be used. Joints should be designed so that proper clearance is maintained during the heating and cooling of the soldering operation so that capillary flow can take place properly.

To maintain alignment of joint components during soldering, special fixtures may be necessary or the units in the assembly can be crimped, clinched, wrapped, or otherwise held together mechanically or by adhesives. Mechanical crimping and clinching enhances the final joint strength by contributing mechanical forces to the bonding forces and in many cases, is the predominant contributor to joint strength. Details of joint design will be discussed in Section 9.7.

9.3.6 Precleaning

An unclean surface will prevent the molten solder from wetting and flowing, making soldering difficult or impossible and contributing to poor joint properties. All metal surfaces to be soldered should be cleaned before assembly to facilitate wetting of the base metal by the solder. Flux should *not* be considered a substitute for precleaning. Precleaning is necessary to remove organic contaminants, like grease, oil, paint, pencil marks, lubricants, coolants, dirt, and inorganic films, like oxides or sulfides or other tarnish layers. The goal of precleaning is to expose atomically clean base metal, and the importance of proper precleaning in soldering, as in welding and brazing, cannot be overemphasized.

Precleaning can involve any or all of three progressively vigorous methods: (1) degreasing, (2) pickling, and (3) mechanical cleaning. Precleaning may also be followed by a fourth step, i.e., precoating. In degreasing, organic contaminants, such as oil and grease, and loosely adhering inorganic contaminants, such as dirt, are removed using either solvents or alkaline solutions. Methods are generally similar to those used for degreasing before adhesive bonding (see Chapter 4, Section 4.4.1). Back-deposition of residue must be avoided by both proper cleaning methods and proper rinsing. All cleaning solutions should be thoroughly removed before soldering.

If the base metals to be soldered have tenacious layers of organic contaminants or inorganic films of rust, scale, oxides, or sulfides, these can be removed by acid cleaning, or *pickling*. Inorganic acids such as hydrochloric, sulfuric, phosphoric, nitric, and hydrofluoric, singly or in combination, can be used. After pickling, parts should be washed in hot water and dried as quickly as possible to stop the action of the acid and remove reaction residues.

Mechanical cleaning may be needed to remove tenacious oxides or other films and to roughen the joint surfaces to improve adhesion through mechanical locking. Methods include grit or shot blasting, mechanical sanding, filing or hand sanding, cleaning with steel wool, and wire brushing or scraping. Mechanical cleaning should be avoided on soft base metals like copper, and care should always be exercised to avoid entrapment of cleaning or cleaner residue.

Precoating may be necessary for metals that are difficult to solder because they oxidize readily and, thus, can reoxidize after precleaning if soldering is delayed too long. Precoating involves coating the base material surfaces to be soldered with a more solderable and more oxidation-resistant metal or alloy before the soldering operation. Tin, copper, silver, cadmium, iron, nickel, and alloys of tin–lead, tin–zinc, tin–copper, and tin–nickel are used as precoats. Sometimes precoating is called *tinning* or *pretinning,* even when tin is not a component of the precoating material. Precoating is essentially mandatory for metals with tenacious oxides (e.g., aluminum, aluminum–bronzes, highly alloyed steels, and cast iron). Precoating produces such advantages as more rapid and uniform soldering and the ability to avoid using strong acid fluxes.

9.3.7 Soldering Process

The soldering process should be selected to provide the proper soldering temperature, heat distribution, and rate of heating and cooling required for the product to be assembled. The number of joints to be soldered, i.e., the production volumes and rates, as well as the joint quality requirements, also influences the process selection. The most important soldering processes are described in Section 9.6.

The specific method for applying the solder and the flux will be dictated by the selection of the soldering process.

9.3.8 Solder and Flux Residue Removal

Generally, flux residues should be thoroughly removed after soldering since such residues are almost always corrosive. The more active the flux, the more important the rapid and complete removal. These residues, or their corrosion products, can also degrade electrical properties by contaminating contacting surfaces. Excess solder, in the form of solder spatter or solder balls, must also be removed, as these can break lose, become trapped at electrical contacts or between circuit paths, and cause electrical shorts. Obviously, solder that bridges across circuit paths (i.e., bridges) must be removed because it also produces electrical short circuits.

Flux residue can usually be removed with hot or warm water (often with sodium carbonate added) but may require organic solvents to hasten dissolution and/or to avoid water damage immediately or due to entrapment. This is a particular concern in electronic assemblies. Ultrasonic assist can facilitate residue removal. Excess solder must usually be removed mechanically.

A growing consideration is the traditional use of chlorinated hydrocarbon or chlorofluorocarbon (CFC) solvents as these are known to cause adverse environmental effects (e.g., ozone depletion in the stratosphere). Non-chlorinated or non-chlorofluoro types are being sought.

9.4 SOLDERS AND SOLDER ALLOY METALLURGY

9.4.1 General Description

Solders are alloys with liquidus temperatures less than 450°C (840°F) and with good fluidity, reasonable strength, and, often, good electrical and thermal conductivity. To satisfy the requirement that liquidus temperatures be below 450°C, most solders are alloys of inherently low melting metals, usually exhibiting an eutectic in their phase diagrams. Good fluidity is no problem with low melting pure metals, since most metals are highly fluid above their melting point, but fluidity of alloys depends on the relative proportions of liquid and solid phases in equilibrium at the temperature at which soldering takes place. Proper alloying, and especially the use of ternary alloying additions, influences the proportions of liquid and solid in equilibrium. Fluidity of alloys was discussed in Section 8.5.1.

The major alloy systems used for solders include (1) tin–lead; (2) tin–antimony; (3) tin–antimony–lead; (4) tin–silver and tin–silver–lead; (5) tin–zinc; (6) cadmium–silver; (7) cadmium–zinc; (8) zinc–aluminum; (9) bismuth–containing, fusible alloys; and (10) indium alloys.

There is growing pressure to find effective Pb-free solders given the growing concern over lead pollution of the environment. Such pollution occurs during lead's production as well as during its deterioration in obsolete, land-filled soldered components and devices.

9.4.2 Tin–Lead Solders

Solders of the tin–lead (Sn–Pb) binary alloy system constitute the largest portion of all solders in use. They are used for joining most metals and have good corrosion resistance to most media. Most cleaning and soldering processes can be used with tin–lead solders, and fluxes of all types can also be used, with the choice of the specific flux depending on the base metal(s) to be joined.

In describing these solders, it is customary to give the tin content first, so a "40-60 solder" is a 40 wt. % Sn and 60 wt. % Pb alloy. Tin–lead solders, like some other solders, are referred to as soft solder alloys, or *soft solders*.[7] Table 9.4 lists several commonly used soft-solder alloys of various compositions and their uses.

The metallurgy of the tin–lead alloy system is fairly representative of all solder alloys and is similar to the metallurgy of braze alloys (see Section 8.5.3), in that it is the metallurgy of a eutectic system. Figure 9.4 shows the constitutional or equilibrium phase diagram for the tin–lead alloy system. It should be immediately apparent that this is a typical diagram for a eutectic system, in this case with a solid solution terminal at the lead-rich end and a much more limited solubility solid solution at the tin-rich end. The highest temperature at which a metal or alloy is completely solid is given by curve ACEDB, called the solidus temperature or solidus curve. The lowest temperature at which a metal or alloy is completely liquid is given by curve AEB, called the liquidus temperature or liquidus curve. At the extreme ends of the diagram are the pure metals lead (on the left) and tin (on the right). As is always the case for pure metals, pure lead and pure tin melt at a specific, singular temperature called the melting point. Pure lead melts at 327°C (621°F), shown by point A, while pure tin melts at 232°C (450°F), shown by point B. Between these two pure metals are solid solution strengthened alloys, which melt over a range of temperatures, beginning at the solidus and ending at the liquidus. The temperature differential between the solidus (ACEDB) and the liquidus (AEB) is

---

[7] Usually *soft solders* are physically soft and low melting. These two characteristics go hand in hand since both hardness and melting point are primarily cohesive properties determined by the nature and strength of bonding.

Table 9.4 Common Soft-Solder Alloys and Their Uses

	Properties			Comparable National Material Standards and Designations				
Uses	Designation	Melting Point or Range °C	Density at 20°C g/cm³	France	Germany	Great Britain	Japan	USA
Solder manufacture, tinning	Sn99.95	232	7.3		Sn99.95 DIN 1704	grade T$_1$ BS 3252	Tin metal class 1a JIS H 2108	Grade AA ASTM B 339-72
Joining electronic components	SnPb40	183–189	8.5	60/40 NF C90-550	L-Sn60Pb DIN 1707	BS solder K and KP BS 219	H60S H60A JIS Z3282	Alloy Grades 60A and 60B ANSI/ASTM B 32-76 SN60 QQ-S-571
Step soldering, tinning of winding wire	PbSn35	183–245	9.5		L-PbSn35(Sb) DIN 1707	BS solder H BS 219	H35S H35A JIS Z 3282	Alloy Grade 35A and 35B ANSI/ASTM B 32-76 Pb65 QQ-S-71
Soldering of mechanical components	PbSn20Sb1	183–277	9.8				H20B JIS Z 3282	Alloy Grade 20C ANSI/ASTM B 32-76 Sn20 QQ-S-571
High service temperature	PbSn5	300–315	11.2	5/95 NF C90-550			H5S H5A JIS Z 3282	Alloy Grade 5A and 5B ANSI/ASTM B 32-76 Sn 5 QQ-S-571
Good strength at elevated temperatures	PbAg2.5Sn1	295–320	11.3	0/97/Ag3 NF C90-550	L-PbAg3 DIN 1707			Alloy Grade 1,5S ANSI/ASTM B 32-76 Ag1,5 QQ-S-571
Solder manufacture	Pb99.99	327	11.3		Pb99.99 DIN 1719	Pb99.99 BS 334 Type A	Pig lead special class JIS H 2105	Corroding lead ASTM B 29-55

Note: It is the custom, in the abbreviated designation of the composition of the solder, to mention the metal with the highest content first.
Example: solder SnPb40, PbSn50, PbSn40; mass per cent of tin 60, 50, 40 respectively.
Reprinted with permission from R. J. Klein-Wassink, *Soldering on Electronics*, Electrochemical Publications Ltd., 1984, p. 85.

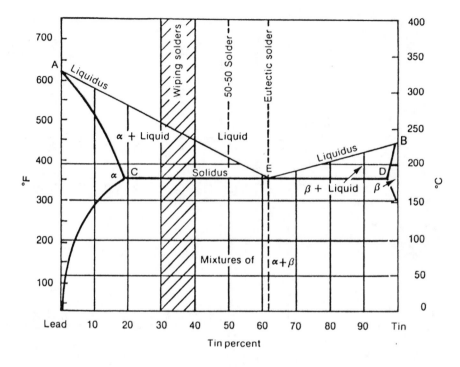

Figure 9.4 Phase diagram for the tin–lead alloy system. (Reprinted with permission from *Soldering Manual*, 2nd Ed., American Welding Society,1978, p. 5.)

called the melting range, and within this range the solder is partially melted. The only exception is at point E, where there is a composition that melts at a single temperature, like a pure material. This, however, is not a pure material but, rather is a microstructural constituent, consisting of intimately mixed lead-rich and tin-rich solid solutions. This constituent is called an eutectic and is the alloy with the lowest liquidus temperature in the alloy system. The eutectic composition is approximately 6 wt. % Sn and 37 wt. % Pb (point E), and the eutectic temperature is 183°C (361°F).

Solders containing 19.5 wt. % (point C) to 97.5 wt. % (point D) tin all have the same solidus temperature; that is, they all start melting at the same temperature, namely 183°C (361°F). Only the eutectic composition, however, becomes completely liquid at 183°C (361°F). All other compositions in this range are only partially melted at 183°C (361°F) so are less fluid than the eutectic at this temperature.

There are many important solder compositions in the tin–lead system, as shown in Table 9.5, which lists Sn–Pb solders from ASTM Specification B32. Some characteristics of some of the more specific alloys of importance follow:

- 5–95 is relatively high melting, has a narrow melting range, and is poor in terms of wetting and flow compared to higher tin alloys. Mechanical properties of this high-tin alloy are better at 149°C (300°F) than most other alloys with more tin.

Table 9.5   List of Important Sn-Pb Solders from ASTM Specification B32.

ASTM Alloy Grade	Fed. Spec. QQ-S-71	Tin % Desired	Lead % Nominal	Antimony % Minimum	Antimony % Desired	Antimony % Maximum	Silver % Desired	Melting Range Solidus °C	Solidus °F	Liquidus °C	Liquidus °F
70A		70	30	-	-	0.12	-	183	361	192	378
70B	Sn	70	30	0.20	-	0.50	-				
63A		63	37	-	-	0.12	-	183	361	183	361
63B	Sn63	63	37	0.20	-	0.50	-				
	Sn62	62	36	0.20	-	0.50	2				
60A		60	40	-	-	0.12	-	183	361	190	374
60B	Sn60	60	40	0.20	-	0.50	-				
50A		50	50	-	-	0.12	-	183	361	216	421
50B	Sn50	50	50	0.20	-	0.50	-				
45A		45	55	-	-	0.12	-	183	361	227	441
45B		45	55	0.20	-	0.50	-				
40A		40	60	-	-	0.12	-	183	361	238	460
40B	Sn40	40	60	0.20	-	0.50	-				
40C		40	58	1.8	2.0	2.4	-	185	365	231	448
35A		35	65	-	-	0.25	-	183	361	247	477

35B	Pb35	35	65	0.20	-	0.50	-	183	365	243	470
35C	Sn35	35	63	1.6	1.8	2.0	-	183	361	255	491
30A		30	70	-	-	0.25	-				
30B	Pb30	30	70	0.20	-	0.50	-	185	364	250	482
30C	Sn30	30	68.4	1.4	1.6	1.8	-	183	361	266	511
25A		25	75	-	-	0.25	-				
25B		25	75	0.20	-	0.50	-				
25C		25	73.7	1.1	1.3	1.5	-	184	364	263	504
20B	Pb20	20	80	0.20	-	0.50	-	183	361	277	531
20C	Sn20	20	79	0.80	1.0	1.20	-	184	363	270	517
15B		15	85	0.20	-	0.50	-	227	440	288	550
10B	Sn10	10	90	0.20	-	0.50	-	268	514	299	570
5A		5	95	-	-	0.12	-				
5B	Sn5	5	95	0.20	-	0.50	-	270	518	312	594
2A		2	98	-	-	0.12	-				

Note: Bismuth content allowed for the above alloys is 0.25% maximum. The allowable copper content is 0.08% maximum, the iron content is 0.02% maximum, the aluminum content is 0.005% maximum, and the zinc content is 0.005%. The arsenic content of solders containing 60 to 70% tin is 0.03%, for 45 to 50% tin the arsenic allowed is 0.025%, and for all solders from 20 to 40% tin the arsenic limit is 0.02%.

Reprinted with permission from *Soldering Manual*, 2nd ed., American Welding Society, 1978, p. 6–7.

- The 10–90, 15–85 and 20–80 solders have lower liquidus and solidus temperatures, a wider melting range, and better wetting and flow than 5-95. All are prone to solidification cracking or hot tearing, however, if movement occurs during cooling (from thermal contraction stresses, for example).
- The 25–75 and 30–70 solders have lower liquidus temperatures than all previous solders but the same solidus as 20–80. The melting range is thus narrower.
- The 35–65, 40–60, and 50–50 solders have low liquidus temperatures and, as a group, have the best combination of wetting, strength, and economy.
- The 60–40 alloy is used wherever exposure temperature restrictions are critical for the assembly or some of its components, since the composition is close to the eutectic.
- The 70–30 alloy is a special purpose solder, used where a high tin content is required, for wetting or other compatibility.

Besides these alloys, pure lead and pure tin can be and are used for soldering, and there are actual eutectic solders with the 63-37 composition.

9.4.3 Tin–Antimony and Tin–Antimony–Lead Solders

Tin–antimony (Sn–Sb) solders are generally stronger than tin–lead solders, since antimony is an effective solid solution strengthener. This is especially true at somewhat higher temperatures than can be tolerated by tin–lead alloys, resulting in good creep strength. The tin–antimony phase diagram is shown in Figure 9.5. The 95 wt. % tin (Sn)–5 wt. % antimony (Sb) solder provides a narrow melting range at a temperature higher than the tin–lead eutectic. For this reason, this alloy is a good choice for mechanical applications such as plumbing, refrigeration and air conditioning. One serious problem is that antimony, like lead, is toxic, so the alloy should be handled carefully and use in certain applications should be avoided.

Antimony may be added to tin–lead solders as a substitute for some of the tin. Additions up to 6 wt. % increase the mechanical properties with only slight impairment of soldering characteristics (i.e., wetting and flow).

9.4.4 Tin–Silver and Tin–Lead–Silver Solders

The tin–silver, lead-free solders are often used for soldering of stainless steel for food processing applications, where lead must be avoided because of its toxicity. Interest in tin-silver alloys is growing as a Pb-free alternative. The tin–silver binary phase diagram is shown in Figure 9.5.

The high lead tin–lead solders with silver added provide higher soldering temperature and exhibit good tensile, shear, and creep strength and are excellent for cryogenic applications. The 62 w/o Sn–36 w/o Pb–2 w/o Ag solder is used to solder to

Table 9.6 List of Various Tin-Bearing Binary and Ternary Solders

ASTM Class.	Composition, wt.%					Solidus		Liquidus		Melting Range	
	Sn	Pb	Sb	Ag	Zn	°C	°F	°C	°F	°C	°F
5	5	95	-	-	-	300	572	314	596	14	24
10	10	90	-	-	-	268	514	301	596	14	24
15	15	85	-	-	-	225	437	290	553	65	116
20	20	80	-	-	-	183	361	280	535	97	174
25	25	75	-	-	-	183	361	267	511	84	150
30	30	70	-	-	-	183	361	255	491	72	130
35	35	65	-	-	-	183	361	247	477	64	116
40	40	60	-	-	-	183	361	235	455	52	94
45	45	55	-	-	-	183	361	228	441	45	80
50	50	50	-	-	-	183	361	217	421	34	60
60	60	40	-	-	-	183	361	190	374	7	13
70	70	30	-	-	-	183	361	192	378	9	17
95TA	95	-	5	-	-	232	450	240	464	8	14
	96	-	-	4	-	221	430	221	430	0	0
	62	36	-	2	-	180	354	190	372	10	18
	5	94.5	-	0.5	-	294	561	301	574	7	13
	2.5	97	-	0.5	-	303	577	310	590	7	13
	1	97.5	-	1.5	-	309	588	309	588	0	0
	91	-	-	-	9	199	390	199	390	0	0
	80	-	-	-	20	199	390	269	518	70	128
	70	-	-	-	30	199	390	311	592	112	202
	60	-	-	-	40	199	390	340	645	141	255
	30	-	-	-	70	199	390	375	708	176	318
96.5TS	96.5	-	-	3.5	-	221	430	221	430	0	0

silver-coated surfaces for electronic applications. The addition of silver retards silver dissolution from the coating, as well as increases creep strength.

9.4.5 Tin–Zinc Solders

A large number of tin–zinc solders are used for joining aluminum. Galvanic corrosion is minimized by their use. The tin–zinc binary phase diagram is shown in Figure 9.5. Table 9.6 lists various tin-bearing binary and ternary solders with antimony, antimony and lead, silver, silver and lead, and zinc, along with their solidus and liquidus temperatures and melting ranges.

9.4.6 Cadmium–Silver Solders

Cadmium–silver solders, whose phase diagram is shown in Figure 9.6, are used where service temperatures will be higher than permissible with lower melting solders. Joint

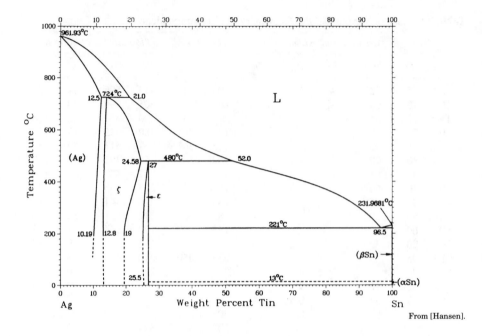

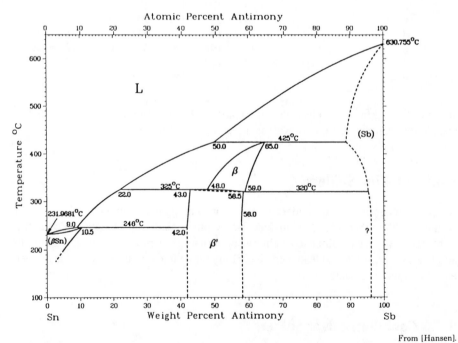

From [Hansen].

Figure 9.5 Some important phase diagrams for various binary alloy systems relevant to soft solders. *Source:* Massalski, Thaddeus B. (ed.), "Binary Alloy Phase Diagrams Vols. 1 & 2, American Society for Metals, Ohio, 1986.

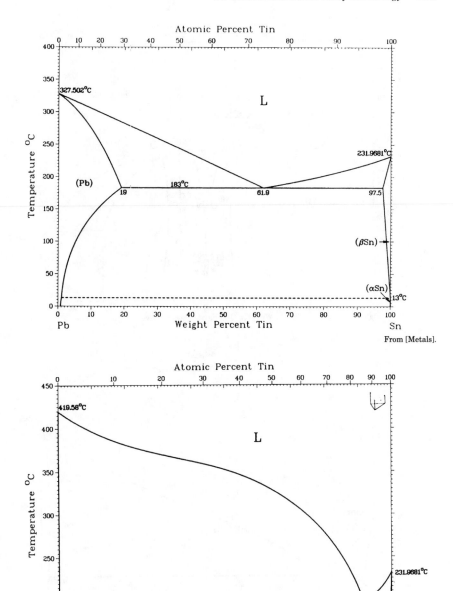

From [Metals].

Z. Moser, J. Dutkiewicz, W. Gąsior, and J. Salawa, 1985.

Figure 9.5 *continued*

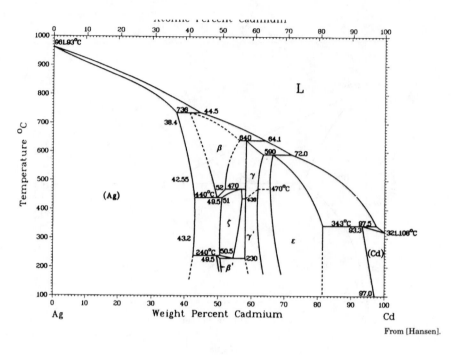

From [Hansen].

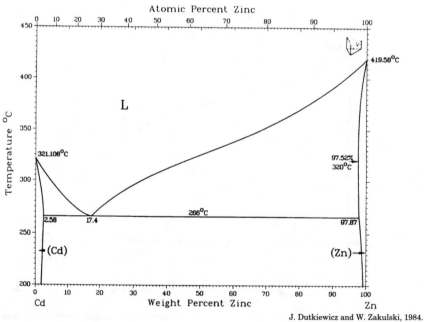

J. Dutkiewicz and W. Zakulski, 1984.

Figure 9.6   Some other important solder alloy phase diagrams. (Reprinted with permission from Thaddeus B. Massalski, (ed.), "Binary Alloy Phase Diagrams Vols. 1 & 2, American Society for Metals, 1986.)

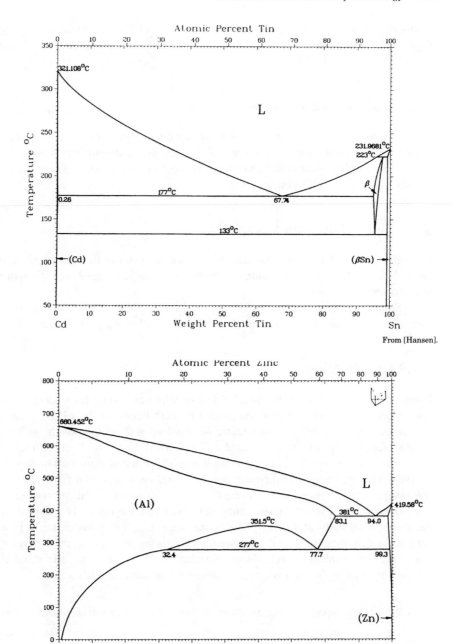

Figure 9.6 *continued.*

strengths can be very high (i.e., upwards of 172.4 MPa or 25,000 psi) when used to join copper. Cadmium, like lead and antimony, is toxic, however, so care must be exercised during both soldering and use.

9.4.7 Cadmium–Zinc Solders

Cadmium–zinc solders are used for soldering aluminum. Strengths and corrosion resistance are intermediate, but cost is considerably lower than cadmium–silver solders. The toxicity of cadmium fumes must again be considered in processing. The cadmium–zinc phase diagram is shown in Figure 9.6.

9.4.8 Zinc–Aluminum Solders

Zinc–aluminum solders, the phase diagram for which is shown in Figure 9.6, are specially designed for soldering aluminum alloys. Resulting joints have high strength and good corrosion resistance.

Some important solder compositions and melting temperatures and ranges for cadmium-silver, cadmium-zinc, and zinc-aluminum solders are given in Table 9.7.

9.4.9 Fusible Alloys

Bismuth-containing solders, or so-called fusible alloys are useful for soldering operations where the soldering temperature must be kept below 183°C (361°F), i.e., the lowest melting (or eutectic) temperature for tin–lead soft solders. Such solders are required for (1) soldering of heat-treated surfaces, where higher soldering temperatures would cause softening; (2) soldering joints where adjacent materials must be kept from overheating (e.g., because of their inherent heat sensitivity or combustibility); (3) *step soldering* to prevent remelting of a nearby solder joint made at a higher temperature; and (4) soldering of temperature-sensing devices, where the device is activated when the fusible alloy melts at relatively low temperature, e.g., fire sprinkler heads. Fusible solders are also sometimes used for mounting or fixturing parts during machining or for producing removable mandrels during forming. In the latter case, the mandrel is removed by melting in cases where it is trapped by the formed part's shape.

Table 9.7   Some Important Cadmium-Silver, Cadmium-Zinc, and Zinc-Aluminum Solders

Composition, Wt. %					Solidus		Liquidus		Melting Range	
Cd	Ag	Zn	Al		°C	°F	°C	°F	°C	°F
95	5	-	-	-	338	640	343	740	55	100
82.5	-	17.5	-	-	265	509	265	509	0	
40	-	60	-	-	265	509	335	635	70	126
10	-	90	-	-	265	509	399	750	134	241
-	-	95	5	-	382	720	382	720	0	0

Table 9.8   List of Low-Melting or Fusible Alloys

Melting Point °C	Composition in Mass Percent					Other Elements	Generic Name
	Sn	Pb	Bi	In	Cd		
16				24		76 Ga	
20	8					92 Ga	
25						95 Ga; 5 Zn	
29.8						100 Ga	
46.5	10.8	22.4	40.6	18	8.2		
47.2	8.3	22.6	44.7	19.1	5.3		
58	12	18	49	21			
61	16		33	51			
70	13.1	27.3	49.5		10.1		Lippowitz' alloy
70–74	12.5	25	50		12.5		Wood's metal
72.4		34	66				
79	17		57	26			
91.5		40.2	51.7		8.1		
93	42		44	14			
95	18.7	31.3	50				Newton's metal
96–98	25	25	50				d'Arcet's metal
103.0	26		53.5		20.5		
96–110	22	28	50				Rose's metal
117	48		52				
125		43.5	56.5				
127.7			75	25			
139	43		57				
144		62	38				
145	49.8	32		18.2			
156.4			100				
170	57					43 Tl	
176	67			33			
178	62.5	36				1.5 Ag	
180	63	34	3				
183	61.9	38.1					

Reprinted with permission from R. J. Klein-Wassink, *Soldering in Electronics,* Electrochemical Publications, 1984, p. 126.

The compositions and melting properties of a selection of low melting, fusible alloys are given in Table 9.8.

9.4.10 Indium Solders

Indium solders offer special properties, namely, very low vapor pressures for use in high vacuum seals and the ability to adhere to a wide variety of metals, as well as to a wide variety of ceramics and glasses. Some nonmetallics that have been successfully soldered with indium-based solders include glass, quartz, marble, granite, mica,

Table 9.9   Some Indium-Based Solders

Composition, Wt. %					Solidus		Liquidus		Melting Range	
In	*Sn*	*Bi*	*Pb*	*Cd*	°C	°F	°C	°F	°C	°F
50	50	-	-	-	117	243	125	257	8	14
25	37.5	-	37.5	-	138	280	138	280	0	0
50	-	-	50	-	180	356	209	408	29	52
19.1	8.3	44.7	22.6	5.3	47	117	47	117	0	0
21	12	49	18	-	58	136	58	136	0	0
4.0	12.8	48.0	25.6	9.6	61	142	65	149	4	7
52	48	-	-	-	117	243	117	243	0	0

porcelain, concrete, brick, aluminum oxide, copper oxide, germanium oxide, iron oxide, magnesium oxide, nickel oxide, titanium oxide, and zirconium oxide.

The 50 wt. % In–50 wt. % Sn alloy shown in Table 9.9 adheres to glass readily and can be used for glass-to-glass or metal-to-glass joining by soldering (see Chapter 13, Section 13.7). Phase diagrams for indium–lead and indium–tin are shown in Figure 9.7.

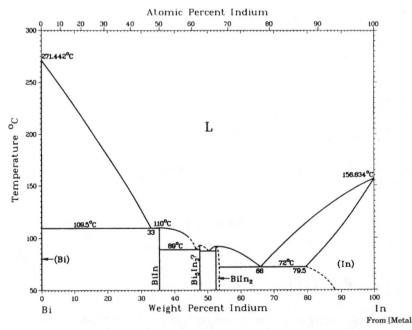

Figure 9.7   Phase diagrams for indium-based binary solder alloys. (Reprinted with permission from Thaddeus B Massalski, (ed.), *Binary Alloy Phase Diagrams,* Vols. 1 & 2, American Society for Metals, 1986.)

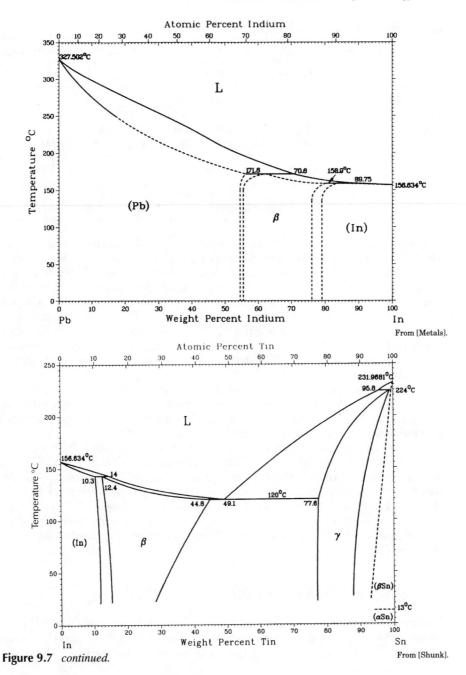

Figure 9.7 *continued.*

From [Shunk].

9.4.11 Other Special Solders

There are, besides these more common solders, many special solders for particular applications. The compositions of some of these special solders and their melting properties are listed in Table 9.10.

Table 9.10   Some Special Solder Alloys

Composition, Wt. %					Solidus		Liquidus		Melting Range	
Au	Sn	Ge	Si	In	°C	°F	°C	°F	°C	°F
80	20	-	-	-	280	536	280	536	0	0
88	-	12	-	-	356	673	356	673	0	0
96.4	-	-	3.6	-	370	698	370	698	0	0
82	-	-	-	18	451	843	485	905	34	62
Sn	Ag	Sb	Ge	Al	Solidus		Liquidus		Range	
65	25	10	-	-	233	451	?	??	?	?
-	-	-	55	45	242	795	424	795	0	0

One particularly important special solder is the gold–silicon (Au-Si) eutectic used for die[8] attachment in hermetic electronic packages, particularly for military applications. The process is often called *eutectic bonding* or *die bonding*. Both gold and silicon melt at temperatures above 1000°C (1800°F), but their phase diagram exhibits a eutectic composition containing 3 wt. % Si, which melts at 363°C (685°F). In the ideal case, a silicon die is placed into an gold-plated ceramic package cavity. The package is then heated to a temperature above that of the Au–Si eutectic, usually above 425°C (795°F). The silicon diffuses into the gold until the eutectic composition is approached and melting begins. The liquid front advances in the gold as silicon continues to diffuse, and an intimate bond is formed.

Several other gold-based solders are shown in Table 9.10, including binaries with tin, germanium, and indium. While obviously expensive, these alloys offer extraordinary corrosion resistance, good wettability, and compatibility with silicon that justify their use in semiconductor device assembly and package sealing.

Two other specialty solders are 65 wt. % Sn–25 wt. % Ag–10 wt. % Sb and 55 wt. % Ge–45 wt. % Al. These alloys were each developed for electronic applications, the first for applications requiring very high strength and the second for applications requiring very high service temperature. The Sn–Ag–Sb alloy has a tensile strength of 124 MPa (18 ksi), while the Ge–Al alloy has a eutectic melting temperature of 424°C (795°F).

As new solders are developed to satisfy the certain requirements to remove Pb, new compositions will undoubtedly appear. Many will probably be based on Sn, such as Sn-Cn, Sn-Mn or others.

9.4.12 Physical Forms of Solder

Solders, like braze fillers, come in many physical forms to satisfy many different applications, including pigs, ingots, bars, solid wires, flux-cored wires, foils, sheets or ribbons, preforms (of all shapes and sizes), segments or drops, and pastes or creams. These types are listed in Table 9.11.

---

[8]*A die*, in this context, is the individual semiconductor element or integrated circuit after it has been cut or separated out of the processed semiconductor wafer, distinct from a completely packaged or encapsulated integrated circuit with leads attached.

Table 9.11   Commercially Available Solder Product Forms

Pig	Available in 25 and 45 kg (50 and 100 lb) pigs
Ingots	Rectangular or round; 1-4, 2-3, and 4-3 kg (3, 5, and 10 lb)
Bars	Available in many cross sections, weights and lengths
Paste or cream	Available as mixture of powdered solder and flux
Foil sheet or ribbon	Available in various thicknesses and widths
Segment or drop	Triangular bar or wire cut into any number of desired pieces or lengths.
Wire, solid	Diameters of 0.25 to 6.36 mm (0.010 to 0.250 in.); spools
Wire, flux-cored	Solder covered with rosin, organic or inorganic fluxes; diameters as above
Preforms	Unlimited range of sizes and shapes to meet need

In addition to these common commercially available forms, some special forms (or means) have come into use in modern electronic assembly for establishing interconnection between devices and the integrating substrate. Two examples are solder bumps and tape automated bonding (or TAB). *Solder bumps* are round solder balls that are bonded to a transistor contact area and used to make connections to a conductor by face-down or "flip-flop" bonding techniques. One solder bump method is IBM's C4, which stands for controlled collapse chip connection. *Tape automated bonding,* or TAB, is a highly automated method for surface mounting packages that can provide interconnection of chips with large numbers (up to 500) of input/output (or I/O) terminals. In TAB, a continuous polymer tape is fabricated with fine-pitched metal lead frames spaced along its length. A window is made in the center of each frame where the chip is to be placed. The leads of the lead-frame are then bonded to the chips, on which bonding platforms, or *bumps*, have been deposited. Bumps serve as bonding platforms for subsequent joining, usually by soldering but possibly by other means (such as adhesive bonding).

9.5 SOLDERING PROCESSES

9.5.1 General Description

The proper application of heat is of paramount importance during any soldering operation. The heat should be applied in such a manner that the solder melts while the surface is heated to permit the molten solder to wet and flow over the surface. Heating can be done manually or by machine (semiautomatically or automatically), locally at the joint or uniformly over the entire assembly, and using any of several sources. The following sections describe some of the more important soldering processes.

9.5.2 Iron Soldering

Iron soldering is a traditional manual process that uses a copper tip that is heated by electrical resistance, gas, or oil to heat the area to be soldered locally (i.e., near the point of solder application). The selection of soldering irons can be simplified by classifying them into four groups, as follows: (1) soldering irons for service or repair; (2) transformer-type, low-voltage pencil type; (3) special quick-heating or plier types; and (4) heavy duty industrial irons.

Regardless of the heating method, the tip performs several important functions. It stores and conducts heat from the heat source to the part(s) to be soldered; it stores molten solder (through wetting and adhesion to the tip); it conveys the molten solder; and it withdraws surplus molten solder (also through wetting of the tip).

Iron soldering has the advantage of being quite inexpensive, but it relies on operator skill and is slow.

9.5.3 Torch Soldering

In *torch soldering,* a combustion flame is used as the source of heat. Heating is usually localized but can be more general, raising the temperature of the entire assembly or a large portion of the assembly. The process can be, and often is, manual, but can be automated. The selection of the torch is dictated by the size, mass, and configuration of the assembly to be soldered.

The temperature of the torch flame and the heat intensity of the process depend on the gas or gases used. Lower temperatures are attained with propane, butane, or natural gas burned with atmospheric or pressurized air. High temperatures are attained with acetylene burned with compressed oxygen. The design of a soldering torch is essentially identical to welding and brazing gas torches (see Chapter 6, Figure 6.9). As in welding and brazing, the fuel gas and air (or oxygen) mixture can be adjusted to render the flame oxidizing (with excess oxygen), neutral (when in proper molar ratios), or reducing (with insufficient oxygen). Proper gas adjustment should generally avoid a sooty flame, which is one starved for air or oxygen, as the soot deposited hinders wetting. Multiple flame tips or burners are often used for large or massive work.

9.5.4 Dip Soldering

Dip soldering uses a molten bath of solder to supply both the heat and the solder necessary to produce soldered joints. Heating is general to the entire assembly. The process is automated and excellent for production, as many joints can be soldered at the same time. Figure 9.8 schematically shows the dip soldering process, as well as some other soldering processes. There is a danger with dip soldering of thermally shocking the assembly being soldered, since heat is applied (from a large heat source) very suddenly. Obviously, all of the materials in an assembly to be dip soldered must be able to tolerate the soldering temperature.

9.5.5 Wave Soldering

In *wave soldering,* the solder is pumped out of a narrow slit to produce a wave or series of waves on the surface of a molten bath of solder. These waves sweep over the assembly, which is suspended above the surface of the bath with just the joints close enough to be

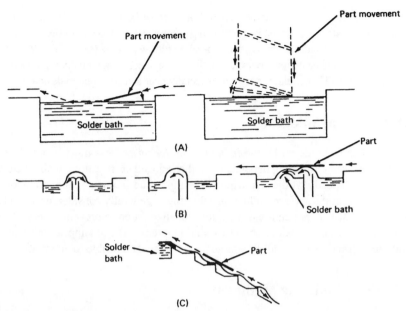

Figure 9.8　(a) Dip, (b) wave, and (c) cascade soldering processes. (Reprinted with permission from *Welding Handbook*, 8th ed., Vol. 2, American Welding Society, 1991, p. 443.)

struck by the wave, providing both heat and solder. Wave soldering systems, such as that shown in Figure 9.8, provide a virtually oxide-free solder surface to the part by breaking any oxide skull on the top of the bath with the wave motion. In addition, the waves act to sweep flux and vapors away after soldering. Heat tends to be localized at the joints, and flux must be preplaced before soldering.

There are, in fact, several variations of wave soldering, including *lambda-wave* and *double-wave*, with the principal differences being the form of the wave or waves.

9.5.6 Oven Soldering

Oven soldering is a good, high-production process that should be considered when the entire assembly can be brought to the soldering temperature without damage, production volume and/or rate justifies the cost of jigs and fixtures (for holding joint components together during soldering), and the assembly is complicated, making other methods impractical. Only fluxes that do not decompose upon exposure to elevated temperature for some time can be used in oven soldering.

9.5.7 Resistance Soldering

In *resistance soldering*, the work is placed either between a ground and a moveable electrode or between two moveable electrodes to complete an electrical circuit. Heat is applied locally to the joint both by the electrical resistance of the metal being soldered and by conduction from the electrode, which is usually carbon. Solder must be fed into the joint during this process or, more often, is supplied by preplaced preforms or solder coatings (i.e., pretinning).

The resistance soldering process is a so-called reflow process (see Section 9.5.10) and is used for its accurate heat input control, short heating cycle, and potentially highly localized heating. These are all beneficial for certain, especially heat-sensitive materials or assemblies.

9.5.8 Induction Soldering

The material to be soldered by the *induction soldering* process must be an electrical conductor. The rate of heating depends upon the induced current flow, while the distribution of heat depends on the frequency. The higher the frequency, the more concentrated the heat is at the surface. Induction soldering is generally considered for (1) large scale production; (2) application of heat to a localized area to prevent damage to surroundings; (3) minimum oxidation to surrounding areas; (4) good appearance and consistently high joint quality; and (5) simple joint designs, amenable to automation.

9.5.9 Other Soldering Methods

There are several other methods of supplying heat to a joint or collection of joints to accomplish soldering. Infrared soldering focuses an infrared light source to accomplish heating locally, in thin-section structures. Hot-gas soldering uses a fine jet of heated inert gas (e.g., nitrogen) for heating. Ultrasonic soldering is really ultrasonically assisted dip soldering, where the ultrasonic energy breaks up tenacious oxides on the part or in the bath. Soldering can also be accomplished using spray guns for applying molten solder. In condensation soldering, the latent heat of a condensing saturated vapor provides heat for soldering. Many of these processes are used in so-called reflow methods of soldering or, simply, reflow soldering.

9.5.10 Reflow Methods of Soldering

In *reflow soldering,* solder is predeposited at each joint as a preform, paste, or solder plating. The solder is then melted in place, or reflowed, by any one of several methods, including resistance or hot gas heating, forced air oven, radiant (IR) heating, liquid immersion, and vapor-phase condensation. More recently, lasers are being used for soldering by reflow methods, with the potential for melting solders with low or high melting temperatures or even for joining by microwelding without the need for solder. Lasers also may permit fluxless soldering, by ablating away oxide layers.

9.6 FLUXES FOR SOLDERING

9.6.1 General Description

A *soldering flux* is a liquid, solid, or gaseous material that, when heated, is capable of promoting or accelerating the wetting of metals by solders. The purpose of the flux is to remove small amounts of oxide and other surface compounds from the surfaces being soldered by the process of chemical reduction and to keep those oxides or compounds from reforming by remaining on the surface until displaced by the molten solder. A fluxing compound or flux should become chemically active just below the melting point (i.e., the solidus of the solder) and remain active to at least the liquidus temperature of the solder. During soldering, the flux should be displaced by the molten solder. This requires the solder to be denser than the flux and reduce the surface energy of the substrate over that of the substrate–flux interface. The mechanism by which fluxes clean oxidized base metal surfaces to promote wetting by molten solder is shown in Figure 9.9.

Fluxes may be classified into three groups: (1) inorganic fluxes, which are the most active and chemically aggressive; (2) organic fluxes, which are moderately active; and (3) naturally occurring rosin fluxes, which are the least active and only mildly corrosive.

9.6.2 Inorganic Fluxes

Inorganic fluxes consist of inorganic acids and salts that are extremely chemically aggressive and highly corrosive. Inorganic fluxes are used to best advantage where conditions concerning the base metal and/or the environment require rapid and highly active fluxing action (i.e., extremely tenacious oxides or tarnishes on metals and alloys considered to have poor solderability). These fluxes are usually formulated to provide activity and stability over the entire soldering temperature range. One distinct disadvantage of these inherently aggressive fluxes is that the residue they produce after reacting with the contaminating layer, if not removed, can cause severe corrosion at the joint. Examples of some important inorganic fluxes are listed in Table 9.12.

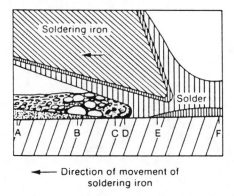

← Direction of movement of
soldering iron

Figure 9.9 The mechanisms by which fluxes clean a substrate surface during soldering. A, flux over oxidized metal; B, boiling flux removes oxide; C, base metal in contact with molten flux; D, molden solder displaces molten flux; E, solder alloys with base metal; F, solder solidifies. (Reprinted with permission from American Welding Society, *Soldering Manual*, 2nd ed., American Welding Society, 1978, p. 14.)

Table 9.12   Inorganic, Organic, and Rosin Fluxes

	Typical fluxes	Vehicle	Use	Temperature Stability	Tarnish Removal	Corrosiveness	Postsolder Cleaning Methods
			Inorganic				
Acids	Hydrochloric, hydrofluoric, orthophosphoric	Water, petrolatum paste	Structural	Excellent	Excellent	High	Hot-water rinse and neutralize; organic solvents; degrease
Salts	Zinc Chloride, ammonium chloride, tin chloride	Water, petrolatum paste, polyethylene glycol	Structural	Excellent	Excellent	High	Hot-water rinse and neutralize, 2% HCl solution, hot-water rinse and neutralize; organic solvents; degrease[a]
Gases	Hydrogen forming gas; dry HCl	None	Electrical	Excellent	Very good at high temperatures	None normally	None required
			Organic: Nonrosin Base				
Acids	Lactic, oleic, stearic, glutamic, phthalic	Water, organic solvents, petrolatum paste, polyethylene glycol	Structural, electrical	Fairly good	Very good	Moderate	Hot-water rinse and neutralize; organic solvents; degrease[a]

Halogens	Aniline hydrochloride, glutamic acid hydrochloride, bromide derivatives of palmitic acid, hydrazine hydrochloride or hydrobromide	Water, organic solvents, polyethylene glycol	Structural, electrical	Fairly good	Very good	Moderate	Hot-water rinse and neutralize; organic solvents; degrease[a]
Amines and amides	Urea, ethylenediamine, mono- and triethanolamine	Water, organic solvents, petrolatum paste, polyethylene glycol	Structural, electrical	Fair	Good	Moderate	Hot-water rinse and neutralize; organic solvents; degrease[a]
Organic: Nonrosin Base							
Acids	Lactic, oleic, stearic, glutamic, phthalic	Water, organic solvents, petrolatum paste, polyethylene glycol	Structural, electrical	Fairly good	Very good	Moderate	Hot-water rinse and neutralize; organic solvents; degrease[a]
Halogens	Aniline hydrochloride, glutamic acid hydrochloride, bromide derivatives of palmitic acid, hydrazine hydrochloride or hydrobromide	Water, organic solvents, polyethylene glycol	Structural, electrical	Fairly good	Very good	Moderate	Hot-water rinse and neutralize; organic solvents; degrease[a]

(continued)

Table 9.12 (continued)

Typical fluxes	Vehicle	Use	Temperature Stability	Tarnish Removal	Corrosiveness	Postsolder Cleaning Methods
Organic: Nonrosin Base						
Amines and amides	Urea, ethylenediamine, mono- and triethanolamine	Water, organic solvents, petralatum paste, polyethylene glycol	Structural, electrical	Fair	Good	Moderate	Hot-water rinse and neutralize; organic solvents; degrease[a]
Organic: Rosin Base						
Rosin, superactivated	Rosin or resin with strong activators	Alcohols, organic solvents, glycols	Structural, electrical	Fair	Very good	Moderate	Water-base detergents; isopropanol; organic solvents; degrease[a]
Activated (RA)[b]	Rosin or resin with activator	Alcohols, organic solvents, glycols	Electrical	Fair	Good	Only to critical electronics	Water-base detergents; isopropanol; organic solvents; degrease[a]
Mildly activated (RMA)[b]	w/w rosin with activator	Alcohols, organic solvents, glycols	Electrical	Poor	Fair	None	Water-base detergents; isopropanol; organic solvents; degrease[a]
Nonactivated (water-white rosin) (R)[b]	w/w rosin only	Alcohols, organic solvents; glycols	Electrical	Poor	Weak	None	Same as activated rosin but normally does not require post-cleaning

[a]For optimum cleaning, follow by wash with demineralized or distilled water.
[b]Follows Federal Spec. QQ-S-571 or MIL-F-4265
Reprinted with permission from H. H. Manko, *Solders and Soldering*, McGraw-Hill, 1964, p. 28–29.

9.6.3 Organic Fluxes

Organic fluxes, while less active than the inorganic types, are effective at soldering temperatures from 90° to 320°C (200° to 600°F). They consist of organic acids and bases and, often, certain of their derivatives such as hydrohalides. Examples are listed in Table 9.12. Organic fluxes are active from just below the soldering temperature into the soldering range, but the period of their activity is short because of their inherent susceptibility to thermal decomposition. When used properly, i.e., to match base metal needs and in the proper amounts, residues are fairly inert and water soluble.

9.6.4 Rosin Fluxes

There are three types of rosin fluxes: (1) nonactivated, (2) mildly activated, and (3) activated. In nonactivated rosin, also known as *water-white*, the active ingredient is abietic acid, which becomes mildly active at 177° to 316°C (350° to 600°F). Many nonactivated rosins are natural in origin, including wood rosin and pine pitch. The residue of such rosins is hard, nonhydroscopic, electrically nonconductive, and noncorrosive. Although it is not essential to remove this residue, it is best to do so. Warm to hot water works well.

Mildly activated rosins were developed to increase the fluxing action of natural rosins, without significantly altering the noncorrosive nature of the residue. Such rosins are commonly used for soldering military and other high-reliability electronic products.

Activated rosins are more active and should be restricted to use in commercial electronics, where residue can be completely removed.

9.6.5 Special Fluxes

Reaction fluxes are a special group of fluxes used for soldering aluminum. Decomposition of the flux when it is heat activated provides a metallic film on the surface of the aluminum in place of the oxide film. Soldering is then performed to this metallic layer, as wetting and spreading occur more easily.

9.6.6 Physical Forms of Fluxes

Flux in most of the aforementioned categories is available in single or multiple cores in wires or in liquid, pastes or dry powders, or preforms. Selection of the form of the flux depends on the joint design, base material, and particular process to be used.

9.6.7 Fluxless Soldering

There is growing interest in being able to solder without requiring fluxes. The reasons are primarily to eliminate the need for subsequent cleaning with chlorinated hydrocarbon solvents or chlorofluorocarbons (CFC's), but also to eliminate the need for flux application, often by screening. One fluxless approach may employ a laser to remove contaminants and oxide by ablation. This is known as laser ablative fluxless soldering (LAFS).

9.7 JOINT DESIGNS AND JOINT PROPERTIES FOR SOLDERING

9.7.1 Solder Joint Designs

The selection of a solder joint design for a specific application depends not only on the service requirements of the assembly, but also on the heating methods to be used, the fabrication techniques before soldering, the number of items to be soldered, and the soldering process to be used.

When the service requirements of a joint are severe, it is generally necessary to design so that the strength of the joint is equal to or greater than the load-carrying capacity of the weakest member of the assembly. Solders have inherently low strength compared to the materials that are generally being soldered, so the soldered joint should be designed to avoid dependence on the strength of the solder alone. The necessary strength can be provided by shaping the parts to be joined so that they engage or interlock, requiring the solder only to seal and stiffen the assembly. This is generally true, whether the soldered joints are intended for use in mechanical systems or electrical systems.

Two basic types of designs are used for soldering of mechanical assemblies: the lap joint and the butt joint, shown in Figure 9.10. The lap joint is, by far, the most common solder joint and should be applied wherever possible, since it offers joints of maximum strength by taking loading in shear. The strength of the joint varies with the size of the overlap and with peeling forces induced at the ends of the overlaps by bending. The tendency toward peeling can be resisted by going to more elaborate double-lap and strapped lap designs. This design is also good for sealing.

The butt joint is the simplest and weakest type joint, as the area of soldered interface is limited by the cross section of the parts. Loading usually results in pure tension, however, and this is good. A scarf joint is a butt joint with a larger cross sectional area produced by cutting the joint at an angle through the thickness. In addition to increasing load-bearing area, the scarf converts tensile loading to shear, which is usually advantageous in solders (as in adhesives).

For the best performance in mechanical applications, solder joints should be mechanically reinforced by any of several schemes as shown in Figure 9.10, Group II or III. In addition to these joint types there are hooked and twisted types, which are used primarily in electrical assembly but also provide a significant mechanical component of strength.

For electrical and electronic assemblies, there are three primary attachment configurations that employ these basic connection types: (1) straight-through attachment,

(2) clinched lead attachment, and (3) surface mounting. The first two methods are referred to as *through-hole* methods. Predominantly used in the past, through-hole joints still find application today. More and more, surface-mounting methods are being used.

In *straight-through attachment,* component leads are inserted into holes drilled in the circuit board and either left straight or hooked or crimped slightly. Actual electrical connection is usually by manual, dip, or wave soldering. Plated-through holes are recommended if the printed wiring has more than one layer, to effect interconnection between layers. The hook or crimp provides a reasonable mechanical component that reinforces the joint and helps prevent creep of the solder in service and eventual failure. In *clinched-lead attachment,* component leads are inserted into circuit holes and are bent by wiping the protruding leads and soldering or, even, by twisting the leads and soldering. Clinched lead attachment is recommended for multiple-lead components, such as integrated circuits, where greater lead strength is available and in systems where the extra lead length will not adversely affect performance. The added strength provided by the mechanical locking is significant to the joint's performance.

In surface mounting, component leads are attached to the circuit by soldering in a lap configuration. Surface mounting is the only attachment method that provides access to conductive joints from the component side. Because components can be attached to both sides of the circuit, surface mounting technology (or SMT) allows maximum packaging density. Connection in SMT is by conventional or reflow soldering, or by adhesive bonding to either or both sides of the circuit. The mechanical integrity of surface-mounted soldered components is quite limited.

When soldering is used principally for mechanical assembly, as opposed to electrical connectivity, different joint designs are required. Often, these joints involve larger bonded areas and, occasionally, mechanical interlocks. Examples of the 21 methods that can be used to make solder joints, as compiled by the American Welding Society, are shown in Figure 9.11(a), while more general approaches for designing solder joints to provide full mechanical security before soldering are shown in Figure 9.11(b).

Irrespective of the specific design, feeding of solder into the joint during soldering is a major consideration during the design process. As in brazing, joint clearance is a critical factor. As a rule, joint clearance should be 0.075 mm (0.003 in.) for optimum strength. In fact, the maximum capillary rise that is achievable in a joint is directly related to the joint gap or clearance given by

$$h = 2\sigma \cos\theta / d\rho g \qquad (9.1)$$

where σ is the surface tension (in ergs or dynes); θ is the contact angle of wetting (in degrees); d is the capillary gap (in meters); ρ is the solder density (in kilograms per cubic meter); and g is gravity (in meters per square second).

Factors in the joint design that relate to the flow of solder include (1) providing a suitable reservoir of molten solder; (2) providing a feed path to the capillary; (3) providing suitable capillary entrance and exit; (4) controlling the gap to provide a suitable capillary driving force; (5) balancing the mass in the joint to assure even heating and to control flow; (6) providing a joint that is suited to the planned method of heating; and (7) providing enough freedom in the joint to prevent entrapment of flux or flux residue.

Group I—No mechanical security prior to soldering						
Butt connections						
No.	Type	Diagram	Controlling formula	Conditions	Fixtures	Current
1	Round to round	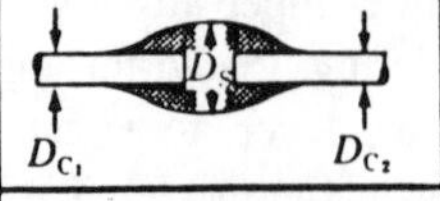 D_{C_1} D_{C_2}	$D_s = \sqrt{\delta D_{C_1}}$	$\rho_{C_1} \geq \rho_{C_2}$ $D_{C_1} \leq D_{C_2}$	Yes	Small
2	Square to square	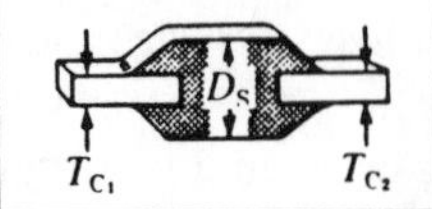 T_{C_1} T_{C_2}	$D_s = \sqrt{\dfrac{4}{\pi}\delta T_{C_1}}$	$\rho_{C_1} \geq \rho_{C_2}$ $T_{C_1} \leq T_{C_2}$	Yes	Small
3	Rectangle to rectangle	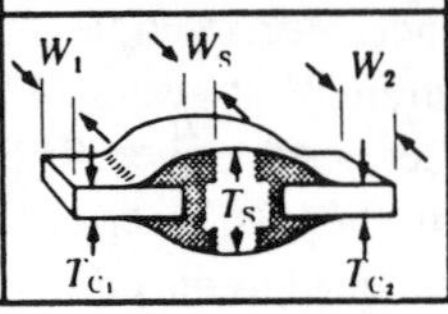 W_1 W_s W_2 T_s T_{C_1} T_{C_2}	$T_s = \delta T_{C_1}$	$\rho_{C_1} \geq \rho_{C_2}$ $W_1 = W_2 = W_s$ $T_{C_1} \leq T_{C_2} \neq T_s$	Yes	Small
Lap connections						
1	Round* to round	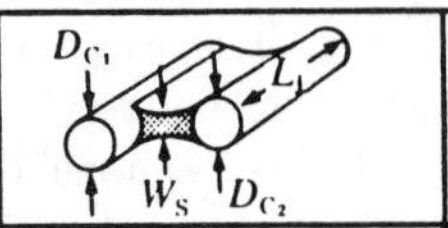 D_{C_1} W_s D_{C_2}	$L_J = \dfrac{\pi}{2}\delta D_{C_1}$	$\rho_{C_1} \geq \rho_{C_2}$ $D_{C_1} \leq D_{C_2}$ $W_s \geq \dfrac{D_{C_1}}{2}$	Yes	Large
2	Round to flat	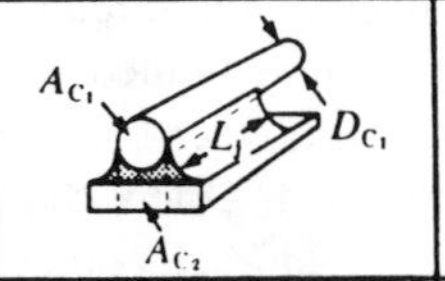 A_{C_1} L_J D_{C_1} A_{C_2}	$L_J = \dfrac{\pi}{4}\delta D_{C_1}$	$\rho_{C_1} \geq \rho_{C_2}$ $A_{C_1} \leq A_{C_2}$	Optional	Large
3	Fat to flat	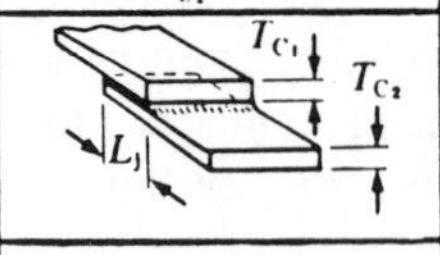 T_{C_1} T_{C_2} L_J	$L_J = \delta T_{C_1}$	$\rho_{C_1} \geq \rho_{C_2}$ $W_1 = W_2 = W_s$ $T_{C_1} \leq T_{C_2}$	Optional	Large
4	·Wire to post	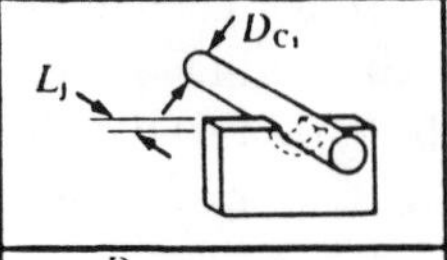 L_J D_{C_1}	$L_J = \dfrac{1}{2}\delta D_{C_1}$	$\rho_{C_1} \geq \rho_{C_2}$ Solder fillet $\geq \dfrac{D_{C_1}}{2}$	No	Medium
5	Wire to cup	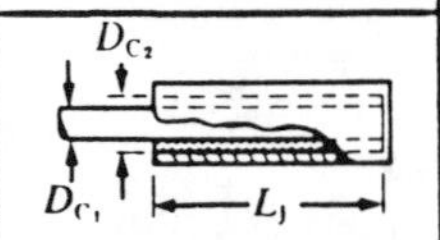 D_{C_2} D_{C_1} L_J	$L_J = \dfrac{1}{4}(\delta - 1)D_{C_1}$	$\rho_{C_1} \geq \rho_{C_2}$	No	Large
6	Wire to hole	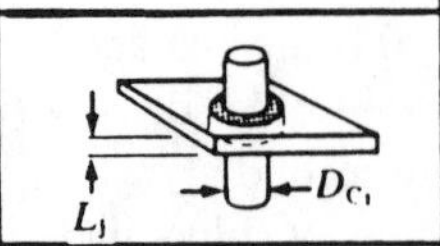 L_J D_{C_1}	$L_J = \dfrac{1}{4}\delta D_{C_1}$	$\rho_{C_1} \geq \rho_{C_2}$	Optional	Medium

[1]H. H. Manko, How to Design the Soldered Electrical Connection, *Prod. Eng.*, June 12, 1961, p. 57.

Figure 9.10   The four principal types of solder connections used in electrical assembly: butt connections, lap connections, hook connections, and wrap connections. (Reprinted with permission from *Soldering Manual*, 2nd ed., American Welding Society, 1978, p. 30–31.)

Group II—Partial mechanical security prior to soldering						
Hook connections						
No	Type	Diagram	Controlling formula	Conditions	Fixtures	Current
1	Round to round		$D_{c_1} = \dfrac{2}{\delta}D_{c_2}$	$\rho_{c_1} \geq \rho_{c_2}$ $D_{c_1} \leq D_{c_2}$ Hook $\geq 180°$	No	Large
2	Round to flat		$D_{c_1} = \dfrac{1}{\pi\delta}(8L_j + 4T_{c_2})$	$\rho_{c_1} \geq \rho_{c_2}$ $A_{c_1} \leq A_{c_2}$ Hook $\geq 180°$	No	Medium
Group III—Full mechanical security prior to soldering**						
Wrap connections						
No	Type	Diagram	Controlling formula	Conditions	Fixtures	Current
1	Round to round		$L_j = \dfrac{\pi}{2}\delta D_{c_1}$	$\rho_{c_1} \geq \rho_{c_2}$ $D_{c_1} \leq D_{c_2}$ $N > 1$	No	Large
2	Round to flat		$D_{c_1} = \dfrac{8}{\pi\delta}(L_j + T_{c_2})$	$\rho_{c_1} \geq \rho_{c_2}$ $A_{c_1} \leq A_{c_2}$ $N = 1$	No	Medium
3	Round to post		$D_{c_1} = \dfrac{4N}{\delta}D_{c_2}$	$\rho_{c_1} \geq \rho_{c_2}$ $D_{c_1} < D_{c_2}$ $N \geq 1$	No	Large

D_{c_1} — Diameter of smaller conductor
A_{c_1} — Area of smaller conductor
S — Solder
W — Width
L_j — Length of joint

T — Thickness
N — Number of Turns
δ — Resistivity Ratio $\dfrac{\rho_S}{\rho_{c_1}}$
ρ — Resistivity (Microhm − cm)

*Use only when large conductor diameter is 3 to 4 times larger than small diameter; otherwise use round-to-flat lap-joint formula.

**In cases where loosening or breaking of the joint would result in a hazardous condition, mechanical security should be specified.

Figure 9.10 *continued.*

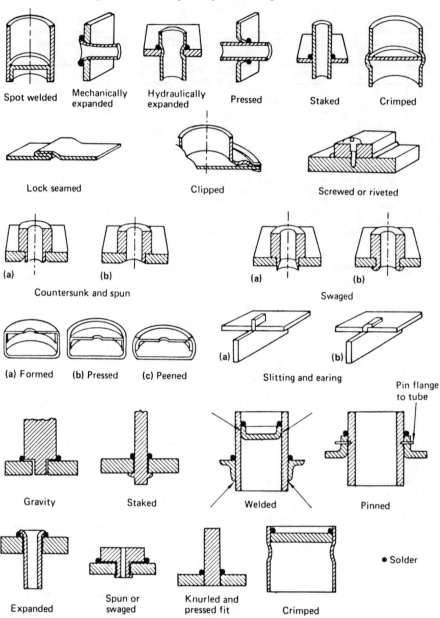

Figure 9.11(a)   Methods for making electrical and mechanical solder joints as compiled by the American Welding Society. (Reprinted with permission from H. H. Manko, *Solders and Soldering*, McGraw-Hill, 1964, p. 142-143.)

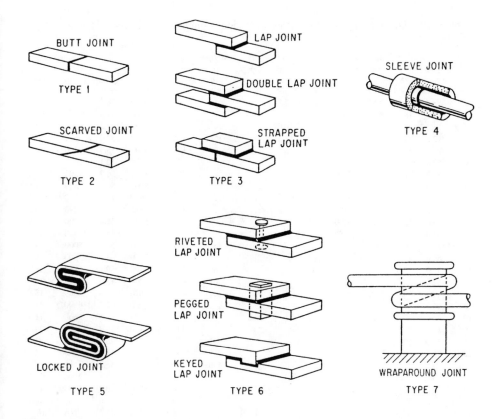

Figure 9.11(b) More general approaches for designing solder joints to provide full mechanical security before soldering. (Reprinted with permission from H. H. Manko, *Solders and Soldering*, McGraw-Hill, 1964, p. 142–143.)

9.7.2 Solder Joint Properties

In general, the shear strength of soldered joints ranges from less than 7 to 49 MPa (1,000 psi to around 7,000 psi). Typical mechanical properties for a variety of soft-solder alloys are shown in Table 9.13.

The strength of solders generally increases with decreasing temperature (i.e., at lower homologous temperatures) and decreases with increasing temperature as creep becomes more significant. As an example, the ultimate tensile strength of 65/35 tin–lead solder increases from 52 MPa (7,260 psi) at room temperature to 96 MPa (13,720 psi) at –180°C. The effect of increased temperature on the shear strength of some solders is shown in Table 9.14.

Table 9.13   Some Mechanical Properties for Various Common Soft-Solder Alloys

	Alloy Composition %							Melting Temp. °F		Tensile Strength		Shear Strength	
No.	Sn	Pb	Bi	In	Sb	Cd	Ag	Solidus	Liquidus	lb/in²	kg/cm²	lb/in²	kg/cm²
1	8.3	22.6	44.7	19.1	...	5.3	...	...	117	5400	38.0	NA	NA
2	12.0	18.0	49.0	21.0	...	...	...	...	136	6300	44.3	NA	NA
3	13.3	26.7	50.0	...	...	10.0	...	...	158	5990	41.5	300[b]	2.1[b]
4	12.5	25.0	50.0	...	...	12.5	...	158	165	4550	32.0	NA	NA
5	11.3	37.7	42.5	...	...	8.5	...	160	190	5400	38.0	300[b]	2.1[b]
6	50.0	...	...	50.0	...	...	...	...	243	1720	12.1	1630[c]	11.5[c]
7	37.5	37.5	...	25.0	...	...	...	...	280	5260	36.9	4300[c]	30.2[c]
8	42.0	...	58.0	...	...	...	...	...	281	8000	56.2	500[b]	3.51[b]
9	...	15.0	...	80.0	...	...	5.0	...	314	2550	17.9	2150[c]	15.1[c]
10	...	...	...	100.0	...	...	...	...	314	515	3.6	890[c]	6.3[c]
11	70.0	18.0	...	12.0	...	...	...	302	345	5320	37.4	4190[c]	29.5[c]
12	63.0	37.0	...	...	...	...	...	...	361	7700	54.1	5400	38.0
13	70.0	30.0	...	...	...	...	...	361	367	7800	54.8	5000	35.2
14	60.0	40.0	...	...	...	...	...	361	370	7600	53.4	5600	39.4
15	50.0	47.0	...	...	3.0	...	...	365	399	8400	59.1	6850	48.2
16	50.0	50.0	...	...	...	...	...	361	417	6200	43.6	5200	36.6
17	...	50.0	...	50.0	...	...	...	...	419	4670	32.8	2680[c]	18.8
18	96.5	...	...	...	...	...	3.5	...	430	8900	62.7	4600	32.3
19	...	...	...	90.0	...	...	10.0	...	448	1650	11.6	1600[c]	11.2[c]
20	...	75.0	...	25.0	...	...	...	...	448	5450	38.3	3520[c]	24.7[c]
21	40.0	60.0	...	...	...	...	...	361	460	5400	38.0	4800	33.7
22	95.0	...	...	...	5.0	...	...	452	464	5900	41.5	6000	42.2
23	95.0	...	...	...	...	...	5.0	430	473	8000	56.2	NA	NA
24	20.0	80.0	...	...	...	...	...	361	531	4800	33.7	4200	29.5
25	...	90.0	...	5.0	...	...	5.0	...	554	5730	40.3	3180[c]	22.4[c]
26	...	97.5	...	...	...	...	2.5	...	579	4400	31.0	2900	20.4
27	1.0	97.5	...	...	...	...	1.5	...	588	4420	31.1	NA	NA
28	...	95.0	...	5.0	...	...	...	...	599	4330	30.4	3220[c]	22.6[c]

[a]Ultimate stress alloy alone, except as noted.   [e]Depends on specimen preparation.
[b]Recommended working stress in joint.   [f]Vickers pyramid diamond.
[c]Ultimate stress in joint.   NA = Not available.
[d]Modified Brinell hardness, using 100-kg load, ½ min.
Reprinted with permission from H. H. Manko. *Solders and Soldering,* McGraw-Hill, 1964, p. 132–133.

Table 9.14   Shear Strengths (kg/km²) of Various Solders as a Function of Temperature

				Temperature 20°C		Temperature 100°C	
Alloy				Strain Rate 0.05 mm/min	Strain Rate 50 mm/min	Strain Rate 0.05 mm/min	Strain Rate 50 mm/min
Sn	Pb	Sb	Ag				
60	40			20	39	13	34
10	90			17	36	11	16
62	36		2	28	52	12	34
40	58	2		24	43	11	25
95		5		28	55	14	29
5	93,5		1,5	18	30	12	21

The indicated strain rate is the rate of movement of the beam of the testing machine; the solder gap between the ring and plug is between 0.005 and 0.25 mm.
Reprinted with permission from R. J. Klein-Wassink, *Soldering in Electronics,* Electrochemical Publications, 1984, p. 120.

Elonga- tion, %	Brinell Hardness	Electrical Conductivity % IACS	Electrical Resistivity, μΩ-cm	Expansion Coefficient μm/(in)(°F)	Specific Gravity	General Notes
1.5	12.0	3.3	51.62	13.9	8.86	Expands, then shrinks to zero in 30 min.
50.0	14.0	3.0	57.47	12.8	8.58	Expands, then shrinks to zero in 60 min.
200.0	9.2	4.0	43.10	12.2	9.38	Expands to 0.0057 in/in permanently
30.0	25.0	3.1	55.61	NA	9.50	Nonelectrical solder for low-ambient temp.
220.0	9.0	4.3	40.38	NA	9.43	Shrinks to 0.0025 in/in, then expands to zero in 60 min.
83.0	4.9^d	11.7	14.74	NA	7.74	Low vapor pressure; good for glass
101.0	10.2^d	7.8	22.10	NA	8.97	Very good resistance to alkaline corrosion
200.0	22.0	5.0	34.48	8.3	8.72	Expands to 0.0007 in/in, then shrinks to 0.0005 in/in
58.0	5.2	13.0	13.26	NA	8.20	Good for thin precoat on ceramics
41.0	Too soft	24.0	7.18	18.3	7.44	Expensive, bonds to nonmetallics
135.5	12.0^d	12.2	14.13	NA	7.96	Good strength, low-cost indium alloy
28–30^e	17.0	11.5	14.99	13.7	8.46	Used where pasty range is intolerable
20.0	17.0	12.5	13.79	12.0	8.17	Good pretinning alloy
27–40^e	16.0	11.5	14.99	13.3	8.52	Good electrical-grade solder
29.0	15.6	9.6	17.96	NA	8.75	Similar to 50/50 Pb–Sn, resists creep well
38–98^e	14.0	10.9	15.82	13.0	8.90	General-purpose solder
55.0	9.6^d	6.0	28.74	NA	9.14	Very good resistance to alkaline corrosion
73.0	40.0	14.0	12.31	NA	10.38	High-temp, electrical solder for instruments
61.0	2.7^d	22.1	7.80	NA	8.10	Solders silver, fired glass, and ceramics
47.5	10.2^d	4.6	37.48	NA	9.80	Very good resistance to alkaline corrosion
39–115^c	12.0	10.1	17.07	13.9	9.28	Inexpensive utility solder
38.0	13.3	NA	NA	15.0	7.20	Lead-free, used in food equipment
30.0	13.7^f	12.6	13.70	NA	NA	High-temp, electrical solder
22.0	11.0	8.7	20.50	14.7	10.04	Cheap solder for body work and plumbing
23.0	9.0^d	5.6	30.79	NA	11.30	Tin-free indium solder
42.0	Too soft	8.8	19.50	NA	NA	Torch solder, poor corrosion resistance
23.0	9.5	NA	NA	NA	11.28	Slightly better corrosion resistance than no. 26
52.0	6.0^d	5.1	33.80	NA	11.35	Zinc-free indium solder

9.8 SOLDERABILITY TESTING

9.8.1 General Description

The ability of a material to be wet with solder is intimately related to its solderability, although there are some additional factors. Solderability depends partly on the inherent character of the base material involved, partly on the degree of cleanliness of the surfaces to be soldered after the fabrication of components, and partly on the aging caused by environmental attack during storage.

Good solderability results from good wetting, which means the formation of a uniform, smooth, unbroken, adherent coating of solder on the base material without the use of highly active fluxes. With poor solderability, poor wetting, nonwetting, partial wetting, or dewetting is observed. These terms refer to different degrees and sources of wetting problems. Poor wetting refers to a high contact angle between the molten solder and the substrate, with the result that spreading is slow or hindered. Nonwetting refers to the extreme case where the angle of contact is essentially 180 degrees, and no wetting occurs. Partial wetting refers to wetting that has occurred over only a portion of the substrate surface, probably because some areas were not properly cleaned or were recontaminated. Dewetting refers to the condition where wetting occurred, the molten solder spread, but subsequently drew back, leaving islands of substrate devoid of solder. The origin of dewetting is complex but related to reactions that occur after wetting and spreading that change (i.e., raise) the surface energy locally.

Although there is no perfect test, the main methods for assessing the solderability or wettability of a material for soldering are the (1) dip-and-look method; (2) wetting balance method; (3) globule method; (4) rotary dip method; (5) meniscus rise method; and (6) timed solder rise method.

9.8.2 Wetting Balance Method

The *wetting balance method* is one of the most common and versatile and, certainly, most quantitative methods for evaluating solderability. In this method, the specimen, which should be representative of the base material (i.e., alloy and form), is suspended from a sensitive balance (typically a spring balance) and immersed endwise to a predetermined depth in molten solder at a controlled temperature. The resultant vertical force from buoyancy and surface tension acting upon the immersed specimen is detected by a transducer. The plotted force trace is compared to one derived from a perfectly wetted specimen of the same nature and dimensions.

There are five stages in the test, as shown in Figure 9.12. In the first stage, just before immersion, there are no vertical forces (other than the downward force of gravity on the specimen). In the second stage, immediately upon immersion of the tip of the

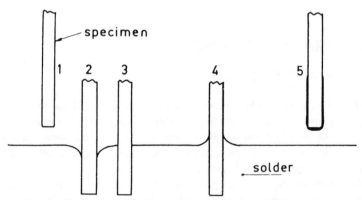

Figure 9.12   The setup and the five stages of the wetting balance method for testing solderability. (Reprinted with permission from R. J. Klein-Wassink, *Soldering in Electronics,* Electrochemical Society, 1984, p. 207.)

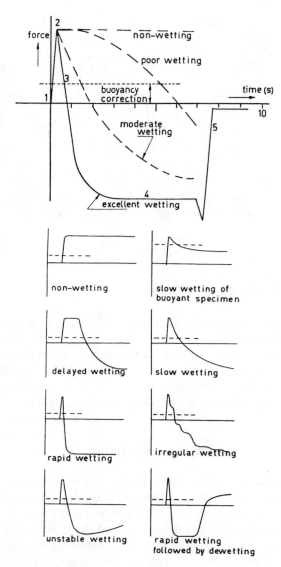

Figure 9.13 Typical traces from the wetting balance test for various wetting conditions. (Reprinted with permission from R. J. Klein-Wassink, *Soldering in Electronics,* Electrochemical Society, 1984, p. 21.)

specimen, the surface of the molten solder bath is pushed down, the meniscus is still curved down, and an upward force results from surface tension. When wetting has occurred, in the third stage, the vertical force of surface tension goes to zero and only an upward buoyancy forces exists. In the fourth stage, the meniscus curves upwards, and a downward force results from the surface tension. The fifth stage is when the specimen is withdrawn from the bath and forces return to zero (or gravity alone). Typical traces for various types or degrees of wetting are shown in Figure 9.13.

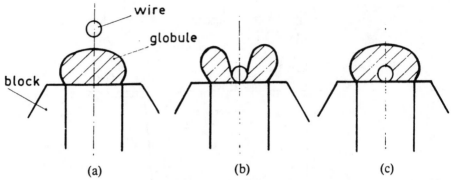

Figure 9.14   The globule method of testing solderability. (Reprinted with permission from R. J. Klein-Wassink, *Soldering in Electronics,* Electrochemical Society, 1984, p. 221.)

9.8.3 Globule Method

The longest established method for assessing solderability or wettability is the *globule method* developed by Philips-Eindhoven (in the Netherlands). The method measures the time required for molten solder at the soldering temperature to wet a circular wire pressed into a molten globule of the solder (see Figure 9.14). The solder glob on an iron post is bisected by the immersed wire test sample. The globule remains bisected until wetting occurs, at which point the globule closes around the wire.

9.8.4 Other Solderability Test Methods

The other common methods for testing solderability are the *dip and look, rotary dip, meniscus rise* and *timed solder rise,* methods. In the dip and look, which is the oldest method, the part is dipped into a molten solder bath under controlled conditions, and, after it is removed, the solder coating is inspected and evaluated for defects (e.g., partial wetting, nonwetting, or dewetting). This evaluation is obviously highly subjective, but the test yields useful results. In the rotary dip method, which was originally developed for printed wire boards to simulate wave soldering, a sample is placed in a holder and, by means of a rotating arm, is passed across the surface of a molten solder bath such that contact time is controlled. The test involves exposing samples for progressively longer times of contact until full wetting is achieved. This test, like the globule test, is semiquantitative. In the meniscus rise and timed solder rise methods, the time for solder to wet a substrate is measured by observing the formation of a meniscus with the substrate (i.e., the formation of a low contact angle).

In soldering, as in brazing and adhesive bonding, achieving good wetting is the key to producing sound joints. It is prudent to assess the wettability of the substrate to be soldered (or brazed or adhesively bonded) by testing before production assembly.

SUMMARY

Soldering, like brazing, enables solid materials (e.g., metals, ceramics, and glasses) to be joined by causing liquid filler to flow into and fill the space between joint faces and, then, to solidify without requiring or causing melting of the base materials. The liquidus of the filler in soldering is below the solidus of the base materials *and* below 450°C (840°F), and wetting and spreading by capillary forces are critical to proper filler distribution. Solder joint strength arises from a combination of metallic or other bonding as well as from a substantial contribution from mechanical interlocking between the filler and microscopic asperities on the base material surfaces. The principal reasons for soldering are to provide electrical connectivity and conductivity and/or leak-tightness or hermeticity, as opposed to providing mechanical strength. The low temperature of the process minimizes thermal damage to substrates, allows many joints to be made at once, and permits disassembly by desoldering.

Soldering depends critically on the wetting or solderability of the substrate by the solder at the soldering temperature. Precleaning and the use of fluxes are necessary to assure proper wetting. After soldering, any residue from fluxes must be removed to avoid corrosion of the substrates or degradation of electrical conductivity or shorting.

Solders are low-melting metals or alloys designed to be fluid at the soldering temperature. Most solder alloy systems exhibit eutectics. Ternary additions are often made to eutectic binaries to alter the eutectic temperature or increase fluidity by increasing the proportion of liquid in liquid-solid mixtures at the soldering temperature. Important alloys are tin–lead (the most common), tin–antimony, tin–silver, tin–zinc, cadmium–silver, cadmium–zinc, zinc–aluminum, very low-melting (i.e., fusible) bismuth-based alloys, and indium alloys. There is growing pressure to find effective Pb-free solders for environmental reasons.

Soldering processes differ in their source of heat, whether heating is localized or general, and whether the process is intended to be practiced in a manual or automated mode. Major processes include iron, torch, dip, wave, oven, resistance, and induction soldering. Modern soldering for electronic assembly often uses reflow methods in which preplaced solder is simply remelted to accomplish joining. All processes require the use of fluxes to help remove organic and oxide contaminants from the surface of the base materials to facilitate wetting. Fluxes vary in chemical aggressiveness or activity, depending on the need, ranging from mild, nonactivated rosin or more aggressive activated rosin, through aggressive organic acids or bases, to highly aggressive inorganic acids or salts.

Solder joints differ for mechanical and electrical assemblies, but all rely on mechanical interlocking to provide strength, and all attempt to minimize loading of the inherently low strength solder. Creep is a particular problem in soldered joints, since the service temperature is frequently high compared to the absolute melting point of the solder.

To assess the solderability of a base material, various testing methods are used to quantify the degree of wetting that can be achieved. Principal methods include wetting balance, globule, rotary dip, and meniscus rise protocols.

PRACTICE QUESTIONS AND PROBLEMS

1.   Define the process of *soldering* and compare it to the process of brazing. How are the two processes similar? How are the two processes different?
2.   How are the processes of soldering, brazing, and adhesive bonding similar in terms of the way coalescence of joint materials is achieved and the role of chemical bonding in producing joint strength? How do the absolute joint strengths and joint efficiencies compare among these three processes?
3.   What is the major advantage of soldering over fusion welding, brazing, and even adhesive bonding? What is the major limitation of soldering compared to many other joining processes?
4.   What is meant by the *solderability* of a base material? What inherent properties of the base material contribute to good solderability? What inherent properties contribute to poor solderability?
5.   What are two fundamentally different ways of improving the solderability of a base material with inherently poor solderability? What are the relative advantages and disadvantages of one approach compared to the other?
6.   What is meant by *aging* as it pertains to soldering and solderability? Why is aging problematic in typical manufacturing environments?
7.   How are specific solders selected for particular applications? What are considerations related to the base materials involved in the joint? What role, if any, does the planned soldering process play in the selection of the solder alloy?
8.   Why is precleaning so important in soldering? What various steps are typically involved in precleaning? What is precoating and why is it used?
9.   What is *solder residue*? Why is solder residue a concern and how can it be handled? What is *flux residue*? Why is it a concern and how can it be handled?
10.   For a typical binary solder alloy exhibiting eutectic and terminal phases with limited solid solubility, describe how solidification would progress, as an alloy with a solute content less than the solubility limit of its terminal solid solution phase was cooled from the melt. Describe how solidification would progress if the solute content of the solder exceeded this solubility limit but was less than the eutectic composition. How would solidification progress at the eutectic composition?
11.   What is the effect of substituting antimony for tin in a tin–lead solder? What is the effect of replacing all of the lead in a tin–lead solder with antimony?
12.   Why is silver added to tin–lead solder? Why is silver sometimes used to totally replace lead in tin-based solders?
13.   What benefits are derived from cadmium-based solders? What limitations are there for these solders?
14.   What is meant by a *fusible alloy*? Give four examples of where these are used.
15.   What are the benefits of indium-based solders? How do gold–silicon eutectic bonding solders work?
16.   What are the special considerations if dip or oven soldering is to be performed? What are the advantages of these processes over many other processes?
17.   What is meant by *reflow soldering*? Give some examples of reflow soldering processes.

18. Discuss the three broad classifications of fluxes used in soldering. Are fluxes always required in soldering? If so, why? If not, why not?
19. What factors must always be taken into account when designing joints for soldering? What special factors must be taken into account in joints to be used in mechanical assemblies versus electrical assemblies?
20. Describe five different methods for assessing the solderability of a particular material or component. Which methods give quantitative measures of solderability?

REFERENCES

1. American Welding Society, *Soldering Manual,* 2nd ed.,Miami: American Welding Society, 1978.
2. American Welding Society, *Welding Handbook, Vol. 2, Welding Processes—Arc and Gas Welding and Cutting, Brazing and Soldering,* 7th ed., Miami: American Welding Society, 1978.
3. ASM International, *Electronic Materials Handbook, Volume 1—Packaging,* Materials Park, Ohio: ASM International, 1989.
4. Combs, C. F., Jr., *Printed Circuit Handbook,* New York: McGraw-Hill, 1988.
5. Hwang, J. S., *Solder Paste in Electronics Packaging,* New York: Van Nostrand Reinhold, 1989.
6. Klein-Wassink, R. J., *Soldering in Electronics,* Ayr Scotland: Electrochemical Publications Limited 1984.
7. Manko, H. H., *Solders and Soldering,* New York: McGraw-Hill, 1964.
8. Manko, H. H., *Soldering Handbook for Printed Circuits and Surface Mounting,* New York: Van Nostrand Reinhold, 1986.
9. Massalski, Thaddeus B. (Ed.), *Binary Alloy Phase Diagrams,* Vols. 1 & 2, Metals Park, Ohio: American Society for Metals, 1986.

Other Joining Processes

10.1 INTRODUCTION

There are two fundamental ways to join materials: chemically or mechanically, with and without the formation of atomic or molecular bonds, respectively. The pure mechanical joining of materials, structures, parts, or components is known as mechanical fastening and relies on mechanical interference or interlocking between joint elements, with varying dependence on the development of friction at the joint faying surfaces to achieve joining or holding. No formation of bonds is required or occurs.[1] Chemical joining of materials (or the structures, parts, or components composed of those materials) relies upon the intentional formation of very large numbers of interatomic or intermolecular bonds. Chemical joining methods include the broad categories of welding and adhesive bonding and the subcategories of brazing and soldering within welding. Welding involves purely chemical bonding, with no dependence at all on mechanical forces. The resulting bonds are almost always primary in nature: metallic bonds between metals or alloys; ionic, covalent, or mixed bonds between ceramics; covalent bonds between glasses; and covalent bonds or, more likely, secondary bonds between thermoplastic polymers. Within the subcategory of brazing, bonding is also primary, with no dependence on mechanical forces for joint strength, while within the subcategory of soldering, bonding can be primary, but there is often a substantial mechanical contribution to joint strength. In adhesive bonding, there is bond formation, but it is often secondary, from either van der Waals induced dipole or permanent dipole attraction between the agent used to facilitate joining, i.e., the adhesive, and the materials being joined, i.e., the adherends. Usually, however, there is also considerable contribution from mechanical forces during adhesive bonding as well.

These processes represent the overwhelming majority of all means for joining and certainly represent all of the fundamental processes. There are, however, some important variations or hybrids[2] of these fundamental processes. *Variations* are literally

---

[1]During mechanical fastening, no bond formation is caused to occur intentionally, although there is, almost certainly, some bond formation as a result of the fact that at least some atoms, ions, or molecules from abutting members of the joint come to their equilibrium interatomic spacing. This low level of bond formation leads to the frictional force.

[2]A *hybrid* is something of mixed origin or composition.

just that; they achieve joining just as one of the fundamental processes does but use some unusual means for doing so. *Hybrids,* on the other hand, combine fundamental processes (or specific processes within a fundamental process) to achieve some unique characteristics or capabilities; often with synergistic[3] results. These variations or hybrids tend to be of more recent origin and have often been developed to overcome particular shortcomings of the basic processes for particularly challenging applications (e.g., advanced aerospace structures). This chapter explores these other processes.

10.2 OTHER OPTIONS FOR JOINING MATERIALS

Other joining processes are either *variations or hybrids* of the fundamental processes of mechanical fastening, adhesive bonding, welding, brazing, or soldering. These currently include (1) thermal spraying; (2) weldbonding; (3) weldbrazing; (4) braze welding; and (5) rivet-bonding. In addition, some important examples of *hybrid welding* processes will also be discussed.

Briefly, *thermal spraying* is a variation of welding or adhesive bonding, depending on whether a metal or ceramic or a thermoplastic is being deposited and for what purpose. *Weldbonding* is a hybrid of resistance spot welding and adhesive bonding. *Weldbrazing* is a hybrid of resistance spot welding and brazing. *Braze welding* is a hybrid of brazing and fusion welding. *Rivet-bonding* is a hybrid of adhesive bonding and mechanical fastening. *Hybrid welding* refers to some relatively new welding processes that are, in fact, hybrids of two or more specific welding processes. Two examples are laser-GTA and plasma-GMA.

10.3 THERMAL SPRAYING

10.3.1 General Description

Thermal spraying (THSP) is a process in which a material is heated and propelled in atomized or particulate form onto a substrate. Initially, the material may be in the form of a continuous solid wire, a rod, or powder, and the source for heating the atomized material to the plastic or molten state can be an oxy-fuel gas flame, electric or plasma arc, or an explosive gas mixture. Material particles are propelled (i.e., given kinetic energy) from the heat source, called a *spray gun* or *torch.* They are heated either during the process of their formation if the starting form is wire or rod, or while immersed in the flame or arc or jet if the starting form is a powder. Bonding is facilitated both by the kinetic energy of the particles and by their heat.

The thermal spraying process is sometimes called *metallizing* or *metal spraying*, but it is really broader than those terms denote. Most metals, oxides, carbides,

---

[3]*Synergism* or *synergy* means the action of two or more substances, organisms, or processes to achieve an effect of which each is individually incapable, with the net effect being better than the simple sum.

ceramic–metal mixtures (called *cermets*) or hard metallic or intermetallic compounds can be deposited by one or more of the thermal spray processes. It is also possible to thermally spray or deposit thermoplastic polymers.

The thermal spray process is a joining process in the sense that it joins one material to another on a macroscopic or bulk scale using microscopic and atomic-level mechanisms. Although usually used for applying protective surface coatings (not unlike welding or brazing for the purpose of hardfacing[4]), thermal spraying can be used for joining in combination with brazing or adhesive bonding. Here, it is used to apply the braze alloy or thermoplastic adhesive by spraying, and then the process of bonding or joining is completed by heating that deposit under some pressure in a process called *fusing*. The process can also be used to produce solid objects using a disposable substrate or form, in which case, the finished object is near-net shape. Here, thermal spraying is a shape welding process.

10.3.2 Mechanism of Coating Bonding

When molten or softened particles strike a substrate at high velocity, they flatten in the direction of motion and form thin platelets that conform to the substrate surface. These platelets rapidly cool and solidify, or simply harden. Successive layers are built up to the desired thickness by having sprayed particles bond to one another, as opposed to to the substrate. Figure 10.1 shows how particles are propelled toward a substrate, deform on impact, giving up their kinetic energy in the form of heat, conform to the contour of the substrate, and adhere. The entrapment of voids, oxides, and other inclusions is also evident in Figure 10.1.

The bond between the sprayed deposit and the substrate may be mechanical, metallurgical, chemical, or a combination of these, depending on the nature of the material being sprayed, the substrate, and the process and spraying conditions. The kinetic energy of the propelled particles brings the materials into intimate contact, ruptures any oxide layers on the particle or droplet or the substrate, thereby enabling bond formation, and uses the heat from initial heating as well as from conversion of kinetic energy to accelerate the kinetics of diffusion and/or reaction with the substrate surface. In some cases, a thermal treatment of the resulting layered composite structure is used to increase the bond strength by diffusion and/or chemical reaction between the deposit and the substrate. As stated previously, this is called *fusing*.

The strength of the bond between the sprayed coating and the substrate depends on the substrate material, the substrate geometry, substrate surface preparation, angle of spray particle impingement with the substrate, presence of preheat, deposit thickness, and postspraying thermal treatment. Sometimes bonding is aided with a special intermediate material applied before the final deposit, called a *bond layer*.

---

[4]*Hardfacing* involves the application of a material, usually a hard metal alloy or ceramic, to the surface of a part for the express purpose of improving the corrosion and/or wear properties of that part. In welding and brazing, this coating is applied in the molten state. In thermal spraying, the material being applied (i.e., the *hardface*) may be molten or only plastic.

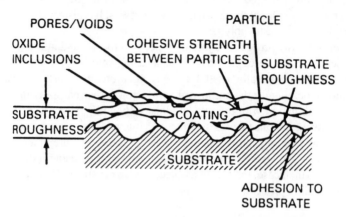

Figure 10.1   The mechanism of particle bonding and coating buildup during thermal spraying. (Reprinted with permission from *Welding Handbook,* Vol. 2, 8th ed., American Welding Society, 1991, p. 865.)

10.3.3 Properties of Bonded Coatings

The properties of the deposited and bonded coating depend on the density of the deposit, the cohesion between the deposited particles within the deposit, and the adhesion of the first particles deposited to the substrate. The density of the deposit produced by thermal spraying depends on the type of material being sprayed, the method of deposition, the specific spraying procedure, and subsequent processing.

Because the process is often performed with the substrate at room temperature or only slightly heated, and since the direct heat and heat from kinetic energy transferred by the particles as they strike the substrate are relatively small (on a bulk scale), there is usually little thermal alteration of the substrate. For this reason, thermal spraying is excellent for restoring dimensions by replacing material lost by wear or corrosion or for applying a protective coating against wear or corrosion. The process is especially good for creating unique surface microstructures, such as composites (i.e., intimate mixtures of metals and ceramics, for example) and amorphous metals.

10.3.4 Applications

Thermal spraying is most widely used for surface alteration. Surface deposits can restore dimensions where material has been lost to wear or corrosion or to apply material to dies or molds or castings to achieve precise dimensions not achieved in initial fabrication. They can also improve surface resistance to wear, corrosion, oxidation, or combinations of these. Zinc, aluminum, stainless steel, bronze, hard alloys, and ceramics are used. Because of its ability to restore dimensions and protect surfaces, thermal spraying is a valuable and widely used process in maintenance and repair, as well as in original equipment manufacture (known as OEM).

Thermal spraying can also be used to provide certain other desired properties at surfaces, such as thermal or electrical conductivity or insulation, high friction for traction (such as on stairs and walking surfaces on the decks of ships, where water can make those surfaces slippery), or lubricity through the application of a polymer to a metal.

An increasingly important application of the process is the spray consolidation of reinforcing materials (e.g., filament wound or preplaced fibers) to produce a metal- or ceramic-matrix composite part. Finally, thermal spraying can be used for joining by applying a braze or solder alloy or thermoplastic adhesive and then fusing that deposit under pressure to effect bonding and joining.

10.3.5 Process Variations

The four variations of thermal spraying are (1) flame spraying (FLSP); (2) plasma spraying (PSP); (3) electric arc spraying (ASP); and (4) detonation flame spraying (DFSP). These variations are based on the method by which the material being sprayed is heated to the molten or softened plastic state and the technique for propelling the resulting hot particles to the substrate.

In *flame spraying,* the material to be sprayed is continuously fed into and melted or softened by an oxy-fuel gas flame. A typical flame spray gun or torch is shown in Figure 10.2. The material feedstock may initially be in a wire, rod, or powder form. Obviously, for wire or rod forms, sufficient heating by the flame must occur to cause melting and atomization into particles. For powders, on the other hand, heating in the flame may, but need not, cause melting. Regardless of the starting form of the feed material, molten or softened particles are propelled onto a substrate by either an auxiliary air jet or the combusted and expanded gases themselves.

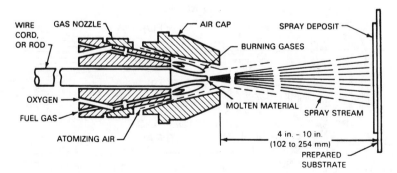

Figure 10.2 A typical flame spray gun. (Reprinted with permission from *Welding Handbook,* Vol. 2, 8th ed., American Welding Society,1991, p. 869.)

The heat value of an oxy-fuel combustion flame is limited compared to other sources (although modern gun design has enabled heat values of greater than 100,000 BTUs to be achieved), so there are practical limits on the types of materials that can be sprayed successfully. Refractory materials (i.e., metals or ceramics) can be difficult or impossible to spray. In addition, the velocity that can be obtained is also limited (although it approaches supersonic velocities) and so the as-deposited density and adhesion can be limited. Table 10.1 lists various materials that can be flame sprayed. This list is not exhaustive.

In *plasma spraying,* the heat for melting or softening the material being sprayed is provided by a nontransferred plasma arc. Such an arc (described in Chapter 6, Section 6.4.3) is maintained between the non-consumable tungsten electrode and the constricting inner nozzle of a torch. A typical plasma spray gun is shown in Figure 10.3. An inert or reducing gas, under pressure, enters the annular space around the electrode, where it is heated to a very high temperature (typically above 10,000°C) by the arc. The extremely hot gas is ionized and expanded tremendously, so that it passes through and exits from the nozzle as a very high (often supersonic) velocity jet. The material being sprayed is usually injected in powder form into the hot gas jet, where it picks up heat and melts or softens while being propelled onto the substrate.

The inherently higher heat value of plasma spray systems allows essentially any material to be sprayed, no matter how refractory. Further, the very high kinetic energy imparted to the spray particles results in more adherent and denser deposits. Table 10.1 lists some important types of materials that can be plasma sprayed.

In *electric arc spraying,* two continuously fed wires are melted by an electric arc sustained between them. This arc heats the wires to melting and atomizes the material into particles. The arc force or arc jet propels the atomized particles onto a substrate, usually with the assist of an auxiliary jet of air. A typical arc spray gun is shown in

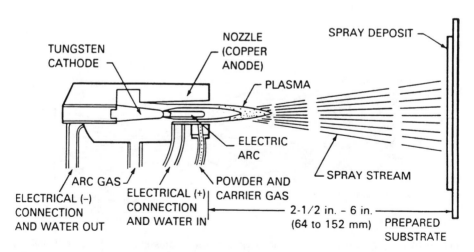

Figure 10.3 A typical plasma spray gun or torch. (Reprinted with permission from *Welding Handbook,* Vol. 2, 8th ed., American Welding Society,1991, p. 874.)

Table 10.1 Materials That Can Be Thermally Sprayed by Various Methods

	Flame Sprayed	
Metal wires	*Ceramic rods*	*Powders*
Aluminum	Alumina–titanium	Hard metals
Copper	Alumina	Carbides
Molybdenum	Zironia	Borides
AISI 1025 steel	Rare earth oxides	Oxides
304 stainless steel	Zironium silicate	Nitrides
Zinc	Magnesium zironate	Thermoplastics
	Barium titanate	
	Calcium titanate	
	Chromium oxide	
	Magnesia–alumina	
	Mullite	

Electric Arc Sprayed

Aluminum	Stainless steel
Copper	Mild steel
Brass or bronze	Babbit metal
Zinc	

	Plasma Sprayed	
Metals	*Ceramics*	*Cermets*
Aluminum	Cr–carbides	Alumina-nickel
Chromium	Ti–carbides	Alumina-nickel aluminide
Copper	W–carbides	Magnesia–nickel
Columbium (Niobium)	Alumina	Zirconia–nickel
Molybdenum	Cr–oxides	Zirconia–nickel aluminide
Nickel	Magnesia	
Nickel–chromium	Titania	
Tungsten		
Tantalum		

Detonation flame Sprayed

Alumina	Tungsten carbide/Co–binder
Alumina–titania	Tungstyen carbide/Ni–Cr binder
Chromium carbide/Ni–Cr binder	Tungsten chromium carbide/ Ni–Cr binder

Figure 10.4. This method is restricted in use to metals that can be produced as continuous wires. Deposited densities are not particularly impressive. Table 10.1 lists some important materials that can be electric arc sprayed.

The *detonation flame spraying* method operates on a very different principle than the other thermal spray methods. Charges of powder are repeatedly heated and projected as molten or plastic particles onto a substrate by rapid, successive detonations of

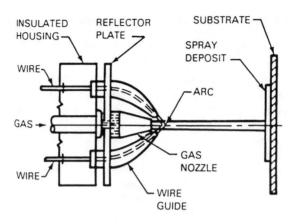

Figure 10.4   A typical electric arc spray gun. (Reprinted with permission from *Welding Handbook,* Vol. 2, 8th ed., American Welding Society,1991, p. 873.)

an explosive mixture of acetylene and oxygen in a gun chamber (see Figure 10.5). The particles leave the gun at much higher velocities than in other processes, thereby producing much denser deposits. The process may not (and usually does not) produce the amount of heating of other processes, however. Table 10.1 lists some materials sprayed by the detonation flame method.

Table 10.2 compares the various thermal spray processes in terms of heat source temperature, particle velocity, and types of materials that can be sprayed.

10.4 HYBRID JOINING PROCESSES

10.4.1 General Description

There are some examples of *hybrid joining processes,* where two different processes are combined to create a new process with extended capability. Sometimes the resulting hybrid simply combines the characteristics of both of the parent processes to obtain the best of both. An example is braze welding. Other times, the hybrid exhibits unique benefits as the result of some synergy between the two parent processes. Three examples are weldbonding, weldbrazing, and rivet-bonding. Not every combination of basic joining processes results in a useful hybrid, however, and some combinations can cause problems.

10.4.2 Weldbonding

Weldbonding, also called *spot-weld adhesive bonding*, is a method of fabricating hardware that uses both welding and adhesive bonding techniques. In its most common

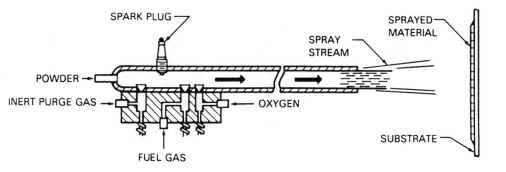

Figure 10.5 A detonation flame spray gun. Source: American Welding Society, *Welding Handbook*, Vol. 2, 8th ed., American Welding Society, 1991, p. 873.

Table 10.2 Comparison of Various Thermal Spray Process Characteristics

arc	Flame Oxyfuel gas Combustion	Detonation Oxyfuel gas Pulsed explosion	Arc Electric arc, Two wires	Plasma Nontransferred Arc	PTA Plasma Transferred
Heat source temperature	4,700–5,600°F	6,000°F +	8,000°F	15,000°F	15,000°F
Particle velocity	800 ft/sec	2500 ft/sec	800 ft/sec	1800 ft/sec	—
Coating Materials					
Wire-metal	Yes	No	Yes	Yes	Yes
Powder-metal	Yes	Yes	No	Yes	Yes
Powder-ceramic	Yes	Yes	No	No	No
Rod or cord ceramic and plastic	Yes	No	No	No	No

Source: Howard B. Cary, *Modern Welding Technology*, 2e, ©1989 pp. 220, 221, 222. Reprinted with permission of Prentice-Hall, Englewood Cliffs, New Jersey.

form, a layer of adhesive, either in paste or film form, is applied to one of the metal members to be joined. The other metal member is placed on top, forming a lap-type joint, and the assembly is then clamped or tack-welded to maintain part alignment. The two metal members are then joined by resistance welding through the adhesive, using a conventional spot or seam welder. The welds are commonly spaced 2.5 to 5.0 cm (1 to 2 inches) apart, center to center. It is also possible, in another variation of the process, to spot weld first and then infiltrate the gap between the faying surfaces with a thinned adhesive, relying on capillary action to cause flow and fill.

Figure 10.6 shows a spot weld in a single-lap weld-bonded joint. Spot welding pressure and heat displace the adhesive and allow metal fusion to form a nugget. A visible mark on the face surface of the metal sandwich sheet denotes the weld location. The inner circle outlines the weld nugget. The area between the two dashed circles in the figure, or the *halo*, is effectively unbonded because of the displacement of adhesive

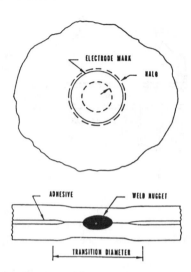

Figure 10.6   A spot weld in a single-lap weld-bonded joint.

and heating effects. Beyond the halo is a region of transition to full adhesive thickness and adhesive bonding.

The weldbonding process offers some important advantages over simple mechanical fastening, as well as simple adhesive bonding. Compared to a mechanically fastened structure, a similar weldbonded structure offers (1) increased static tensile and/or shear strength (by increasing the total area of joining from just the area of the fasteners to the area of the welds and the adhesive); (2) increased fatigue life (by spreading the loading through the adhesive and minimizing stress concentrations); (3) gas-tight and/or fluid-tight joints (through the sealing action of the adhesive); (4) increased rigidity (by preventing slip at fasteners or buckling between points of fastening or welding); (5) improved corrosion resistance (through sealing); (6) inexpensive tooling; (7) weight savings; (8) smooth, hermetically sealed inner and outer surfaces (e.g., for aerodynamics); and (9) complete interface bonding to improve load transfer.

Compared to pure adhesive bonding, weldbonding dramatically increases peel and/or cleavage strength, as well as pure tension, compression (due to buckling), and shear. There is evidence that static and especially fatigue strength is better than the purely added strengths of the adhesive and the spot welds. This probably has to do with uniform load or stress distribution versus stress concentration.

Weldbonding is competitive in static strength with titanium fasteners in sheet titanium up to 4 mm (0.156 in.) thick. Beyond this point, spot-welding limitations and static strength capacity make weldbonds less desirable. Epoxy and polyimide adhesives are typically used for titanium, often using fillers of silica (7 wt%), 3% strontium chromate (for corrosion resistance), and metal powder. Aluminum alloys are also weldbonded, using epoxy, modified-epoxy, or elastomeric urethane adhesives, sometimes with fillers. When conductive metal fillers are used, welding parameters are nearly the same as those used for welding without adhesive.

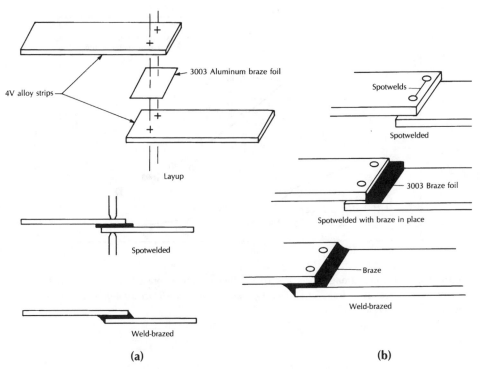

Figure 10.7 Weldbrazing with capillary flow and by spot welding through prepunched braze filler foils. (a) Weldbraze by capillary flow; (b) weldbraze through preplaced prepunched foil. (Source: T. T. Bales *et al., Development of the Weld-Braze Joining Process,* NASA Technical Note D-7281, 1973.)

Welding through an adhesive causes a higher percentage of irregularly shaped nuggets, but the strength of the joints is not adversely affected. There is, however, also a tendency for a higher percentage of spot welds that exhibit expulsion, or *spitting,* resulting in lower quality and lower strength welds. The approach of welding first and then back-infiltrating with a low-viscosity adhesive opens up more options for the welding processes that can be employed. For example, GTA-spot, laser-spot, and EB-spot become possible.

10.4.3 Weldbrazing

NASA and the Air Force have had some limited experience in which resistance spot welds have been made and then the joint back-infiltrated with braze filler metal (see Figure 10.7a). Braze alloy preforms have also been used where there are prepunched holes through which spot welds are first made, and, then, the braze alloy is fused, as shown in Figure 10.7b. In either variation, the process is called weldbrazing.

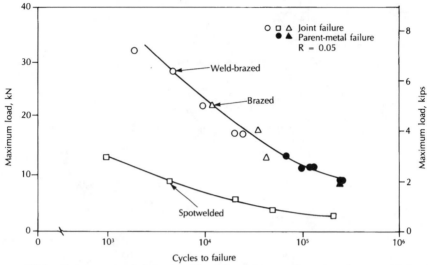

Figure 10.8   Fatigue strength behavior of weldbrazed versus brazed and spot welded only specimens. (Source: T. T. Bales *et al, Development of the Weld-Braze Joining Process,* NASA Technical Note D-7281, 1973.)

Two systems that have been weldbrazed to advantage are titanium and aluminum alloys, both using aluminum alloy filler. The process of weldbrazing results in improvements in static shear and, especially, peel and fatigue strength. There is not much evidence of a synergistic effect in weldbrazing, however. Static shear strength seems to be simply additive, and there appears to be no gain in weld-brazed joints over brazed joints (see Figure 10.8). Strength-at-temperature, on the other hand, is definitely improved somewhat, mostly by extending service temperatures slightly beyond those for brazing alloys alone. Figure 10.9a shows the tensile shear strength as a function of test temperature, while Figure 10.9b shows stress-rupture behavior, both for single-overlap specimens.

10.4.4 Braze Welding

Braze welding is an odd process—neither simply a variation of welding or brazing nor a true hybrid. It is accomplished by the use of a filler metal having a liquidus above 450°C (840°F) and below the solidus of the base metals to be joined, fulfilling one of the key requirements of brazing. However, unlike brazing, in braze welding the filler is *not* distributed in the joint by capillary action but, rather, is added to the joint as a welding rod (analogous to gas-tungsten arc welding using filler) or as a consumable arc welding electrode (analogous to shielded metal arc welding). Typical of brazing,

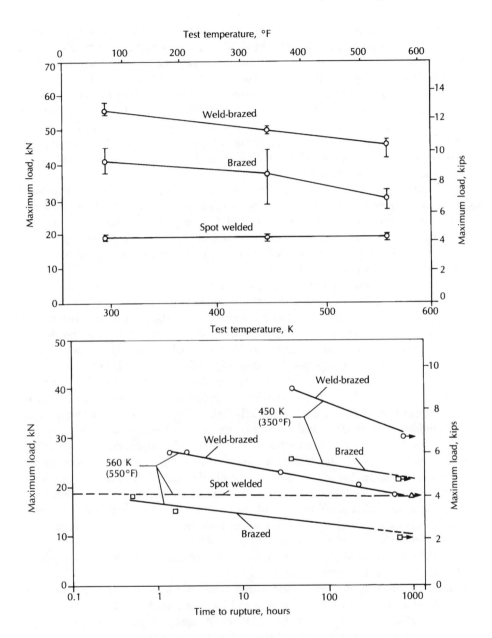

Figure 10.9 Tensile-shear strength and stress-rupture behavior of single-overlap weldbrazed specimen as a function of temperature. (*Source*: T. T. Bales, *et al*, *Development of the Weld-Braze Joining Process*, NASA Technical Note D-7281, 1973.)

however, the base metals being joined are not melted, only the filler. Bonding in braze welding takes place between the deposited filler and the hot, unmelted base metals in the same manner as in conventional welding or brazing, i.e., through the formation of metallic bonds during interdiffusion or reaction at faying surfaces and filler.

Joint designs for braze welding are similar to those used for oxyacetylene welding and are typically grooved. To obtain a strong bond between the filler metal and the unmelted base metal, the molten filler must wet the hot base metal. The process involves heating the base metal to the required temperature, adding flux (for oxy-fuel processes), precoating the joint faces with filler alloy (known as *buttering*), and adding filler until the joint is completely filled.

Most braze welding is done using an oxy-fuel gas welding torch, but the process can be done using a nonconsumable carbon arc, gas-tungsten arc, or plasma arc torch. Carbon arc is often used with galvanized steel. Gas-tungsten and plasma arcs are used with inert gas shielding and with filler metals that have relatively high melting temperatures. Specially formulated shielded consumable electrodes can also be used with shielded metal arc welding or SMAW.

The braze welding process has some distinct advantages over conventional fusion welding processes. First, less heat is required to accomplish bonding than conventional welding, thereby permitting higher joining speeds, less energy consumption, and less distortion. Second, the deposited filler metal is relatively soft and ductile, thereby providing good machineability and low residual stresses. Third, joints can be produced with adequate strength for many applications. Fourth, brittle metals (e.g., gray cast iron) can be braze welded without extensive preheating and minimal tendency toward cracking. Fifth, the process provides a convenient way to join dissimilar metals (e.g., copper to steel or cast iron and nickel–copper alloys to cast iron or steel). Sixth, the equipment required for braze welding is relatively inexpensive and easy to use.

Despite these numerous advantages, there are disadvantages. First, the strength of braze welds is limited to that of the filler metal, which, since melting temperatures are low, tends to be relatively low. Second, permissible service temperatures for braze welded assemblies are lower than those of fusion welded assemblies, again because of the lower melting fillers employed. Third, the joint may be subject to galvanic corrosion and differential chemical attack owing to the filler-base metal mismatch. Fourth, the color of the braze weld may not match the base metal.

Filler metals used for braze welding tend to be commercial torch brazing brass filler alloys containing approximately 60 wt% copper and 40 wt% zinc, with small additions of tin, iron, manganese, and silicon to improve flow characteristics, decrease volatilization of zinc, scavenge oxygen, and increase strength and hardness. Nickel additions of up to 10 wt% whiten the color and increase strength and corrosion resistance. Some typical braze welding fillers are listed in Table 10.3.

Fluxes for braze welding are compounded for the base metal and the brass filler metal. They tend to remain active for longer periods of time at higher temperatures than fluxes designed for capillary brazing, since the nature of the process demands that.

Joints for braze welding are usually of groove, fillet, or edge types, similar to those used for conventional fusion welding. Assemblies can be simple or complex and can be made up of sheet, plate, pipes, tubes, rods, bars, forgings, or castings. For best

Table 10.3 Some Typical Braze Welding Fillers

AWS Classification	Chemical composition (wt %)					Minimum tensile strength		Liquid temperature	
A 5.7 and A 5.8	*Cu*	*Zn*	*Sn*	*Fe*	*Ni*	*MPa*	*ksi*	*°C*	*°F*
RBCuZn-A	60	39	1	—	—	275	40	900	1650
RBCuZn-B	60	37.5	1	1	0.5	344	50	890	1630
RBCuZn-C	60	38	1	1	—	344	50	935	1630
RBCuZn-D	50	40	—	—	10	413	60	935	1715

strength, the bond area between the brazing alloy and the base metal should be as large as possible.

10.4.5 Rivet-Bonding

Like weldbonding, *rivet-bonding* is a hybrid of adhesive bonding but where mechanical fastening (by rivets) is combined with adhesives to create a structural joint with improved properties. Like weldbonding and weldbrazing, rivet-bonding offers (1) reduced joint stress, (2) longer fatigue life, and (3) improved fracture characteristics and a degree of redundancy in the load paths. Figure 10. 10 shows a schematic representation of a rivet-bonded joint, as well as a weldbonded joint.

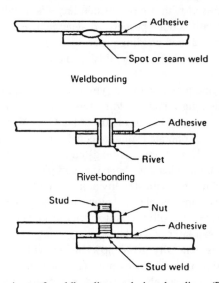

Figure 10.10 Comparison of weldbonding and rivet-bonding. (Reprinted with permission from *Welding Handbook,* Vol. 3, 7th Ed., American Welding Society, 1980, p. 359.)

10.5 HYBRID WELDING PROCESSES

10.5.1 General Description

Just as it is possible to combine different basic joining processes to create new, hybrid joining processes (such as weldbonding), it is possible to combine two different welding processes to create a new, hybrid welding process. This is usually done to create a new hybrid process that combines the best characteristics of each parent process or to obtain some synergistic benefit. Two examples of hybrid welding processes are laser-gas tungsten arc (or GTA) welding and plasma-gas metal arc (or GMA) welding.

10.5.2 Laser-GTA Welding

The highly monochromatic, collimated light beam from laser sources permits the surface heat treatment, surface alloying, cutting, drilling and welding of materials, almost without limit and, essentially, independent of melting point. One of the principal disadvantages of the laser as a welding heat source is the low process efficiency, or the limited amount of energy from the laser beam that actually enters the workpiece to effect a weld. Efficiencies of 10 to 15% are common, and even lower values prevail for high conductivity or reflective materials (e.g., aluminum). Although most metals are weldable by the laser-beam process, high thermal conductivity materials, such as copper and aluminum, are difficult to weld because the surface absorptivity of a material decreases as its conductivity increases.

The thermal coupling and penetration of a laser can be dramatically enhanced when the laser beam is used with a gas-tungsten arc. The resulting process is called *laser-GTA welding*. When it is done properly, a synergistic effect occurs in which the joint penetration of the combined welding processes is greater than the simple sum of the two processes. In addition, the arc stability of the GTA process is also improved markedly.

To work properly and give rise to the synergistic effects, the laser beam must be placed on axis with a gas-tungsten arc welding torch. On-axis means the laser beam is positioned to within 3.2 mm (1/8 in.) of the tip of the GTAW electrode on a line such that, when both processes are operated simultaneously, a combined plasma results. This is shown in Figure 10.11. In fact, the gas-tungsten arc can be placed either above or below the workpiece, while the laser is above the workpiece. In the first case, the laser and arc form a combined plasma with stability at low GTAW currents and high rates of travel. When the GTAW torch is placed below the workpiece, the arc preheats the material, increases surface absorptivity, and results in higher process efficiency and joint penetration for a given laser power level.

Joint penetration is generally 20 to 50% higher with laser-GTA than straight GTA welding. Penetration improvements are even more dramatic for higher power lasers. The stability of the arc is also increased, especially at lower currents. Three mechanisms have been proposed to account for these observed phenomena in laser-GTA welding. The first is stabilization of the anode spot. Unlike the normal situation where the anode spot (located at the work in DC-electrode negative) motion does not

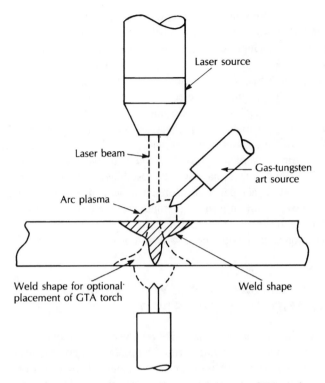

Figure 10.11 The laser-GTA welding process. As an option, the GTA torch can be located on the back side of the worpiece, altering the weld as shown.

correspond to torch motion but tends to lag and then snap back under the electrode erratically, in laser-GTA welding the anode spot seems rooted directly under the electrode at all times. This has been proposed to improve current flow and heating efficiency at the weld. Second, laser beam absorption is promoted. Laser absorption is improved in the presence of the GTAW arc because the superheated molten region at the surface of the workpiece under the electrode dramatically increases absorptivity. This is true since absorptivity is directly related to the temperature of a body. Third, the direct interaction of photons with gas atoms could be enhancing ionization. The collision of photons with excited gas atoms could be a minor contributor to increased penetration and increased arc stability.

Whatever the precise mechanism or mechanisms involved, the combining of laser-beam and GTA welding to create laser-GTA has interesting and beneficial results.

10.5.3 Plasma-GMA Welding

In 1972, engineers at Philips Research Laboratories in Eindhoven, the Netherlands, developed an electrode holder that combines the features of plasma arc and gas-metal arc welding (see Figure 10.12). When operated at certain power levels, the system produces welds that differ from those of either of the component processes used separately. The resulting plasma-GMA process is a hybrid welding process.

For the plasma, an inert gas (e.g., argon) is used, while for shielding, either argon or carbon dioxide can be used. The resulting plasma arc is of the transferred type, the workpiece being connected on the negative side of the circuit.

In operation, the gas-metal arc electrode is fed through the center of the shielded plasma stream. Under DCRP (electrode positive), the filler metal consumable electrode establishes its own arc while being preheated by the surrounding plasma arc. Under these conditions, two modes of operation have been discovered that have widely different metal deposition characteristics. First, when comparatively low current is passed through the filler metal, a firm, narrow arc is established. This is good for high speed welding of thin sheet or deep penetration of plates. Second, when the current in the filler metal rises to above a certain transition value, the mode of deposition changes, with swirling of the electrode arc in the plasma producing a shower of fine drops. A flat, shallow penetration results, with very low heat input.

10.6 OTHER COMBINATIONS: WHAT MAKES SENSE AND WHAT DOES NOT?

The combination of welding with adhesive bonding in weldbonding, with brazing in weldbrazing, and of adhesive bonding with riveting in rivet-bonding were all described as resulting in interesting hybrid joining processes, but the possibility of creating useful hybrids by combining fundamental processes is not endless. Some combinations make sense, in that the combination offsets weaknesses of one process with the strengths of the other, and the possibility exists that there could be some synergistic effect. Other combinations do not make sense, in that the strengths and weaknesses of the two parents not only do not offset, but the weakness may add. It is worth looking at other combinations to see what makes sense and what does not.

Another combination of adhesive bonding with mechanical fastening that is not usually thought of as a hybrid is referred to simply as *sealing*. It is done to prevent leaking or provide fluid-tightness or to prevent corrosion due to intrusion of moisture, salt, salt water, solvents, or corrosive chemicals or agents in the crevice created by fastening. If the sealant is a nonstructural adhesive, which is usually the case, it is used strictly as a gasket,[5] and does not provide much, if any, improvement in static or fatigue strength. If a structural adhesive is used with mechanical fasteners, it is probably being used primarily for its strength properties and secondarily for its sealing properties. In this case, the sealant does increase strength by increasing the load-bearing area, especially in shear, and distributing or spreading the load to reduce fatigue. The use of a structural adhesive usually provides some synergism because the adhesive takes some of the load off the fasteners and redistributes the remaining load more uniformly. The fasteners, in turn, also take some load off the adhesive, particularly out

---

[5]A *gasket* is any of a wide variety of seals or packings used between matched machine parts (or around pipes) to prevent the escape or intrusion of a gas or fluid.

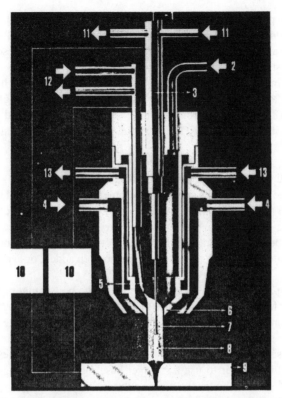

Figure 10.12 The plasma-GMA welding torch. 1, bar wire filler metal; 2, plasma gas; 3, filler metal contact tube; 4, shielding gas; 5, nonconsumable tungsten electrode; 6, water-cooled copper nozzle; 7, plasma arc; 8, gas metal arc; 9, workpiece; 10, power source, 11, 12, and 13, water cooling system. (Reprinted with permission from Philips Research Laboratories, "A New Process: Plasma-GMA Welding," *Welding Journal,* Vol. 51, No. 8, p. 561.)

of plane. While the actual analysis of load distribution between the fasteners and the adhesive is complicated because of their drastically different moduli (or strain response to stress), the contribution of a structural adhesive to a mechanically fastened joint can be considerable.

Example 10.1: Stress in the Adhesive of a Bolted Joint

Consider the joint analyzed in Chapter 2, Example 2.2. Five 3/4-in. (19.1 mm) diameter ASTM A325 steel bolts in a double-lap shear joint under a 38,250 lb. (170.1 kN) load each experience a shear stress of 9,646 psi (67.5 MPa). If an adhesive were applied to this joint, it would only experience a shear stress of

$$\tau = 38{,}250/[(7.75 \times 7.75) - (3.14 \times 0.3875^2)]$$
$$= 641.88 \text{ psi } (4.5 \text{ MPa})$$

simply because of the larger area of bonded surface.

In fact, the presence of the bolts would limit the amount of load transferred to the adhesive even more, as the bolts, having a much higher modulus of elasticity, would strain less under a given load than the adhesive. For joints such as these, analysis has to be based on the strain rather than stress caused by an applied load.

For this joint, a shear stress of 9,646 psi (67.5 MPa) in the steel bolts would result in a strain of approximately 0.0008, given a shear modulus of approximately 12,000,000 psi (83,000 MPa), from $\gamma = \tau/G$. This strain level would result in a shear stress of approximately 296 psi (12.3 Pa) for an adhesive with a shear modulus of 370,000 psi (15.4 MPa).

As pointed out earlier, not every combination of joining processes provides any benefit, and some combinations can lead to problems. The "belts and suspenders" philosophy of some so-called conservative engineers can work. Spot welds, for example, provide a degree of improvement or protection to adhesively bonded or brazed joints when peeling loads are encountered. However, sometimes two different joining methods cannot operate at the same time, so provide no such added, or even redundant, protection.

One example is bolting a welded joint. Here, the bolts cannot develop a friction force at the joint faying surfaces because there are no faying surfaces. The weld, in creating metallic bonds, creates a very high strength and efficiency at the joint. Bolts with a preload could add very little to further improve strength. If bolts are used to provide redundancy for safety, should the weld fail, it is questionable that they would work as planned. It would be wiser to use a stronger base material, a stronger filler, a better welding process, and/or a better (i.e., more substantial) joint design.

In conclusion, it is important to consider the combination of joining processes carefully before combining them. Most combinations make sense, but some do not.

SUMMARY

It is possible to modify or combine the basic joining processes of mechanical fastening, adhesive bonding, or welding (including brazing and soldering) to create useful variations and hybrids. A variation of a basic process usually involves simply a unique means of accomplishing the process. Thermal spraying is an example, as is braze welding. A hybrid process is an entirely new process that combines the best features of two (or more) parent processes, often with synergistic results. Examples of hybrid joining processes include weldbonding (a combination of spot welding and adhesive bonding), weldbrazing (a combination of spot welding and brazing), and rivet-bonding (a combination of mechanical fastening with adhesive bonding).

Thermal spraying applies material in particulate form by atomizing a wire or rod or simply heating a powder to the molten or soft plastic state and propelling those heated particles onto a substrate. Bonding occurs between the particles and the substrate by adhesion and between the particles by cohesion through the formation of chemical bonds. Methods include combustion flame, electric arc, nontransferred plasma arc, and detonation flame. The degree of adhesion, cohesion, and the density of the deposit vary directly with the kinetic energy of the particles during spraying. The process can create shapes, restore lost material and dimensions, apply protective coating (not unlike weld hardfacing), produce joints by applying braze alloys or adhesives, and consolidate reinforcements by building a matrix around them.

Weldbonding and rivet-bonding and weldbrazing all improve the out-of-plane (i.e., peel) static strength of the adhesive or braze. They also improve the fatigue strength of the joint by distributing the load more uniformly, reducing stress concentrations. Braze welding enables the joining of heat-sensitive materials in thick sections by filling a groove or producing a fillet with braze alloy without relying on capillary action. Filler can be applied by oxy-fuel or arc methods.

In addition to combining the various basic joining processes to create hybrids, it is possible to combine specific welding processes to create hybrid welding processes. Two examples are laser-GTA and plasma-GMA welding. Both of these processes exhibit improved deposition rates, improved penetration, lower heat input, greater arc stability, and other unique features.

Not every combination of basic joining processes makes sense, in that the strengths of the parents may not add and may not even offset weaknesses. Care must be exercised in deciding when such combinations are useful to improve properties or to provide redundancy.

PRACTICE QUESTIONS AND PROBLEMS

1. Differentiate between special joining processes that are variations of the fundamental processes of mechanical fastening, adhesive bonding, and welding, including brazing and soldering and processes that are hybrids of these. Cite two examples of each.

2. How does the process of thermal spraying operate? How is thermal spraying considered a joining process?

3. There are four major variations of thermal spraying. Briefly describe how each operates, emphasizing differences. Rank order these four variations in terms of heat source intensity and in terms of particle velocity.

4. What are five different applications of thermal spraying?

5. Describe the process of "weldbonding". What are the advantages of this process over its parent processes?

6. Describe the process of weldbrazing. What are its advantages over its parent processes?

7. Differentiate among braze welding and fusion welding and brazing. Give some advantages of braze welding.

8. Describe the process of rivet-bonding. Where would you recommend using this process?

9. Describe the two major hybrid welding processes. Give some advantages of these processes over the parent processes.

10. For the rivet-bonded joint shown in the illustration of Figure 10.10, calculate the shear stress in the rivets and in the adhesive. Assume the rivets are made from a

titanium alloy with a shear strength of 210 MPa and a shear modulus of 50 GPa and the adhesive is an epoxy with a shear strength of 20 MPa and a shear modulus of 2,464 MPa .

REFERENCES

1.   American Welding Society, *Welding Handbook, Volume 2—Welding Processes,* 8th ed., Miami: American Welding Society, 1991.

2.   Bales, Thomas T., Royster, Dick M., and Arnold, Jr., Winfrey E., *Development of the Weld-Braze Joining Process,* NASA Technical Note D-7281, Washington D.C.: National Aeronautics and Space Administration, June 1973.

3.   Dalley, J. W., and McClaren, S. W., "Aerospace Weld-Bonding and Rivet-Bonding," in *Advances in Joining Technology,* J.J. Burke, A.E. Gorum, and A. Tarpinian (eds.), Chestnut Hills, Massachusetts: Brook Hill Publishing, 1976, pp. 615–622.

4.   Diebold, T. P., and Albright, C. E., "Laser-GTA Welding of Aluminum Alloy 5052," *Welding Journal,* June 1984, Vol. 63, No. 6, pp. 18s–24s.

5.   Houck, D. L., "Thermal Spray Technology: New Ideas and Processes," *Proceedings of ASM's 1988 Conference*, Metals Park, Ohio: American Society for Metals, 1989.

6.   Landrock, Arthur H., *Adhesives Technology Handbook,* Park Ridge, N. J: Noyes Publications, 1985, pp. 218–223.

7.   Philips Research Laboratories, "A New Process: Plasma-GMA Welding," *Welding Journal,* Vol. 51, No. 8, pp. 560–563.

8.   Schwartz, M. M., *Brazing,* 2nd ed., Metals Park, Ohio: ASM International, 1990.

9.   U.S. Air Force, *Manufacturing Methods Development of Spot-welded Adhesively Bonded Joining for Titanium,* AFML-TR-71-93, 1971.

PART 2

Joining of Specific Material Types and Combinations

11

Joining Advanced Metals, Alloys, and Intermetallics

11.1 INTRODUCTION

11.1.1 Joining Process Options for Metals and Alloys

Metals, more than most other materials, are used in structural applications, so joining of metals and their alloys is an extremely important process in fabrication and assembly. Much of the popularity of metals is related to their general high strength, ductility, toughness, and fabricability, including castability, formability, and machinability. Other desirable characteristics are electrical and thermal conductivity, resistance to many corrosive environments, and, for many, serviceability at elevated temperatures. The predominant method for joining metals and alloys is welding, which includes the subcategories of brazing and soldering, but essentially all other methods of joining, including mechanical fastening, adhesive bonding, thermal spraying, weldbonding, weldbrazing, and rivet-bonding are also possible and used.

The reason for the popularity of welding, brazing, and soldering is that these processes result in strong joints through the creation of primary metallic bonds. Metals readily form metallic bonds with other metals, as long as intimate contact is achieved between clean faying surfaces. For this reason, joining under the action of heat and/or pressure is relatively easy, and the resulting joints are sound. Resulting structural integrity is high, structural efficiency (i.e., strength-for-weight) and joint efficiency (i.e., joint to base material stress level ratios) are generally high, and physical properties such as electrical or thermal conductivity are essentially unaffected by the joint. Continuous welds also provide excellent hermeticity and, done properly, provide superb environmental durability.

Also popular as a joining method for metals and their alloys is mechanical fastening. Here, the good ductility and fabricability of most metals and alloys permits the use of fasteners since fastener holes can be produced easily. The toughness of metals, combined with their high strength, gives tolerance to the stresses associated with hole production and fastener installation and to the stress concentrations associated with holes developed under service loads. Ease of disassembly is a major benefit of mechanical fastening, as is the inherent damage tolerance of the assembly afforded by physical barriers (i.e., joint faying surfaces) to crack propagation out of one structural member into another. The shortcomings compared to welding are principally lower structural efficiency (i.e., joint strength to weight), loosening, and leaking.

Less popular, but also possible and growing, is adhesive bonding. Load-carrying capacity can be quite high using large bonding areas, and damping characteristics can be excellent. Structural integrity can be high, but structural efficiency (i.e., strength for weight) and joint efficiency for some applications (e.g., bonding thick sections) can be quite limited. The single greatest shortcoming is that environmental durability can be a problem if temperatures are high, very low, or if there is prolonged exposure to water, moisture, solvents, weather, and so on.

Thermal spraying of metals onto similar or dissimilar metals, onto ceramics, or even onto polymers works well, producing adherent coatings. Most hybrid joining processes can also be used, often to special advantage. This includes weldbonding, weldbrazing, braze welding, and rivet-bonding.

11.1.2 Special Challenges of Joining Advanced Metals and Alloys

The special challenges of joining metals and metallic alloys to one another are largely associated with extremes. Joining of metals and alloys with extreme melting temperatures, extreme reactiveness, particular sensitivity to heat, highly incompatible chemical compositions or physical properties, extremes of geometry, or extremes of environment in which they must be joined and/or survive all present special challenges.

Refractory metals and alloys (e.g., tungsten, molybdenum, tantalum, and hafnium) can be difficult to join because of their extremely high melting and use temperatures. Melting with fusion welding processes can be difficult. Susceptibility to embrittlement from contamination by oxygen, hydrogen, or other interstitials is high. Suitability to elevated temperatures limits the utility of brazing and, especially, soldering and adhesive bonding. The body-centered structure of the refractory metals and their most common alloys leads to a ductile-brittle transition at relatively high temperatures compared to other metals, and grain growth at elevated temperatures can be severe, leading to loss of ductility and toughness.

Reactive metals and their alloys (e.g., titanium, zirconium, niobium, columbium, and beryllium) are challenging to join by welding, brazing, and soldering because of their extreme reactivity. These metals tend to form tenacious oxides, which in turn impede wetting. They also tend to form embrittling compounds (e.g., carbides, nitrides, and hydrides) during welding or intermetallics during brazing or soldering that can degrade performance in service, especially under fatigue or impact loading.

Many metals alloys are very sensitive to heat input. Low-melting metals and alloys, such as aluminum, magnesium, and zinc; age-hardenable alloys, such as heat-treatable aluminum alloys and nickel–base superalloys; transformation hardened (i.e., quenched and tempered) or hardenable alloys, such as medium- or high-carbon and low-alloy, alloy and tool steels; and inherently brittle alloys, such as cast irons, are all extremely prone to thermal damage (e.g., cracking) or property degradation (e.g., softening or embrittlement) during joining by welding.

Dissimilar metals and alloys can be incompatible in terms of their chemical and/or physical properties, leading to the formation of embrittling phases or the development of deleterious internal stresses. Chemical incompatibility can severely limit mutual solubility and can frequently cause embrittlement through the formation of intermetallic compounds at interfaces (e.g., copper to stainless steel or tin–lead solders to copper). The problem of incompatibility is again most serious when extremes are involved, such as extreme differences in chemical properties (e.g., mutual solubility or electronegativity), physical properties (e.g., melting temperature or coefficient of thermal expansion), or even mechanical properties (e.g., strengths or elastic moduli).

Sometimes the extremes that cause difficulty in joining are not simply the material's nature but are related to the geometry of the assembly or to the joining or service environment. Examples include (1) joining of very thin sections; (2) joining of very thick sections; (3) joining of very small components; (4) joining of very large components; and (5) joining in hostile environments such as in space, under water, or in areas of high radioactivity.

11.1.3 The Challenge of Joining Intermetallics

Joining of intermetallic compounds, in both monolithic and composite forms, is becoming important, as these materials are considered for applications in severe environments of temperature or corrosiveness. Aluminides, silicides, and borides of iron, nickel, titanium, and refractory metals are the subject of extensive interest and development. As a group, intermetallics tend to be brittle and refractory, making joining, especially by welding, a challenge. Many of these materials derive their unique and attractive elevated temperature properties (e.g., increasing strength with increasing temperature) from the fact that they exhibit long-range ordering. Welding can also disrupt this desirable order and, thus, desirable behavior.

The following sections will address the joining of advanced metals, alloys, and intermetallics.

11.2 JOINING REFRACTORY METALS AND ALLOYS

11.2.1 General Comments

Refractory metals and metallic alloys are those that have very high melting points and, thus, can function and survive in high temperature environments. Tungsten, molybdenum, and tantalum are the most common, but others include niobium or columbium, iridium, ruthenium, osmium, and rhenium. All melt over 2205°C (4000°F). Other metals and their alloys are also sometimes considered refractory, including hafnium, chromium, vanadium, and rhodium. These all have melting points over 1650°C or 3000°F. By far the most important refractory metals are tungsten, molybdenum, tantalum, and niobium.

Table 11.1 lists the refractory metals and their most common alloys, along with their melting temperatures or ranges, use temperature ranges, and strengths, as well other pertinent data.

As a group, these metals have body-centered cubic (BCC) crystal structures, high to very high densities, low specific heats, and low coefficients of thermal expansion. Mechanical strength and structural integrity at elevated temperatures are excellent. Like most body-centered cubic metals, the refractories exhibit a ductile-to-brittle transition in impact behavior, often at surprisingly low temperatures compared to their high melting temperatures (see Table 11.1).

The high melting temperature of these materials makes casting difficult, impossible, or impractical, and deformation processing can be difficult because of their typical body-centered cubic structures. Most refractory metals and alloys are thus produced by pressing and sintering powders to produce billets or net-shaped parts, or they are produced in arc cast forms. Complex structures from refractory metals and alloys tend to be produced by powder metallurgical techniques, such as hot compacting and forging, hot isostatic pressing, and, more recently, powder injection molding. Large and/or complex structures frequently require joining because of net-shape processing limitations, however.

As a result of the methods of metal or alloy production, defect (e.g., porosity) and contamination (e.g., absorbed gases or oxides or inclusions) content can be high. Working to create the desired finished shape is usually done warm versus hot, not much above the recrystallization temperature (which tends to be high for these materials). The result of all of this is that the final microstructure can be quite heterogeneous, further complicating processing and welding.

Refractory metals and their alloys tend to be susceptible to contamination by interstitial elements, especially oxygen, nitrogen, and carbon, and sometimes hydrogen (e.g., tantalum and niobium), both during processing and in service. As a result, contamination must be avoided by excluding these elements during processing, including joining, and through the use of coatings for sustained elevated temperature service.

Since these materials tend to be selected and used for their refractoriness, joining options are usually limited to those methods that can tolerate elevated temperatures. Mechanical fastening, welding, and brazing tend to predominate. While possible, soldering and adhesive bonding are generally not practical since the service temperatures for which the refractory metal is usually selected in the first place are probably too high for joints made by these methods to provide meaningful properties. Some adhesive bonding is performed with refractory metals and alloys but usually for joining to ceramics using inorganic refractory ceramic cements (see Chapter 15).

11.2.2 Mechanical Fastening

Refractory metals and their alloys can be joined by mechanical fastening, using either integral design features (e.g., interlocks) or fasteners. Methods are typical of those described in Chapter 3, Sections 3.3, 3.4, and 3.6.

The key consideration in mechanically fastening refractory metals and alloys, especially with the use of fasteners, is thermal, including differential expansion and con-

Table 11.1 Refractory Metals and Their Most Common Alloys

Metal or Alloy	Melting Point °C (°F)	Density g/cm³	Recrystal- lization Temp °C (°F)	Transition Temp °C (°F)	Elastic Modulus GPa (10⁶psi)	C.T.E. 10⁻⁶m/m °C
Tungsten (W) 25 Re-W	3390 (6130)	19.3	1200-1650 (2190-3000)	260-370 (500-700)	345 (50)	4.5

Molybdenum (Mo)
  0.5Ti-0.08 Zr-Mo (TZM)
  20ᵃ% Re-Mo
	2620
	(4750)
		10.2	1150-1200
			(2100-2190)
				150-260
				(300-500)
					324
					(47)
						4.8

Tantalum (Ta)
  2.5 W-Ta (KBI-10)
  10.0W-Ta (Ta-10W)
  2.5W-0.15 Nb-Ta (FS-63)
  8W-2Hf-Ta (T-111)
  10W-2.5Hf-0.01C-Ta(T-222)
  8W-1Re-0.7Hf-0.025C-Ta
    (Astar 811C)
	2996
	(5425)
		16.6	1100-1400
			(2010-2550)
				<-195
				(<-320)
					189
					(27)
						6.5

Niobium (Nb)
  1 Zr-Nb (Nb-1 Zr)
  5 Mo-5 V-1 Zr-Nb (B-66)
  10 Hf-1 Ti-Nb (C-103)
  10W-1 Hf-0.1 Y (C-129Y)
  10W-2.5 Zr (Nb-752)
  28 Ta-11 W-2.5 Zr-Nb
    (FS-85)
  10 Ta-10 W-Nb (SN 6-291)
	2468
	(4474)
		8.57	985-1150
			(1805-2100)
				−200- −75
				(−330- −105)
					105
					(15)
						7.1

Reactive Metals: Titanium	1668 (3034)	4.5	-	-	117 (16.8)	8.4
Zirconium	1852 (3350)	6.5	-	-	101 (14.4)	5.9
Beryllium	1290 (2355)	1.85	-	-	300 (44.5)	11.5
Other Refractories: Hafnium (Hf)	2230 (4046)	13.1	-	-	138 (19.8)	5.9
Iridium (Ir)	2454 (4459)	22.4	-	-	530 (76.0)	6.5
Ruthenium (Ru)	~2500 (4532)	12.2	-	-	418 (60.0)	9.6
Osmium (Os)	~2700 (4900)	22.5	-	-	565 (81.0)	6.0
Rhenium (Re)	3167 (5732)	20.53	-	-	465 (66.7)	12.45 ‖c-axis 4.67⊥c-axis
Vanadium (V)	1700 (3092)	5.96	-	-	132 (19.0)	-
Rhodium (Rh)	1966 (3560)	12.4	-	-	296 (42.5)	8.5
Chromium (Cr)	1920 (3490)	7.1	-	-	251 (36.0)	8.2

traction, thermal fatigue, and thermal stability. These problems arise from the fact that refractory materials are generally used at high service temperatures, so the severity of thermal excursions gives rise to considerable changes in dimensions and can cause degrading reactions. The relatively low thermal coefficients of expansion for the refractory metals means mismatch with other metals and alloys is likely. The thermal expansion coefficients of the fastener material and the refractory metal or alloy joint members should match as closely as possible, ideally no more than 1 to 1.5 x 10^{-6} mm/mm/°C or in./in./°F different.[1]

For the best compatibility, refractory metal or alloy fasteners should be used with like refractory metal or alloy joint elements. Another approach, however, is to use slip-joints, which allow relative movement of the joint elements during thermal excursions. These joints can use loose-fitting slots rather than close-tolerance, round holes to allow parts to move.

Thermal stability of both the joint elements and the fasteners is usually dealt with by using appropriate coatings (i..e, thermal barriers) to prevent oxidation or other adverse chemical reactions.

11.2.3 Welding

Refractory metals and alloys can be welded using either fusion or nonfusion welding processes. The two keys to successfully welding the refractory metals and their alloys are achieving melting or softening for coalescence and preventing embrittlement from absorbed interstitial elements. For fusion welding, the higher energy density processes are generally more successful and satisfactory as they are better able to cause melting. Electron beam, laser beam, and plasma arc welding processes are commonly used, but so is gas-tungsten arc. Welding can be and often is accomplished autogenously, but fillers, usually of closely matching composition to the base materials, can also be used. For electron beam welding, protection from contamination is provided by the vacuum employed with the beam. For the other processes, inert shielding gases (e.g., argon or helium) are used.

The problem of obtaining melting can be overcome by using processes that require little or no melting, such as resistance or nonfusion welding. Resistance welding, using both spot and seam methods, and nonfusion processes, especially friction or diffusion processes, all work well. These latter processes accomplish joining with little or no melting, relying predominantly on pressure to effect bond formation. Diffusion bonding is generally difficult to accomplish because diffusion rates tend to be slow at practical bonding temperatures, however. As mentioned above, embrittlement must be prevented by using proper shielding methods and procedures.

Another problem encountered in welding the refractory metals is related to their BCC structure. Body-centered cubic metals tend to form coarse-grained microstructures during both solidification and recrystallization in the heat-affected zone, and they exhibit

---

[1] The only exception is when a temperature gradient will exist in the part or structure during normal service. In this case, rather than matching the coefficients of thermal expansion, materials are selected and parts and joints are designed to minimize the stresses that will arise from the gradient.

significant strain-rate sensitivity. The strength, ductility, and toughness of BCC metals falls with increasing grain size, so any tensile stress in service or during postweld cooling or processing concentrates strain in the heat-affected zone. Strain-rate sensitivity is thus increased, the tendency toward embrittlement is accentuated, and cracking may occur during cooling or in service upon repeated thermal cycling (i.e., thermal fatigue).

The risk of cracking due to strain in these metals and alloys can be reduced by using preheat (but not so much as to cause excessive grain growth in the heat-affected zone), using postweld heat treatment to reduce stresses, selecting a filler that is softer than the base metals, and, above all else, avoiding contamination.

When filler is required for fusion welding, it is generally selected to match the composition of the base material as closely as possible or be of a softer composition. As an example, Mo-50Re or Mo-20Re filler is often used with pure molybdenum. Specific recommendations for specific refractory metals or alloys should be sought in appropriate references (such as the American Welding Society's 8th edition of its *Welding Handbook,* or the ASM International's *Metals Handbook,* Volume 6, on "Welding, Brazing and Soldering"). While fillers are not normally used during nonfusion or resistance welding, brittleness may occur in spot-welded refractory metals like molybdenum. It can be minimized by inserting ductile foils (e.g., zirconium, nickel, or copper) or metal fiber between the surfaces to be spot welded. Diffusion bonding can also often be enhanced by using intermediate materials such as nickel.

11.2.4 Brazing

The technology of brazing the refractory metals and their alloys is still in the development stage, with new techniques and brazing materials being reported every day. Most refractory metals and alloys can be brazed to themselves and to other refractory metals and alloys. Care must be exercised in (1) selecting a compatible filler alloy, (2) relieving stresses before brazing, and (3) minimizing the time at brazing temperature to avoid embrittling reactions from absorption of gases or interdiffusion to produce intermetallics.

As general rules, for tungsten and its alloys, (1) avoid brazing alloys with excessive nickel to prevent recrystallization in the base metal due to the high brazing temperatures required, and (2) avoid contact with graphite to prevent carbide formation.[2] For molybdenum and its alloys, (1) prevent oxidation by using protective coatings, (2) prevent contamination by interstitials, (3) prevent recrystallization by careful alloy selection or brazing cycle selection (limiting time over 1090°C or 2000°F), and (4) use barrier layers (e.g., chromium) to avoid diffusion-induced embrittlement by intermetallic compound formation with filler components (e.g., Ni). For tantalum and columbium and their alloys, (1) remove *all* reactive gases (e.g., O_2, CO, NH_3, H_2, N_2, and CO_2) and (2) electroplate with copper or nickel to prevent oxidation, and then use an appropriate filler that is compatible with the plating.

Table 11.2 lists some of the more common brazing fillers used with various refractory metals and alloys.

---

[2] This latter requirement may seem odd, but it is not. Graphite being itself highly refractory is a likely candidate for use with tungsten and tungsten alloys.

Table 11.2   Common Braze Fillers for Use with Refractory Metals and Alloys

| Filler | Liquidus | | For use with | | | |
	°C	°F	W	Ta	Mo	Nb
Ag	963	1765	✓		✓	
Au	1066	1950	✓		✓	
Cb or Nb	2468	4475	✓		✓	
Cu	1083	1981	✓		✓	
Ni	1453	2647	✓		✓	
Pd	1552	2826	✓		✓	
Pt	1769	3217	✓		✓	
BAg Series	618-971	1145-1780	✓		✓	
BAu Series	891-1166	1635-2130	✓		✓	
BNi Series	877-1135	1610-2075	✓		✓	
BCo-1	1149	2100	✓		✓	
Au - 6 Cu	991	1815	✓		✓	
Au-50 Cu	971	1780	✓		✓	
Au-35 Ni	1077	1970	✓		✓	
Au-8 Pd	1241	2265	✓		✓	
Au-13 Pd	1304	2380	✓		✓	
Au-25 Pd	1410	2570	✓		✓	
Au-25 Pt	1410	2570	✓		✓	
Cr-25 V	1752	3185	✓		✓	
Pd-35 Co	1235	2255	✓		✓	
Pd-40 Ni	1235	2255	✓		✓	
Au-15.5 Cu-3 Ni	910	1670	✓		✓	
Au-20 Ag-20 Cu	835	1535	✓		✓	

| Filler | Liquidus | | Remelt | For use with | | | |
	°C	°F	°C (°F)	W	Ta	Mo	Nb
10 Ta-40V-50 Ti	~1760	~3200	~2400 (4350)		✓		
20 Ta-50 V-30 Ti	~1760	~3200	~2400 (4350)		✓		
25 Ta-55 V-20 Ti	~1843	~3350	~2200 (4000)		✓		
30 Ta-65 V-5 Ti	~1843	~3350	~2400 (4350)		✓		
5 Ta-65 V-30 Nb	~1816	~3300	~2300 (4170)		✓		
25 Ta-50 V-25 Nb	~1871	~3400	~2500 (4530)		✓		
30 Ta-65 V-5 Nb	~1871	~3400	~2300 (4170)		✓		
30 Ta-40 V-30 Nb	~1927	~3500	~2000 (3630)		✓		
93 Hf-7 Mo	~2093	~3800	~2250 (4060)		✓		
60 Hf-40 Ta	~2193	~3980	~2150 (3880)		✓		
66 Ti-34 Cr	~1482	~2700	~2080 (3780)		✓		✓
66 Ti-30 V-4 Be	~1316	~2400	~2090 (3800)		✓		✓
48 Ti-48 Zr-4 Be	~1050	~1920	-				✓
75 Zr-19 Nb-6 Be	~1050	~1920	-				✓
91.5 Ti-8.5 Si	~1370	~2500	-				✓
73 Ti-13 V-11 Cr-3 Al	~1620	~2950	-				✓
90 Pt-10 Ir	~1815	~3300	-				✓

Table 11.2   *(continued)*

Filler	Liquidus		Remelt	For use with			
	°C	°F	°C (°F)	W	Ta	Mo	Nb
90 Pt-10 Rh	~1900	~3450	-				✓
Nb	2416	4380	-	✓	✓	✓	
Ta	2996	5425	-	✓			
Ti	1816	3300	-	✓	✓	✓	✓

Filler	Liquidus		For use with			
	°C	°F	W	Ta	Mo	Nb
Nb-2.2 B			✓			
Nb-20 Ti			✓			
Pt-3.6 B (+11 w/o W powder)			✓			
W-25 Os			✓			
W-50 Mo-3 Re			✓			
Mo-5 Os			✓			
V-35 Nb					✓	
Ti-30 V					✓	
Ti-8.5 Si (+ Mo powder)					✓	
Ti-25 Cr-13 Ni					✓	
Ti-65 V					✓	
V					✓	
V-50 Mo					✓	
Mo B-50 Mo C					✓	

11.3 JOINING REACTIVE METALS AND ALLOYS

11.3.1 General Comments

The reactive metals, which can include some of the refractory metals, present some difficult joining problems when welding or brazing is used, as well as when mechanical fastening is used. These metals (e.g., beryllium, titanium, zirconium, as well as the reactive refractories niobium or columbium, molybdenum, tantalum, and tungsten) have in common an extraordinarily high affinity for oxygen and other elements in their gaseous, liquid, and even solid forms. Some of the important properties of reactive metals are listed in Table 11.1.

In welding or brazing, which are often the preferred methods for joining, cleanliness of the workpieces is of special importance. Fluxes (especially welding fluxes) must be avoided or very carefully selected, and exposure to the atmosphere must be avoided when the base metal is hot or, especially, molten. The consequence of improper procedures is poor wetting and/or embrittlement of the resulting joint.

Mechanical fastening and adhesive bonding can be used, but adhesive bonding is generally far less commonly used. When mechanical fastening is done, the fasteners must be selected or specially fabricated from materials that will not galvanically corrode in the presence of these active materials. Coatings are often employed on mechanical fasteners used with reactive metals to provide an electrically insulating barrier as well as a general reaction barrier. Adhesive bonding has been used most notably with titanium and its alloys, especially for joining lightweight structures such as honeycomb core sandwiches for use in aircraft and spacecraft. Adhesives are selected based on the adherends to be joined and the environment for service (see Chapter 5). Care must be taken to avoid reactions between adhesive and adherend, and during cleaning prior to bonding.

11.3.2 Welding

The processes that have been applied experimentally and in production are gas-tungsten arc (especially direct current straight polarity or direct current electric negative), gas-metal arc, plasma arc, electron beam, resistance spot or seam, flash butt, pressure and explosive, ultrasonic, and diffusion welding. The most common processes are, by far, GTAW, PAW, and EBW without filler.

The principal problems encountered in welding the reactive metals and their alloys are: (1) embrittlement due to contamination, (2) embrittlement due to recrystallization, and (3) porosity.

Beryllium,[3] titanium, zirconium, niobium and tantalum all react rapidly at temperatures well below their melting points with all the common gases except the inert gases. Contamination of the weld pool by oxygen and nitrogen absorbed from the atmosphere, for example, results in an increase in strength and hardness and a reduction in ductility and toughness, even under fairly effective shielding. Besides reacting with gases, many of these metals react with other elements, both metallic and nonmetallic in ionic, liquid, or even solid forms (e.g., Ti with Fe).

For the most effective shielding, with gas-tungsten arc or plasma arc welding, (1) blanket both the torch and the underside of the joint with shielding gas, (2) consider a trailing shield to allow cooling under an inert gas shield (for GMAW use a leading shield as well), and (3) consider employing a dry box or tent containment.[4] For the most stringent applications, where prevention of contamination is critical, a vacuum atmosphere of 10^{-3} to 10^{-5} Torr is used along with an electron beam (EB) source. Resistance or flash welds can normally be made in air because the time cycle (i.e., time at temperature) is too short for any substantial degree of gas pickup.

Porosity can be a serious problem in welding reactive metals and their alloys. Porosity can arise from the release of entrapped gases in the metal (e.g., introduced from processing by powder metallurgy) or in tooling (i.e., incipient leaks), from dis-

---

[3] Beryllium is extremely toxic in many forms, including vapor, powder, dust, and chips. Consequently, extreme care should be exercised in welding, brazing, and soldering beryllium.

[4] A *dry box* is a chamber for allowing welding (or other processing) under inert shielding. Rubber gloves, attached and sealed to the box, allow the operator to gain access to the welding torch inside the shielded environment to perform welding.

solved gases (e.g., from secondary processing by plating or chemical milling or acid cleaning or pickling), or from other sources (e.g., poor cleaning). Hydrogenation of titanium and zirconium is a particular problem, resulting in both severe and persistent porosity and/or embrittling hydrides. Effective precleaning is critical to the successful welding of reactive metals and is accomplished by machining, scraping, grit-blasting, or grinding followed by degreasing or careful pickling.

Cracking of reactive metals and their alloys can be a problem during welding. The unalloyed reactive metals are not especially prone to hot cracking nor are their alloys that are commonly welded. The brittle metals beryllium, molybdenum, and tungsten, however, may crack at low temperatures due to strain sensitivity if welded under restraint. Postweld stress relieving and proper joint design help.

The reactive metals and their alloys are fusion welded either autogenously or using fillers that closely match the base metal or alloy in composition and properties. Specific filler recommendations are available in the AWS and ASM handbooks.

11.3.3 Brazing

Brazing is an attractive alternative for joining most reactive metals and alloys and is the preferred method for metallurgically joining beryllium (since dangerous Be vapor is not generated). Because of the chemical and metallurgical reactivity of most reactive metals and their alloys, brazing techniques must be highly specialized. Filler alloys must be selected carefully to avoid brittle intermetallic formation, and fluxes must be chosen carefully to avoid undesired reactions.

For beryllium, the preferred brazing alloys are zinc (with a melting range of 427° to 454°C or 800° to 850°F), aluminum–silicon (566° to 677°C or 1050° to 1250°F), silver–copper (649° to 904°C or 1200° to 1660°F), and silver (882° to 954°C or 1620° to 1750°F).

For titanium and zirconium, embrittlement by surface contaminants, atmosphere, and even flux components must be avoided by very careful cleaning, immediate brazing, and careful flux selection and use, respectively. Acceptable fillers for titanium are (1) 3003 Al foil for thin, lightweight structures (e.g., honeycomb core); (2) electroplated metals such as copper, iron, or nickel on the titanium to react *in situ* during brazing to form a eutectic with the titanium; (3) silver-based alloys with lithium, copper, aluminum, or tin, including an especially good Ag–9Pd–9Ga alloy for filling large gaps; (4) 48Ti–48Zr–2Be; (5) 43Ti–43Zr–12Ni–2Be; and (6) various titanium–nickel–copper alloys.

A big problem with selecting an appropriate filler for titanium and its alloys is achieving reasonable corrosion resistance in the brazed joint, since titanium is often chosen for its excellent corrosion resistance. The Ti–Zr–Be and Ti–Zr–Ni–Be fillers offer excellent corrosion resistance.

For zirconium and its alloys, many good braze fillers have long been sought. Some successes are 95Zr–5Be (1004°C or 1840°F), 50Zr–50Ag (1521°C or 2770°F), 71Zr–29Mn (1371°C or 2500°F), and 76Zr–24Sn (1732°C or 3150°F), as well as Zr-4-5% Be, Ni-7% P or Ni-20% Pd-3% In.

11.4 JOINING HEAT-SENSITIVE METALS AND ALLOYS

11.4.1 General Comments

Some metals and alloys are quite sensitive to heat, so joining must be done either without using heat, using adhesives or mechanical fastening, or using heat input that is very carefully controlled and/or localized during fusion or nonfusion welding, brazing, or soldering. Sensitivity to heat in metals and alloys can be manifested in any one of several ways. First, localized melting or burn-through can occur in low-melting and/or high thermal conductivity metals or alloys such as aluminum, magnesium, zinc, tin, lead, and their alloys, copper-based brasses and bronzes, and even the precious metals silver, gold, platinum, and palladium. Second, softening or loss of strength can occur in quenched and tempered (i.e., transformation hardened) steels with medium to high carbon and/or low to high alloy content, in age-hardened alloys (e.g., aluminum, nickel, cobalt based or PH stainless steels), or in cold-worked metals and alloys. Third, loss of ductility, embrittlement, and/or cracking can occur in alloys that undergo certain cooling rate related phase transformation (e.g., hardenable steels) or that cannot tolerate thermal shock because of their inherent brittle nature (e.g., cast irons).

Another possible manifestation of heat sensitivity is adverse reactions with the surrounding atmosphere (e.g., embrittlement by absorption of gases, such as oxygen or hydrogen, interstitially) or with adjoining materials (e.g., iron near titanium or copper near austenitic stainless steels). Since this occurs most commonly with the refractory and reactive metals and alloys described previously, it will not be discussed again here.

If welding or brazing, and, to a far lesser extent, soldering are chosen for joining heat-sensitive metals and alloys, care must be exercised to minimize the amount, the concentration, and the duration of heating. Obviously, mechanical fastening and adhesive bonding may be preferred joining options for these materials.

11.4.2 Welding

The welding of heat-sensitive cold-worked, age-hardened, or transformation hardened alloys was discussed in Chapter 7, Section 7.5, so will not be repeated here. Suffice it to say that some heat-sensitive alloys of these types include but are not limited to (1) cold-worked brasses, bronzes, and stainless steels; (2) age-hardenable aluminum alloys, including Al–Cu, Al–Cu–Mg, Al–Mg–Si (2xxx and 6xxx) types; Al–Mg–Zn (7xxx) types; and new Al-Li types; (3) age-hardenable magnesium alloys (including Mg–Al–Mn types, such as A8, A10, and A12; and Mg–Al–Zn–Mn types, such as AZ63, AZ92, and AZ101); (4) age-hardenable nickel–base alloys (including those with controlled additions from among magnesium, beryllium, aluminum, silicon, titanium, and molybdenum, such as the "K," "KR," and "S" Monels and the Hastelloy A, B, and C types, as well as various superalloys such as Inconel's, Udimet's, Niomonic's, Rene's, and Waspaloy); (5) age-hardenable cobalt–base alloys (including those strengthened by complex carbides, such as the Co–Cr–C and Co–Cr–Mo–C types, and those hardened by intermetallics, such as the Co–Cr–Mo with Co_3Mo, Co–Cr–Ta, W, or Nb and Co–Ti or Co–Cr–Ni with Ni_3Ti types) like the various Haynes and Haynes Stellite alloys; (6) age-hardenable steels (such as maraging and precipitation-hardenable stainless steels such as 17-4 PH and 15-5 PH); (7) hardenable carbon and alloy steels, including heat-resistant, corrosion-resistant, and tool steels; and (8) cast irons.

Some general guidelines for welding heat-sensitive metals and alloys follow. First, use a process that minimizes heating and melting, such as nonfusion friction or pressure welding or resistance welding. Second, use fusion welding processes that minimize linear heat input (i.e., joules per meter or joules per inch), by using high welding speeds with GTAW, GMAW or, especially, by using high energy density (inherently high-speed) processes like EBW or LBW. Third, use pulsed-current modes with GTAW, GMAW, or PAW to lower the average current and power. Fourth, use electrode-positive DC or AC modes to shift the bulk of the heat from the workpiece to the electrode. Fifth, use preheat with transformation hardenable alloys to reduce the cooling rate after welding. Sixth, minimize joint thickness and the number of passes. Seventh, maintain low interpass temperatures. Finally, eighth, use intermittent, or stitch, welding techniques, to reduce the total heat input.

The welding and brazing of low melting as well as high thermal conductivity metals and alloys present special problems. Aluminum, magnesium, lead, tin, and zinc and their alloys have low melting temperatures, as do certain copper-based alloys such as the brasses, i.e., copper–zinc alloys, and bronzes, i.e., copper–tin alloys (see Table 11.3). In addition, aluminum, magnesium, copper, and the precious metals silver, gold, platinum, palladium, and others, have very high thermal conductivities compared to most other structural metals and alloys (e.g., steels, nickel– or copper–base alloys, and titanium alloys).

A low melting metal or alloy can cause problems because many such metals and alloys do not give any visual indication of how hot they are, i.e., they do not emit visible radiation or change color as do steels, for example. This presents a problem to a

Table 11.3 Low-Melting or High Conductivity Metals and Alloys

Metal (or Alloy)	Melting Temp °C (°F)	Boiling Temp. °C (°F)	Density g/cm³	Thermal Conductivity cal/cm³/sec /°C/cm
Aluminum	660.2 (1220)	~2500 (4530)	2.7	0.57
Copper	1084 (1983)	~2500 (4530)	8.92	0.92
Brasses				
Bronzes				
Gold	1063 (1945)	~2950 (5350)	19.3	0.70
Lead	327.4 (621)	1750 (3182)	11.34	0.082
Magnesium	650 (1202)	1103 (2017)	1.74	0.35
Silver	960.5 (1761)	2160 (3920)	10.5	1.001
Tin	231.9 (449)	~2400 (4350)	7.30	0.147
Pewter				
Babbitts				
Zinc	419.5 (787)	907 (1665)	7.14	0.265

manual welder. Also, obviously, they can only tolerate so much heat input before melting occurs, and, with many of the higher intensity welding heat sources (e.g., arcs and beams), excessive melting can easily occur or vaporization can quickly follow melting, leading to burn-through. To add to the problem, many of these metals and alloys oxidize easily and form a refractory oxide layer (e.g., aluminum, magnesium). Consequently, enough heat must be applied to disrupt the oxide but not so much as to burn through the weldment.

The property of high thermal conductivity (e.g., in aluminum, copper, and the precious metals) means that heat from the welding source disperses quickly in the weldment, so there is a tendency, especially with manual processes, to dwell too long to compensate. This often results in overheating and burn-through. If one thinks about welding aluminum versus steel, for example, aluminum has a thermal conductivity that is three to five times greater than steel but with half the melting temperature. Thus, heat input must be kept low when welding aluminum and its alloys.

Aluminum and magnesium and their alloys are typically welded at high speeds with GTAW for thinner sections, GMAW for thicker sections, or EBW for very thick sections or special, precision applications. Zinc and lead and their alloys are usually welded with GTAW or oxy-gas (such as oxyacetylene). The precious metals (silver, gold, platinum, palladium) and their alloys are usually welded with oxy-gas processes but can be welded with GTAW using DCSP. To help keep heat input low, smaller diameter tungsten electrodes tend to be used with GTAW when welding these metals and alloys. Resistance welding also works well with this entire group of materials except zinc and lead and their alloys.

Resistance and various nonfusion processes are often good choices for welding heat-sensitive metals and alloys, in general. Resistance welding of aluminum and magnesium was mentioned previously. Friction and pressure welding, including diffusion bonding, have all been successfully used with at least some of the heat-sensitive materials. A requirement is that the material have enough ductility to permit some essential upset during nonfusion welding. Consequently, cast iron is not a viable candidate for these processes. Nickel– and cobalt–base superalloys, however, as well as aluminum, and magnesium, and some copper alloys are joined by these processes, including DBW.

11.4.3 Brazing and Soldering

Brazing and soldering are good choices for joining heat-sensitive metals and alloys. Strong bonds can be obtained with brazing at moderate temperatures and heat inputs, while sound bonds can be obtained with soldering at low temperatures and heat inputs. Age-hardenable alloys, such as those of aluminum, magnesium, nickel, or cobalt, can all be brazed with minimal effects on properties if heating is properly localized through proper process selection. Aluminum and magnesium and their alloys present a problem in wetting, because of their inherent and tenacious refractory oxides. Torch, furnace, and dip processes are commonly used, with induction, infrared, and resistance used for special situations. Aggressive fluxes are often required for achieving good wetting. Brazing cycles, i.e., time at the brazing temperature, should be kept short to minimize penetration of the filler components by diffusion, leading to adverse brittle compound or low melting eutectic formation.

For aluminum and its brazeable alloys, BAlSi fillers are used. The high-strength wrought alloys and certain cast alloys have high alloy contents and low melting temperatures, so generally are not brazed. The 2xxx and 7xxx Al alloys, commonly used in aerospace applications, fall into this category. For magnesium and its alloys, BMg fillers are used, including BMg1 (9Al–2Zn–0.1Mn–0.00005Be) and BMg2 (12Al–0.5Zn–0.00005Be). Two other solders are used for filling in defects in magnesium castings; 60Cd–30Zn–10Sn (with a 150°C or 300°F melting temperature) and 90Cd–10Zn (with a 300°C or 500°F melting temperature). No fluxes are usually required or used with magnesium or its alloys.

The age-hardenable, heat-sensitive nickel– and cobalt–base alloys, including most of the so-called superalloys used in high-temperature applications (e.g., aircraft gas turbines) can be successfully brazed; the nickel alloys with BNi or BAu fillers and the cobalt alloys with BCo fillers.

For applications where mechanical strength is not critical, or where the melting temperature of the metal or alloy to be joined is especially low (e.g., tin, lead, tin–lead alloys, zinc or zinc alloys), soldering can be an excellent choice. Aluminum, brasses, bronzes, and zinc die-casting alloys are frequently soldered very successfully. Magnesium is essentially impossible to solder because of the extreme difficulties in achieving wetting. For aluminum and its alloys, aggressive inorganic fluxes are required, and solder fillers include Sn–Pb, Sn–Zn, Pb–Bi, Zn–Cd, and Pb–Cd–Sb–Ag types. Silver is added to many solders to improve the generally poor corrosion resistance of solder joints in aluminum. One example is 80.1Pb–18Sn–1.9Ag. The precious metals and their alloys can also be successfully soldered.

11.4.4 Mechanical Fastening and Adhesive Bonding

Obviously, mechanical fastening methods and adhesive bonding are excellent choices for joining heat-sensitive metals and alloys. For mechanical fastening, the same general principles apply to avoid overloading the fastener or the joint element. For the soft and/or low-melting metals and alloys (such as tin, lead, and precious metals), deformation of the fastener hole by bearing and/or fastener tear-out must be carefully considered. For the inherently brittle alloys, such as zinc die-cast alloys, or cast irons, care must be exercised in hole preparation and loading. In addition, care should be exercised to avoid problems with thermal mismatch due to severe differential thermal expansion/contraction, especially with high thermal conductivity materials like aluminum, magnesium, copper or precious metal alloys, or with materials that are intended to function over wide temperature ranges such as the nickel– and copper–base superalloys. Finally, normal precautions should be taken regarding potential corrosion (i.e., galvanic) couples, especially with the noble metals.

Adhesive bonding can be very successful with heat-sensitive metals and alloys, but it is usually restricted to use with the lower melting metals and alloys that tend to fulfill applications at lower service temperatures. Adhesive bonding is also extremely useful for joining thin section and/or lightweight structures. The best examples are probably the bonding of aluminum and magnesium alloys, including aluminum–lithium. It is also not uncommon to find weldbonding being used with these lightweight materials, especially when out-of-plane loading is anticipated or required. Adhesive bonding is usually not used with the heat-sensitive alloys intended for elevated temperature service (e.g., nickel– and cobalt–base superalloys), for reasons of environmental degradation.

Table 11.4 lists the types of adhesives commonly used with various lightweight and/or low-melting metals and alloys of aluminum, brass, bronze, gold, lead, magnesium, silver, and zinc.

Table 11.4   Adhesives Recommended for Use with Various Lightweight or Low-Melting Metals and Alloys

Aluminum (2.70 g/cm^3)
    Modified epoxies   2nd generation acrylics
    Modified acrylics   Cyanoacrylates
    Epoxy–phenolics   Silicones
    Neoprene–phenolics

Beryllium (1.82 g/cm^3)
    Epoxies   Nitrile–phenolics
    Epoxy–phenolics   Polyimides
    Epoxy–nylons   Polybenzimidazoles
    Epoxy–nitrile   Polyurethanes

Magnesium (1.74 g/cm^3)
    Epoxies   Cyanoacrylates
    Epoxy–phenolics   Polyvinyl acetates
    Epoxy–nylons   Polyurethanes
    Vinyl–phenolics   Silicones
    Nitrile–phenolics

Titanium (4.5 g/cm^3)
    Epoxies   Epoxy–phenolics
    Nitrile–epoxies   Polymides
    Nitrile–phenolics

Tin (231.9 °C or 449°F)
    Epoxies   Polyacrylate
    Polyvinyl alkyl ether   Styrene-butadiene

Zinc (419°C or 787°F)
    Nitrile–epoxy   Cyanoacrylates
    Epoxies   Rubber-based adhesives
    Silicones

11.4.5 Welding, Braze Welding, and Brazing Cast Irons

Cast irons are very high carbon iron–carbon alloys, typically with 2-5 wt% carbon. As a group, cast irons are hard and brittle and highly susceptible to thermal shock. Thus, they are often considered unweldable. Cast irons come in many different grades, with many common characteristics, especially low melting temperature and high fluidity, leading to good castability. Various types or grades differ principally in their chemistry and the morphology of their graphite or cementite phases. Major types include white cast irons (where carbon exceeds 1.7 wt% and is all combined in the form of iron carbide or cementite), gray cast irons (where carbon ranges from 2 to almost 5 wt% and is in the form of flakes of graphite imbedded in a steel matrix), mottled cast irons (which are a cross between white and gray cast irons, obtained by composition and cooling rate control from the melt), malleable or ductile cast irons (where the cementite has dissociated into small particles or nodules, of graphite in the solid state, usually by heat treatment), and nodular or spheroidal graphite (SG) cast irons (where graphite nodules are formed directly during the process of solidification and are embedded in a matrix of steel). White cast iron, for all intents and purposes, is not weldable but can be brazed, if done properly. The other grades are weldable if proper procedures are used, and, can be brazed as well.

The detailed metallurgy of various grades of cast iron is left to other sources (e.g., Reference 1). Only the welding, braze welding, and brazing of these alloys will be discussed here.

Gray cast iron is the most common and least costly of all the cast materials, so it is very widely used. It is usually composed of 2.5 to 3.5 wt% C in iron, in which most of the carbon is present as free graphite flakes. These flakes promote brittleness and sensitivity to thermal shock, while providing excellent damping characteristics. Additions of phosphorus to enhance fluidity for casting and impurities of sulfur promote hot shortness and hot cracking. The spheroidal graphite (or nodular) and ductile (or malleable) grades are rendered much more ductile by causing the graphite to form spheres or nodules by special heat treatment and/or the addition of special nucleation additives like magnesium or nickel or rare earth elements.

The metallurgical changes that take place in the heat-affected zone of fusion welds in these two materials, i.e., gray and spheroidal cast iron, are basically the same. In the region heated above the eutectoid temperature, the ferrite is transformed to austenite. Above 800°C (1370°F), graphite starts to go into solution and, simultaneously, cementite is precipitated; first at grain boundaries and at higher temperatures, when more graphite is dissolved, within the austenite grains. At still higher temperatures, some melting occurs partially by constitutional liquation. On cooling, the cementite network remains, but the austenite transforms: high carbon regions to martensite, and low carbon regions to pearlite.

Thus, the heat-affected zone of fusion welds in cast irons has a complex structure; comprising melted or remelted regions, undissolved graphite, martensite, fine pearlite, coarse pearlite, and some ferrite. This structure is very hard and very brittle, and, when tested in tension or bending, usually fails through the FZ-HAZ boundary.

Welding

During fusion welding, the effects described previously can be mitigated in several ways. First, by preheating to about 300°C or 570°F, and using low-hydrogen electrodes, hard zone formation and cracking will be minimized and hydrogen cracking will be avoided. Second, postweld heat treatment at 650°C (1200°F) or full annealing will reduce HAZ hardness even more. Third, when preheating is not possible or practical (e.g., due to size and/or location), heat input should be reduced to the lowest possible level by operating in DCRP, by making short runs (i.e., stitch welding), or by allowing cooling between each run (to minimize the interpass temperature). This technique reduces graphite dissolution. Fourth, after welding, the cast iron weldment should be cooled slowly to avoid thermal shock. Insulating blankets or even resistance heated blankets can be used.

Cast iron can be welded with various fillers, using the oxy-gas, SMAW, FCAW, GTAW, GMAW, PAW, or SAW processes. In gas welding, a cast iron filler rod is often used. The two most common fillers for arc welding are pure Ni and 55Ni–45Fe. These are available in both electrode and wire forms. The relatively low strength and ductility of these fillers allows stresses generated from the heat of welding to concentrate in the weld proper. The low melting point of these fillers minimizes hardening effects.

A common procedure for welding cast irons is known as *buttering*. Here, a soft Ni or Ni–Fe (or even an austenitic stainless steel) is applied in thin, narrow, short passes on each faying surface. Then the joint is filled to complete the weld, using 55Ni–45Fe, austenitic stainless steel, or even mild steel. The nickel in the butter layer prevents carbon pickup in the main weld area.

Braze Welding

Braze welding, sometimes called *bronze welding* with cast irons, combines brazing with fusion welding using a bronze filler metal. The process (described in Chapter 10, Section 10.4.3) is like brazing since little or no melting of the cast iron substrate occurs with the low melting bronze alloy. Also, the filler is obviously very different in composition than the substrate (i.e., is heterogeneous). The process is like normal fusion welding in that the joint is filled by melting the consumable bronze filler wire, rod, or electrode in place and *not* relying on capillary flow.

The advantage of braze welding with cast iron is that only the region immediately around the joint needs to be heated to allow filler alloy melting. This localization of heating minimizes adverse heat effects.

The most popular filler used in braze welding cast irons is 39Zn–1%Sn, balance Cu.

Brazing

The brazing of cast irons, like welding, requires special consideration. Before brazing, the faying surfaces of gray, malleable (ductile), or nodular (spheroidal) cast irons are generally treated by electrochemical surface cleaning, seared with an oxidizing flame, grit-blasted, or chemically cleaned. This is to remove graphite by decarburization and, thereby, promote critically important wetting.

Brazing with silver brazing alloys should be done below 760°C (1400°F). Copper braze alloys can be used at higher temperatures if heating and cooling are done carefully. It is often desirable to preheat the entire casting during brazing to minimize gradients between the braze area proper and the body of the casting.

11.5 JOINING HEAT-SENSITIVE STRUCTURES

11.5.1 General Comments

Not all of the problems associated with the heat of joining metals and alloys by welding, brazing, and, to a lesser extent, soldering, are due to the inherent sensitivity or susceptibility of the material's composition and/or microstructure to thermally induced changes. Sometimes the geometry or the geometric arrangement of the material elements in a structure or an assembly give rise to a sensitivity to heat-induced damage.

Several types of situations can give rise to geometrically related heat-induced damage, including (1) joining of very thin sections to other very thin sections (e.g., thin sheet or foil welding or honeycomb core or core-to-face sheet welding) or to thick sections (e.g., flexible thin sheet metal bellows to end machined fittings or housings); (2) joining of very small details (e.g., fine wire leads to chips or pads in electronic packages or assemblies); (3) joining metals or alloys in close proximity to heat-sensitive, non-metallic materials (e.g., near plastics or organic-matrix composites) or brittle materials (e.g., ceramics or glasses); and (4) joining metals or alloys in the vicinity of combustibles (e.g., fuels or fuel lines; lubricants; coated, painted, or plated areas).

Often the type or form of heat-induced damage is distortion or warpage. In other instances, the damage can be related to melting or combustion. Figure 11.1 illustrates some of these heat-sensitive structural situations. The following sections briefly describe some considerations and techniques for avoiding or minimizing heat-induced damage in heat-sensitive structures.

11.5.2 Joining Very Thin Sections

Thin sheets or foils of metals or alloys occasionally have to be joined to other thin sheets or foils or to thicker components such as cast or machined frames or housings or stiffeners. Obviously, these thin sections are quite susceptible to heat damage, either by overheating to cause melting or burn-through or by distortion that is so severe as to prevent proper fit and/or function. Some examples where very thin metallic elements must be joined are (1) thin, formed sheet metal diaphragms to support frames or housings (e.g., in hermetically sealed pressure, vacuum, or environmental systems); (2) thin, formed sheet metal bellows joining to one another or to end fittings or housings (e.g., in ducting or vacuum systems); (3) welded or brazed metal honeycomb core to create the honeycomb core proper or to attach the core to face sheets to produce a sandwich panel

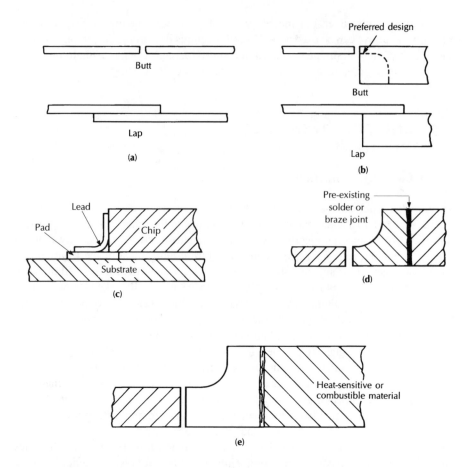

Figure 11.1   Various heat-sensitive structural conditions or arrangements. (a) Thin to thin; (b) thin to thick; (c) fine wire bonding; (d) step brazing or soldering; (e) joining near heat-sensitive or combustible material.

or structure, where service temperatures preclude the use of adhesives; and (4) seam welding (or soldering) of thin steel or aluminum can stock.

If adhesives can be used to bond these thin sections to one another or to thicker sections, that should often be the choice. In fact, adhesive bonding is widely used in joining thin sections, as exemplified by the adhesively bonded honeycomb sandwich panels shown in Figure 11.2. If the service environment is hostile to adhesives because of elevated temperatures, moisture, solvents, or biological agents, welding, brazing, or soldering are the logical joining alternatives. Mechanical fastening is essentially impossible because the thin sections are unable to tolerate bearing and/or fastener tear-out or pull-through loads.

The key to joining very thin sections is keeping the temperature of the joint as low as possible and minimizing the net linear heat input by welding at very high speeds,

Figure 11.2   Adhesively bonded thin-section honeycomb sandwich panels. (Courtesy of Grumman Aircraft Systems Division of Grumman Aerospace Corporation.)

for example with EBW or LBW. Great care must also be taken to concentrate the heat on the specific area to be joined, minimizing the exposure of surrounding areas to potentially damaging heat to prevent metallurgical degradation or distortion, but without concentrating the heat so much that overheating or burn-through occurs. This involves delicately balancing heat input.

Soldering, because of its requirements for melting only the filler and not approaching the melting point of most metal substrates, is obviously the preferred choice over welding and brazing if solder strength is sufficient. For very thin sections, with sufficient overlap, joint strengths can easily be considered structural as well as hermetically tight. In addition, proper joint design can enhance the joint strength (see Chapter 9, Section 9.7.1).

For more demanding structural applications, brazing is the next best choice. Here, however, the entire structure is normally heated to just below its solidus temperature (unless special processes are used, such as induction brazing, infrared brazing) and the filler alloy is melted to create the joint by capillary flow. Once again, the preferred joint configuration is a lap. Rarely would a gas torch be used because of the risk of rapidly overheating accidentally. Structural metal honeycomb, such as titanium honeycomb for use in aerospace applications, is usually created by brazing the stack-up used to produce the honeycomb upon expansion, as well as to attach the sandwich panel cover or face sheets.

Fusion welding can be used to join very thin sections, provided the heat input is kept to a minimum to cause just the melting that is desired and needed *and* the source is precisely located and properly concentrated over the area to be joined and that area only. Preferred processes for welding very thin sections include laser-beam and electron-beam for precise control of energy input and position and to minimize heat input through high welding speeds; gas-tungsten arc and plasma arc using very small torches, small diameter electrodes, and very low currents (e.g., less than a few amperes) called needle arcs; and resistance spot or seam or ultrasonic friction welding. The last two processes minimize heat input by minimizing or avoiding melting and, in the case of resistance welding, using very high heating rates. Diffusion bonding is also a viable option to producing sound lap joints in appropriate materials (e.g., titanium, nickel, copper).

For those special cases where very thin sections are to be joined to much thicker sections, special care must be exercised to properly distribute or balance the heat input to each joint portion. Techniques that can help are preheating the heavier section to reduce the heat needed to raise its temperature sufficiently to cause melting of it or a filler and/or applying cooling (e.g., moistened pads, or a heat-blocking compound) to the thin section to extract excess heat and prevent unwanted heat spreading.

11.5.3 Joining Very Small Components, or Microjoining

Joining very small joint components (in a process often referred to as *microjoining*) can cause difficulty in welding, brazing, or soldering, as well as in mechanical fastening and adhesive bonding, from several standpoints. First, and most common, is the difficulty of aligning the joint components to create the proper joint and assembly. This is predominantly a *fixturing* problem, but may also involve inherent problems in directing the heat source and controlling its heat input. It is often possible to tend to the alignment or fixturing problem through careful design of the joint in the first place. For example, joining of leaded or unleaded (or leadless) electronic packages to electrically conductive pads on a substrate should allow sufficient play or tolerance to produce a satisfactory joint, even if there is some relative misalignment during assembly. Some mechanical alignment aids such as stops or grooves or slots can facilitate alignment.

Second, it can be difficult to focus the energy (i.e., heat) of a welding source just on the element to be joined and nowhere else. Here, processes like laser or electron beam are attractive for three reasons. First, these beams are highly focused. Second, they can be precisely directed or positioned by optical or electromagnetic as well as mechanical means. Third, the energy in these beams can be precisely controlled.

The final point is also the final point important to microjoining in general. That is, it can be difficult to control the amount of energy (i.e., the heat input) needed to create just the amount of melting required to form the needed and desired joint. Again, for welding, brazing or soldering, laser or electron beam sources are attractive.

Usually for brazing or soldering microcomponents the filler metal is preplaced on or at the joint components before attempting to apply heat. Plated or dipped coatings of the filler (often called *tinning* or *pretinning* in soldering) or preforms are employed. Joints are formed by reflowing the filler by applying a general rise in temperature of the assembly using waves, furnaces, etc.

11.5.4 Special Situations

There are many situations in which a joint must be made between two or more pieces of metal using welding, brazing, or soldering and where some adjoining joint element or neighboring component in the assembly must be protected from the heat used to make the joint. This is a common requirement when repairs are being made on structures or assemblies that cannot be disassembled (e.g., exhaust systems on cars).

The techniques used often involve shielding the heat-sensitive material or area. Some specific techniques are (1) use of a *thermal barrier* (such as a chill bar of copper or aluminum or other metal or a moistened barrier compound[5] or moistened rag) as a heat sink, and (2) use of a coolant (e.g., water spray or impinging gas stream).

For all these special situations, safety is a major concern and consideration. Do not use water indiscriminately around electric arc processes. Be very careful welding near combustible fluids or coatings or neighboring parts.

Another approach to joining components in the vicinity of adjacent heat-sensitive materials or structure is known as *step-joining*. Two specific examples are *step-brazing* (Section 8.4.8) and *step-soldering* (Section 9.5.9). In this approach, progressively lower melting fillers are used to prevent remelting of adjacent or previously made joints.

11.6 JOINING VERY THICK SECTIONS OR VERY LARGE STRUCTURES

11.6.1 Joining Very Thick Sections

Structures that consist of very thick elements requiring joints pose challenging joining problems. Almost all heavy section joints are required to provide high strength and high quality, often for highly demanding and even hostile environments. Heavy-section armor where ballistic impact resistance is required; thick-walled pressure vessels or

---

[5] These compounds were formerly composed of asbestos, but now contain asbestos alternatives.

containers where high static and dynamic loads, hermeticity, cryogenic or elevated temperatures, and corrosion resistance are required; submarine or submersible hulls where high compressive loads, detonation impact loads, and corrosion resistance are required; and nuclear reactor components that experience high temperatures, high pressures, high static and dynamic loads, corrosive agents, and radiation are examples of where very thick sections must be joined to result in high integrity but with economy. Sometimes these joints (as in submarines) must be aerodynamically or hydrodynamically smooth or should be as lightweight as possible (e.g., for shipboard nuclear reactors).

When joints exceed a certain thickness and structural demands become paramount, adhesive bonding is usually no longer a viable joining option. While mechanical fastening is possible, and is practiced, there is a practical limit. This limit is dictated by the difficulty of hole preparation, the difficulty of fastener installation, and the ability to achieve desired and necessary preloads or joint clamping forces. For practical assembly reasons, for the most stringent pressure requirements, and for absolute fluid tightness, welding is generally the only viable option.

Normally, very thick sections are welded using many passes with heavy deposition processes like gas-metal arc welding (GMAW), submerged arc (SAW), electroslag (ESW), or electrogas (EGW). Other options are GTAW at very high current (i.e., several thousand amperes), hot-wire GTAW (i.e., HWGTAW), or various narrow-gap welding processes, such as a multiwire variant of the GMAW process. Joints for such processes are usually prepared with wide openings to provide access to the welder. Single V, U or J or double V, U or J preparations are common. Figure 11.3 shows some typical thick-section joint options. Figure 11.4 shows a typical thick-section welded assembly.

While multipass, heavy-deposition rate welding processes can produce sound welds, the total amount of filler metal required and the tremendous total heat input leads to significant shrinkage, distortion, and residual stresses. In addition, the skill level required of welders or operators, the high labor intensity, and overall cost are usually high. Furthermore, inspection can be quite challenging and repair can be extremely difficult and costly. Finally, multiple passes result in overlapping MAZ's, which can greatly complicate the effects on microstrucutre of these regions.

To minimize the problems associated with the large mass or volume of a multipass weld, welding procedures must be carefully controlled. Welds are sequenced to balance heat input. Alternating passes are often made from opposite sides of the joint to offset any shrinkage, and not all passes are made in one joint until it is finished, but, rather, a pass is made in one joint and then another somewhere else on the weldment to offset any possible distortion. In hardenable alloys, each pass is used to temper previous passes to some degree, and subsequent passes always provide some stress relief in previous passes. Finally, overall temperature, i.e., interpass temperature, is controlled to prevent metallurgical degradation and distortion or residual stresses.

An attractive alternative to multipass welding is to use a fusion welding process that is capable of producing thick-section or deep-penetration welds in single, or at least fewer, passes. Candidate processes must be capable of producing welds with very large depth-to-width ratios. Example processes are electron-beam, high-power laser-beam, and certain narrow-gap processes. Narrow, single-pass welds can be produced by these processes as follows:

- up to 25 to 35 cm (or 10 to 14 inches) in steels or titanium or aluminum by EBW with 100 KVA (kilovolt-amperes) systems
- up to 10 to 15 cm (4 to 6 inches) by narrow gap processes in steel or nickel-base alloys
- up to 5 cm (or 2 inches) by high-power (e.g., over 10 kW) LBW in certain metals with low reflectivity, like steel.

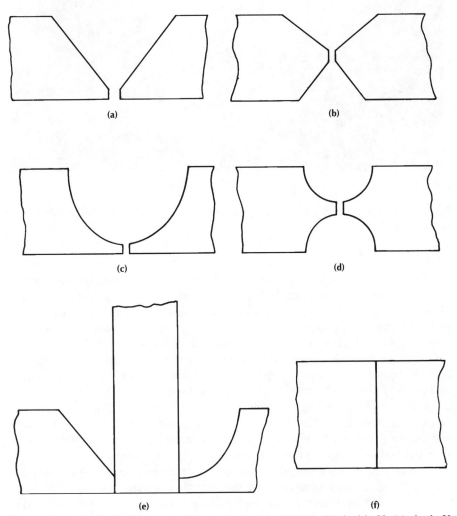

Figure 11.3 Typical thick-section joint options: (a) single V; (b) double V; (c) single U; (d) double U; (e) half V and J; (f) straight or square butt. Joint (e) is used almost exclusively with EBW or LBW processes.

Figure 11.4   Typical thick-section welded assembly. (Courtesy of Newport News Shipbuilding and Dry Dock Company.)

For the most part, these processes (i.e., EBW and LBW) have very high energy density heat sources (greater than 10^9 watts per square centimeter), and, thus, operate in a keyhole mode. The result is deep penetration, minimal weld fusion zone width, minimal volumetric shrinkage, minimal distortion, high welding speeds, low linear heat input, minimal heat-affected zone, and minimal distortion and residual stresses. Also, complications from overlapping MAZ's are avoided. An example where single-pass EB welding enabled precision assembly without distortion is the F-14 fighter aircraft's all Ti-6Al-4V wing carry-through structure (see Figure 11.5). This 22-foot-long, 4-foot-wide, 1½-foot-deep box assembly was entirely EB welded, using single pass welds, with joints varying from approximately 1 inch to over 2 inches. EB welding enabled more than 300 pounds to be saved over mechanical fastening. Welded joints were designed to demanding fatigue and fracture toughness criteria.

Usually these single-pass, deep-penetration welding processes require fairly precise joint fit-up to help reduce shrinkage and distortion. Square- or straight-butt preparations are common (Figure 11.3f) , and welding is usually done autogenously (i.e., filler is not used) because of the difficulties of getting the filler down into the deep, narrow weld. Resulting welds are usually fairly parallel sided, so the heat distribution through the thickness is fairly uniform. This further reduces the distortion typically found with tapered, multipass welds.

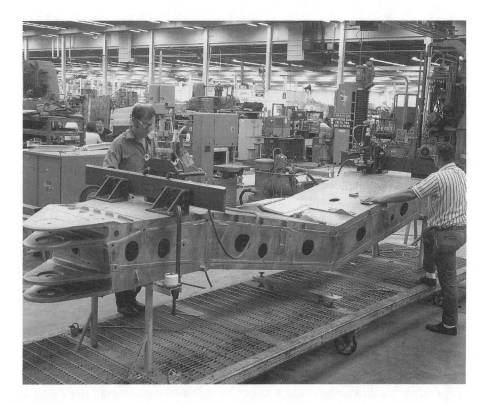

Figure 11.5   EB-welded F-14 aircraft wing carry-through structure. (Courtesy of Grumman Aircraft Systems Division of Grumman Aerospace Corporation.)

Because the level of residual stresses that can develop can be very high in thick section or heavy structures, special care must be given to relieving these stresses after welding. Thermal treatments, i.e., stress-relief annealing, are preferred, but size and location often make such treatment difficult or impractical. Localized stress relief is often practiced through the use of high-intensity quartz heat lamps, resistance-heated blankets, gas torches, etc. If thermal stress relief is impossible, welding procedures should be carefully selected and practiced to minimize residual stress development, and vibration stress relief or postweld mechanical treatment by shot-peening or other means should be attempted.

11.6.2 Joining Very Large Structures

The challenge of welding very large structures is primarily the sheer labor intensity involved, but there are also challenges posed by having to produce welds in all positions and, usually, outdoors. Examples of such large welded structures are ships, oil or gas transmission pipelines, chemical reaction vessels, storage tanks, and off-shore drilling platforms. A typical large welded structure is shown in Figure 11.6.

The challenge of making welds in other than a horizontal plane with the welding heat source in the downward position is to deal with the effects of gravity. Gravity pulls on the molten weld pool and can make it unstable. The result can be spill-out of the molten metal, resulting in gross underfill, massive voids, and sloppy crown and root beads. Special technique, special fillers, and special operating modes must usually be used when welding in the so-called out-of-position directions, i.e., vertically up, vertically down, or horizontally in a vertical plane or overhead in the horizontal plane.

Generally, weld heat input must be kept even lower during out-of-position welding than for downhand (or in-position) welding, minimizing the size of the weld pool and the volume of molten weld metal. This is true for shielded metal arc, flux-cored arc, and gas-metal arc processes. For processes like submerged arc and electroslag welding[6], weld pools are so large or the volume of molten metal so great that welding out-of-position is impossible.

In addition to the effect of basic process and heat input or power level within a process, the mode of molten metal transfer in consumable arc welding processes can also have a pronounced effect on the ability to weld out of position. The preferred mode for metal transfer when welding with GMAW is short-circuit arc, as molten droplets are drawn directly off the tip of the melting electrode by the surface tension forces of the molten weld pool whenever the droplet makes contact. The globular mode is usually out of the question as transfer of the large glob of molten metal would not be aided by gravity for out of position welds. Spray mode is usually too hot to facilitate out-of-position welding, resulting in too large a molten pool. Pulsed-current spray or P-GMAW does allow out-of-position welding, because the periodic drop in current allows intermittent cooling and solidification of the weld metal.

The problems associated with welding out of doors are maintaining proper shielding in high winds, keeping prepared weld joints dry during rainy, snowy, or icy weather, and maintaining proper preheat or interpass temperatures to prevent undesirable untempered martensite formation in hardenable steels. The problem of shielding can be dealt with by using protective shelters or tents, or using the flux-cored or open-arc welding process. Recall that in this process the protective gases generated by the decomposition of the core fluxing ingredients must pass through the arc before being exposed to any convective currents (such as wind). Thus, by the time the protective gases can be blown away, they have done their job of shielding the molten weld metal.

Keeping the weld joint dry during welding and controlling the joint temperature are much more challenging. Again, protective shelters can be used to keep off rain, and preheating can be accomplished using resistance heating blankets, quartz lamps, or gas torches. Preheating also helps dry the joint before welding.

The final problem associated with welding large structures is stress relieving the joints after welding. Obviously, extremely large structures are not going to be placed in heat treating furnaces. Instead of bringing the structure to the heat treating furnace, heat treating is brought to the structure. Localize heating using resistance heaters, quartz lamp

---

[6]Joint design in electroslag welding allows welding vertically up, but not horizontally.

Figure 11.6   Typical large welded structure: an offshore oil drilling platform. (Courtesy of the Chicago Bridge & Iron Company.)

heaters, or even torches is employed. Another approach is to mechanically relieve the most severe residual stresses by peening, burnishing the welds, or using local vibration.

11.7 JOINING DISSIMILAR METALS AND ALLOYS

11.7.1 General Comments

Special design requirements sometimes suggest or require that dissimilar metals or alloys should be employed in different, adjoining, or abutting areas of a structure or assembly. Despite the possible dissimilar chemistries and crystal structures of these metals or alloys, a sound structural joint is required. This can be difficult if the different metals or alloys are incompatible chemically or metallurgically or if they are incompatible in terms of a key physical property like thermal expansion coefficient.

Chemically or metallurgically incompatible materials may not permit mixing because of mutual insolubility or very limited solubility, resulting in severe macroscopic segregation and gross heterogeneity. Even if mixing occurs, chemical or metallurgical incompatibilities may result in severe embrittlement upon mixing because of the formation of intermetallic compounds. Finally, there could be an electrochemical incompatibility that would lead to severe galvanic corrosion in service. Incompatibility of thermal expansion coefficients can give rise to severe thermal stresses, which, in turn, give rise to severe distortion, residual stresses, cracking, or fracture.

Often, dissimilar metal or alloy combinations are needed to meet special environmental demands, such as corrosion resistance (e.g., using cladding), wear resistance (e.g., using hard coating or wear facing), temperature resistance, and electrical or thermal conductivity requirements. The requirement for successfully joining dissimilar metals or alloys that are chemically or metallurgically incompatible is to prevent or minimize mixing or to use intermediate materials (or intermediaries) in the joint. The requirement for successfully joining dissimilar metals or alloys with mismatching coefficients of thermal expansion (i.e., C.T.E.s) is proper joint design and/or use of intermediaries.

11.7.2 Avoiding or Minimizing Melting

There are several options for avoiding or minimizing melting to enable the successful joining of chemically or metallurgically incompatible metals or alloys. First, use mechanical fastening methods. These preclude the need for melting and mixing and do not alter the chemical composition or microstructure of any joint element. The only potential problems are electrochemical (i.e., galvanic) incompatibility, which can be handled by using suitable gaskets or coatings, or thermal expansion mismatch, which may be handled by careful joint design or by the use of intermediaries (see Section 11.7.3). Second, adhesively bond the joint. This could provide suitable joint strength, efficiency, and durability, provided the proper adhesive is selected and proper preparation and application procedures are used. Adhesive bonding precludes the need for mixing, usually does not alter the chemistry or microstructure of metallic adherends, and can overcome electrochemical and thermal expansion mismatch incompatibilities by keeping joint elements electrically separated and by providing flexibility or strain compliance in the adhesive. A third approach is to avoid substrate melting and mixing by using brazing or soldering methods. This requires selecting a suitable, mutually compatible braze or solder filler alloy.

If welding is felt to be needed for joint strength and serviceability, the amount of melting can be minimized or avoided by using a nonfusion welding process. Examples are friction or inertia welding, roll bonding, forge welding, flash welding, explosive welding or bonding, diffusion welding or bonding, and, to a lesser extent, resistance welding (actually, a fusion process). Brittle intermetallics can still result from some of these processes, either because some melting and mixing does occur or because of interdiffusion. Obviously, electrochemical and C.T.E. compatibility must be acceptable for these processes to work.

11.7.3 Using Intermediate Layers or Intermediaries

Undesirable mixing, or gross incompatibilities of physical properties, such as electrochemical potential or coefficient of thermal expansion, can be overcome by separating the two different materials with some intermediate material or intermediary. This intermediate material serves to transition or *grade* the properties from one joint element to the other by gradually changing the composition and/or properties or by serving as a mutually compatible bridging agent (i.e., the intermediary is compatible with each joint material).

Intermediaries can take several forms. First, coatings could be used. Metallic or nonmetallic (e.g., ceramic or polymeric) coatings can act to separate or insulate dissimilar metals and alloys from one another or can be used to bridge from one material to another. Metallic electrolytic (e.g., hard chromium) or electroless platings (e.g., electroless nickel), sputtered coatings, chemical or physically deposited coatings, or clad layers all could be used to transition between metallic materials. These layers are usually quite thin, however, and may not tolerate grossly different physical property disparities. Second, interlayers could be used. These metallic layers can have substantial thickness and can be applied by plating, cladding, thermal spraying, fusion welding (i.e., weld cladding), brazing, etc. One or more interlayers can be used to gradually change the composition and/or properties from one joint element to the other. Finally, transition pieces could be used. These are usually fairly thick and are often welded or brazed or soldered, layer by layer, piece by piece. Obviously, each piece must be compatible to the substrate to which it is joined. The number of transition pieces (or interlayers) depends on how different the two base joint materials are. For C.T.E. mismatch, 1 to 1.5×10^{-6} per °C (or °F) differences between adjoining materials are usually considered reasonable to bridge two very different C.T.E. materials. (Again, recall that in the presence of a severe temperature gradient, C.T.E. matching may not be what is needed but rather selection of C.T.E.s to minimize the thermally induced stresses caused by the gradient.)

After an intermediary is applied, joining by brazing or soldering or, even, welding, is then accomplished between one joint component and the intermediary, choosing a suitable filler for that couple. Similar joints are made between abutting pairs of intermediaries, until the entire joint—base element to base element—is bridged. In brazing, the use of intermediaries is called *sandwich brazing*.

The trick in joining dissimilar metals and alloys is to make the joint components look less different chemically and in terms of physical properties. A similar *trick* is used for joining dissimilar material types, such as metals to ceramics, metals to glass, and metals to polymers. This is discussed in Chapter 15.

11.8 JOINING IN HOSTILE ENVIRONMENTS

Another challenge posed to joining of metals and alloys is when that joining must be accomplished in a hostile environment. Hostile environments may mean hostile to the welder or welding operator, thus requiring that welding be done automatically and, often, remotely, or it may mean hostile to the welding process or the weld proper. The options for joining processes other than welding are extremely limited. Three examples of hostile environments for welding are in space, under water, and in high radiation areas. Each of these requires highly specialized approaches and techniques, so will be described here only in the most general terms.

The challenges of welding in space are fourfold. First, the absence of air, i.e., the presence of a vacuum, limits process options. Arc welding can be difficult as ionization is difficult. The result is arc instability. Electron-beam and laser-beam welding, on the other hand, become very attractive. Second, and related to the presence of a vacuum, evaporative loss of high vapor pressure solutes and even solvents could pose severe problems in achieving desired composition and density (because of outgassing of the fusion zone). Third, the absence of gravity can make molten metal transfer and the precise placement of weld metal difficult. For example, globular transfer, which depends on the force of gravity to detach the molten metal at the tip of the melting electrode, would be impossible, while spray transfer, through the action of the operative electromagnetic force, or short-circuit transfer, through the action of surface tension forces, would work well. Fourth, and common to all hostile environments, manual welding becomes difficult, impractical, or impossible. Thus, welding must be done automatically and, usually, remotely using robotic systems equipped with elaborate sensors (e.g., vision systems). Obviously, the environment of space provides extremely good shielding.

As for other options for joining in space, most organic adhesives are out of the question as they would volatize. Mechanical fastening, while possible, poses some challenging logistic problems, i.e., how to prepare the fastener holes and how to store, handle, and install fasteners. The opportunity to use interlocking design features to accomplish mechanical fastening is intriguing, however.

The challenges of underwater welding are the obvious problem of removing water from the welding envelop to prevent hydrogen embrittlement, the less obvious problem of extreme pressure on arc action and the forcing of gases into the molten weld pool, the potential problem that rapid cooling could cause, the safety hazard posed with arc welding processes, and the frequent need to accomplish welding remotely. Special welding consumables and procedures have been developed for welding underwater, but much work needs to be done given increasing needs in nuclear reactors, on offshore drilling platforms, etc. Adhesive bonding with suitable adhesives is not out of the question, and mechanical fastening using divers or robots to install the fasteners is possible.

Finally, the challenge of welding in a highly radioactive area is primarily associated with the need to weld automatically and remotely, with the added problems associated with having to prepare joints remotely during repair welding. The use of most adhesives is precluded by the severe degradation of many polymers by high radiation, while the use of mechanical fastening is largely precluded by the difficulties associated with preparing fastener holes and installing fasteners remotely.

11.9 JOINING OF INTERMETALLICS

11.9.1 General Comments

As application opportunities continue to emerge that require materials to perform at higher and higher temperatures for sustained periods of time, traditional metals and alloys of construction, such as iron, nickel, and cobalt base superalloys, reach their usable strength and thermal stability limits. This happens at about 950-1050°C (1750-1925°F) for 1000 hours at a stress of 150 MPa (22 ksi) for many superalloys and only slightly higher for some specific alloys and forms (e.g., wrought versus cast versus single crystal). Admittedly, refractory metals (e.g., tungsten, molybdenum, tantalum, and niobium) and their alloys can function at higher temperatures, but their high densities ($10.2 \ g/cm^3$ for molybdenum to $19.3 \ g/cm^3$ for tungsten versus $7.9 \ g/cm^3$ for iron to $8.9 \ g/cm^3$ for nickel and cobalt) and lower specific strengths, as well as their higher reactivity, often preclude their use in many applications (e.g., gas turbine components).

There are thus three options for higher temperature service: (1) monolithic or reinforced ceramics; (2) graphite or carbon-based (i.e., carbon–carbon composites); and (3) monolithic or reinforced intermetallics. Each of these material types has its advantages and shortcomings, but all need to be capable of being joined to enable advanced structures to be produced. Thus, weldability is an especially desirable characteristic. For weldability, intermetallics offer the most promise and, so, are addressed here. Joining of ceramics, graphite, and carbon–carbon composites will be discussed in Chapters 12 and 14.

11.9.2 A Description of Intermetallics

Intermetallics or *intermetallic compounds,* as the name implies, are compounds, or alloys of narrow ranges of solubility, composed entirely of metallic elements. Many intermetallic compounds, such as those formed in brazed or soldered joints at interfaces (e.g., Cu_3Sn), are extremely brittle and have no engineering value. In fact, they are troublesome. There are some intermetallics, however, that are attractive as engineering materials because they are strong, stable at elevated temperatures, and can be made reasonably ductile by appropriate alloying and/or processing. Some particularly attractive groups of intermetallics are:

- metal (e.g., Fe, Ni, Ti or refractory metal, Hf) aluminides (e.g., Ni^3Al or NiAl)
- metal (e.g., Ti, Zr, V, Nb, Ta, Cr, Mo, W) silicides (e.g., $MoSi^2$)
- metal (e.g., Ti, Zr, V, Nb, Ta, Cr, Mo, W) borides (e.g., TiB or TiB^2)
- metal (e.g., Fe, Ni, or Co) compounds or long-range ordered alloys with Ti and V (e.g., $[Fe,Ni]^3[V,Ti]$).

Some also consider at least some of the metal carbides as intermetallics, such as SiC. Table 11.5 shows a tabulation of various refractory metals and alloys, ceramics, and intermetallics for comparison.

As a group, intermetallics have a narrow range of compositions, based on stoichiometric atomic ratios, and have structures that usually exhibit long-range order. Thus, intermetallics and long-range ordered (LRO) alloys are a unique class of materials with an atomic arrangement distinctly different from that of conventional, disordered alloys. This distinctly different structure gives rise to distinctly different properties.

Below a certain critical ordering temperature, T_c, alloying atoms in LRO alloys arrange themselves on specific lattice sites and form an ordered crystal structure. This order persists over many, many unit cells, so is called long-range order.[7] As a result of long-range order, these materials exhibit unique properties because of the effect of that order on atomic mobility and dislocation dynamics. Most significant, elevated temperature performance is superior. The yield strength of many LRO alloys increases rather than decreases with increasing temperature. Elevated temperature creep and fatigue strength are also high, and many intermetallic materials (like the aluminides) have outstanding corrosion and oxidation resistance. The one shortcoming is a tendency toward brittle behavior in their ordered state, especially at lower temperatures. This can make fabrication difficult and can limit structural utility.

There are several intermetallics or LRO alloys that are receiving particular attention, including studies of their weldability. These include, but are not limited to

- $(Fe,Ni)^3(V,Ti)$, e.g., $(Fe50Ni50)^3(V98Ti2)$ with a $Tc = 670°C$ (1238°F)
- $(Fe,Co)^3(V,Ti)$, e.g., $(Fe22Co78)^3(V98Ti2)$ with a $Tc = 950°C$ (1742°F)
- Nickel-aluminides, Ni^3Al: with up to 13% of a ternary alloy addition of Co, Cu, Fe, Ti, or V for improved ductility; or with 0.1 at.% B or 0.5 at.% Hf for improved ductility
- Titanium-aluminides, Ti^3Al: with ternary or quaternary additions of Nb, V, Mo, or W for increasing ductility.

The titanium aluminides are particularly interesting because of their inherently lower density (due to their Ti base). Examples of two alloys receiving attention for their weldability are: Ti-13.5 wt% (24 at.%) Al-21.5 wt% (11 at.%) Nb and Ti-48 at.% Al-6.5 vol.% TiB_2.

11.9.3 Weldability of LRO Alloys and Intermetallics

LRO alloys and intermetallics, such as those described above, exhibit a range of weldability, but most have been welded autogenously using GTAW, EBW, and LBW (pulsed Nd:YAG) with Ar shielding. Inertia friction welding has also been used successfully with the Ti_3Al alloys, as has diffusion welding.

---

[7] Some short-range order almost always exists in crystalline materials, but this order, unlike long-range order, does not persist very far and so does not have much of an effect on properties.

Table 11.5 Refractory Metals, Ceramics, and Intermetallic Compounds

Material	Melting Point °C	Density g/cm³	Modulus GPa	Hardness (Annealed)	UTS @RT MPa	UTS @Temp. MPa
Tungsten	3,380	19.3	345	450 V	345-1400	0.5X @ 600°C
Tantalum	3,000	16.6	189	-	520	520 @ 1000°C
Molybdenum	2,600	10.2	324	200 V	560	275 @ 1000°C
Niobium	2,500	8.6	105	160 V	275	-
Hafnium	2,130	13.4	138	190 DHH	450	110 @ 375°C
Zirconium	2,130	6.5	101	180 DPH	344	138 @ 315°C
Thorium	1,830	11.7	59	37 V	218	-
Titanium	1,730	4.5	117	-	235	-
Vanadium	1,700	6.0	132	85 R_B	517	-
Uranium	1,100	18.7	167	(Soft)	386	-
Rhenium	3,170	20.5	465	270 KHN	1157	-
Osmium	2,700	22.5	565	800 DPH	-	-
Iridium	2,450	12.2	530	170 V	623	331 @ 1000°C
Ruthenium	2,430	22.4	418	220 V	540	247 @ 1100°C
Rhodium	1,970	12.4	296	135 BHN	958	-
Platinum	1,780	21.4	148	100 V	124	-
Palladium	1,560	12.0	118	38 V	145	-
Carbon	4,730	2.25				
Boron	2,050	2.3				
Silicon	1,410	2.4	11 A			
				kg/mm³		
HfC	3,890	12.7	-	-		
TaC	3,880	14.5	-	1,800		
NbC	3,500	7.8	342	2,400		
Ta_2C	3,400	15.2	-	-		
TiC	3,250	4.9	314	3,200		
ZrC	3,175	6.4	-	2,600		
VC	2,830	5.8	272	2,800		
WC	2,630	15.8	715	2,400		
W_2C	2,700	17.3	422	3,000		
MoC	2,695	8.4	-	>1,800		
Mo_2C	2,690	9.2	228	1,800		
ThC_2	2,655	9.6	-	-		
ThC	2,625	10.7	-	-		
UC	2,475	13.6	-	-		
UC_2	2,375	11.7	-	-		
Cr_3C_2	1,895	6.7	-	1,300		
U_2C_3	1,800	12.9	-	-		

(continued)

Table 11.5   *(continued)*

Material	Melting Point °C	Density g/cm³	Modulus (GPA)	Hardness kg/mm² (unless noted)	UTS @RT (Annealed)	UTS@Temp. MPa
HfN	3,310	14.0	-	-		
TaN	3,090	14.1	-	+ Mohs		
ZrN	2,980	7.3	-	1510 KHN		
TiN	2,930	5.4	80	9-10 Mohs		
NbN	2,050	7.3	-	+ 8 Mohs		
NbN	2,050	7.3	-	+ 8 Mohs		
VN	2,050	6.1	-	-		
CrN	1,500[a]	6.1	-	-		
HfB$_2$	3,250	11.2	-	-		
ZrB$_2$	3,040	6.1	-	2,200		
TaB$_2$	3,000[a]	12.6	-	-		
TiB$_2$	2,980	4.5	-	3,400		
NbB$_2$	>2,900[a]	7.2	-	+8 Mohs		
WB	2,860	16.0	-	9 Mohs		
W$_2$B	2,770	16.7	-	-		
MoB$_2$	2,100	7.8	-	1,280		
VB$_2$	2,100[a]	5.1	-	8-9 Mohs		
MoB	2,180	8.8	-	1,570		
Mo$_2$B	2,000[a]	4.3	-	-		
CrB	1,550	6.2	-	8.5 Mohs		
TaSi$_2$	2,400	8.8	-	1,560		
WSi$_2$	2,150	9.3	-	1,090		
NbSi$_2$	1,950	5.3	0	1,050		
MoSi$_2$	1,870[a]	6.1	-	1,290		
VSi$_2$	1,750	4.7	0	1,090		
CrSi$_2$	1,570	4.4	-	1,150		
TiSi$_2$	1,540	4.4	-	870		
ZrSi$_2$	1,520[a]	4.9	-	1,030		
ThO$_2$	3,050	9.7	139		97	-
MgO	2,800	3.5	170		210	-
ZrO$_2$	2,690	5.8	167		145	90 @ 1000°C
BeO	2,530	3.0	300		138	-
Al$_2$O$_3$	2,050	3.9	350	2,100 KMN	262	234 @ 1000°C
BN	3,000	2.25	-	-	-	-
SiC	2,700[b]	3.2	470	2500 KHN		
B$_4$C	2,350	2.5	293	2800 KHN		
Si$_3$N$_4$	1,900[b]	3.4	-	-	-	-

[a] Decomposes
[b] Sublimes
Adapted from Paul Schwartzkopf and Richard Kieffer, *Refractory Hard Metals,* Macmillan, 1953.

There are two predominant problems in trying to weld the LRO alloys or intermetallics. First, hot cracking tends to occur in some intermetallics in both the heat-affected and fusion zones because of microsegregation of certain phases. Thermal stresses, aggravated by higher welding speeds, promote such cracking. The tendency toward hot cracking can be lowered by keeping welding speeds below 13 mm/sec (30 ipm) during EBW of thin sheets of Ti_3Al alloys. Lowering the cooling rate in the HAZ to below 300 K/sec by preheating also has been shown to help. Finally, additions of certain ternary elements, such as 10 at.% Fe to Ni_3Al or 1.7 wt% Hf to B-doped Ni_3Al, definitely reduces the hot cracking tendency. Second, the loss of long-range order in some alloys has been shown to reduce strength, hardness, and ductility in the FZ and HAZ slightly. One example is in $(Fe,Ni)_3(V,Ti)$ LRO alloys. Order can be restored by a postweld reordering heat treatment if it is essential to obtain optimum properties.

During the continued development of LRO alloys and intermetallics, the best way to assess weldability is to conduct continuous cooling studies of their microstructure development and cracking tendency. Such studies are useful for selecting welding processes and parameters expeditiously.

11.9.4 Other Options for Intermetallics

The use of an exothermic brazing process, SHS or pressurized combination synthesis offers potential for joining refractory intermetallics for one thing, joining could be done simultaneously with material synthesis and shape production; for another, joint strength would be essentially identical to the substrates. This process is referred to in Chapter 8.4.8.

11.10 THERMAL SPRAYING OF METALS, ALLOYS, AND INTERMETALLICS

As described in Chapter 10, it is possible to apply material to a substrate by propelling that material in its molten or heat-softened (i.e., plastic) state at high kinetic energies. The process is known as *thermal spraying*. Thermal spraying is a joining process in the sense that it unites one material to another, as opposed to one part to another. As opposed to producing an assembly of parts, thermal spraying produces an assembly of materials, often a composite.

Metals, alloys, and intermetallics can all be deposited or joined by thermal spraying. Examples are shown in Table 10.1 of Chapter 10. By thermal spraying, it is possible to restore material lost by wear or corrosion or to apply a protective layer or coating. Thermal spraying is widely used in the original fabrication and subsequent repair of gas turbine blades, vanes, discs, and other components. The resulting adhesion is excellent, often equalling the cohesive strength of the substrate material itself.

Often called *metallizing* in the past, thermal spraying of metals can be accomplished using any of the four possible methods: (1) combustion flame spraying, (2) electric arc spraying, (3) plasma arc spraying, or (4) detonation-flame spraying. The temperature and/or kinetic energy of the particles being sprayed increases in the order these processes are listed. Consequently, more refractory metals and alloys, as well as intermetallics and ceramics, are usually sprayed using the plasma arc or detonation-flame methods.

Thermal spraying can be used to deposit a surface layer of essentially any material on any material, provided there is proper match of C.T.E.s. Pure metal, alloy, or intermetallic coatings (as well as ceramic coatings) can be applied to pure metals, alloys, or intermetallics (or even ceramics) in any combination.

SUMMARY

Metals and metallic alloys are overwhelmingly the most widely used materials in the fabrication of assemblies and structures because of their inherently high strength, ductility, toughness and fabricability by casting, deformation processing, and machining. Joining of these materials is critically important. Welding, brazing, and, to a lesser extent, soldering, are the most popular joining processes, in that order, since they result in the formation of metallic bonds that offer high joint integrity and continuity of important physical properties (e.g., electrical or thermal conductivity). Mechanical fastening is also widely used, especially where disassembly is required, but also because of its ease and economy. Ease of fastener hole production and good tolerance to bearing loading and inevitable stress concentrations make mechanical fastening of metals and alloys attractive and practical. Although less popular, adhesive bonding is an excellent option, especially where light weight and ease of assembly are desired. Limited environment durability is the principal drawback for severe service conditions. Both mechanical fastening and adhesive bonding are excellent choices for joining dissimilar metals and alloys, provided proper consideration is given to electrochemical and thermal expansion incompatibilities.

Joining of metals and alloys becomes difficult when extremes are involved—either material extremes or geometric extremes. Material extremes include extreme refractoriness, extreme reactivity, and extreme sensitivity to heat (due to low melting temperatures or susceptible microstructures). Geometric extremes include very thin sections, very small components, very thick sections, or very large components. Other extremes are dissimilar chemistries and/or properties and hostile environments, such as space or under water.

For refractory metals and alloys, welding and brazing are preferred, with the major consideration being how to achieve bonding without contamination by interstitial pickup. For fusion processes, achieving the degree of required melting can be difficult. Adhesives are rarely used because of severe use temperature limitations. For reactive metals and alloys, welding and brazing are again preferred, but extreme care must be taken to provide adequate shielding from oxygen, nitrogen, hydrogen, and carbon, as well as other materials. For heat-sensitive metals and alloys, extreme care must be exercised during welding to limit heat input. Brazing and soldering, as well as adhesive bonding, are often preferable. Examples of heat-sensitive metals and alloys are those that are low melting, cold worked, age hardened, transformation hardened or hardenable, or inherently brittle and prone to thermal shock (like cast iron).

Some metal structures are sensitive to heat because they are very thin in section, very small, or in close proximity to materials that cannot tolerate heat (such as plastics or combustibles). If welding, brazing, or soldering is chosen, heat input must be carefully directed and controlled in magnitude. For very thick sections, welding is the joining method of choice almost by default. Single-pass processes (like EBW or LBW) are generally preferable to high-deposition rate, multipass processes like GMAW, SAW, ESW, and EGW. Total heat input from the higher energy density processes is lower. Regardless of the choice of process, welding heat input must be carefully balanced to prevent distortion and severe residual stresses.

Dissimilar metals and alloys can be readily joined by adhesive bonding or mechanical fastening, provided electrochemical and thermal coefficient incompatibilities are properly addressed. Welding, and to a lesser extent, brazing and soldering, require consideration of chemical, as well as electrochemical and C.T.E. compatibility. Chemical incompatibility is often overcome through the use of coatings, interlayers, or transition pieces to bridge the dissimilar joint elements.

Intermetallic compounds and long-range ordered (LRO) alloys, such as aluminides, silicides, borides, and some carbides, are most preferably joined by welding since they are often intended for extreme high temperature service. Hot cracking, due to inherent low ductility, and loss of long-range order and desirable properties are the two principal concerns.

Finally, thermal spraying can be used to join surface layers of metals, alloys, or intermetallics to other metals, alloys, intermetallics, or even ceramics, in any combination.

PRACTICE QUESTIONS AND PROBLEMS

1. What properties of metals and alloys make these such attractive materials for use in such diverse and demanding structural applications?

2. What is the predominant method for joining metals and alloys and why? What is the next most popular method for joining metals and alloys and why?

3. What are the six special challenges for joining metals and alloys that involve extremes?

4. Name the four most common refractory metals. What are some secondary refractory metals? What are the challenges associated with joining the refractory metals and their alloys? Give at least three challenges that are generic to all refractory metals and alloys.

5. What particular problems are associated with welding the refractory metals? What special problems are presented by the crystal structure that is characteristic of the refractory metals?

6. What are the considerations in mechanically fastening the refractory metals and alloys? Give at least two problems and their solutions.

7. What processes are typically used to weld the refractory metals and alloys? Explain why each of these is popular. If fillers are used, how are they generally selected?

8. What are the challenges associated with brazing the refractory metals and alloys? Propose some preferred brazing processes for use with the refractory metals.

9. Name the three most common reactive metals. What other metals are especially reactive? What causes metals to be classified as reactive?

10. What are the special problems associated with welding the reactive metals and alloys? What processes can be used successfully? What special modifications need to be made to conventional processes, if any?

11. What are the special problems associated with brazing the reactive metals and alloys? How are these problems resolved?

12. Do mechanical fastening or adhesive bonding overcome all of the problems associated with the reactive metals and alloys? If not, why not? What are the problems? If so, explain how.

13. What is meant by "heat-sensitive" metals and alloys? Give several different generic examples of heat sensitivity as well as metals or alloys that fall under each example.

14. What joining methods are preferred for use with heat sensitive metals and alloys? Explain.

15. If welding is used to join heat-sensitive metals and alloys, what processes are preferred and why? What special procedures need to be employed?

16. Why is the joining of cast irons a special challenge? What joining processes would you recommend as most preferred? Describe any special procedures that should be used.

17. What is meant by "heat-sensitive structures"? Give several examples of such structures.

18. How would you recommend joining very thin metals or alloys to other very thin metals or alloys? How, if at all, would you change your answer if you had to join very thin metals to thick metals?

19. What are the challenges associated with joining very small metallic components? How would you meet these challenges?

20. What are the challenges associated with joining very thick sections of metal? Rank order the processes of welding, mechanical fastening, adhesive bonding, and brazing for joining such structures. Give some specific problems with each general joining approach.

21. What are the special challenges of welding very large structures? Explain your answer. Give some solutions.

22. What problems are associated with welding dissimilar metals and alloys? What welding approaches are used to overcome these problems? What other joining approaches might you use?

23. What are considered "hostile environments" for joining? Why? What are the problems associated with mechanical fastening in such environments? What about with adhesive bonding?

24. What are "intermetallics"? Why are these materials important in modern engineering? What special problems arise in attempting to join these materials?

25. Why is thermal spraying attractive for applying metals and alloys to surfaces? What types of applications require thermal spraying with metals and alloys? How would you recommend spraying intermetallics?

REFERENCES

1. American Welding Society, *Welding Handbook,* Vol. 2, 8th ed., Miami: American Welding Society, 1991.

2. Cary, Howard B., *Modern Welding Technology,* 2nd ed., Englewood Cliffs, N.J: Prentice-Hall, 1989.

3. Messler, Robert W., Jr., and Paez, Carlos A., "Welding for Low-Cost Advanced Titanium Aircraft Structures," *Welding Technology for the Aerospace Industry,* Miami: American Welding Society,1980.

4. Klein-Wassink, R.J., *Soldering in Electronics,* Ayr, Scotland: Electrochemical Publications, 1984.

5. Landrock, Arthur H., *Adhesives Technology Handbook,* Park Ridge, N.J.: Noyes Publications, 1985.

6. Schwartz, Mel M., *Brazing,* Metals Park, Ohio: ASM International, 1990.

7. Schwartzkopf, Paul and Kieffer, Richard, *Refractory Hard Metals,* New York: Macmillan, 1953.

Joining of Ceramics and of Glasses

12.1 INTRODUCTION

12.1.1 Ceramics and Glasses Defined

Ceramics and glasses, like some metals (e.g., gold, platinum, silver, and copper) occur naturally, i.e., in a natural state. Most, however, are synthesized, especially those offering the highest strength, highest hardness, best toughness, greatest resistance to corrosion and heat, and the most unusual electrical, magnetic, and optical properties. As materials, ceramics and glasses have the longest history; beginning almost exclusively as clay, clay mixtures, and clay derivatives, then moving to refractory oxides such as dolomite, a mixture of magnesium oxide and calcium oxide, and within the last 30 to 40 years to so-called advanced ceramics, including structural ceramics, dielectrics, and electronic and magnetic ceramics.

It can be difficult to define ceramics because of their variety and diversity. Most generally, ceramics are all those solid materials that are *not* metals or polymers, i.e., neither metallic nor organic. More specifically, ceramics are usually defined as materials composed of nonmetallic, inorganic compounds, singly or in combinations as mixtures or alloys. They can be conveniently divided into two groups: oxides and nonoxides. The oxides are of metallic or metalloid (e.g., silicon, germanium, boron) elements, and include alumina (Al_2O_3), beryllia (BeO), magnesia (MgO), silicon dioxide (SiO_2), thoria (ThO_2), titania (TiO_2), urania (UrO_2), and zirconia (ZrO_2), as well as many less common types. The nonoxides include carbides (e.g., B_4C, SiC, TiC, WC), borides (e.g., TiB_2, MoB), nitrides (e.g., BN, AlN, Si_3N_4, TiN), silicides (e.g., $MoSi_2$, WSi_2), and beryllides (e.g., $ZrBe_{13}$, $MoBe_{12}$)[1].

Ceramics and glasses are distinguished from one another based on their methods of processing and their final structures. *Ceramics* are nonmetallic, inorganic solids of various compositions that have attained a crystalline state by the firing of crystalline

---

[1] Within this group there are nonmetallic, inorganic compounds of metals or metalloids with carbon (i.e., carbides), nitrogen (i.e., nitrides), and hydrogen (i.e., hydrides), as well as intermetallic compounds, such as the borides, silicides, and beryllides. Intermetallic compounds are sometimes considered ceramics, and sometimes are considered as a distinct and separate material group. In this book, intermetallic compounds are primarily addressed in Chapter 11, as a separate group along with metals and alloys.

inorganic, nonmetallic starting materials.[2] Ceramics are crystalline materials usually made from crystalline materials. *Glasses,* on the other hand, are noncrystalline solids produced by the firing of crystalline inorganic, nonmetallic materials (e.g., oxides, fluorides, borides, nitrides, and silicates). By using fairly recently developed techniques, some glasses can be converted to crystalline ceramics when certain nucleating agents are added to the glass constituents. Thus, glasses are noncrystalline, while ceramics are crystalline. Some consider glasses a subcategory of ceramics, while others consider glasses an entirely separate category of materials, since they can, in fact, be produced from starting materials that are polymers or metals as well as ceramics.[3]

Ceramics and glasses have many unique properties compared to metals and polymers because of their atomic structures and bonding. Ceramics consist of crystalline arrays of ions with bonds that are either ionic or covalent or somewhere in between.[4] Bonds are usually very strong, and electrons are, for the most part, associated with, or *tied,* to individual atoms in the crystal. This gives rise to the properties that make ceramics so attractive for so many applications where metals cannot perform as well or at all. Ceramics have high cohesive strength if microflaws induced during processing can be avoided[5] and extraordinarily high hardnesses up to the 8000 kg/mm² of diamond, the 5000 kg/mm² of cubic boron nitride, or the 2800 kg/mm² of silicon carbide. They also have high melting temperatures, i.e., they are refractory; have excellent chemical stability (i.e., they have good resistance to corrosion and oxidation); and are usually electrically and thermally insulative since electrons are tied up. Table 12.1 lists some of the properties of some important ceramics versus certain metals.

The unique, noncrystalline or amorphous structure of glasses gives them unique properties compared to crystalline ceramics and metals. Bonding within glasses is predominantly covalent or ionic or mixed, as in ceramics, so they share some properties in common with ceramics. The individual atomic species in glasses, however, are not arranged in regular, three-dimensional crystalline arrays, so they exhibit some unique properties. Common properties include high strength, especially in compression (again, due to the existence of microflaws from processing); high hardness; good chemical stability; and, usually, electrical and thermal insulating abilities (i.e., low electrical and thermal conductivity). Unique properties include the absence of a distinct melting point or range but, rather, a continuous softening with increasing temperature, and optical transparency or translucency, due to the absence of or lesser degree of scattering by periodic charge centers. Other attractive properties of ceramics and glasses include

---

[2] Some modern ceramics are produced by other means, i.e., without firing. Example processes are chemical or vapor deposition used in producing silicon carbides, silicon nitrides, and many of the borides.

[3] Metal glasses are produced by cooling metals from their molten state extremely rapidly, to prevent the rearrangement of the metal atoms into the preferred crystalline array. These metal glasses are highly metastable, while other glasses are less so.

[4] Recall that in ionic bonding, electrons are transferred from one atom (i.e., the positively-charged cation) to another (i.e., the negatively charged anion) to allow each atom to achieve a stable electron configuration. Bonding occurs by Coulombic or electrostatic attraction. In covalent bonding, atoms (or ions) share electrons in pairs or small groups (of 3 or 4 or so) and, in this way, achieve stable electron configurations, reducing overall system energy. In mixed ionic-covalent bonding, there is some exchange and some sharing.

[5] Because of the existence of microflaws in most ceramics produced in practice, compressive strengths are usually much higher than tensile strengths.

Table 12.1 Properties of Some Important Ceramics versus Some Metals for Comparison

Material	Melting Point (°C)	Modulus E (GPa)	Hardness (KHN) 100 g	K_{Ic} (MPa√m)	CTE (×10⁶/°C)	Resistivity (Ω-cm)	Thermal Conductivity (J/s-m·°K) 100°C	1000°C
Alumina-crystals	2050	380	2200	3-5	8-8	10^{10}-10^{12}	30	6.3
porcelain	"	370	"	"	"	"	"	"
sintered	"	370	"	"	"	"	"	"
Beryllia-sintered	2530	310	9.0 MOH	-	9.0	-	219	20
Boron carbide (B_4C)	2350	290	2800	3-4	4.5	0.5	-	-
Boron nitride (BN)	(3000)	83	230	-	-	-	-	-
Borosilicate glass	850	69	-	-	-	10^{13}	-	-
Graphite	3652	-	-	1-2	7.8	-	180	63
Magnesia-sintered	(2800)	210	5.5	3	13.5	10^{13}	38	7.1
Mullite^a-porcelain	1870	69	-	2-3	5.3	-	5.9	3.8
Silicon carbide (SiC)	2650	340-470	2500	3	4.7	10	-	-
Silicon nitride (Si_3N_4)	1900	-	9.0 MOH	4-5	2.1	-	-	-
Silica glass	1700	72	800	0.5	0.5	>10^{14}	2.0	2.5
Soda glass	700	70	550	0.7-0.8	9.0	-	1.7	-
Thoria (ThO_2)	-	-	640	-	9.2	-	10	2.9
Titanium carbide (TiC)	3250	310	3200	3-4	7.4	10^2-10^3	25	5.9
Zirconia (stabilized)	2690	150	7.0 MOH	9	6.5	10^{12}	2.0	2.3
Al	660.2	70	-	20-45	23.5	1.5×10^{-6}	221	-
Co	1495	185	-	300	12.3	5×10^{-6}	69	-
Ni	1453	180	-	350	13.3	3.5×10^{-6}	92	-
Steel	1536	200	-	60-140	12.1	10×11^{-6}	75	-
Ti	1668	117	-	55-115	8.4	-	-	-
W	3390	345	-	-	4.5	5.5×10^{-6}	201	-

^{a} $3Al_2O_3 \cdot 2 SiO_2$

retention of mechanical properties at high temperatures (in ceramics, but not in glasses), low coefficient of thermal expansion, and low density.

The same structures that impart useful properties to these materials, however, impose certain limitations or complications in processing. For example, ceramics are difficult to melt cast, generally impossible to shape by plastic deformation, difficult to machine, and difficult to join. Likewise, glasses pose some similar as well as some different problems.

As a group, ceramics and glasses fill an extremely important niche in materials engineering because of their exceptional or unique properties compared to metals, in particular. For this reason they are seeing ever increasing use in both traditional and advanced structural, electronic, magnetic, and optical applications. Joining, therefore, has become a major processing need. Once again however, it is extremes in properties that tend to make it difficult to join ceramics or glasses.

12.1.2 The Drivers for Joining Ceramics and Glasses

Joining of ceramics and of glasses to themselves, to one another, and to other materials, especially metals, is vitally important for many advanced, high-performance applications. These materials are often used for their unique optical, thermal, electrical, magnetic, chemical, or mechanical properties but rarely have the toughness to constitute an entire structure.[6] Therefore, components made from these materials must usually be structurally integrated into an overall system, which is often predominantly of metallic construction.

Ceramic-to-ceramic or glass-to-glass joining is also important for three general reasons. First, joining is needed to overcome processing size limitations. Both physical (i.e., material and facility) and economical constraints in producing ceramics and, to a lesser extent, glasses limit the size of components that can be made. Inherent brittleness in ceramics at all temperatures, and in glasses as they cool, leads to the formation of microflaws that degrade tensile properties and toughness in service and can cause gross failure during processing. Severe thermal gradients and fast cooling rates either during processing or later on in service can lead to gross failure. Thus, fabricating larger objects from ceramics or glasses requires joining smaller, easier-to-fabricate components.

Second, joining is needed to overcome processing shape limitations. Many ceramic parts, especially those having the most attractive engineering properties, can only be made in relatively simple shapes. This again relates to the inherent susceptibility of these brittle materials to process-induced flaws (e.g., from shrinkage from various sources, from differential thermal expansion, or from thermal shock). Thus, fabrication of more complex shapes requires machining, which is difficult because of the inherent high hardness of these materials, or joining. Obviously, joining is an attractive option.

Third, joining of ceramics and glasses enables material optimization. Some applications require more than one type of ceramic or glass to be combined in the design to obtain the optimum desired properties in the assembly. Often, it is preferable to have these properties in an integral (i.e., one-piece) component. Thus, joining of dissimilar ceramics or glasses becomes important.

Joining of ceramics or glasses to other materials or to ceramics or glasses of other compositions is usually motivated by technological needs, while joining of these materials to make larger or more complex shapes is usually forced by economic considerations in addition to or instead of technological needs. Figure 12.1 shows how ceramics and glasses are commonly joined to produce large, complex structures from smaller, simpler shapes or for producing hybrid structures from ceramics of different compositions.

12.1.3 Basic Joining Techniques for Ceramics and Glasses

There are essentially four basic joining techniques used with ceramics and glasses: one exclusively for use during the initial production of the ceramic or glass article and three for use in secondary processing or assembly. The first technique can only be used for joining a ceramic to another ceramic of the same or different composition, while the last

---

[6] Modern, so-called structural ceramics are being developed to have better toughness, in and of themselves (i.e., as monolithic materials), and through reinforcement by dispersed particles or second phases, by fibers or by laminates of other ceramics or materials (e.g., as ceramic-matrix composites).

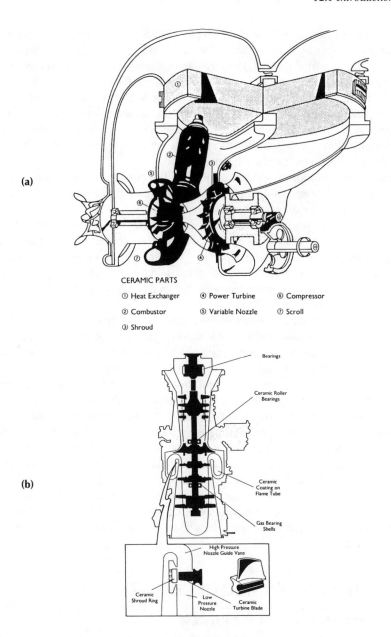

Figure 12.1 Schematic showing how ceramics are joined to produce larger, more complex structures or hybrid structures. (a) Automotive gas turbine and (b) helicopter gas turbine. (Reprinted with permission from Mel M. Schwartz, *Ceramic Joining*, ASM International, 1990, p. 167–169.)

three techniques can be used to join ceramics to ceramics, glasses to glasses, or ceramics or glasses to other materials.

The first technique is *sinter bonding,* a process that is almost exclusively restricted to the joining of ceramics. During the initial production of ceramic articles, it is possible, and often the practice, to join smaller, simpler shapes together to form larger and/or more complex shapes by cofiring them. The joint is created by diffusion, often, but not necessarily, during partial melting. Primary chemical bonds of the same type as found in the parent ceramics (i.e., ionic or covalent or mixed) are formed. The growth of grains across the initial interface between the abutting pieces obliterates the interface if the process is done properly. This growth is the result of ordinary sintering. This so-called sinter bonding can occur directly, without the aid of any intermediate material such as a glassy frit or a slurry of the powdered crystalline ceramic or mixed ceramics, or may require the use of such an intermediate material.[7]

The second technique is mechanical joining. It is possible to join ceramics to other ceramics or to other materials through the use of mechanical interlocking (i.e., physical features) or, to a lesser extent, mechanical fasteners.

The third technique is called *direct joining*. It is possible to join ceramics to other ceramics (or glasses to other glasses) and, to a far lesser extent, to other materials by welding, either by fusion or nonfusion (i.e., solid-phase) processes. The process is in complete analogy to metals. In such direct joining, no intermediate material is required.[8] This is the most common technique for joining glasses to other glasses of similar or different composition.

The fourth technique involves *indirect joining*. This is the most common technique for achieving high-integrity joints between ceramics or between ceramics or glasses and other materials. An intermediate bonding material is absolutely required and can be an organic adhesive (in adhesive bonding), glasses or glass-ceramic combinations (in frits), oxide mixtures (in cementing or mortaring), a ceramic (in ceramic brazing), or metals (in metal brazing or soldering, or even in solid-state diffusion bonding).

Elevated temperature serviceability is often a major factor leading to the selec-tion and application of a ceramic. In general, service temperature capability increases for ceramics joined by the preceding processes in the following, ascending order: organic adhesives, solders or mortars, mechanical fasteners and metal brazes, inorganic cements and metal or ceramic brazes, mechanical interlocks, welding, and sinter bonding.

In the following sections, the joining of ceramics to ceramics will be treated first, and then the joining of glasses to other glasses will be treated. Many of the techniques apply to both. The joining of ceramics or glasses to other materials (e.g., metals) is treated in Chapter 15.

12.2 MECHANICAL JOINING OF CERAMICS

12.2.1 Characteristics of the Process

Mechanical joining, as seen in Chapters 3 and 4, is basically quite simple and efficient and should, in principal, be capable of providing reasonably high strength joints that

---

[7] This technique for joining ceramics or glasses during their initial production could be grouped under either the *direct joining* or *indirect joining* secondary processing techniques, depending on how it is actually carried out, i.e., without or with an intermediate material.

[8] In actual practice, an intermediate is sometimes used in solid-phase joining by diffusion, i.e, diffusion bonding. This could be considered indirect joining but usually is not.

can tolerate sustained high temperatures. The characteristic lack of ductility in ceramics (and glasses) until quite high temperatures are reached, however, severely limits the temperature range over which this method of joining is applicable because of the severe stress concentrations that can arise and lead to catastrophic failure. The inability of ceramics and glasses to tolerate strain also severely limits the tensile stresses that can be sustained by mechanical joints.

In mechanical joining applications involving tensile loads, grommets or shims of a suitable soft material (e.g., soft metals like lead or copper) may be used to minimize stress concentrations by distributing the load and preventing damage to the ceramic's surface. Flow (i.e., creep) of such materials, however, limits the stresses that can be supported. High compressive loads can usually be tolerated much better than tensile loads, but considerable care must be taken to avoid the development of tensile components from bending or other complex loading conditions.

Widely different thermal expansion coefficients for ceramics (which have low C.T.E.s) and metals (which have high C.T.E.s) must be carefully considered to avoid the development of thermally induced stresses during temperature excursions. Mechanical connectors or fasteners must be selected to minimize differences in C.T.E.s, softening grommets or shims must be used, or the joint must be designed in such a way as to allow some slippage to accommodate any mismatch.

12.2.2 Mechanical Joining Methods

Despite the limitations just described, a variety of methods can be used to join ceramics mechanically to themselves or to other materials. Examples include (1) mechanical interlocking (e.g., tongue-and-groove, dovetails, "dogbone" connectors); (2) press or shrink fits (for close tolerance assemblies) that develop compressive residuals in the ceramic; (3) metal hangers or brackets or clamps that allow play or slip; (4) mechanical fasteners (e.g., bolts, screws, pins); and (5) integral threads on the body of the ceramic part. Some of these techniques are shown in Figure 12.2.

It should be noted that, as with refractory metals and alloys, the major problem in mechanical fastening ceramics is accommodating large dimensional changes caused by thermal expansion and contraction over wide temperature excursions in processing or service. For this reason, both mechanical interlocking schemes and schemes using fasteners must be able to accommodate these sometimes large movements. One approach is to use loose-fitting joints, with built-in play, or slop. Another approach is to design the joint and the mechanical interlock or fastener to allow slip. These are called *slip joints* and are similar in concept to the joints found on roadways over bridges or on railroad tracks, where expansion and contraction from temperature changes can cause gross movements. Some examples of slip joints are loose-fitting keyed joints (e.g., dovetails or dogbones); slotted (versus round) fastener holes, to allow movement in at least one direction; spring-loaded interlocks or fasteners, to take up dimensional changes; and inherently flexible joints, such as ball-and-socket types.

Mechanical fasteners must be used quite carefully to avoid excessive loading in bearing and to avoid further aggravating normal stress concentration effects. Tensile loads must be kept low at all cost.

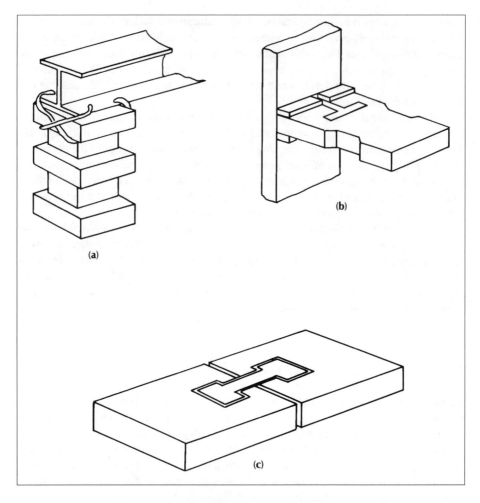

Figure 12.2   Illustration of various mechanical interlocking and fastening methods for ceramics and glasses. (a) Metal hangers; (b) metal bracket; (c) refractory or metal dog bone.

12.3 ADHESIVE BONDING, CEMENTING, AND RELATED JOINING OF CERAMICS

12.3.1 Adhesive Bonding or Joining of Ceramics

The indirect joining of ceramics with organic-based materials, such as sealants, elastomers, or adhesives, is one of the simplest and most widely used methods of joining ceramics to other ceramics of like or unlike composition, as well as to organic materials (such as plastics) or to metals. These processes offer manufacturing advantages of low cost, high speed, and simplicity, as well as two very important mechanical advantages: (1) adhesives distribute applied loads, thereby minimizing potentially de-structive stress concentrations, and (2) most adhesives can accommodate large strains without significant development of stress in the ceramics because of their low elastic moduli.

Despite these advantages, two serious shortcomings are associated with organic adhesives. First, the strengths of organic adhesives, and especially sealants and elas-tomers, which are generally nonstructural adhesive types, are low. While true structural adhesives can achieve strengths of 35 MPa (or 5,000 psi), sealants and elastomers are commonly an order of magnitude weaker. This may not be a particular problem if either the total loading is low or sufficient bond area is available to allow the required load to be carried. Second, the service temperature of an adhesively bonded assembly is severely limited for most ceramics. Organic adhesives can typically tolerate only a few hundred degrees Celsius, while the useful service temperatures of ceramics can be sev-eral thousand degrees. Thus, organic adhesives are normally used to join ceramics when elevated temperature service is not the reason for the ceramic's selection but rather special electrical or wear properties are needed.

Adhesives for use with ceramics (or glasses) include polyvinyl butyral, phe-nolic butyral, neoprene, polysulfide, silicone, vinyl acetate, epoxies (usually with about 200°C or 400°F limits), and epoxies modified with elastomers. For high tem per-atures (i.e., 300°C or 575°F), polyimides, polysulfones, polyphenylquinoxalines, and polytriazines are used. One important application of organic adhesives is for joining glasses to one another and to metals, as in forward-facing windows in commercial air-craft.

12.3.2 Cement and Mortar Joining of Ceramics

Cement-mortar type interlayers are widely used for the simple and relatively inexpensive indirect joining of ceramics to one another (as for joining bricks or cement blocks). Cements and mortars are inorganic materials consisting of various mixtures of glasses, glasses and crystalline ceramics, or just crystalline ceramics. They are inorganic adhesives.

Cementing and *mortaring* are also known generically as *ceramic sealing* or *ceramic sealant bonding*. As a group, cements and mortars offer higher temperature serviceability than organic adhesives and still offer most of the advantage of load spreading but little or no strain accommodation offered by their organic counterparts. The strengths of cements and mortars can be substantially lower than the strengths of adhesives, however.

Cements and mortars fall into two broad categories: (1) unfired ceramic cements, and (2) fired ceramic cements. The unfired ceramic cements and mortars offer good bonding and insulating and sealing capabilities from room temperature to temperatures approaching refractories. As stated previously, they are basically inorganic and provide good adhesive bond strength, good corrosion resistance, and electrical and thermal insulation. There are three basic types of unfired cements: (a) hydraulic set, (b) air set, and (c) chemical set. Some useful properties of these, as well as some fired cements, are given in Table 12.2.

Table 12.2   Properties of Unfired and Fired Ceramic Cements

Property	Adhesive Cement	Electric-Heater Cement	Hi-Temp Cement	Electrotemp Cement
Type set	Air	Air	Air	Chemical
Density (lb/in^3)	110	110	110	110
(gm/cc)	(1.76)	(1.76)	(1.76)	(2.56)
C.T.E.[a] in./in./°F	6.2×10^{-6}	6.5×10^{-6}	6.0×10^{-6}	2.6×10^{-6}
(cm/cm/°C)	(11.2×10^{-6})	(11.3×10^{-6})	(10.8×10^{-6})	(4.68×10^{-6})
Comp strength, psi	3900	2700	3500	5000
(kg/cm^2)	(274)	(189)	(246)	(316)
Dielectric constant	3.5–6.0	5.0–7.0	3.5–6.0	3.0–4.0
Dielectric strength, volts/mil (volts/mag)				
70°	12.5–51.0	12.5–51.0	12.5–51.0	76.0–101.5
(21°C)	(490–2000)	(490–2000)	(490–2000)	(2900–3900)
750°	15.0	15.0	15.0	25.0–38.0
(399°C)	(588)	(588)	(588)	(980–1490)
1475°F	1.3	3.8	1.3	12.5–25.0
(801°C)	(51)	(149)	(5.1)	(490–980)
Max service temp. °F	1800	2500	2500	2600
(°C)	(982)	(1370)	(1370)	(1426)
Modulus of rupture, psi	460	320	420	450
(kg/cm^2)	(32)	(22)	(29)	(31)
Shear strength, psi	710	300	600	365
(kg/cm^2)	(49)	(21)	(42.2)	(26)
Tensile strength, psi	410	285	400	250
(kg/cm^2)	(28)	(20)	(28.1)	(17)
Vol resistivity ohm-cm				
70°F (21°C)	10^8–10^9	10^7–10^8	10^8–10^9	10^{10}–10^{11}
750°F (399°C)	10^4–10^5	10^4–10^5	10^4–10^5	10^9–10^{10}
1475°F (801°C)	10^2–10^3	10^2–10^3	10^2–10^3	10^8–10^9
Thermal conductivity, BTU/ft^2/hr/°F/in.	4.7	5–8	7.0	10–12
$\frac{gm\text{-}cal}{sec\text{-}cm\text{-}°C}$	(1.6×10^{-3})	$(1.7\text{–}2.7 \times 10^{-3})$	(2.4×10^{-3})	$(3.4\text{–}4.1 \times 10^{-3})$
Resistance				
Acids	Yes except hydrofluoric	Yes except hydrofluoric	Yes except hydrofluoric	No
Alkalies	No	No	No	No
Water	Wash with acid	Wash with acid	Wash with acid	No
Oil	Yes	Yes	Yes	Yes
Electricity	Yes	Yes	Yes	Yes

[a]Coefficient of thermal expansion.
Reprinted with permission from C. E. Zimmer, "Engineer It with Cement," *Materials Engineering,* Sept. 1984, pp. 34–35.

Hydraulic-set cements contain various anhydrous compounds that react with water to form hydrated compounds. The major types are portland, calcium aluminate, natural lime-silica, and combinations thereof. Portland cement functions to about 260°C (500°F), while calcium-aluminate is a more refractory mortar. Air-set cements are inorganic compounds, often mixed oxides, that are blended with water and sodium or potassium silicate.

Electrical-Insulating Cement	Low-Expansion Cement	Acidproof Cement	Electrical-Resistor Cement	Electrical Refractory Cement
Chemical	Chemical	Chemical	Air	Hydraulic
160	141	121	116	156
(2.56)	(2.25)	(1.94)	(1.85)	(2.5)
3.1×10^{-6}	4.6×10^{-6}	6.3×10^{-6}	13.0×10^{-6}	8.0×10^{-6}
(5.58×10^{-6})	(8.28×10^{-6})	(11.3×10^{-6})	(23.4×10^{-6})	(1.43×10^{-6})
4000	3900	2200	3300	3500
(281)	(274)	(154)	(232)	(246)
3.0–4.0	5.0–7.0	5.0–7.0	3.5–4.5	-
-	25.0–51.0	12.5–38.0	12.5–51.0	-
-	(980–2000)	(490–1490)	(490–2000)	-
-	12.5–25.0	12.5–38.0	12.5–25.0	-
-	(490–980)	(490–1490)	(490–980)	-
-	1.3	2.0	7.5	-
-	(51)	(78)	-	-
2200	1550	1750	2600	2600
(1204)	(843)	(954)	(1429)	(1429)
510	460	455	460	500
(35)	(32)	(32)	(32)	(35)
400	430	430	375	210
(28)	(30)	(30)	(26.4)	(15)
290	425	430	325	400
(20)	(29)	(28)	(22.8)	(28)
-	10^7–10^9	10^9–10^{11}	10^8–10^9	-
-	10^4–10^6	10^7–10^8	10^5–10^6	-
-	10^2–10^3	10^2–10^3	10^3–10^4	-
8–11	8	4–6	11	12
$(2.7–3.8 \times 10^{-3})$	(2.7×10^{-3})	$(1.4–2.0 \times 10^{-3})$	(3.8×10^{-3})	(4.1×10^{-3})
No	Yes except hydrofluoric	Yes except hydrofluoric	Yes except hydrofluoric	Above pH 5
No	No	No	No	Yes
Yes	Yes	Yes	Wash with acid	Yes
Yes	Yes	Yes	Yes	Yes
Yes	Yes	Yes	Yes	Yes

Setting occurs by evaporation of the water, and service temperatures for cementing applications approach 1100°C (2000°F). The chemical-set cements involve a broad range of chemical reactions but predominantly acid-alkali types to form gels. Various siliceous compounds as well as phosphates, chlorides, and sulfates are used. With all these types, shrinkage can be 1–5% and must be accommodated by design and processing.

Fired ceramic cements include (1) vitreous ceramic sealants composed of glazes and other ceramic materials that form stable glasses on firing, and (2) crystalline ceramic sealants comprising recrystallized glasses and crystalline compounds that contain little or no glassy phase. Both of these materials require high maturing temperatures for joining shapes but offer excellent refractoriness. Service temperature capability is excellent, however.

Some considerations need to be taken into account with cements and mortars. First, thermal expansion mismatch between the cement or mortar and the ceramic can be an important factor limiting service temperature. This must be taken into account during design of the joint. Second, thermal decomposition of the cement or mortar (e.g., lime-based mortar) can be a problem. This decomposition is usually what limits service temperature. Third, chemical interaction between the cement or mortar and the ceramic can cause problems. Degradation of the joint strength is most common, but degradation of the base ceramic can also occur. Fourth, unfired ceramic sealants may not provide hermeticity, as they tend to be porous and permeable to gases.

Cements and mortars are typically selected to have compositions that are similar to the ceramics they are to bond. (An excellent reference is "Cementitious Bonding in Ceramic Fabrication," by J.F. Wygant, Reference 10.)

12.4 BRAZING AND SOLDERING OF CERAMICS

12.4.1 General Comments

As defined in Chapters 8 and 9, and by the American Welding Society, soldering and brazing are joining processes that create bonds through the capillary flow of molten fillers between solid substrates. Soldering refers to those processes that are carried out with fillers that melt and flow below 450°C (840°F), while brazing refers to those processes that are carried out with fillers that melt and flow above this temperature. Based on this definition, most indirect joining processes using metal intermediate materials applied to ceramics are brazing rather than soldering, but the latter term is appropriately used for sealing with glasses (see Section 12.8.4), even though it is occasionally done over 450°C.

A number of brazing processes for ceramics apply a metal layer to the ceramic, referred to as *metallizing*, with subsequent brazing to join the metallized ceramic to another metallized ceramic or to a metal.

12.4.2 Characteristics of Brazing Methods for Ceramics

There are fundamentally two general methods for brazing ceramics: (1) those that use metallic materials as the intermediate, called *metal brazing*, and (2) those that use ceramic materials as the intermediate, called *ceramic brazing*.

For metal brazing, there are three specific methods that are not necessarily fundamentally different in terms of the mechanism of bonding but are differentiated by practice. These are (1) noble-metal brazing, (2) active-metal brazing, and (3) refractory-metal brazing. Metal brazing of ceramics is one of the most common methods of joining ceramics, especially for high-performance applications. The basic advantage of metal brazing is that a variety of materials can be joined together by generally simple procedures, producing vacuum-tight joints that have modest to substantial strength.

For ceramic brazing, the ceramic filler actually fuses or melts and distributes in the joint by capillary action. This distinguishes it from ceramic cementing, where no such overt melting and no capillary flow occurs. An advantage of ceramic brazing over metal brazing is the closer match of properties (especially, physical properties like C.T.E.) between the filler and the substrates.

12.4.3 Metal-Brazing of Ceramics

The obvious problem in trying to braze or solder ceramic substrates is achieving wetting by the molten filler. When a metal filler is used, the problem is overcome either by altering the surface of the ceramic to make it like, or make it act like, a metal or by getting the filler to act like a ceramic. The foregoing three major methods of brazing ceramics using a metal filler or intermediate layer are described in more detail in this section. It can be seen that in noble-metal brazing and in one form of active-metal brazing, the metal filler is made to act like a ceramic by having its components oxidize, while in another form of active-metal brazing and in refractory metal brazing, the surface of the ceramic is chemically altered by the metallic braze filler to act like a metal. By metallizing the surface of the ceramic by depositing or embedding metal by electroplating, sputtering, ion-implanting, or some other means, brazing with a metal filler can be accomplished as it normally is accomplished with metal substrates, that is, by simply selecting a filler that is compatible with the metallized surface material.

Noble-Metal Brazing

Noble-metal brazes are most commonly based on silver or platinum and their alloys with other noble metals (e.g., palladium and gold) but frequently include copper or nickel as base metals. Such brazing is normally done in air, with evidence that noble-metal oxides form and bond with the ceramic substrate(s). Typical ceramics that have been brazed with platinum, palladium, gold, or silver, with little or no pressure include MgO, Al_2O_3, ZrO_2, UO_2, BeO, ferrite, SiO_2, glasses, and graphite.[9]

Noble-metal braze joints have strengths that are approximately 50–100% of the strength of epoxy-bonded joints. Strengths ranging from 23 MPa (3,400 psi) for lead to 252 MPa (36,000 psi) for platinum in Al_2O_3 have been reported, however.

Active-Metal Brazing

Active-metal brazing is most commonly based on the use of titanium but may also be based on zirconium, niobium, chromium, or yttrium, particularly for Al_2O_3. Iron,

---

[9] Some consider graphite a ceramic, while others do not. In this book, the joining of graphite is discussed in chapter 12 and in chapter 14.

cobalt, and nickel have also been used with certain oxide ceramics. Two procedures have been used with traditional titanium-based active brazing. In the first, the titanium (or other active metal) is incorporated into the brazing alloy to aid wetting by forming an oxide or other compound (e.g., carbide, boride) through reaction of the active metal with the ceramic. In the second, the surface of the ceramic to be brazed is coated with the active metal or a compound that decomposes to that metal (e.g., a metal hydride, such as TiH_2). In this second approach, done in a vacuum, the metal layer is referred to as *metallization*. This layer is then wet by an appropriate metal braze alloy.

Another method of applying active metals (per Pattee *et al.*, Reference 5) is to use molten alkali halides salts (e.g., alkali or alkaline earth halides on titanium) to prevent oxidation while the active metal is deposited on the ceramic substrate. In a diffusion brazing approach, brazing is accomplished below the liquidus of the braze filler alloy, through the formation of a transient liquid phase.

For graphite, a particularly difficult material to join, Ag-Cr, Ag-Ti, Ag-Zr, Au-Zr, and Cu-Cr alloys worked well by forming carbides with the active component. Two commercially available braze fillers for use with graphite are 68.8Ag-26.7Cu-4.5Ti (with a melting range of 830–850°C or 1525–1560°F) and 70Ti-15Cu-15Ni (with a melting range of 910–960°C or 1670–1760°F). Because graphite reacts readily with oxygen, oxygen must be excluded, usually by brazing under a vacuum of approximately 10^{-4} Torr. The reaction of an active component and subsequent diffusion is a common means for achieving wetting and bonding by brazing difficult-to-wet substrates.

Refractory Metal Brazing and Metallization

Although some brazing is actually done using refractory metals and their alloys (e.g., Al_2O_3 or Si_3N_4 to themselves using niobium and zirconium, respectively), these materials are usually used to coat or "metallize" a ceramic with subsequent brazing to the metallized layer (often with a second metal applied by plating). Tungsten and molybdenum are normally used as the refractory metal metallization. The process requires high temperatures (e.g., 1400–1600°C or 2600–2900°F) and a hydrogen atmosphere. The refractoriness of the resultant braze, along with high strength and reliability, make this an attractive option, especially for Al_2O_3 and BeO. Strengths are typically 105 to 150 MPa (15,000 to 30,000 psi) for molybdenum brazing of Al_2O_3 at 1500–1600°C (2750–2900°F).

The operative refractory metal needed to metallize the ceramic's surface can be embedded as a powder during initial processing, mixed with the parent ceramic and fired to apply a coating, chemically or vapor deposited, or sputtered on. Another common method of metallizing with a refractory is called the *molybdenum-manganese (MoMn) process*. Here, a paint of molybdenum and manganese metal or their oxides is applied to the ceramic as a powder slurry. The assembly is fired in hydrogen with a controlled dew point so that the manganese is converted to its oxide (MnO) while the molybdenum remains as a metal. The MnO then reacts with the ceramic grains and any liquid glassy phase to form a controlled amount of glassy phase. The molybdenum sinters to form a porous coating into which the glassy phase penetrates and interlocks mechanically. In addition, the glass at the interface reacts with the molybdenum to form MoO, thus forming a chemical bond to the ceramic grains since they are compatible. An electrodeposited coating (e.g., nickel) is often plated over this metallized layer to further facilitate brazing. Alumina and beryllia substrates have been metallized this way using either molybdenum with manganese or tungsten with manganese.

12.4.4 Ceramic-Brazing of Ceramics

Ceramics can be joined to themselves (or, as will be seen in Chapter 15, to metals) by brazing with ceramic fillers or intermediate materials instead of metals. Glasses are common ceramic braze filler materials, but mixtures of glasses and crystalline phases, or all crystalline phases, often as eutectics, can be used. Ceramic brazes can be applied and processed in the same fashion as metal brazes, i.e., in the solid state by diffusion brazing or in the fluid or liquid state by conventional brazing. While ceramic brazes tend to be preplaced before rather than applied during brazing, they do distribute uniformly within the joint by capillary action, just as all brazes must by definition.

Ceramic brazes provide good environmental compatibility (e.g., service temperature and corrosion resistance); often better than most metals. Unfortunately, ceramic brazes tend to be less tolerant of thermal expansion mismatch than most metals, so care must be exercised in their selection and use and in joint design.

One common ceramic braze filler is Pb-Zn borosilicate glass. Some refractory ceramic brazes are manganese pyrophosphate and MnO eutectic (30:70), Al_2O_3-MnO-SiO_2 for Al_2O_3, and Al_2O_3-CaO-MgO-SiO_2 for Al_2O_3.

Normally, ceramic brazing is accomplished using the furnace brazing method with appropriate atmosphere control, which may include vacuum.

12.5 WELDING OF CERAMICS

12.5.1 General Comments

Welding is really the oldest technique for joining a major group of ceramics, namely silicate glasses. The joining of two or more glass shapes to one another in the practice of glass working, though not usually referred to as such, is really welding. Welding of ceramics, in general, and very refractory, single-phase crystalline ceramic materials, in particular, is relatively new.

The basic requirements for welding ceramics are twofold. First, the ceramics involved in the joint must be chemically compatible with each other and with the environment in which they will be joined. Second, the ceramics being joined must be mechanically compatible (i.e., have reasonably comparable strengths and coefficients of thermal expansion (i.e., within 1 to 1.5×10^{-6} mm/mm/°C or 0.5 to 0.8 in./in./°F).

The two general methods for welding ceramics are (1) in the solid-state using nonfusion processes, and (2) using melting with fusion processes.

12.5.2 Solid-State Welding of Ceramics

Solid-state or nonfusion welding is accomplished by heating the components to be joined while they are held in contact, usually under pressure. For metals, which are often inherently ductile, this pressure causes plastic deformation of surface asperities or localized points of contact, bringing more points into contact. With more points in contact, there are more paths for solid-state diffusion. A filler is rarely needed or used. For ceramics, such plastic deformation is difficult or impossible, with only limited elastic deformation able to contribute to increasing the number of points of contact for subsequent diffusion. Thus, in the solid-state welding of ceramics, an intervening layer of ceramic in powdered form is often sandwiched between the joint components. The powder can be of the same composition as the two bodies being joined if they are of the same composition, or of one or the other or a mixture of both if the two bodies are of different compositions. In extreme cases, where the two materials being joined are chemically or physically incompatible, a series of thin layers of powder, grading from one composition to the other, can be used. Figure 12.3 shows a schematic concept for grading powdered fillers between two incompatible ceramics to be solid-state welded.

Bonding during solid-state welding occurs by creation and growth of a reaction zone between the two materials to be joined and/or by sintering, both of which rely on diffusion. When dissimilar materials are being joined, bonding will usually depend on the reaction and/or interdiffusion of both components. With similar materials, it depends primarily on the sintering ability of the materials, i.e., diffusion and grain growth. The principal process used for accomplishing solid-state welding of ceramics to ceramics is diffusion bonding, in the form of hot-pressing and isostatic pressing. The process of dif-

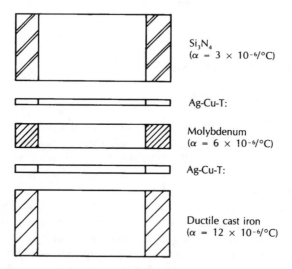

Si_3N_4
$(\alpha = 3 \times 10^{-6}/°C)$

Ag-Cu-T:

Molybdenum
$(\alpha = 6 \times 10^{-6}/°C)$

Ag-Cu-T:

Ductile cast iron
$(\alpha = 12 \times 10^{-6}/°C)$

Figure 12.3   A graded seal assembly or graded joint for solid-state welding between two incompatible ceramics.

fusion bonding for ceramics is essentially identical to that for metals (Chapter 6, Section 6.5.5). Various friction welding processes have also successfully joined ceramics as well.

A wide variety of oxide and nonoxide ceramics have been solid-state welded to themselves, to one another, and to other metals. Table 12.3 summarizes solid-state welding of oxide and nonoxide ceramics.

12.5.3 Fusion Welding of Ceramics

As with metals, fusion welding of ceramics is achieved by filling the joint between parts with molten material obtained by melting the edges of the components making up the joint while in contact, i.e., autogenous welding, or with melt from a filler of similar or compatible materials. In the fusion welding of dissimilar materials, their melts must also be compatible with one another. Besides chemical compatibility between substrates and any filler, the ceramics being fusion welded must be compatible with the welding environment and must be physically compatible with one another.

While compatibility with the welding environment depends somewhat on the particular fusion technique being used, an overall requirement is that the ceramics melt properly and, then, solidify properly. There are several problems associated with attempting to fusion weld ceramics. First, some ceramics (e.g., BN, Si_3Ni_4, and SiC) vaporize without melting (i.e., they sublime), so, at normal pressures, cannot be fusion welded. This problem may be overcome by solid-state welding. Second, some ceramics (e.g., MgO) have very high vapor pressures at their melting points, so they vaporize after they melt, making fusion welding difficult. This problem may be overcome by solid-state welding. Third, the very high temperatures involved in the fusion of many ceramics can cause problems with phase transformations in surrounding heat-affected areas, leading to fracture or severe property degradation. This problem may once again be avoided by solid-state welding. Fourth, thermal stress fractures, resulting from severe temperature gradients around the fusion zone or from thermal shock on heating or cooling, have long been co sidered a major obstacle to fusion welding of ceramics. Supplemental heating around the weld, to reduce thermal gradients, helps greatly. This problem may not be overcome by solid-state welding, depending on the process.

The principal processes for fusion welding ceramics, in descending order of popularity, are (1) laser-beam welding (primarily CO_2, but also Nd:YAG), (2) electron-beam, and (3) arc welding (especially plasma arc and gas-tungsten arc). While laser- and electron-beam welding can be attempted with any ceramic, arc welding is restricted to ceramics that are electrically conductive, at least at elevated temperatures. Plasma arc welding can be performed on nonconductive ceramics, provided the ceramic weldment is not part of the electrical circuit, i.e., the arc is nontransferred. Table 12.4 gives a brief summary of some successfully fusion welded ceramics and ceramic combinations.

Microwave joining of ceramics is also showing promise. Here, microwave energy at very high frequencies (>2.45 GHz) is absorbed by the ceramic, raising its temperature to allow joining by reaction and sintering, with or without filler, with or without fusion.

Table 12.3 Summary of Solid-State Welding of Oxide and Nonoxide Ceramics

Materials Welded	Type[a]	Temperature (°C)	Pressure (1000 psi)	Time (min)	Ambient Weld Strengths (1000 psi)
Oxide					
MgO-MgO	C to C	1750–1950	3–5	5	7–17
Al$_2$O$_3$-Al$_2$O$_3$	C to C / C to C / C to P / P to P	1700	1	30–40	-
Al$_2$O$_3$-Al$_2$O$_3$	C to C	1900–2000	0.1–1	60–70	-
Al$_2$O$_3$-Al$_2$O$_3$, Al$_2$O$_3$ + 1 wt-% MgO Al$_2$O$_3$ + 1 wt-%MgO	C to C	1400	2	60	-
Al$_2$O$_3$-Al$_2$O$_3$	C to C	1600	0.6	180	-
Al$_2$O$_3$-Al$_2$O$_3$	P to P	-	-	-	up to 54
Al$_2$O$_3$-Al$_2$O$_3$	P to P	1300–1400	8–10	20–30	30–45
Al$_2$O$_3$-Al$_2$O$_3$	P to P	1500	0.7	60	-
Al$_2$O$_3$-Al$_2$O$_3$ + H$_3$PO$_4$ or AlF$_3$	P to P	900–1200	0.7	60	20-50 estimated
NiO-NiO	P to P	1000	10	90	-
MgO-MgO	P to P	800–1100	10–12	30	1–15
MgO-MgO	P to P	1300–1400	5	15	-
CaO-CaO (with 2 wt-% LiF)	P to P	1000-1100	5	15	-
CeO$_2$-CeO$_2$	P to P	1000	-	-	-
Al$_2$O$_3$-Nb	P to P	-	-	-	up to 11.7
Al$_2$O$_3$-Cu	P to P	1000	0.3	10	10–28
Al$_2$O$_3$-Ni, Fe, etc.	P to P	1250–1300	0.3	10	7–28
Al$_2$O$_3$-Ni	C or P to P	1200	-	-	-
Al$_2$O$_3$-Nb	P to P	1500	2.1	17	11–13
Al$_2$O$_3$-ZrC, ZrN, or ZrB$_2$	P to P	1750	3	60	-
ZrO$_2$ or ZrB$_2$	P to P	1800	3	60	-
Nonoxide					
LiF-LiF	C to C	840–860	0		
B$_4$C-Si					
TaC-TaC	P to P				
(ZrC, NbC, TaC)-(Nb, Ta, Mo, W)	P to P	1200–2000	0.7	5–60	
ZrC-ZrB$_2$					
ZrN-ZrB$_2$	P to P	2100	4	30	
ZrC-ZrN					
TiC-TiB$_2$					
TiC-TiN					
TiB$_2$-TiN	P to P	2100	4	60	
TiC-ZrB$_2$					
C-C	P to P				

[a]C, crystal; P, polycrystal

Reprinted with permission from R. W. Rice, "Joining of Ceramics," *Advances in Joining Technology*, John J. Burke *et al.,* (eds.), Brook Hill Publishing, 1976, pp. 97–98.

Table 12.4   Summary of Successfully Fusion Welded Ceramics and Ceramic Combinations

Welded Parts	Weld Method	Representative Weld[a] Strengths Achieved (1000 psi)
Al_2O_3-Al_2O_3	Electron beam	15–30
Al_2O_3-Ta	Electron beam	5–10
Al_2O_3-Al_2O_3	Laser	25
ZrB_2-ZrB_2	Arc	20–60
TaC-TaC	Arc	20–30
ZrB_2 + C or SiC to itself	Arc	~10
W-graphite	Arc	3–5
ZrB_2-Mo, Nb, or Ta	Arc	15–30
ZrB_2 + SiC-Mo or Ta	Arc	15–30
ZrB_2-graphite	Arc	~2
Fireclay, brick to itself	Arc (with feed material)	0.2–1.5

[a]These strengths generally represent only preliminary trials, and may often be limited by the quality of the bodies to be welded.

Reprinted with permission from R. W. Rice, "Joining of Ceramics," *Advances in Joining Technology,* John J. Burke *et al.,* (eds.), Brook Hill Publishing, 1976, pp. 97–98.

12.6 OTHER METHODS FOR JOINING CERAMICS TO CERAMICS

The hybrid joining processes of weldbonding, weldbrazing, and rivet-bonding have not been applied to the joining of ceramics to other ceramics or to other materials (such as metals). The process of thermal spraying, however, has been used successfully. Thermal spraying is used to apply ceramic coatings to other ceramics or to other materials (principally metals), or to restore dimensions to ceramic parts by depositing the same ceramic. While this is not a joining process in the normal sense of the word, i.e., to assemble, attach or unite parts, it is joining in the sense of uniting materials. Obviously, the thermal spraying process can be used to apply metallization to ceramic surfaces to facilitate metal brazing.

Normally, the higher heat-intensity thermal spraying methods are used with ceramics because of their inherent refractoriness. Plasma spraying is most commonly used, but the detonation-flame process has also been used, to take advantage of the extraordinary high kinetic energy achieved. With both processes, bonding results from the impact of molten or softened particles of ceramic with a clean, heated substrate. Adhesive strength can be comparable to the inherent cohesive strength of the base ceramic.

An ever increasing variety of ceramics is being applied by thermal spraying, including alumina, zirconia, yttria, and silicon carbide. Table 10.1 in Chapter 10 lists some of the ceramics that have been successfully sprayed.

A final method worth mentioning is SHS joining or pressurized combustion synthesis joining. This is an exothermic brazing process (see Section 8.4.8) that can operate in the solid-state or with some liquid phase formation. While still very much the subject of research, SHS joining is attractive for joining simultaneously with material synthesis and shaping (by reactive sintering) or as a secondary operation.

12.7 COMPARISON OF JOINING TECHNIQUES FOR CERAMICS

It is worth reviewing the various methods for joining ceramics to one another, as well as the relative merits of each method. Table 12.5 gives a summary comparison of the various methods for joining ceramics to ceramics.

Briefly, the primary advantage of organic adhesive bonding, inorganic cementing or mortaring, and, to some extent, mechanical fastening is their ease, versatility, and low cost. Strength is good. Temperature capability is extremely limited for organic adhesives but not bad for some inorganic cements and for mechanical interlocking or fastening. Sealing, for hermeticity, can be a problem with porous, unfired ceramic cements. Low-stress, low-temperature service is good.

Brazing, especially with active-metal type fillers, is most widely used for demanding applications. The obvious challenge is achieving good wetting. Strength is quite good. Sealing is excellent. Temperature stability is good, and chemical resistance is fair to good. Soldering proper is generally not performed.

Welding offers the ultimate in temperature capability, strength, and environmental stability. Solid-state processes, because of their lower temperatures, offer the greatest strength, i.e., up to 420 MPa (60,000 psi).

With every joining process, great care must be given to matching the low thermal coefficient of expansion of most ceramics with other materials. The C.T.E.s of dissimilar ceramics must also be closely matched to avoid failure by fracture.

Table 12.5   Summary of Various Methods for Joining Ceramics

					Welding	
Requirement	*Adhesive*	*Cement*	*Mechanical*	*Brazing*	*Diffusion*	*Fusion*
Strength (numbers in 1000 psi)	Low, e.q. < 5	Usually low, often below 1	Low-medium, e.g., 1–10	Good, e.g., 10–40	Often best, e.g., 20–60+	Good, e.g., 10–50
Compatibility with severe environments	Generally poor[a]	Poor-medium	Poor-medium	Medium-good	Medium-good	Generally best
Vacuum tight	Questionable[b]	Generally not	Generally not	Usually	Frequently[c]	Usually
Cost	Low	Low	Low-medium	Medium-high	Medium-high	Low-high[d]

[a]e.g., due to temperature and some chemical limitations.
[b]e.g., due to outgassing of organic constituents.
[c]May not always be as reliable as brazing or fusion welding.
[d]Can depend substantially on the number of parts, material, and type of weld, e.g., low for considerable arc welding but would be high for small numbers of electron beam welds.
Reprinted with permission from R. W. Rice, "Joining of Ceramics," *Advances in Joining Technology,* John J. Burke *et al.,* (eds.), Brook Hill Publishing, 1976, p. 107.

12.8 JOINING OF GLASSES

12.8.1 General Comments

Glasses are amorphous solids that exhibit no distinct melting point nor associated discontinuous volumetric change at some particular temperature but, rather, a continuous decrease in specific volume and increase in viscosity with decreasing temperature. The rate of decrease of specific volume with decreasing temperature decreases at some temperature, known as the *glass transition temperature, T_g*. Below the T_g, a glass behaves like a brittle solid. Bonding within glasses is primarily covalent, although some ionic and mixed primary bonding and some secondary bonding also occurs. While strength can be quite high, easily exceeding 700 MPa or 100 ksi, it is far higher in compression than in tension. Unfortunately, ductility is quite low. This inherent brittle behavior is related to the presence of microflaws induced during processing, similar to the those found in crystalline ceramics.

The properties of glasses are as unique as their structure. The lack of periodicity of atoms (i.e., location on precise lattice sites) drastically reduces scattering of electrons and photons, so optical properties are unique, usually resulting in transparency or translucency in the visible or near-visible range. Electrical properties can also be unique, although glasses are usually dielectric or insulative. Glasses generally exhibit good chemical and environmental stability and are usually thermally insulating.

Like ceramics, described earlier, glasses can be joined by several methods, including, in approximate decreasing order of frequency, (1) welding or *fusing*, (2) cementing using glassy frits, (3) adhesive bonding using organic adhesives, (4) soldering to dissimilar materials, and (5) mechanical fastening. The amorphous structure of glasses particularly favors welding or fusing, while their property of increasing workability (i.e., decreasing viscosity) with increasing temperature enables some special mechanical interlocking methods, especially useful for joining to metals. In fact, glasses are only rarely mechanically joined to other glasses and are much more commonly mechanically joined or sealed to dissimilar materials. For this reason, the mechanical fastening of glasses will be discussed only in Chapter 15, although the methods shown in Figure 12.2 for ceramics generally apply.

12.8.2 Welding or Fusing of Glasses

Glasses are readily fused or welded to other glasses. In fact, glassworkers and glassblowers practice such fusing or welding all the time. Such welding is accomplished by heating the two glasses to be joined into a temperature range where they deform or flow easily, known as the working range.[10] Once at the proper temperature, the glasses are placed in contact, pressed together, and held under pressure to allow flow across the

---

[10] The *working range* is just one of several designated ranges of temperatures useful in characterizing the behavior of glasses. These ranges, which also include *annealing*, *softening* and *melting* ranges, characterize the viscosity range of the glass at some temperature to allow certain actions to take place, like shaping, stress relieving, etc.

interface. The mechanism involved is bulk viscous flow rather than simply diffusion, although interdiffusion certainly occurs at the same time. The resulting joint is usually indistinguishable from the original joint elements, and joint properties, including strength and physical properties, are excellent. Heating sources for welding glasses are usually gas flames but can be furnaces using combustion or electrical sources, principally relying on convection and radiant heating, respectively. Carbon dioxide lasers have also been used successfully.

In order to successfully weld two glasses, they must be compatible chemically, which is usually not a problem, and in terms of thermal expansion coefficients, which can be a problem. When there is C.T.E. mismatch, so-called matched joints are made using several other glasses with graded coefficients of thermal expansion as intermediates.

12.8.3 Cementing and Adhesive Bonding of Glasses

Glasses can be bonded adhesively with either inorganic or organic adhesives. When inorganic adhesives are used, the process is typically called *cementing*. When the adhesive is organic, the process is simply called adhesive bonding. In both types of bonding, bonding forces are a combination of chemical (from secondary van der Waals forces) and mechanical (from interlocking). Because glasses are generally nonporous, mechanical roughening is usually required to improve adhesion. This can be accomplished using abrasive particles (e.g., fine alumina or carborundum in a water slurry) in a grit-blast or acid etching with a mixture of sodium dichromate and concentrated sulfuric acid in water. Degreasing with acetone or detergent is always required, followed by drying. Glasses intended for optical use should *not* be acid etched, and abrasive treatment must be limited.

Inorganic cements used to join glass to glass include (1) soluble silicates of potassium or sodium (such as sodium silicate water glass); (2) basic salts of heavy metals (known as Sorel cements), including magnesium oxychloride; and (3) lithargic cements, which are mixtures of glycerine and lead oxide (PbO). Glass frits can also be used as cements, but because firing is required to cause bonding, these are more akin to fillers for welding or brazing if melting (i.e., softening) of the substrate glasses is limited.

Organic adhesives for use in glass-to-glass bonding include a variety of transparent, heat-setting resins, usually exhibiting water resistance to provide environmental stability for outdoor service. Example adhesives include polyvinyl butyral, phenolic butyral, nitrile-phenolic, neoprene, polysulfide, silicone, vinyl acetate, and clear epoxies.

12.8.4 Soldering of Glasses and Solder Glasses

Glasses can be soldered to other glasses as well as to other materials, including, most importantly, metals. What is especially interesting is that the fillers used can be either metallic or special composition glasses. When they are metals (or, more properly, metallic alloys) they are called *solders*. When they are glasses, they are called *solder glasses.*

Metallic solder alloys for use with glass are based on indium, since these alloys, almost uniquely, are able to wet glasses. The most common alloy used is 50 wt% In-50 wt% Sn, with a melting range of 117-125°C (243-257°F). Joint strengths are quite good, thermal match is good, and excellent hermetic seals can be achieved. Indium solders also have extremely low vapor pressures, so they perform well in vacuum applications where glasses are frequently used.

Particular glasses have been developed specifically for soldering purposes between glasses or between glasses and other inorganic materials (such as ceramics or mica) or metallic alloys. The name *solder glass* has been given to this group of glasses on the basis of their technological similarity to metallic solders, i.e., low melting or softening temperature, wetting of glass surfaces, and adhesion to base material.

In conventional glass working, two glass articles or parts are fused together (i.e., welded) and reformed at relatively high working temperatures. In contrast, solder glasses are used to join two glass parts together at temperatures low enough to avoid any significant distortion due to flow or damage due to thermal shock. Excellent strength and sealing are obtained. Because of the low sealing temperatures they allow, solder glasses are widely used in hermetically sealing electronic devices.

The most difficult part of the development or formulation of a solder glass is to adjust its properties for a particular application. Usually its thermal expansion characteristics should match those of the parts to be sealed, its viscosity should remain low within the soldering temperature range, and its chemical durability should not be lowered beyond acceptable limits. Generally, however, the lower the viscosity of a glass, the higher is its thermal expansion rate and the poorer its chemical durability.

To fulfill specific requirements with respect to the foregoing properties, a wide variety of solder glass compositions have been explored and developed. Table 12.6 lists some of the more important solder glasses and their properties. Types include PbF_2-B_2O_3-ZnO systems, Sb_2O_3 and As_2O_3 types, phosphate (P_2O_5)-based types, and CuO- and Cu_2O-containing types. Some solder glasses must be used at temperatures that exceed the normal limit of metallic solders (i.e., 450° or 840°F), but are considered solders nevertheless.

SUMMARY

Ceramics and glasses are fundamentally different from metals, alloys, and polymers, and even differ from each other, so joining presents some special challenges. Ceramics are crystalline, inorganic nonmetallic solids, while glasses are noncrystalline or amorphous, inorganic solids (except for glassy polymers). Both ceramics and glasses are characterized by strength that is quite high in compression and, usually, much lower in tension because of the existence of processing-induced microflaws. In ceramics, ductility is generally quite limited even to the highest temperatures at which the material remains solid, and both ceramics and glasses are extremely brittle at normal (i.e., room) temperatures. Chemical and other environmental durability of both materials is excel-

Table 12.6   Some of the More Important Solder Glasses and Their Properties

Composition (mole %) or (wt. %*)

Sb_2O_3	As_2O_3	B_2O_3	SiO_2	ZnO	PbO	PbF_2	P_2O_5	K_2O	Na_2O	Li_2O	C.T.E. $(\times 10^{-7}/°C)$	Softening Temp (°C)
Solder Glasses in the PbF_2-B_2O_3-ZnO System												
		35		35		30					75	440
		27.5		32.5		40					70	470
		40		37.5		32.5					63	490
		42.5		32.5		25					66	490
		42.5		40		17.5					58	510
Solder Glasses Containing Large Amounts of Sb_2O_3 and As_2O_3*												
90.3								9.7			185	260
86								14.0			194	280
41	27.3				31.7						128	287
43.8	44.8							11.4			212	252
73.4					18.7			7.9			192	250
80.6		4.8		2.8	7.7			2.2	1.2	0.7	150	290
65.0		5.0	3.0		20.0			5.0	1.0	1.0	133	312
87.3			3.0		20.0			9.7			182	292
45.1	30.6	5.4		3.1	8.0			7.3			184	255
68		15.0			17.0						122	342
Phosphate-Based Sealing Glasses												
					30		70				131	300
					41.2		58.8				153	255
					50		50				171	310
			50				50				84.5	380
			20	40			40				132	345
			20	30			50				121	325
			30	20			50				116	330
			40	20			40				119	367

Data from Takeshi Takamori, "Solder Glasses," *Treatise on Materials Science and Technology,* Vol. 17, Minuro Tomozawa and Robert H. Doremus, (eds.), New York: Academic Press, 1985, pp. 186-88.

lent, and electrical, magnetic, thermal, and optical properties can be unique among all material types. Thus, joining to produce larger, more complex shapes than can be readily produced by normal processing, or to produce hybrid structures with optimum properties achieved through the mixing of materials, becomes critical for both economic and technical reasons.

During their initial production, ceramics can be direct bonded by sintering, with or without some liquid phase or bonding aid present, with great ease and success. The process is called *sinter bonding* and is almost exclusive to ceramics. Once ceramic articles have been produced, joining becomes more difficult but can be accomplished by, in decreasing order of use: direct joining by solid-state or fusion welding; indirect joining by adhesive bonding with organic adhesives or cementing or mortaring with inorganic adhesives; by brazing with either metallic or ceramic fillers, or by soldering; or by mechanical fastening.

Direct joining by welding, either in the solid state (almost exclusively by diffusion bonding, with or without an intermediate, but also by friction) or by fusion (e.g., using an electron-beam, laser-beam, or an arc) offers the optimum in temperature capability, strength, and environmental durability. During fusion welding, sublimation, rapid decomposition, and evaporation from the melt, or thermal shock on heating or cooling must be considered, avoided, or overcome. Indirect joining with inorganic or organic adhesives is simple, versatile, and relatively low cost. Strength is good, but temperature capability, especially for organic adhesives, can be extremely limited. Sealing for hermeticity can be difficult, unless the inorganic adhesive is fired. Brazing with noble-metal or active-metal fillers, or to refractory metal metallized layers works well, providing good strength and temperature capability. With all indirect joining methods using thin-layer, intermediate materials, thermal mismatch (i.e., differential coefficients of thermal expansion, or C.T.E.s) must be dealt with. This is usually accomplished through the use of intermediates that gradually step-change the critical properties from one side of the joint to the other.

Mechanical fastening by either interlocking or through the use of fasteners works well, provided bearing forces and/or clamping forces do not cause intolerable stress concentrations that lead to brittle fracture. Flexibility in mechanically fastened joints must usually be provided to accommodate severe differential C.T.E.s.

Glasses can be joined directly by welding or fusing, indirectly using organic adhesives or inorganic cements or using solders, or, to a far lesser extent, mechanically using either interlocking techniques or fasteners. Solders for glasses are either indium-based metallic alloys or special, low-melting, low-viscosity glasses called solder glasses.

PRACTICE QUESTIONS AND PROBLEMS

1. Define what is meant by "ceramics". How is a glass different from a ceramic? How are these materials related?
2. What are some of the properties of ceramics that set them aside from metals and better suit them to certain applications?
3. What are some of the properties that give glasses a unique niche among materials? Why do glasses display these properties?
4. Why is joining so important to ceramics and glasses? What makes joining of these materials so challenging?
5. What is meant by "sinter bonding" as it pertains to joining of ceramics?
6. Explain what is meant by "direct joining" of ceramics and glasses. Give some examples of direct bonding processes.
7. Explain what is meant by "indirect joining" of ceramics and glasses. Give some examples.
8. What are the advantages and disadvantages of mechanically joining ceramics or glasses? What are some of the methods used? What special caution must be taken?
9. Differentiate between "adhesive bonding" and "cementing" or "mortaring" as these pertain to ceramics and glasses. What are the advantages of each method? What are the shortcomings of each?
10. What are the two general methods for brazing ceramics? What are the relative advantages and disadvantages of each method?
11. Describe noble-metal, active-metal, and refractory-metal brazing.

12. Describe how ceramics can be welded. Give examples of specific welding processes that can be used successfully.
13. How is thermal spraying considered a process for joining ceramics? Explain. Give some examples of where thermal spraying might be used to advantage.
14. Describe how glasses can be welded.
15. Describe how glasses can be soldered. What types of solders are used? Differentiate between glass soldering and solder glasses.

REFERENCES

1. Houck, D.L., *Thermal Spraying Technology: New Ideas and Processes,* Proceedings of the ASM's 1988 Conference, Materials Park, Ohio: ASM International, 1989.
2. Kelley, J.E., Sumner, D.H., and Kelly, H.J., "Systems for Uniting Refractory Materials," *Advances in Joining Technology,* Chestnut Hill, Mass: Brook Hill Publishing, 1976, pp. 155–183.
3. Landrock, Arthur H., *Adhesives Technology Handbook,* Park Ridge, N.J: Noyes Publications, 1985.
4. Moore, Thomas J., "Feasibility Study of the Welding of SiC," *Journal of the American Ceramic Society,* Vol. 68, No. 6, 1985, pp. C-151–C-153.
5. Pattee, H.E., Evans, R.M., and Monroe, R.E., "Joining Ceramics and Graphite to Other Materials," Washington, D.C.: NASA Report No. SP-5052, 1968.
6. Palaith, David, and Silberglitt, Richard, "Microwave Joining of Ceramics," *Ceramic Bulletin,* Vol. 68, No. 9, 1989, pp. 1601–1606.
7. Rice, R.W., "Joining of Ceramics," *Advances in Joining Technology,* ed. by John J. Burke *et al.*, Chestnut Hill, Mass: Brook Hill Publishing, 1976, pp. 69–111.
8. Schwartz, Mel M., *Ceramic Joining,* Materials Park, Ohio: ASM International, 1990.
9. Takamori, Takeshi, "Solder Glasses," *Treatise on Materials Science and Technology,* Vol. 17, Glass II, Minuro Tomozawa and Robert H. Doremus, (eds.), New York: Academic Press, 1985, pp. 173-251.
10. Wygant, J.F., "Cementitious Bonding in Ceramic Fabrication," *Ceramic Fabrication Processes,* W.D. Kingery (ed.),Cambridge, Mass: MIT Press, 1958, pp. 171-188.

Joining of Polymers

13.1 INTRODUCTION

13.1.1 Polymers Defined

Polymers, or what are commonly referred to as plastics, are materials that are composed of long-chain molecules that can occasionally be arranged in regular arrays to render the material totally crystalline, frequently into regions (or packets) of regular arrays to render the material partially crystalline or semicrystalline, or sometimes randomly to produce a completely amorphous structure. The degree to which these long-chain molecules can take up regular, crystalline arrangements depends on their size and structural complexity or bulkiness. The larger and more complex the molecule, the less crystalline the polymer.

Within the polymer chains, bonding is covalent and, characteristically, strong. Between chains, bonding can vary from weak secondary bonding to strong covalent or even ionic bonding. When covalent or ionic bonding occurs between chains, the polymer is said to be *cross-linked,* and the resulting structure is more rigid and less flexible. When the bonding between chains is strictly secondary, the polymer is typically weaker and more flexible, or *plastic,* from which the common name derives.

The molecular chains of polymers are usually based on carbon (e.g., polymeric hydrocarbons) but can also be based on silicon. The basic building block of the central chain molecule of a polymer is called a *mer*. A mer has the basic chemical formula and structure that is repeated in the chain; creating the *poly- mer* from the mers. While in this treatment only carbon-based polymers will be considered, in all cases, the chain molecules are extremely long, consisting of hundreds, thousands, or even tens of thousands of mers.

As a result of their being built up of many repeating units or mers, polymers typically have very high molecular weights. Thermodynamics, specifically the desire to maximize entropy, tends to cause these very long and often complex chains to coil, twist, or kink, as well as intertwine or entangle, i.e., to tend toward disorder. The result is unusually high elasticity, the ability to stretch under tensile loading and then recover upon unloading; viscoelasticity, the characteristic of continuing to elongate or strain for some time after the application of an instantaneously applied load or stress; and stress relaxation, the ability to reduce the stress produced by a particular level of strain by having molecules adjust by unkinking, uncoiling, or sliding along one another. This combined behavior is what gives polymers their unique properties, e.g, extraordinary elasticity, resilience, and formability. These properties in the bulk polymer arise from the behavior of the individual chain molecules and the interactions between these chain molecules.

Since there are many, many basic building blocks (i.e., mers or monomers) from which polymer molecules can be produced, and these building blocks can be arranged in many different ways within and along the chain, there are a phenomenal number of possible polymeric materials. Chains can be of various lengths (i.e., molecular weights), each with slightly different properties (e.g., melting point or glass transition temperature). Furthermore, depending on whether chains can align and pack efficiently to produce crystalline or semicrystalline structures, or whether the structure of the chain is so complex that such alignment and packing are impossible, in which case an amorphous structure results, properties can also differ even for the same basic composition.

Chain molecules can consist entirely of one type of mer, different mers arranged along the central spine or backbone of the molecule in clusters or groups (i.e., block copolymers) or randomly (i.e., random copolymers), or along secondary spines or branches (i.e., branched or grafted polymers). Furthermore, various organic radicals or side-groups (e.g., methyl groups, benezene rings, etc.) can link to the central spine in different ways to produce different properties (i.e., stereoisomerism): randomly on both sides of the chain (i.e., atactic), regularly along one side (i.e., syndiotactic), or regularly and symmetrically on both sides of the chain (i.e., isotactic). There can also be alternative structures of the same basic mers known as geometric isomers (i.e., *cis-*, *trans-*, and *gauche-*), analogs to alloying (i.e., substitution of one element in a mer for another, such as one chloride ion for one hydrogen ion in producing a vinyl chloride mer from an ethylene mer), and many other variations. Figure 13.1 illustrates some of the many ways that basic polymer structures can vary.

Fundamentally, there are two types of organic polymers: (1) *thermosetting polymers* and (2) *thermoplastic polymers*. Briefly, thermosetting polymers tend to be cross-linked (i.e., have primary bonds between chains). Further, they cross-link or *cure* from a melt or liquid phase during synthesis to become irreversibly rigid. As a group, the thermosets tend to be higher strength, less elastic, more thermally durable, more chemically resistant, and nonrecyclable. The thermoplastics, on the other hand, tend to have simpler molecules, exhibit far less cross-linking (if any), soften on heating and harden on cooling reversibly, exhibit greater elasticity, have more limited thermal durability and chemical (i.e., solvent) resistance, and are recyclable. Within these primary groups are found secondary classifications such as *elastomers* or *elastomerics,* which exhibit unusually high recoverable elongation upon loading, and *rigid foams,* which contain open or closed cells or pores that are produced during polymer processing to lighten structural weight and improve thermal insulating qualities. Table 13.1 lists some of the more important and common polymer materials of various types: thermosetting, thermoplastic, elastomeric, and rigid foam.

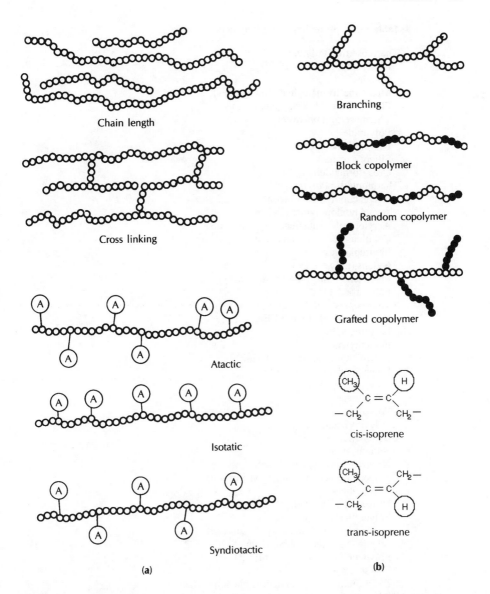

Figure 13.1 Some of the many ways the basic structure of a polymer can vary. (a) Stereo-isomerism; (b) geometric isomerism.

13.1.2 The Challenge of Joining Polymeric Materials

The need to produce larger, more complex, and higher-performance parts from polymers or plastics has created an increased need for joining these materials. The need for a greater vari-

Table 13.1   Some Important Polymers

Thermosetting Polymers:
  Diallyl phthalate (DAP)
  Epoxies
  Melamine-formaldehyde (melamines)
  Phenol–formaldehyde (phenolics)
  Thermosetting polyester
  Polyimide
  Polyurethane
  Silicone resins
  Urea–formaldehyde
Thermoplastic Polymers:
  Acetal copolymer (Celcon)
  Acetal homopolymer (Delrin)
  Acrylonitrile-Butadiene-Styrene (ABS)
  Cellulous acetate, acetate butyrate, nitrate
  Fluoroplastics
  Nylons (polyamides)
  Phenylene-oxide based resins (Noryl)
  Polycarbonate
  Thermoplastic polyester
  Polyetheretherketone (PEEK)
  Polyethersulfone
  Polyethylene
  Polymethylmethacrylate (PMMA)
  Polyphenylene sulfide (PPS)
  Polypropylene
  Polystyrene
  Polysulfone
  Polyvinyl chloride
Elastomeric Polymers:
  Natural rubber
  Neoprene (polychloroprene)
  Silicone rubber (polydimethylsiloxane)
  Butyl rubber
  Nitrile rubber (butadiene-acrylonitrile)
  Polyurethane elastomers
  Synthetic natural rubber (polyisoprene)
  Styrene-butadiene rubber (Buna S)
  Polybutadiene (butadiene rubber)
  Polysulfide rubber
  EPDM (ethylene-propylene-diene terpolymer)
Rigid, Structural Foams:
  Phenylene oxide-based resins (Noryl) ⎫
  Polyethylene   ⎪
  Polystyrene   ⎬ T/P's
  Polyvinyl chloride   ⎪
  Polycarbonates   ⎭
  Epoxy   ⎫
  Phenolics   ⎪
  Polyurethanes   ⎬ T/S's
  Silicones   ⎪
  Urea-formaldehydes   ⎭

ety of joining methods and for improved joint properties and efficiencies is also increasing because, perhaps more than any other material, polymers are being engineered to improve strength or other properties and, thereby, permit applications demanding higher performance.

The diversity of types within polymers makes the selection of an appropriate joining process more challenging than for most other materials, i.e., the extreme associated with the class of materials is diversity. All metals are fundamentally the same. They are all metallically bonded on an atomic scale, they are all crystalline in nature,[1] they all soften and then melt upon heating, and they are all simple in structure. Therefore, joining of metals is fundamentally simpler than for other materials. Likewise, ceramics are all ionically and/or covalently bonded, are all crystalline in nature, and are only slightly more complex in structure than metals. Their joining is more complex by certain processes because of their greater refractoriness, greater brittleness, and lower chemical activity. Glasses, close cousins of ceramics, are covalently bonded but amorphous and soften progressively on heating. Thus, joining is generally easier than for ceramics for many joining processes. Polymers, on the other hand, are always complex and of greater variety, so joining can be more complex.

13.2 GENERAL METHODS FOR JOINING POLYMERS

Polymeric materials, or plastics, can be joined fundamentally by mechanical fastening, adhesive bonding including solvent cementing, and welding or fusion bonding. Brazing, soldering, and weldbrazing are not (or, at least, have not been) used.[2] Thermal spraying is used to enable adhesive bonding by applying thermoplastic adhesives to polymeric or other substrates. Weldbonding is theoretically possible, at least for thermoplastics but seemingly has not been used commercially and may offer no distinct advantage, as will be explained later.

To date, most semistructural and structural parts (i.e., secondary and primary structures, respectively) have been made from thermosetting polymers, with or without reinforcement (e.g., epoxy-graphite aircraft parts). Mechanical fasteners and adhesives have been used for joining these materials. However, greater impact resistance, easier processing (including assembly), and potential for recycling[3] are increasing the interest in parts made from thermoplastics. While adhesives and mechanical fasteners have

---

[1] Metals can be made non-crystalline or amorphous by causing them to solidify extremely rapidly, but this is an extreme and must be done very intentionally.

[2] Some adhesive bonding methods resemble brazing and soldering in that soft fillers are used to join solid substrates, but these are still not considered brazing or soldering, but, rather, are still considered adhesive bonding.

[3] There are really at least two types or levels of recycling that are important for all materials, and particularly for plastics. In primary recycling, materials are recovered, reclaimed and reused (i.e., are recycled) within the manufacturing environment, as part of the overall manufacturing process. This prevents waste, by maximizing material utilization and minimizing waste or scrap. In "secondary" recycling, materials are recovered, reclaimed or reused (i.e., are recycled) from products which have lived out their useful lives. In fact, within this category for plastics, there are two levels, one called secondary, which recovers waste plastics specie by specie, and ternary which recovers plastics as a group or commingled.

been used for joining thermoplastics also, these polymers, because of their fundamental character (i.e., the ability to reversibly soften and harden upon heating and cooling), offer an alternative joining possibility that is still in its infancy, namely, welding or *fusion bonding.*

13.3 JOINING THERMOSETTING POLYMERS

13.3.1 General Comments

Thermosetting polymers or thermosets are those polymers that become rigid upon curing after synthesis and remain irreversibly rigid, i.e., *set*. It is this fundamental behavior of thermosets that precludes joining by welding. Heating thermosets and applying pressure will not cause melting or even softening, so bonds cannot be created across an interface in this way. The predominant methods for joining thermosetting polymers are thus mechanical fastening and adhesive bonding, with adhesive bonding being far more common.

13.3.2 Mechanical Fastening

Thermosetting polymers can be mechanically fastened using both threaded and unthreaded fasteners as well as by employing integral, interlocking design features. When fasteners are used, they are typically made from metals, but they could be made from polymers, especially thermoplastics.

When using mechanical fasteners or integral, interlocking design features, for that matter, the viscoelastic deformation behavior of polymers must be considered to avoid deformation and/or relaxation. In viscoelastic behavior (as shown in Figure 13.2), an applied load, P, causes an instantaneous strain, ε, and a time-dependent strain, $\varepsilon(t)$. While the specific form of the strain response differs with specific polymer types, all polymers exhibit some form of viscoelastic behavior. The time-dependent deformation is often called *cold flow*. Cold flow in polymers can cause distortion of a fastener hole or design feature responsible for interlocking and can lead to fastener loosening, to fastener pull-out or loss, and/or to joint failure.

Besides this macroscopic deformation, which can cause loss of function of the fastener, joint, part, or assembly, this time-dependent strain can also lead to the relaxation of stress. This too can cause loosening of fasteners and loss of fastener effectiveness for fasteners that rely on clamping force or preload to achieve needed joining forces. Polymeric fasteners would also be prone to such viscoelastic deformation and stress relaxation, so they too could loosen. Fastener tear-out or pull-through can also be a problem because of a polymer's inherently low strength. This is especially a problem in thin material.

Mechanical fasteners used with polymers should have large bearing areas at their heads and tails or feet and/or should be used with washers to spread loading to avoid

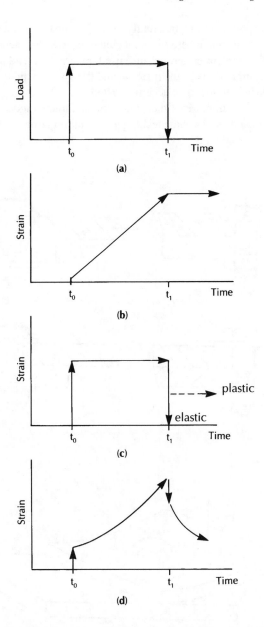

Figure 13.2 Applied load and strain response in viscoelastic polymers. (a) Applied load verses time; (b) viscous behavior; (c) elastic (or elastic-plastic) behavior; (d) viscoelastic behavior.

viscoelastic deformation problems. Fastener holes can also be sleeved to help reduce localized bearing deformation. Loading must be limited when using mechanical fasteners or, to a lesser extent, with interlocks since stress concentrations arise and aggravate viscoelastic deformation and tear-out.

Because polymers are inherently corrosion resistant, there is no problem with electrochemical mismatch between joint elements or with fasteners. Thermal mismatch is usually not a problem either, because use temperatures tend to be limited as a result of the inherent limitations of the material and the viscoelasticity of the polymer allows thermally induced strains to be accommodated.

Figure 13.3 summarizes the predominant failure modes for mechanically fastened polymers and shows some techniques for minimizing such failures.

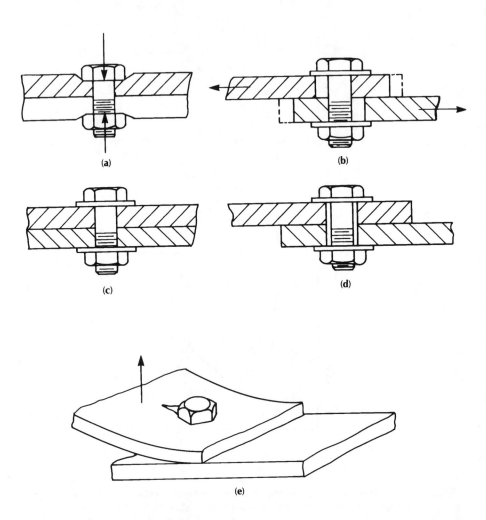

Figure 13.3 Predominant failure modes for mechanically fastened polymers and some means to minimize such failures. (a) Cold flow under clamping load; (b) cold flow under bearing loads; (c) use washers or large-area heads or feet; (d) use sleeves or bushings; (e) fastener pull-through; avoid or resist peel loading.

13.3.3 Adhesive Bonding

Most thermosetting polymers are relatively easy to join or bond with adhesives, so adhesive bonding tends to be the most popular joining method used with these materials. Because thermosets are not particularly soluble, solvents are of little use in causing softening to aid adhesives or for solvent cementing (defined later).[4]

In general, conventional adhesive bonding using compatible adhesives is the most practical way to join a thermoset to a thermoset or to a nonpolymeric material, for that matter. The best adhesives for bonding thermosetting polymers are thermosets themselves. Some examples of adhesives used for bonding thermosetting epoxies are modified acrylics, epoxies, polyesters, phenol-formaldehydes, and cyanoacrylates. For joining phenolics or phenol formaldehyde, neoprene and urethane elastomers, epoxies and modified epoxies, cyanoacrylates, and phenolic adhesives are used. For joining polyurethanes, elastomeric adhesives, epoxies and modified epoxies, and neoprene adhesives are used, while for silicone resins, silicone rubber and silicone resin adhesives are used. Adhesive alloys also work well with thermosets, especially when they are being joined to other polymer types (e.g., thermoplastics). Reinforced thermosets (e.g., with glass fibers or synthetic high-strength fibers) are bonded by considering the base (matrix) resin (see Chapter 14).

Adhesively bonded thermosetting polymers, whether monolithic or reinforced, can generally carry greater loads than mechanically fastened thermosets because loading is uniformly distributed over a larger area, resulting in lower stresses. Maximum stress is limited to the strength of the adhesive used, provided bonding is performed properly.

13.4 JOINING THERMOPLASTIC POLYMERS

13.4.1 General Comments

Thermoplastic polymers are those polymers that can be reversibly softened and hardened by heating and cooling after they have been synthesized. Thermoplastics can be joined by mechanical fastening, adhesive bonding (including solvent cementing), and by welding or fusion bonding.

13.4.2 Mechanical Fastening

Thermoplastic polymers can be mechanically fastened using both threaded and unthreaded fasteners and, to a far lesser extent, by integral, interlocking design features.

---

[4] Solvents may, however, be useful for joining thermosetting to thermoplastic polymers.

Because thermoplastics are often lower in strength than thermosets, they are more prone to viscoelastic deformation or cold flow. Thus, fastener design must be carefully considered, both for the fastener and for the fastener hole (e.g., hole placement relative to edges).

As for thermosets, fastener loading should be distributed as much as possible to avoid stress concentration that will cause deformation. Heads and feet of fastener systems should be as large as practical, or washers should be employed, and holes ought to be sleeved. While not common, especially for true structural applications, fasteners that are themselves made from thermoplastics might be ideal for mechanically fastening thermoplastics.

Failure modes and remedies are typical of those shown in Figure 13.3.

13.4.3 Adhesive Bonding and Solvent Cementing

Thermoplastics can be adhesive bonded after softening by a solvent or by heating using so-called hot-melts. Unlike thermosetting polymers, thermoplastics ordinarily require that their surfaces be physically and/or chemically modified to produce acceptable adhesion. This is especially true for crystalline thermoplastics such as the polyolefins (e.g., polyethylene and polypropylene), linear polyesters, and fluoropolymers (e.g., Teflon). Once treated or modified, bonding can be accomplished with the aid of adhesives or without, depending on the particular polymers involved in the joint and on the application requirements.

Methods for activating the surface of thermoplastic polymers include (1) oxidation by means of chemical or flame treatment, (2) electric discharge to leave a more reactive surface, (3) ionized inert gas to strengthen the surface by cross-linking and leave it more active, or (4) metal ion treatment. Adhesives for use with thermoplastics are themselves thermoplastics and include cellulous acetate, cellulous acetate butyrate, cellulous nitrate, polyvinyl acetate, polyvinyl vinylidene, polyvinyl acetals, polyvinyl alcohols, polyamides, acrylics, and phenoxys. Reinforced thermoplastics are joined by the same adhesives as used for the matrix resin species in its monolithic form, as will be discussed in Chapter 14.

A special option available for joining thermoplastics adhesively is *solvent cementing*. Solvent cementing is a process in which thermoplastics, usually amorphous, are softened by the application of a suitable solvent or mixture of solvents and, then, pressed together to effect a bond. The thermoplastic resin of the joint element itself, after evaporation of the solvent, acts as the adhesive. Any bonds produced by this method are principally primary covalent bonds, but most of the bonding arises from entangling of the long-chain molecules. Joint strength can be very high compared to adhesive joints in which only secondary bonding occurs.

Many thermoplastic resins in either monolithic or reinforced, composite form are easier to join effectively by solvent cementing than by conventional adhesive bonding. Mixed solvents often give the best results. Often, small amounts of the polymer or polymers to be cemented are dissolved in the solvent or solvents to form a "bodied" cement. These aid in gap filling, accelerate setting, and reduce shrinkage and internal stresses.

13.4.4 Welding or Fusion Bonding of Thermoplastics

Just as in the welding of metals, thermoplastic polymers can be joined by softening or melting the base material through the application of heat and effecting bond formation by applying pressure. Fusion processes for joining thermoplastics and thermoplastic composites involve heating the polymer to a viscous state and physically causing polymer chains to interdiffuse, usually by pressure-induced flow. The resulting adhesion is largely the result of chain entangling but can include secondary and even primary covalent bonding. Consequently, joint strength is typically quite high compared to adhesive bonding and can easily approach the strength of the base material.

Welding or *fusion bonding* processes for thermoplastics can be divided into two groups: (1) processes involving external heating (e.g., hot plate, infrared, hot gas, resistance, and radio-frequency heating), and (2) processes involving mechanical movement to produce heating (e.g., frictional heating by vibration, spin, and orbital or by ultrasonic processes). All of these processes have several attractive features, including speed; strong, highly efficient joints; tolerance of contaminated surfaces; suitability to difficult-to-bond substrates; excellent bondline; and improved recyclability.

Processes Involving External Heating

Contact with a heat source results in melted or softened thermoplastic polymer surfaces in the weld area. Forging these surfaces together results in interdiffusion of the molecular chains and a weld.

In *hot plate welding,* heated metal or ceramic platens or bars clamp and heat the thermoplastic, pressing the heated samples together to a preset pressure for a preset length of time to create an upset and bond (see Figure 13.4a). The process is capable of producing weld strengths equal to the parent polymer and can be used with dissimilar thermoplastics provided they are chemically compatible. With dissimilar plastics, each platen heats each plastic to its proper softening point and not necessarily to the same temperature. Filler material may or may not be needed and used. Obviously, heating platens or bars need to be shaped to the contour of the parts being joined. This process has been used for fusion bonding or welding plastic battery cases, fuel tanks, fuel filler pipes, and large pipes for water, gas, sewage, and chemical transport.

In *hot-gas welding,* a stream of heated gas (e.g., nitrogen, carbon dioxide, hydrogen, and oxygen) or air is directed at the joint area to fuse the surfaces (Figure 13.4b). A filler rod may also be used. When a filler is used, it should have the same composition as the substrate, although dissimilar thermoplastics can be joined using a filler that is compatible with each. Joints are usually V-grooved to provide access or fillet welds are made. Gas temperatures depend on the polymer being welded but typically fall in the range of 200 to 300°C (392 to 592°F), and gas flow rates vary from 15 to 60 liters/min. (31.78 to 127.13 ft^3/hr.). Typical plastics welded include: PVC, polyethylene (using nonoxidizing gas), polypropylene, polymethacrylate, polycarbonates, and nylons. Hot-air hair dryers (or higher wattage industrial dryers) are commonly used to join PVC piping.

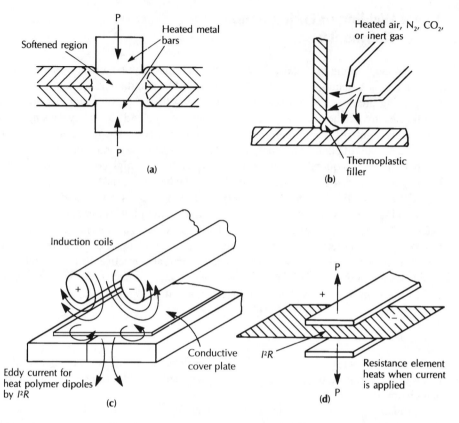

Figure 13.4   Various thermoplastic welding processes relying on external heating sources to accomplish bonding. (a) Hot-bar welding; (b) hot-gas welding; (c) induction welding; (d) resistance welding.

Infrared heating is sometimes used as an alternative to hot plate heating. An infrared radiation source, focused on the weld joint face, is used for heating. *Radio-frequency welding* uses radio-frequency radiation (27 MHz) to heat polymers that have inherent dipolar molecules. Unlike induction welding, no conductor is involved or required. The process has been used for sealing PVC, nylons, and other thermoplastics with strong dipoles. Induction welding is similar to radio-frequency welding, except a conductive material is heated by the radio frequency radiation through I^2R eddy current effects. This heat is used to heat the thermoplastic indirectly. Tapes of thermoplastics filled with iron oxide powder work well. The mechanism of induction heating is shown in Figure 13.4c.

In a final method, *resistance welding,* the resistance to electrical current is used to heat a conductive element that is coated with a plastic and welded (Figure 13.4d). The process has been used for making complicated joints in vehicle bumpers and panels, plastic pipes, containers, and medical devices.

Processes Involving Frictional Heating

In processes involving frictional heating, the heat needed to soften the thermoplastic is provided by simple friction or hysteresis heating. In *friction welding,* a spin or orbital welding process was first reported in the 1930s and remains essentially unchanged today. It involves fixing one substrate while the other is given a controlled angular velocity, is brought into contact to cause frictional heating, and is upset to effect a bond. The process is shown in Figure 13.5a. Parameters include rotational speed (1 to 20 m/sec or 3.28 to 65.6 ft/sec); frictional pressure (80 to 150 kPa or 11.6 to 21.7 psi); forge pressure (100 to 300 kPa or 14.5 to 43.5 psi); and weld time (1 to 20 sec). Because heating depends on relative surface velocity, heating varies across a cross section so can lead to weld zone residual stresses. While a drill press can be used, a controlled energy source, like a flywheel, is better. Advantages of the spin welding process, when it is done properly, are high weld quality, simplicity, speed, and reproducibility. The principal disadvantage is that only limited, rotationally symmetric geometries can be welded.

Another form of friction welding is *vibration* or *linear friction welding.* This involves rubbing two thermoplastics together under at a suitable pressure, frequency, and amplitude to generate enough heat energy to melt the polymers. At this point, vibration is stopped and the parts are aligned and allowed to solidify into a weldment. The process (shown in Figure 13.5b) is like spin or orbital welding, except that the motion is linear rather than rotational. The vibration welding process is suitable for most thermoplastics, including amorphous and semicrystalline forms. Welds are produced in seconds using vibrational amplitudes of 0.5 to a few millimeters and frequencies ranging from 100 to 500 Hz. The process is especially useful for welding crystalline thermoplastics such as acetal, polyethylene, nylons, and polypropylenes that are not easily welded by solvent cementing or other welding methods.

In *ultrasonic welding,* small amplitude vertical oscillations at frequencies of 10 to 50 kHz (usually 20 to 40 kHz) are passed through polymers to be dissipated at a joint or bond interface as frictional heating. Heat is generated by a combination of friction and hysteresis.[5] Oscillatory frequencies are generated by piezoelectric or magnetostrictive crystals mounted between two blocks of metal (called *converters*). Because the amplitudes of vibration in the crystal are usually quite small, they are boosted by mass and geometric effects in a booster and horn (Figure 13.5c). At the horn, amplitudes are usually 20 to 60 micronmeters. Joints often contain projections to accentuate frictional heating in this process.

Ultrasonic welding is currently the most common welding process used with thermoplastics. It is fast, occurring in seconds, results in little upset or flash, and is easy to automate. Applications include joining of plastic flash cubes, dashboard assemblies for cars, audio- and videotape cassette bodies, and luggage tags.

---

[5] *Hysteresis* is the loss of strain energy that can occur on reversed mechanical loading from various sources such as internal friction. In mechanical load reversals, hysteresis causes heat generation.

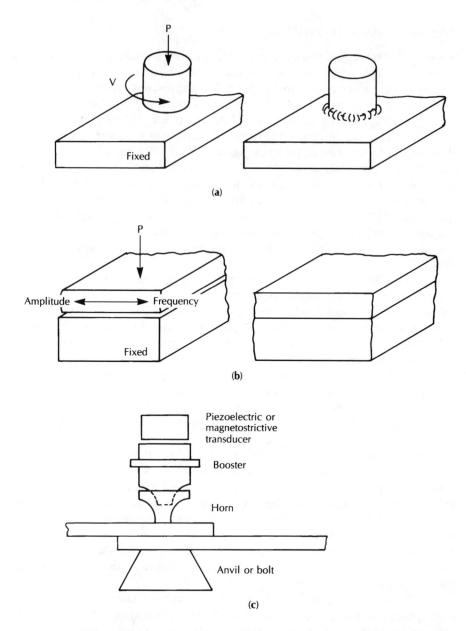

Figure 13.5   Various thermoplastic welding processes relying on friction to accomplish bonding. (a) Spin or orbital friction welding; (b) linear vibration welding; (c) ultrasonic welding.

13.5 JOINING ELASTOMERIC POLYMERS OR ELASTOMERS

An *elastomer* or *elastomeric polymer* is defined as a macromolecular material that, at room temperature, is capable of recovering substantially in size and shape after the removal of a deforming force. Elastomers are composed of very long, very coiled, kinked or tangled polymer chains. Usually, although long, the molecules of elastomers are simple, with few bulky side groups or radicals and little or no branching or cross-linking. Applying a load tends to straighten or align these molecules to cause a change in macroscopic size or shape. When the applied load is removed, the molecules attempt to return to their higher, tangled and kinked entropy state. Elastomers can be thermo-setting or thermoplastic, but are more often thermoplastic.

Joining these highly pliable materials is always accomplished by adhesive bond-ing. To be successful, elastomeric polymers are best bonded using pressure-sensitive adhesives derived from an elastomer similar to the one or ones being bonded. Some typical adhesives used for bonding elastomers to themselves or to other materials are various rubbers, neoprene, acrylics, cyanoacrylates, and epoxies. The key need in bonding elastomeric materials is to have the joint also be pliable.

13.6 JOINING STRUCTURAL OR RIGID PLASTIC FOAMS

Plastic foams can be produced from thermosetting or thermoplastic polymers by creat-ing either open or closed cells, either during initial polymer synthesis or as a secondary foaming operation. Usually gas producing reactions are used. The resulting foams can be soft and easily compressible or rigid. Rigid foams can be used for structural appli-cations where light weight is critical. Rigid foams are often called *structural foams.*

Rigid, structural foams are always joined by adhesive bonding. For joining ther-moplastic foams, solvent cementing is usually preferred over conventional adhesives. Care must be taken not to collapse the cells of structural foams during bonding, since these cells impart the desirable properties of light weight, good thermal insulation, and impact protection. When adhesives are used, water-based types based on polyvinyl acetate and neoprene are frequently used. For thermosetting foams, epoxies and a vari-ety of other adhesives normally used with dense thermosets can be employed success-fully. When adhesives are used, whether they are thermoplastic or thermosetting, is it usually preferable to thin them with a solvent or diluent.

13.7 JOINING DISSIMILAR POLYMERS

Dissimilar combinations of polymers requiring joining can include thermosetting types to thermoplastic types or thermosets or thermoplastics to elastomers or rigid foams. For all these combinations, adhesive bonding is the preferred, and almost exclusive, method of joining.

Table 13.2   Adhesives and Adhesive Alloys for Joining Various Combinations of Polymers

Thermosetting Polymers:
  For epoxies
    Modified acrylics
    Epoxies
    Polyesters (thermosetting types)
    Phenol–formaldehyde
    Cyanoacrylates
  For phenolics
    Neoprene elastomers
    Urethane elastomers
    Epoxies and modified epoxies
    Cyanoacrylates
    Phenolics
  For melamines
    Epoxies
    Polyurethane
    Neoprene
    Cyanoacrylate
    Urea–formaldehyde
  For polyurethanes
    Polyurethanes
    Polyurethane elastomers
    Epoxies and modified epoxies
    Neoprene
  For silicone resins
    Silicone rubbers
    Silicone resin adhesives
Thermoplastic Polymers:
  Solvent cementing
  For cellulosics
    Natural rubber
    Neoprene rubber
    Nitrile rubber
    Resorcinol–formaldehyde
    Cyanoacrylates
  For polyamides
    Phenolics
    Resorcinol–formaldehyde
    Epoxies and modified epoxy
    Polyurethane
    Cyanoacrylates
  For polycarbonates
    Epoxies
    Urethanes
    Silicones
    Cyanoacrylates
    Hot Melts

Table 13.2 *(continued)*

For polyvinyl chloride
 Nitrile rubber
 Polyurethanes
 Neoprenes
 Modified acrylics
 Anaerobics
Elastomer Polymers:
 Pressure-sensitive adhesives
 Neoprene
 Nitrile rubber
 Polyurethanes
 Natural, reclaimed, butyl, nitrile, butadiene, and other rubbers.
Rigid Foams:
 Solvent cementing
 Nitrile rubber
 Flexible epoxy
 Rubber-based adhesives

For most of these dissimilar combinations, adhesive alloys composed of appropriate combinations of the basic materials are recommended. For joining thermosetting and thermoplastic types, adhesive alloys containing mixtures of thermosetting and thermoplastic adhesives are recommended. For joining thermosetting polymers to thermosetting elastomers, adhesive alloys containing thermosetting adhesives and thermosetting elastomeric adhesives are recommended. Likewise, for thermoplastic polymers to thermoplastic elastomers, thermoplastic/thermoplastic elastomers are recommended. For thermoplastics or thermosets to be joined to elastomerics of the opposite fundamental polymer type, alloys of thermosetting-thermoplastic-elastomeric adhesives are recommended.

Table 13.2 lists suggested adhesives or adhesive alloys for various combinations of thermosetting, thermoplastic, and elastomeric polymers and foams.

For joining dense forms of thermosetting and thermoplastic polymers to one another, mechanical fastening methods can be used, with the usual precautions.

SUMMARY

Polymeric materials are composed of large, complex molecules, often based on organic hydrocarbons. There are innumerable varieties due to the variety of molecular weights (i.e., chain lengths or degree of polymerization), chain configurations (i.e., combinations and arrangement of monomers within the chain), cross-linking between or networking among chains, type and arrangement of side-groups or radicals on chains (i.e., stereoisomerism), geometric arrangement of elements within the monomer building blocks (i.e., geometric isomerism), degree of amorphism or crystallinity, etc. Two primary types characterize all organic polymers, however: (1) thermosets that cure upon synthesis into an irreversibly rigid structure and (2) thermoplastics that can be reversibly softened and hardened by heating and cooling. Within each of these primary types are two secondary types: elastomerics that exhibit unusually high recoverable elasticity and rigid foams that contain open or closed cells that lighten the structure and impart high thermal insulating properties.

Thermosetting polymers can be joined by mechanical fastening or adhesive bonding. Mechanical fastening can use fasteners or integral, interlocking design features. In either case, loading must be limited to prevent time-dependent viscoelastic deformation or cold flow, especially in regions of high stress concentration. Adhesive bonding offers the advantage of load spreading, thereby avoiding cold flow. Thermosetting-type adhesives are used to join thermosetting-type polymers.

Thermoplastic polymers can also be joined by mechanical fastening and adhesive bonding, as well as by two other methods, i.e., solvent cementing and fusion bonding or welding. Within adhesive bonding is a special process called solvent cementing, where a solvent is used to soften the thermoplastic substrates and cause intermixing, setting by evaporation or diffusion, and bonding. Welding or fusion bonding is also uniquely possible with thermoplastics, using either external heat sources such as hot plates, hot-gas, infrared, induction, radio-frequency or resistance or internal mechanical heating by friction through spin or orbital, linear vibrational, or ultrasonic methods. In the mechanical fastening of thermoplastics, the same precautions must be taken to avoid cold flow as in thermosets. In adhesive bonding one thermoplastic to another thermoplastic, adhesives based on thermoplastics must be used.

Elastomeric polymers are special, very long-chain molecules that are kinked or coiled in their relaxed state, stretch to become straight under loading, and contract back to their kinked and coiled state upon load removal. These extremely flexible polymers are joined exclusively by adhesive bonding using adhesives that are of the same basic type as the polymer(s) being joined and with elastomeric properties themselves, either inherently or through adhesive alloying.

Rigid plastic foams are always adhesively bonded, and they must be bonded very carefully to prevent collapse of the foam. Solvent cementing is preferred, although thinned adhesives of the appropriate composition can also be used successfully.

Dissimilar combinations of polymers require the use of adhesive alloys, or mixtures of different types of polymers, if they are to be adhesively bonded. For example, when bonding thermosets to thermoplastics, an alloy consisting of a thermoplastic and thermosetting type polymer adhesive are used. Mechanical fastening is also possible for combinations of dense thermosets and thermoplastics.

PRACTICE QUESTIONS AND PROBLEMS

1. Define polymers as a material. What characteristics of polymers lead to their almost unparalleled diversity as materials, allowing them to be so effectively engineered?
2. Differentiate between thermosetting and thermoplastic polymers. Give some of the relative advantages and disadvantages or limitations of each type.
3. What makes the joining of polymers so challenging?
4. In the broadest or most general sense, what are the options for joining polymers?

5.　What are the special challenges of joining thermosetting polymers? What must be considered when such polymers are to be mechanically fastened? What must be considered when such polymers are to be adhesively bonded? Which process is generally preferable? Why?

6.　What are the special challenges of joining thermoplastic polymers? What must be considered when such polymers are to be mechanically fastened? What must be considered when such polymers are to be adhesively bonded? Which process is generally preferable? Why?

7.　Define the process of solvent cementing. Compare this process to more conventional adhesive bonding processes, and categorize it within the SME classification scheme presented in Chapter 5. Compare this process to welding or fusion bonding.

8.　Describe why thermoplastic polymers can be welded but thermosetting polymers cannot be welded.

9.　What are the two broad divisions of welding or fusion bonding processes for joining polymers? Give some specific examples of processes within each division.

10.　What are the special considerations when elastomeric polymers are to be joined? What process would you recommend? How would you perform the process differently than for more conventional thermosetting or thermoplastic polymers?

11.　What are the special considerations when structural foams are to be joined? What process would you recommend?

12.　Explain how dissimilar polymer types must be joined.

REFERENCES

1.　Gauthier, Michelle M., "Sorting Out Structural Adhesives," *Advanced Materials and Processes,* Vol. 138, No. 1, 1990, pp. 26–35.

2.　Gauthier, Michelle M., "Clearing Up Adhesives Confusion," *Advanced Materials and Processes,* Vol. 138, No. 2, 1990, pp. 41–49.

3.　Grimm, R.A., "Fusion Welding Techniques for Plastics," *Welding Journal,* Vol. 69, No. 3, 1990, pp. 23–28.

4.　Landrock, Arthur H., *Adhesives Technology Handbook,* Park Ridge, N.J.: Noyes Publications, 1985.

5.　Shields, J., *Adhesives Handbook,* 3rd. ed., London, Butterworths: 1984.

6.　Skeist, I., *Handbook of Adhesives,* 3rd. ed., New York, Van Nostrand Rheinhold: 1989.

Joining Composite Materials

14.1 INTRODUCTION

14.1.1 General Comments

The performance of a structure or an assembly is critically dependent on the behavior of any joints it contains, and most contain joints. All too often, a weight or strength advantage brought about by clever design or optimal use of materials is lost because the characteristics of the associated joints were not properly understood. Perhaps nowhere is proper joining more important than in structures or assemblies composed of composites or composite materials, where the composite is almost certainly being used to enable the attainment of high strength and low weight beyond that attainable by other, monolithic materials. Examples abound in military, commercial, and civilian aircraft and in missiles, satellites, and spacecraft, and more examples are appearing every day, such as automobiles, appliances and business machines, construction, consumer products, machine tools, marine applications, medical equipment and implants, and sports equipment.

The joining of composite materials presents a very special challenge.

14.1.2 Composites Defined

Early in the history of humankind, it was realized that, for some uses, combinations of materials often produced properties in the mixed material, or *composite,* that were superior to those of the components themselves, for example, mud bricks reinforced with straw in ancient Egypt, layered iron and steel in ancient Damascus, and steel-reinforced concrete in modern construction worldwide.

All these materials could be considered composites in the broadest sense, since they all consist of two or more identifiable component materials. This definition, however—mixtures of two or more identifiable materials—is too broad. A more useful, but narrower, definition of a composite is a combination of a reinforcement material in a binder or matrix material. The key to this definition is that the component materials act together, in concert, to help one another, often synergistically.[1] Often the matrix pro-

---

[1] *Synergistic* means the action or interaction of two or more substances, organs, or organisms to achieve an effect of which each is individually incapable.

vides protection from damage and gives stability to the reinforcement under loading; at the same time the reinforcement enhances strength or hardness or stiffness. As used here, the term *composite* also implies that the component materials are macroscopically identifiable and mechanically separable, not merely different at the molecular level.

Composites or composite materials occur naturally but can be purely synthetic. Some natural composites include wood, bone, and celery. Obviously, these are structural, for their particular need, i.e., to hold up a tree, an animal, or a plant respectively. Modern synthetic composites include almost every imaginable combination of metals, ceramics, glasses, polymers, and intermetallics. A simple pictorialization of possible combinations is shown in Figure 14.1. In addition to these, there are composites that consist of (1) one metal reinforcing another in certain metal-matrix composites, or MMCs (e.g., boron fibers in aluminum or refractory metal particles or fibers or wires in copper); (2) one ceramic reinforcing another in ceramic–ceramic composites, or CCCs (e.g., SiC or Si_3N_4 in Al_2O_3 or partially stabilized zirconia or PSZ in SiC); and (3) one polymer reinforcing another (e.g., aramid fibers in rubber).

In general, synthetic composites are categorized by their matrix: organic, polymer-matrix composites, with either thermosetting or thermoplastic matrices; metal-matrix composites; ceramic-matrix composites; intermetallic-matrix composites; and a special category often included under ceramic composites, carbon–carbon composites, where the matrix and reinforcement are simply different forms or grades of carbon. As shown in Figure 14.1, almost every possible combination exists as an engineering material: (1) polymer matrices with metal reinforcements (e.g., epoxy–boron); (2) polymer matrices with ceramic reinforcements (e.g., epoxy–glass); (3) metal matrices with ceramic reinforcements (e.g., Al–SiC); (4) metal matrices with polymer reinforcements (e.g., aluminum–aramid, called Arall); (5) ceramic matrices with metal reinforcements (e.g., steel reinforced concrete and *cermets*); and (6) ceramic matrices with polymer reinforcements (e.g., talc or $CaCO_3$-filled polymers, such as epoxies). Reinforcements can take the form of randomly dispersed particles (i.e., in particle-reinforced or particulate composites), random or aligned chopped fibers or whiskers (i.e., in fiber-reinforced or whisker-reinforced composites), continuous, aligned fibers

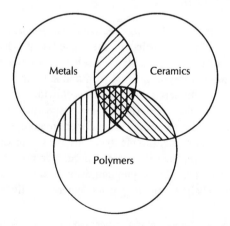

Figure 14.1   Venn diagram showing the possible combinations of materials capable of producing synthetic composites.

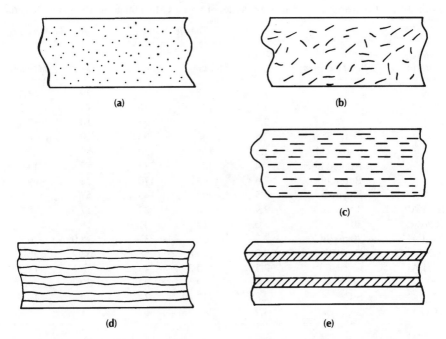

Figure 14.2   Forms of reinforcements in composites. (a) Particle-reinforced or particulate composite; (b) random, chopped fiber; (c) aligned, chopped fiber; (d) continuous fiber composite (unidirectional); (e) laminated composite.

or wires (i.e., in continuously reinforced composites), and layers or laminations (i.e., in laminate-reinforced composites). These are shown schematically in Figure 14.2. Each imparts different degrees of reinforcement, enhances different properties, and leads to different degrees of anisotropy or directionality of properties (i.e., greater anistropy with increasing alignment).

The advantages derived from combining different types of materials to form synthetic composites are enhanced properties, beyond those attainable in monolithic materials, or unique properties, unattainable from single, monolithic materials. Often, the properties being sought are mechanical and related to strength and include tensile yield or ultimate strength, specific tensile strength,[2] tensile modulus or stiffness, or specific modulus. Composites, however, can be synthesized to obtain other enhanced or unique properties, such as improved wear resistance; hardness with impact toughness; controlled thermal expansion (even, zero coefficient of thermal expansion); improved thermal or electrical conductivity or resistivity (depending on need); and enhanced or extended elevated temperature strength.

Some of the enhanced or unique properties obtainable in composites versus monolithic materials are shown in Table 14.1 and in Figures 14.3 through 14.6.

---

[2]*Specific strength* and *specific modulus* take the density of the material into account, to relate the relative strength-to-weight or stiffness-to-weight. See Brett [1989].

Table 14.1   Comparison of Strength Properties of Various Monolithic and Composite Materials

Matrix Material	Reinforcement	Density (g/cm³)	Tensile Strength		Tensile Modulus	
			ksi	ksi/g/cc	Msi	Msi/g/cc
Steel		7.8	145	19	29	3.5
Ti		4.5	134	30	16	3.5
Al		2.8	67	24	10	3.5
	Glass fiber	2.5	246	98	10	4
	Carbon fiber	1.9	228	120	55	29
	Aramid fiber	1.4	385	275	19	14
	Boron fiber	2.6	443	170	23	9
	SiC fiber	3.5	500	143	57	16
6061 Al	30v/o SiC	2.95	80	27	17.5	5.9
7090 Al	30v/o SiC	3.0	112	37	18.5	6.2
2124 Al	20v/o SiC[a]	2.9	108	37	18.0	6.2
AZ91CMg	Gr (P100)[b]	2.0	116.8	58	52.2	26.1
6061 Al	Gr (P100)[b]	2.5	120.2	48	48.6	19.5
Cu	SiC[b]	6.4	150	23	30	4.7
Al	SiC[b]	2.85	212	74	30	10.5
Ti	SiC[b]	3.9	240	62	21	5.4

[a]Whisker
[b]Unidirectional fiber; all others are particulate

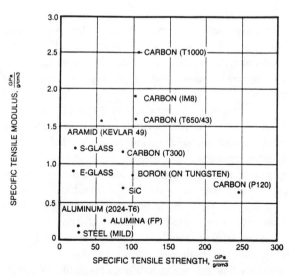

Figure 14.3   Specific tensile strength versus specific tensile modulus for various commercial organic-matrix composites and steel and aluminum. (Reprinted with permission from A. Brett Strong, *Fundamentals of Composites Manufacturing: Materials, Methods and Applications*, Society of Manufacturing Engineers, 1989.)

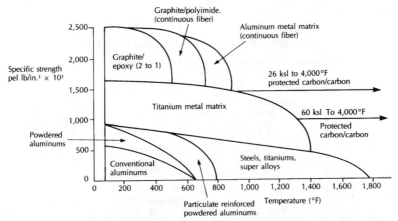

Figure 14.4 Elevated temperature performance of a number of advanced monolithic and reinforced (composite) materials. (Reprinted with permission from A. Brett Strong, *Fundamentals of Composites Manufacturing: Materials, Methods and Applications*, Society of Manufacturing Engineers, 1989.)

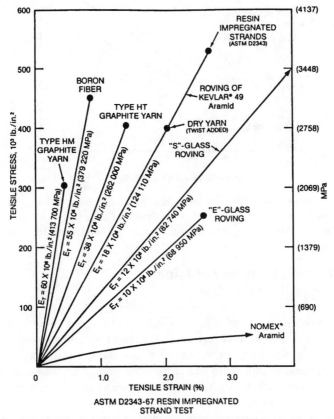

Figure 14.5 Typical stress versus strain behavior of various reinforcing fibers. (Reprinted with permission from A. Brett Strong, *Fundamentals of Composites Manufacturing: Materials, Methods and Applications*, Society of Manufacturing Engineers, 1989.)

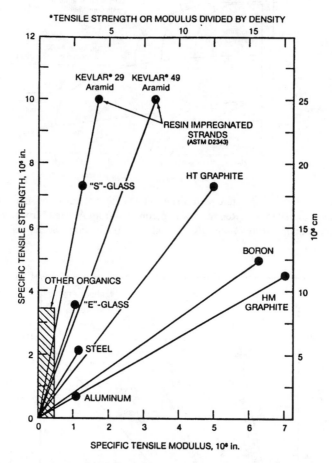

Figure 14.6 Specific tensile strength versus specific tensile modulus for various reinforcing fibers. (Reprinted with permission from A. Brett Strong, *Fundamentals of Composites Manufacturing: Materials, Methods and Applications*, Society of Manufacturing Engineers, 1989.)

14.1.3 Special Challenges of Joining Composite Materials

Joining of composite materials poses a special problem, i.e., how to achieve joint strength or other properties anywhere close to those of the parent composite since the integrity or continuity of the reinforcement across the joint is difficult or impossible to retain or reestablish. Usually to join composites, one attempts to cause bonding of the matrix materials, with as little disruption of or damage to the reinforcing phase as possible.

Depending on how reinforcement is achieved in a particular composite (e.g., by dispersed discrete particles, chopped, randomly distributed fibers or whiskers, or continuous, oriented fibers or wires) and on the specific joining technique, some continuity of reinforcement across the joint or within the joint may be possible.

14.2 OPTIONS FOR JOINING COMPOSITES

14.2.1 General Methods

Depending on the type (i.e., the matrix) of the composite(s) to be joined, mechanical fastening, adhesive bonding, welding, brazing, or soldering, as well as some of the hybrid processes (such as rivet-bonding, weldbonding, or weldbrazing) may be possible.

Generally, for organic polymer-matrix composites or so-called fiber-reinforced plastics or (FRPs), there are two basic options, in decreasing order of usage: adhesive bonding and mechanical fastening. For those organic-matrix composites where the matrix is thermoplastic, welding or fusion bonding is also possible and is growing in usage. Weldbonding of thermoplastic composites is also a viable option, while rivet-bonding is possible with either thermosetting or thermoplastic composites.

For metal-matrix composites, the most common options for joining are, in approximate decreasing order of current usage, brazing, adhesive bonding, mechanical fastening, and welding. As techniques improve, welding is growing in popularity because of the superior properties possible. The hybrid processes weldbonding and weldbrazing are also quite common, while rivet-bonding is possible but less common. Intermetallic-matrix composites tend to be brazed or welded since elevated temperature service is the usual application.

For ceramic-matrix composites, including carbon–carbon composites, the primary options are, in decreasing order of preference, brazing and adhesive bonding. Welding of ceramic-matrix composites is possible but the subject of development only at the present time. Mechanical fastening is occasionally used. Table 14.2 summarizes the various options for joining various composites by type.

Regardless of the specific joining process used, in composites, perhaps more than any other material, joint design is critical. This is because loading must be carefully controlled to take advantage of any anisotropy and to avoid any problems (e.g., delamination through the thickness).

14.2.2 Mechanical versus Adhesive Joining of Composites

For many composites, mechanical fastening and adhesive bonding have been and still are the most common methods of joining. Although adhesive bonding is the principal method for joining composites to other composites, to metals, or to various other materials (such as sandwich honeycomb core), mechanical fastening is used in specific applications where adhesive bonding is not appropriate or optimum.

As usual in joining, the function of the joining material or device (e.g., fastener or interlocking feature) is to transfer load from one joint member to the other. The ability to transfer loads depends on the type of load, the environmental conditions, and the materials being joined. Mechanical fastening and adhesive bonding each offer different advantages and suffer from different disadvantages.

In mechanically fastened composite joints, advantages relative to adhesively

Table 14.2   Summary of the Various Options for Joining
Various Composites by Type (in Descending Order of Use)

Organic- or Polymer-Matrix Composites:
  1. Adhesive bonding
  2. Mechanical fastening
  3. Rivet-bonding
  4. Fusion bonding (of Thermoplastics)
Metal-Matrix Composites:
  1. Welding
  2. Brazing
  3. Weldbrazing
  4. Mechanical fastening
  5. Adhesive bonding
Ceramic-Matrix Composites:
  1. Brazing (ceramic fillers; metal fillers)
  2. Cementing (inorganic adhesive)
  3. Adhesive bonding
Intermetallic-Matrix Composites:
  1. Welding
  2. Brazing
  3. Mechanical fastening
Carbon–Carbon Composites:
  1. Bonding (pyrolyzing organic adhesive)
  2. Mechanical fastening
  3. Brazing
  4. Adhesive bonding

bonded joints include (1) no special surface preparation of components is needed; (2) there is less critical mating or fit-up requirements than for bonding; (3) disassembly is possible without component damage; (4) inspection is possible by direct observation; (5) resistance to peel or other out-of-plane loading is greater; and (6) sensitivity to thermal, water, and other environmental degradation is lower. On the other hand, relative disadvantages include (1) unavoidable stress concentration from holes for fasteners; (2) possible leaks through open joints or fastener holes; and (3) weight penalty due to the weight of the fasteners, required inserts, and/or structural stiffeners or doublers.

For adhesively bonded joints, advantages compared to mechanically fastened joints include (1) little or no undesirable stress concentration but, rather, favorable load spreading; (2) little or no weight penalty with thin adhesive bond lines; (3) smooth external surfaces at the joint to improve aerodynamic or hydrodynamic flow; (4) freedom from galvanic corrosion in assemblies of dissimilar materials; (5) facility with thin sections and no risk of fastener tear-out or pull-through; (6) lower sensitivity to cyclic loading due to lower stress concentrations and better strain accommodation; and (7) freedom from fastener bearing problems in matrices prone to cold flow. On the other hand, relative disadvantages include (1) difficult or impossible disassembly without damage to components; (2) susceptibility to environmental degradation by high or low temperatures, water, humidity, salt or salt spray, solvents, biological agents, radiation, vacuum, and ultraviolet light; (3) requirement for special surface preparation to chemically and physically activate the bonding surface; (4) difficult inspection except by indirect, nondestructive means; and (5) difficult repair of processing or service induced defects.

14.3 JOINING OF ORGANIC (POLYMER) MATRIX COMPOSITES

14.3.1 General Mechanical Fastening

The use of mechanically fastened joints in fiber-reinforced plastics (FRPs) or organic- or polymer-matrix composites is a logical carryover from the fastening of isotropic materials such as metals, where a wealth of experience and understanding exists. The enthusiasm of designers to use mechanical fasteners with such composites has been tepid due partly to a lack of confidence in the ability of composites to tolerate holes and cutouts and partly to designers extending joining techniques used for isotropic materials like metals to composites without understanding or thinking about the anisotropic nature and failure mechanisms of composites. This is not so much the case today because of new understanding.

It is true that FRP (and other directionally reinforced) composites can be considerably weakened by the introduction of holes for fasteners. This is in part a result of the large tensile elastic stress concentration (K) that occurs in the region of such discontinuities. These can be as high as $K = 8$ versus $K = 3$ for isotropic materials. It is also partly a result of a lack of plasticity in composites.

In most isotropic materials, plasticity allows yielding to take place at regions of high stress, and the effects of stress concentrations on the final net failing stress is thus small. This is not the case for unidirectionally reinforced composites, which are essentially elastic in their behavior to failure, so the effect of stress concentration is to give rise to a low net tensile strain.

One technique used to offset the effect of low inherent plasticity is to reduce the degree of anisotropy in the vicinity of a hole, introducing some softening, or pseudo-plastic behavior, thereby increasing efficiency. One specific approach is to incorporate fibers that are oriented in different directions around the holes, another approach is to employ doublers to reduce the net stress in the section.

Mechanically fastened joints in composites display the same basic failure modes as do metals, i.e., tension or shear or bearing. In addition, however, two other modes are possible in composites, namely, cleavage and pull-out. These modes are shown in Figure 14.7. The failure mechanisms in FRPs are complex and varied compared to metals and are dependent on fiber or reinforcement type and orientation, surface treatment, joint design, hole quality, and matrix type.

There are several considerations in mechanical joining of composites, some of which are unique to these special, anisotropic materials. These include (1) the materials of the fastener (to avoid galvanic or anodic corrosion when coupled to highly cathodic

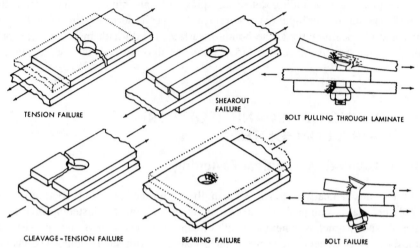

Figure 14.7   Modes of failure for mechanically gastened joints in composites. (a) Tension failure; (b) shearout failure; (c) bold pulling through laminate; (d) cleavage-tension failure; (e) bearing failure; (f) bold failure. (Reprinted with permission from L.J. Hart-Smith, "Design of Adhesively Bonded Joints," *Joining Fiber Reinforced Plastics* F.L. Matthews (Ed.), Elsevier Applied Science, 1986, p. 230.)

fibers, like graphite, in a dielectric epoxy); (2) the shape of the fastener head and foot (to spread loading and avoid excessive through-the-thickness loading, which could promote delamination); (3) hole size (to assure proper fit with the fastener); (4) the strength of the fastener; (5) loss of strength because of material being removed by drilling (i.e., fiber destruction); and (6) joint design. The methods used in hole preparation are usually the same as for monolithic materials, i.e., drilling and countersinking, but inserts are frequently used. Also, certain reinforcements (e.g., glass, Kevlar, and graphite) can be highly abrasive to cutting tools, including drills. Often, holes are sized to produce interference with the fastener, i.e., the hole is made slightly smaller in diameter than the fastener. This causes the fastener to introduce a compressive residual stress around the hole, which counteracts detrimental tensile stress concentration from applied loads.

The types of fasteners used with composites are essentially the same as those used with monolithic materials, such as metals, but with some special features. Self-tapping screws are rarely used because they are susceptible to thread stripping and pull-out from the matrix, which is low strength. Rivets are widely used, but with large heads and feet to distribute loading and avoid cold flow. Bolts are widely used, but with wide heads and nuts or with washers, and, often, with bushings or inserts, all to avoid cold flow from clamping and/or bearing. Finally, pins are used, but usually with hole inserts or bushings or sleeves to avoid bearing-induced cold flow. As mentioned previously, the type of material used in the fastener must be chosen carefully when graphite-reinforced composites are being joined to avoid preferential galvanic corrosion of the fastener.

The most common types of joint designs used for mechanically fastening composites are shown in Figure 14.8. Various fastener designs and inserts are also shown.

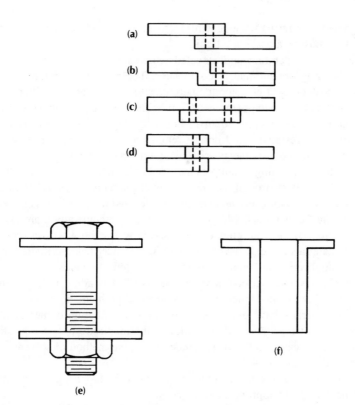

Figure 14.8 Common joint designs, fasteners, and inserts used for mechanically fastening composites. (a) Straight lap; (b) offset lap; (c) butt lap; (d) double lap; (e) large head bearing areas or washers; (f) sleeves.

14.3.2 General Adhesive Bonding

The adhesives most used for bonding organic, polymer-matrix composites are polymeric materials that are generally similar to the matrix (or matrices) of the composite(s) themselves. Thus, thermosetting polymeric adhesives are generally used for adhesively bonding thermosetting-matrix composites (e.g., epoxy–glass), while thermoplastic polymeric adhesives are generally used with thermoplastic-matrix composites (e.g., polyetheretherketone-, or PEEK-, graphite). Solvent cementing can also be used for bonding thermoplastic-matrix composites, just as it can be used for bonding monolithic thermoplastics.

Just as epoxies have been the most common matrix for organic-matrix composites, epoxy adhesives have also been the most common types used in adhesive joining. As described in Chapter 5, epoxy adhesives can be either one-component or two-component types. In the one-component type, the curing agent is already mixed in, and curing is initiated with heat. In the two-component type, the user mixes in the chemical curing agent just before use, and curing occurs by catalysis. The form of the one-component adhesives is usually a sheet, very much like a *prepreg* (i.e., a fibrous material preimpregnated with resin) without reinforcement, or it can be a paste. The two-component system must be a paste or a liquid.

Both room temperature and elevated temperature curing systems are used in adhesively bonding organic-matrix composites. The room temperature curing adhesives require postcuring at elevated temperatures to develop good mechanical properties for service at elevated temperatures. Curing times can range from a few minutes for simple, noncritical parts to more than 12 hours for larger, critical-performance parts.

Epoxy adhesives have good bond strengths and environmental resistance. Peel strengths are usually poor, however, so elastomeric materials (e.g., rubbers) are often added for rubber-toughening. Examples include butadiene, polyurethane, silicone, and polysulfides. While peel strength is increased over plain epoxy, shear strength is lowered, and sensitivity to moisture and creep (i.e., elevated temperature) is increased.

Thermoplastic materials, which are solid at room temperature, are also used as adhesives for joining organic composites, especially for those with a thermoplastic matrix. The materials are called *hot melt adhesives* since they are melted to become fluid for easy application and then reharden when cooled. Hot melts are most often used with thermoplastics and thermosets when the speed of bonding is important. These adhesives should be avoided when creep is a potential problem, however. Thermoplastic adhesives generally have good peel strength and good environmental resistance, except against solvents. As with monolithic thermoplastics, thermoplastic-matrix organic composites can be adhesively bonded using the method of solvent cementing (see Chapter 13, Section 13.4.3).

Specific adhesives for use with thermosetting or thermoplastic composites can be found in Chapter 13 and appropriate references.

The effectiveness of an adhesive bond depends on several factors, including (1) the polymeric composition of the adhesive; (2) surface preparation of the adherend; (3) adhesive lay-up procedure; (4) fitting of parts to be joined; (5) tooling; and (6) the curing process. In surface preparation, as discussed in the chapter on adhesive bonding procedure (Chapter 5), cleanliness (i.e., freedom from contaminants) is critical, and some surface treatment to increase the surface energy and area (i.e., by roughening) is useful.

In critical bond assemblies, a technique called a *mock bond* can be used to assess bond integrity. In this technique, the parts are prefitted and put through a simulated or mock bond cycle using adhesive sandwiched between release films (called verafilm). The parts are then disassembled and the verafilm is inspected to identify areas of poor fit or where extra adhesive may be required.

An adhesive bonding technique that has been proven to be very useful in bonding thermosetting-matrix composite materials to other composites or to metals is to apply the adhesive to the uncured thermosetting composite. Provided the adhesive is chosen to have a cure that is compatible with that of the composite matrix, the composite and the adhesive can be cured together, or *cocured*. This has the obvious advantage of allowing the adhesive and matrix materials to flow together somewhat, making the bond stronger. It also has the obvious advantage of combining processing steps for improved productivity.

The most common types of joints for adhesively bonding organic composites or other materials are shown in Figure 14.9. Note that all joint designs for adhesive

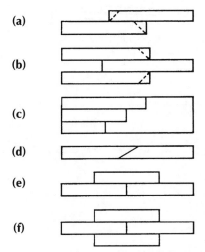

Figure 14.9   Common joint designs used in adhesively bonded composites. (a) Single lap; (b) double lap; (c) stepped lap; (d) scarf; (e) single overlay; (f) double overlay.

bonding attempt to force loading to occur in shear to take advantage of the typically higher shear versus peel strength of the adhesive. Whenever possible, simple lap-type joints are preferable, since no machining of the composite pieces is required. This joint type is not as strong as some other types, however, and can result in increased weight due to excess overlap with very little real improvement in strength. As shown in Figure 4.8, more complicated joint geometries, with multiple bond-lines, are employed for more demanding applications, including either higher loading or thicker adherends. These more complicated designs basically increase the total bond area and, unfortunately, the manufacturing cost.

One of the problems encountered in adhesively bonded joints is joint distortion due to deformation (i.e., bending) of thin adherends under eccentric loading and/or inability of the adhesive to comply elastically or plastically (i.e., behave in a brittle manner). This is shown in Figure 14.10. This problem can be overcome by using more complicated joint designs or stiffeners (see Chapter 4, Section 4.6).

In all composite joint types, proper composite design would orient the surface fibers in the joint area parallel to the direction of loading. The joint itself should be placed and designed to load in shear—avoiding peel or cleavage loading both of the adhesive being used for joining and between plies.[3] Tapering the edges of joint components helps eliminate stress risers under loading, thus reducing the tendency to peel, especially in lap joints. Special joint designs using stiffeners also are useful. Again, some of the schemes described in Chapter 4, Section 4.6 may be useful.

Although not the subject of this book, inspection of joints is key to their proper performance. As pointed out under the disadvantages of adhesive bonding compared to

---

[3]Recall that organic matrix composites are usually built up from many layers of unidirectionally oriented fiber-reinforced layers or plies. Bonding between plies is, in fact, accomplished by adhesives and is adhesive bonding.

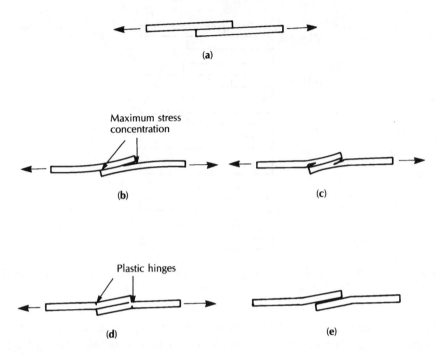

Figure 14.10   Joint distortion of simple, single-lap geometries due to eccentric loading. (a) Low load level; (b) deformation of adherends under moderate load; (c) interlaminar failure of filamentary composite adherends; (d) high load level (metal adherends); (e) permanently deformed (metal) adherends after failure of adhesive. (Reprinted with permission from L.J. Hart-Smith, "Design of Adhesively Bonded Joints," in *Joining Fiber Reinforced Plastics* F.L. Matthews (Ed.), Elsevier Applied Science, 1986, p. 291.)

mechanical fastening, inspection of adhesively bonded joints can be difficult. First, inspection must be indirect, and, second, joint areas are extensive. Table 14.3 summarizes various nondestructive evaluation (NDE) techniques for assessing the quality of adhesively bonded composite (or other) joints. For each technique, the principal characteristic detected, the relative advantages and limitations of the technique, and the types of defects that can be detected are listed.

14.3.3 Fusion Bonding Thermoplastic Composites

When the matrix of an organic composite is thermoplastic, joining can be accomplished by welding or fusion bonding as well as by mechanical fastening and adhesive bonding. Fusion bonding methods are described in Chapter 13, Section 13.5.3), so will not be described here. Suffice it to say welding is accomplished by cleaning the faying surfaces with an organic solvent and/or water and detergent, heating the surfaces to the point of melting, and then pressing the surfaces together to effect the bond. In addition

Table 14.3 Summary of NDE Techniques for Quality Assessment in Adhesively Bonded Joints

	Radiography	Ultrasonics	Acousto-Ultrasonics	Acousto-Emission	Thermography	Optical Holography	Eddy Current
Principle/characteristic detected	Differential absorption of penetrating radiation	Changes in acoustic impedance caused by defects	Uses pulsed ultrasound stress wave stimulation	Defects in part stressed generate stress waves	Mapping of temperature distribution over the test area	3-D imaging of a diffusely reflecting object	Changes in elec. cond. caused by material variations
Advantages	Film provides record of inspection; extensive data base	Can penetrate thick materials; can be automated	Portable, quantitative, automated, graphic imaging	Remote and continuous surveilance	Rapid, remote measurement; need not contact part; quantitative	No special surface preparation or coating required	Readily automated; moderate cost
Limitations	Expensive; depth of defect not indicated; radiation safety	Water immersion or couplant needed	Surface contact, surface geometry	Requires application of stress for defect detection	Poor resolution for thick specimens	Vibration—sensitive if not coupled	Limited to elec. cond. materials; limited penetration depth
Defects detected							
Voids	Yes	Yes	Yes	No	Yes[b]	Yes	Yes
Debonds	Yes[a]	Yes	Yes	Yes	Yes[b]	Yes	Yes
Delaminations	Yes[a]	Yes	Yes	Yes	Yes[b]	Yes	Yes
Impact damage	Yes[b]	Yes	Yes	Yes	Yes[b]	Yes	Yes
Density variations	Yes	Yes	Yes	No	No	No	Yes
Resin variations	Yes	Yes	Yes	No	No	No	Yes
Broken fibers	Yes[c]	Yes	Yes	Yes	No	Yes	Yes
Fiber misalignment	Yes	Yes	Yes	Yes	No	No	Yes
Wrinkles	Yes	Yes	Yes	Yes	Yes	Yes	Yes
Resin cracks	Yes[b]	Yes	Yes	No	Yes	Yes	Yes
Porosity	Yes	Yes	Yes	No	Yes[b]	Yes	Yes
Cure variations	No	No	No	No	No	No	Yes
Inclusions	Yes	Yes	Yes	Yes	No	No	Yes
Moisture	No	Yes	Yes	No	Yes	Yes	Yes

[a]Should be physically separated. [b]Minor damages may not be detected. [c]Might give problems.

Reprinted with permission from A. Brett Strong, *Fundamentals of Composites Manufacturing: Materials, Methods and Applications*, Society of Manufacturing Engineers, 1989.

to conventional heating (e.g., by hot gas, hot platens, and infrared sources), ultrasonic heating, friction or vibrational heating, and induction heating (which requires a metal screen insert called a susceptor) can be used to effect welding.

Welding or fusion bonding of thermoplastic-matrix composites may or may not require the use of a filler. If the entire laminate gets hot enough for the surface fibers to deform and intermingle with those of the mating surface, excess resin is usually not required. If this does not occur, additional resin (usually of the same type) may be added by any one of several means (e.g., sheets, powder, solvent-softened paste). Additional resin is often required in structurally reinforced composites during ultrasonic, friction, vibration, or induction welding.

Joint strengths achievable by welding or fusion bonding can be quite high, approaching the strength of the parent material, even when that material is a composite.

14.3.4 Joining Thermosetting to Thermoplastic Types

For joining organic-matrix composites or organic-matrix composites to monolithic polymers that are fundamentally different, e.g., thermosetting and thermoplastic, mechanical fastening and adhesive bonding are still viable options. The same practices described in Chapter 13, Section 13.7 apply. This challenge will be discussed in Chapter 15.

14.4 JOINING OF METAL-MATRIX COMPOSITES

14.4.1 Metal-Matrix Composites Defined

Metal matrix composites (MMCs) consist of high-performance reinforcements in a continuous metallic matrix. The reinforcements can be in the form of particles, whiskers,[4] chopped (i.e., discontinuous) or continuous fibers, or, even, laminates and can be metallic, intermetallic, ceramic, glass, or polymeric. Whiskers and discontinuous fibers can be random or oriented, while continuous fibers and laminates are always oriented. Metal matrices can be almost anything, but some typical ones are aluminum (e.g., with SiC or graphite), titanium (e.g., with SiC or B), magnesium (e.g., with graphite), and copper (e.g., with Mo, W or WC). Table 14.4 lists some common and emerging metal-matrix composites by matrix.

Metal matrix composites can offer a variety of significant property advantages over unreinforced, monolithic metals, depending on the material and form of the reinforcement. Modulus and strength, in both tension and compression, can be increased up to 100% with discontinuous reinforcements and much more with unidirectional, continuous reinforcements. Dynamic damping capacity is almost always increased (up to

---

[4]*Whiskers* are single-crystal forms of metals or ceramics, often containing a single screw dislocation down the central axis. As single crystals, even with a single screw dislocation, they have phenomenally high strength and modulus, approaching the theoretical maximum strength of the particular material.

Table 14.4   Common and Emerging Metal-Matrix Composites

Reinforcement \ Matrix	Al	Mg	Ti	Zr	Be	Cu	Ag	Ni	Co	SA[c]	Nb	Mo
Al_2O_3 (p, w, f)[a]	•	•	•	•		•						
B (f)	•	•				•						
BN (f)	•											
BorSiC (f)	•		•									
Carbon (f)	•	•				•	•	•	•			
Graphite (f)	•	•						•				
SiO_2 (p, w, f)	•											
Vitreous silica (f)	•	•										
Glass (f)	•											
SiC (p, w, f)	•	•	•	•	•	•		•	•			•
NbC[b]								•	•			
TiC[a]								•	•			
TaC[b]								•	•			
HfC[a]								•				
WC (pt.)	•					•	•	•				
Ni_3Al (f)								•				
Ni–Cr–Al–Y (f)								•				
Be (f)		•	•	•	•							
Stainless steel (f)	•							•				
W (f)						•		•			•	•
Mo (f)			•							•		
Ta (f)			•							•		
Nb (f)						•	•	•	•	•		

[a]p, particle; w, whisker; f, fiber or wire.   [b]In situ.   [c]SA= superalloy.

five times) because of scattering and attenuation, regardless of the form of the reinforcement. The coefficient of thermal expansion can be tailored to need (i.e., adjusted to a particular value), including being made zero or even negative in one particular direction. Wear resistance is usually increased but frequently at the cost of machinability. Fatigue strength and material damage tolerance are usually increased by absorbing the energy at the tip of a propagating crack, relieving the stress concentration, and arresting further growth. Anisotropy of most mechanical and physical properties can be controlled, either to minimize directional effects (i.e., approach isotropy) or to tailor effects for specific directional properties or damage tolerance. Finally, use temperature is increased through the use of refractory (i.e., metal, intermetallic, or ceramic) reinforcements (see Figure 14.4).

14.4.2 General Requirements for Joining MMCs

The high inherent thermal conductivity and serviceability of metals and the extension of their strength to higher temperatures through reinforcement by appropriate materials often dominate property requirements when MMCs are selected and used. Thus, when MMCs need to be joined, the joining process must result in bonding that can tolerate elevated temperatures and preserve the reinforcement across the joint as much as possible. Of course, there are applications for MMCs that do not require elevated temperature serviceability, so other joining options open up.

Preferred methods for joining metal-matrix composites are, in approximate decreasing order of preference for elevated temperature serviceability (1) welding, (2) mechanical fastening, (3) brazing, and (4) adhesive bonding (for polymer-reinforced metals or for nonthermal applications, especially). Welding allows service temperatures to approach the melting point of the lowest melting joint component or the mixed weld metal. Brazing allows service temperatures up to the limiting service temperature of the braze alloy. If diffusion brazing is used, this can be quite high, approaching the melting point of the base materials. Mechanical fastening can allow service to near the melting temperature of the matrix of the joint components, or to the limiting service temperature of the reinforcement, or to the limiting service temperature of the fastener, whichever is lowest. Adhesive bonding would rarely be used with MMCs unless they were selected for one of their special physical properties other than melting point or service temperature (e.g., controlled C.T.E.).

14.4.3 Welding MMCs

Whether a metal-matrix composite is being welded to another metal-matrix composite of the same or different type or composition or to a monolithic metal of similar or compatible composition to the matrix, the preferred processes will (1) preserve the integrity of the reinforcement, often by avoiding or minimizing melting or fusion; (2) minimize exposure of the reinforcement to elevated temperatures during welding to prevent the occurrence of degrading reactions between the reinforcement and the matrix; and (3) minimize forces that could compromise the integrity of whiskers or continuous-fiber reinforcements (e.g., high localized pressure, say from resistance welding electrodes, or from plastic upset, say from friction or inertia welding). The trick is to cause primary metallic bond formation between matrices or between the matrix and a metal component, with as little melting and at as low a temperature for as short a time as possible. Thus, preferred welding processes are, in descending order of preference, nonfusion welding processes, resistance welding processes, high energy density, low linear heat input processes, and low heat input, rapid cooling fusion processes. Viable options are inertia, friction, ultrasonic, and diffusion welding (all nonfusion processes); resistance spot and seam welding; laser- and electron-beam welding (two high energy density processes); and capacitor-discharge welding (a rapid heating and cooling resistance process). Gas-tungsten arc and plasma arc have also been successfully used, if heat input is kept low.

In the first two types of processes (i.e, nonfusion and resistance), the amount of melting is minimal and, for all except diffusion bonding (DB), time-at-peak temperature is short and postweld cooling is rapid. In diffusion bonding, the bonding temperature may be able to be kept low enough to avoid excessive thermal damage to the reinforcements, especially if special diffusion barrier coatings are used on fibers (e.g., silicon carbide on boron in borsic fibers). In resistance welding, the heating cycle is specially designed to melt a nugget and create a bond with minimal melting *and* minimal upsetting to avoid physically breaking brittle fibers. In laser-beam welding, the matrix can be made to melt rapidly, and melting or vaporization of the fiber reinforcement may be avoidable if short, high-energy pulses are used. In electron-beam and in continuous laser-beam, the intensely high energy density of the process usually causes vaporization (i.e., keyholing), so fiber damage is to be expected in the actual region of fusion. Total heat input is low, however, so thermal damage to reinforcements outside the fusion zone is minimized.

Capacitor-discharge welding has produced some excellent welds in some MMCs, most notably SiC-reinforced aluminum alloys. In the capacitor-discharge process, rod-to-rod or rod-to-plate welds are made by causing arcing and subsequent melting by discharging a bank of capacitors across a gap between the parts to be welded. Gravity is used to move the parts together during arcing. As the components meet, the arc is extinguished, solidification occurs, and excess molten metal is expelled. Heating and subsequent cooling are very rapid, minimizing time at temperature.

Exactly how susceptible the reinforcement of a metal-matrix composite is to damage by the heat and pressure of the welding process depends on the material and form of the reinforcement. If the difference in melting temperatures between the matrix and the reinforcement is great, with the reinforcement being much more refractory, the heat of welding for many processes (including certain fusion processes, such as GTAW or PAW) will likely cause little or no problem. When the melting temperatures are closer, or the reinforcement is inherently highly reactive with the matrix (or vice versa), the heat of welding can cause degradation of the reinforcement properties outright or through adverse chemical reactions between the reinforcement and the matrix. The reaction product is often inherently brittle, so degrades adhesion between the matrix and the reinforcement. Another common problem encountered during fusion welding of MMCs is severe outgassing and porosity formation in both the fusion zone and the high-temperature heat-affected zone (HAZ) of the weld. If the reinforcement is inherently resistant to thermal degradation of its properties but is reactive with the matrix, a diffusion barrier coating can be applied to the reinforcement. This is done on silicon carbide fibers intended for use in reinforcing titanium, using proprietary coatings, and on boron fibers using silicon carbide in so-called borsic fibers.

Susceptibility to damage by the pressure associated with some welding processes, such as resistance, inertia, or friction welding, is greatest with continuous fibers, troublesome with whiskers, less serious with chopped fibers, and essentially no problem with particulates or dispersoids. Short, discontinuous fibers and particulate reinforcements, if suitably refractory, may survive some fusion processes, provided vaporization from keyholing does not occur.

14.4.4 Brazing MMCs

Brazing can be an attractive option for joining MMCs, because no melting of the matrix occurs. Thus, degradation of reinforcements is usually minimal, especially if the reinforcement is reasonably refractory. Brazing MMCs to one another or to monolithic metals is accomplished using fillers that are compatible with the metals or alloys involved. In most cases, time at temperature should be minimized, so processes that localize heating to the area being brazed are preferable. For maximum temperature serviceability, provided the reinforcement is suitably refractory and nonreactive, diffusion brazing offers the best choice.

It is conceivable to add particulate or even discontinuous fibers or whiskers to the braze filler so that the braze itself is also reinforced, although this has not been widely done in practice.

14.4.5 Fastening MMCs

Metal-matrix composites can be mechanically fastened using threaded or unthreaded fasteners or by using inherent interlocking design features. As with all composites, care must be taken in placement of fastener holes to avoid or minimize damage to continuous fiber reinforcements. Reinforcing materials, in any form, can be abrasive, making hole drilling difficult and accelerating tool wear. Often, special joint designs soften the areas to be drilled and fastened, eliminating reinforcement in those areas or employing inserts. Deciding on preload can be made expecially difficult by the composites properties.

14.5 JOINING OF CERAMIC-MATRIX COMPOSITES

14.5.1 Ceramic-Matrix Composites Defined

A ceramic-matrix composite (CMC) has a matrix that is neither organic and resinous nor metallic. Generally, ceramic matrices are solid, inorganic, nonmetallic materials that contain positive and negative ions that are either ionically bonded or atoms that are covalently bonded. Mixed ionic and covalent bonding is also possible. The solid material may be crystalline, noncrystalline (i.e., vitreous or glasslike), or mixed (i.e., partially crystalline with a glassy phase mixed in). CMCs are characterized by high strength, refractoriness, and exceptional chemical and thermal stability, with improved toughness over monolithic ceramics, even over the so-called advanced or structural ceramics. Toughness is improved through the incorporation of the reinforcing phase, which can be metallic, ceramic, intermetallic, or even polymeric. Reinforcements can be in the form of particles, discontinuous or continuous fibers, or laminates.

Ceramic-matrix composites often behave quite the opposite of organic- or metal-matrix composites. The fibers in organic- and metal-matrix composites have a lower strain to failure than the matrix material and, thus, serve as the site for failure initiation. The matrix provides overall damage tolerance by yielding and remaining ductile. In ceramic/ceramic fiber composites, on the other hand, the matrix fails first, having the lower strain to failure. It is the fibers that impart toughness by various means, most particularly by absorbing crack energy along the fiber-matrix interface.

Table 14.5 lists some important metal- and ceramic-reinforced ceramic-matrix composites.

Table 14.5   Important Metal- and Ceramic-Reinforced Ceramic Composites

Reinforcement \\ Matrix	Al_2O_3	$(Al, Cr)_2O_3$	B_4C	CeO_2	Glass	$2Al_2O_3\cdot2SiO_2$	Si	SiO_2	SiC	Si_3N_4	ThO	TiC	ZrO_2	NiO
Al_2O_3 (f)[a]					•		•	•				•		
Carbon (f)					•									
Graphite (f)					•		•	•						
CaO														•
MgO							•	•						
Mullite											•			
SiC							•	•					•	
Si_3N_4									•			•		
BeO							•							
BN	•		•		•									
ZrO_2	•													
PSZ	•									•			•	
Cr		•		•									•	
Ni										•				
Nb		•		•									•	
Tn		•		•										•
Mo	•		•	•		•							•	
W	•		•	•		•								•

[a]f, fiber.

14.5.2   General Methods for Joining CMCs

The preferred methods for joining ceramic-matrix composites depend principally on the material of reinforcement, since this material largely affects temperature serviceability because of its inherent melting temperature or its reactivity with the matrix upon heating. Another important consideration, however, is matching of coefficients of thermal expansion (i.e., C.T.E.s).

The lowest temperature serviceability is for polymer-reinforced ceramics, which are far and away the rarest type. In fact, these composites, to date, are more properly considered ceramic-filled polymers. Examples are talc or calcium carbonate filled epoxies. For this group, adhesive bonding using organic adhesives is preferred. The reason is obvious—the polymeric reinforcement cannot tolerate elevated temperature during processing or in service. Adhesive types used are the same as those recommended for monolithic ceramics of the same basic composition as the matrix.

Metal-reinforced ceramics, formerly and still commonly called cermets, are usually designed for wear resistance with toughness. They have found wide application in cutting tools but are being considered for other demanding applications in heat engines and electronics. The preferred methods for joining are, in decreasing order, brazing, cementing or mortaring with inorganic adhesives, and adhesive bonding with organic adhesives. Brazing generally offers the highest temperature serviceability, with the limit being set by the metallic braze filler alloy or thermal mismatch between the metallic filler and the predominantly ceramic substrate. Bonding with inorganics, as in cementing or mortaring, can offer excellent service temperatures, depending on the specific material used. Fired cements offer the best service. Bonding with organic adhesives is most temperature limited but can be useful for nonthermal applications (e.g., electronics).

Ceramic-reinforced ceramic–ceramic composites (or CCCs) offer the greatest refractoriness and are often designed and selected for such applications. As a result, the preferred joining processes are welding, brazing, and cementing or mortaring. Welding can be by either nonfusion (e.g., friction welding or diffusion bonding) or fusion processes (e.g., LBW or EBW). Joint strength and temperature serviceability can both be excellent. Brazing can be accomplished using either a metallic or a ceramic filler, with the greater temperature serviceability for the ceramic filler. As for metal-reinforced ceramics, cementing and mortaring can produce excellent joints of quite good temperature serviceability if the cement is fired.

Two other options actually exist for joining CCCs during their fabrication: (a) sinter bonding and (b) self-propagating high-temperature synthesis (SHS) or combustion synthesis or thermite[5] synthesis. Both of these rely on bonding by diffusion, with varying amounts of liquid phase present. Sinter bonding was described in Chapter 12. Self-propagating, high-temperature synthesis or combustion synthesis is a process that relies on the high heat of formation of certain compounds from their constituent elements. Examples include oxides, carbides, nitrides, borides, silicides, and aluminides. The heat of reaction is so high that it literally propagates the reaction through the volume of reactants, leaving the reaction product in its wake. While not yet developed for use outside the laboratory, the potential exists for using SHS or combustion synthesis for joining difficult-to-join refractory materials, like CCCs.

14.5.3 Direct Bonding of CCCs

Without question the highest joint integrity in ceramic–ceramic composites, or CCCs, is obtained by direct bonding during the initial production of the composite. Two methods are available, depending on the method used to produce the composite: (1) self-propagating high-temperature synthesis (SHS) and (2) sinter bonding or sinter-HIP bonding.

In self-propagating high-temperature synthesis, or SHS, the CCCs of oxide–carbide types are made by reducing oxides in the presence of carbon. The resulting reaction is highly exothermic, and the heat generated propagates through the ceramic composite body. Joining can be accomplished by placing two prereacted shapes in contact and triggering the reaction. The process is also called combustion synthesis or thermite synthesis. Examples of such CCCs are: Al_2O_3/TiC, Al_2O_3/SiC, and MgO/SiC. The carbide phase appears as *rivers* in a matrix of the oxide.

In sinter bonding, various ceramic composites can be joined by diffusion, with various amounts of liquid phase present. Many possibilities exist, including Al_2O_3/SiC, Al_2O_3/Si_3N_4, SiC/PSZ, and sialon/TiN.

---

[5]The classic *thermite* reaction involves the highly exothermic reaction between Fe_3O_4 and Al to produce Al_2O_3 and molten Fe.

With these processes, overall shape complexity and component size is limited, but not as much as for individual molded ceramics.

14.5.4 Welding of CMCs

Ceramic–ceramic composites should be able to be welded to produce good joint integrity and excellent temperature serviceability using either nonfusion diffusion bonding or certain fusion welding processes such as LBW or EBW or even GTAW. The diffusion bonding process will closely resemble the direct bonding process of sinter bonding, although it may be practiced after initial sinter densification, relying on grain growth to effect the bond. In fusion welding, the integrity of the reinforcing phase may be lost, depending on its nature and form. Metal-reinforced ceramics can also be welded by solid-state diffusion bonding, usually with the aid of a metallic intermediate layer.

Specific methods for welding ceramics are covered in Chapter 12, Section 12.5.

14.5.5 Brazing of CMCs and CCCs

Ceramic-matrix composites can be joined to one another, to a monolithic ceramic, or to a metal using brazing. For ceramic–ceramic types, ceramic fillers would almost certainly be preferred. For metal-reinforced ceramics, metal fillers would likely be preferred, but ceramic fillers are also possible. The general methods of brazing ceramics that can be applied are described in Chapter 12, Section 12.4. Naturally, with brazing, joint design is important to offset the loss of continuity of reinforcement across the joint.

14.5.6 Bonding CMCs by Adhesives, Cement, or Mortar

Like monolithic ceramics, CMCs or CCCs can be bonded to other CMCs or CCCs or to monolithic ceramics using selected organic adhesives for nonthermal applications or using inorganic adhesives (i.e., cements or mortars) for elevated temperature applications. Specific methods are described in Chapter 12, Section 12.3.

14.6 JOINING CARBON, GRAPHITE, OR CARBON–CARBON COMPOSITES

14.6.1 General Comments

Carbon, graphite, and carbon-matrix or carbon–carbon composites vary widely in the degree of crystallinity, in the degree of orientation of the crystals (for crystalline types), and in the size, quantity, and distribution of porosity in their microstructure. They also all contain various impurities, especially oxygen or water. These factors are strongly dependent on the starting materials from which final materials are produced and on processing. The physical and mechanical properties of these products are, in turn, also strongly affected by starting materials and processing.

Carbon, graphite, and carbon–carbon composites offer exceptional elevated temperature strength and toughness when properly protected from oxidation. Carbon, after all, has the highest melting point of all materials, at 3,550°C (6,420°F). Carbonaceous materials also have the highest energy absorption or specific heat capacity of any known substance, so they resist heating. The coefficient of thermal expansion of crystalline graphite is highly anisotropic, and this property can be used to advantage to tailor the C.T.E. of composites containing graphite. It is for these properties, primarily, that these materials are employed. Some important properties for some common forms of carbonaceous materials are given in Table 14.6.

Joining of carbon, graphite, and carbon–carbon composites is especially challenging, but there are several options including, in decreasing order of usage, (1) mechanical fastening, (2) brazing, and (3) bonding, in a process related to adhesive

Table 14.6   Some Important Properties of Some Common Monolithic and Composite Carbon Materials

Material	Density (g/cc)	Tensile Strength (MPa)	Compressive Strength (MPa)	Modulus (GPa)	C.T.E. (per °C)
Low strength fiber, graphite felt[a] (un-carbide WDF), random (50–60%) in carbon	0.92–1.15	26–38	-	3.4–5.2	0.6–4.3
Low strength fiber, graphite cloth (un-carbide WCB), unidirectional (60–70%) in carbon	1.17–1.26	60–93	-	3.4–5.2	"
High mod. fiber, Modur Type I, chopped in carbon (70–75%)	1.58–1.62	43–100	-	3.13–3.8	"
Carb. fiber low strength cloth (Carb-I-Tex 100, 500, 700)	1.38–1.41	76–121	-	9.0–19	"
High mod. fiber, Modur Type I, unidir. (50–65%)	1.63–1.69	345–524	-	138–172	"
High strength fiber, Grafol Type II, unidir. (62–65%)	1.47–1.49	1034–1241	-	152–172	"
Fe	7.87	240	same	200	11.7
Ni	8.9	345–655	same	180	13.3
W	19.3	375–400	same	345	4.5
Ti	4.54	235	same	117	8.4
Cu	8.96	220-234		102	9.2
Al_2O_3	3.9–4.0	207	2588	380	8.8
ZrO_2 (PSZ)	-	270	1863	150	6.5
SiC	3.2	180	-	340–470	4.7
Si_3N_4	3.2	250	3447	-	2.1

[a]Flexural.

bonding. Brazing generally offers the best joint strength if the fundamental problem of achieving wetting can be overcome. Carbon and graphite are among the most difficult materials to wet with metals. Mechanical fastening must be done carefully to prevent damage to the inherently soft and sometimes brittle material and to prevent galvanic corrosion, as well as oxidation of the metallic fasteners. Carbon and graphite are extremely cathodic, so they render almost all metals highly anodic, most notably, magnesium, aluminum, zinc, steels, and nickel alloys. Since graphite and carbon–carbon composites are often used for their refractoriness, metallic fasteners can be subjected to severe oxidation. Nevertheless, mechanical fastening generally offers the highest temperature serviceability. Adhesive bonding, using organic adhesives, also presents challenges for achieving wetting but can be done. Obviously, service temperature is severely limited by the organic adhesive, unless, as part of the bonding process, the resinous adhesive is pyrolyzed, in which case the joint is as refractory as the base material. If this is the case, this approach clearly offers the highest temperature serviceability.

A special challenge in joining the carbonaceous materials to other materials is to match or otherwise deal with the coefficient of thermal expansion. The C.T.E. for these materials ranges from 2×10^{-6} mm/mm/°C to 8×10^{-6} mm/mm/°C but can be negative in some directions for some anisotropic forms of graphite.

14.6.2 Joining by Mechanical Fastening

Probably mechanical fastening is the most common method of joining graphite and carbon–carbon composites, particularly to other materials and especially to metals. The ablative tiles used to protect NASA's space shuttle from the severe heating encountered during reentry to the atmosphere from outer space are mechanically fastened to the underlying metallic structure. These tiles are used on the leading edges of wings and other control areas and are attached mechanically to allow expansion as well as replacement when needed.

Three problems must be dealt with when mechanically fastening carbon, graphite, or carbon–carbon composites to one another or to other materials: (1) physical damage to the carbonaceous substrate by fastening loads; (2) thermal stresses from C.T.E. mismatch; and (3) corrosion or oxidation of the fastener.

Carbonaceous materials can be mechanically fastened using interlocking part features, clamps, brackets, or fasteners (usually bolts). When bolts are used, clamping loads must be kept to reasonable levels, and stress concentration must be minimized. Large bearing areas under heads and nuts, often achieved through the use of washers but possibly by integral design features, are used. To minimize thermally induced stresses from the combination of extreme excursions of temperature (often from room temperature or below (in space) to over 3000°C (5400°F) and/or differential coefficients of thermal expansion, joints are designed to allow slip. Fastener holes are loose fitting rather than tight and slotted rather than round. Clamps and brackets and interlocking features are also designed to allow movement in at least one direction.

To prevent galvanic corrosion and/or high-temperature oxidation, fasteners are usually fabricated from titanium alloys or graphite or carbon–carbon composite. Oxidation-resistant coatings are also often used.

14.6.3 Joining by Brazing

The two major difficulties to be overcome in brazing carbonaceous materials to themselves or other materials are achieving wetting and dealing with thermal expansion.

Wettability

The wetting characteristics of all carbons, graphites, and carbon–carbon composites are generally poor but are strongly influenced by such impurities as oxygen or water that are either adsorbed on the surface or absorbed in the bulk of the material. Moisture absorption always occurs to some extent, with levels as high as 0.25 wt.%. Brazeability also depends on the size and distribution of pores, which can vary significantly from one grade to another, depending primarily on manufacturing method. Some graphites are so porous that all available filler metal is drawn into them, resulting in filler-starved joints. Others are so dense and impervious that adherence of filler by mechanical locking is extremely poor.

In order to promote wetting, surface activity must fundamentally be increased, and problems of impurities and porosity must be dealt with. Specific approaches for doing this can be found in references on brazing carbon and graphite (e.g., L.T. Anikin, *et al.*, "The High Temperature Brazing of Graphite," *Weld. Prod.*, Vol. 21 (No. 3), 1977, pp. 39–41; I. Amato, *et al.*, "Brazing of Special Grade Graphite to Metal Substrates," *Welding J.*, Vol. 53 (No. 10), Oct. 1974, pp. 623–628).

Thermal Expansion

A major consideration when brazing carbon, graphite, and carbon–carbon composites is the effect of the coefficient of thermal expansion of these materials. In these materials, C.T.E.s may be less than, equal to, or greater than those of the reactive or refractory metals (e.g., titanium, zirconium, nobium, tungsten, molybdenum, and tantelum) but are always less than the common structural metals (e.g., iron, nickel-, or cobalt-base alloys). In some highly anisotropic forms of graphite or carbon–carbon composites, the C.T.E. can even be negative in certain directions. Table 14.6 lists the C.T.E.s for various monolithic and composite forms of carbon compared to various other materials.

The mismatch with metals used as the bases for brazing alloys (e.g., titanium, copper, silver, gold, and platinum) is a particular problem. Joint failure can occur, particularly during thermal cycling (even during brazing), if too great a difference in C.T.E.s exists between the carbonaceous material and the brazing filler metal. Differential C.T.E.s can cause problems in getting the molten brazing filler metal to flow in the joint gap if the gap opens up too much or squeezes closed too much.

Brazing to a Dissimilar Material

Brazing carbonaceous materials to dissimilar materials can cause problems because the carbonaceous materials have little or no ductility and can be weak under tensile loading. These adverse conditions are usually compensated for in graphite-to-metal joints by brazing the graphite (or carbon–carbon composite) to a transition piece of a metal with a C.T.E. near that of the graphite. Examples of suitable metals for such transitions are molybdenum, tantalum, and zirconium. The transition piece can be subsequently brazed to a structural metal, if required. This technique minimizes shear cracking in the graphite by transposing the stresses resulting from the large difference in thermal expansion to the metallic components. Thin sections of metals, such as copper or nickel, that deform easily when stressed have also been successfully used for brazing graphite to dissimilar metals.

Brazing Techniques for Carbonaceous Materials

Graphite and carbon–carbon composites are inherently more difficult to wet with the more common brazing filler metals. Most merely ball up at the joint, with little or no wetting action. Two techniques are used to overcome this wetting difficulty. First, the graphite or carbon–carbon composite is coated with a more readily wettable layer, such as a 0.008 to 0.31 mil (0.0002 to 0.006 mm) thick layer of molybdenum or tungsten applied by chemical vapor deposition (CVD). This coated or metallized material can then be brazed using something like BCu-1. Second, brazing filler metals that contain strong carbide-forming elements, such as chromium, titanium, or zirconium, can be employed. Such elements react with the carbon to form carbides, and bonding is achieved.

Another brazing option, in theory, is to employ exothermic brazing or SHS joining. By using fillers containing materials that react to form carbides with substantial exothermy, joining should be possible.

Brazing Filler Metals for Carbonaceous Materials

Several brazing filler metals have been experimentally developed for brazing carbonaceous materials to themselves or to refractory metals. These include (1) 48Ti-48Zr-4Be; (2) 49Ti-49Cu-2Be; (3) 35Au-35Ni-30Mo; and (4) 70Au-20Ni-10Ta. Commercially available braze alloys that work are 68.8Ag-26.7Cu-4.5Ti and 70Ti-15Cu-15Ni. For joining graphite to steel, 80Cu-10Ti-10Sn works well.

14.6.4 Joining by Adhesive Bonding

Carbonaceous materials can be joined by the use of adhesives. Because carbonaceous materials (e.g., graphite and carbon–carbon composites) are usually chosen for their refractoriness in the first place, and service will inevitably involve extreme temperature exposure, the adhesives that are applied are usually pyrolyzed to affect the bond. This process changes the organic resin to a carbonaceous material itself by thermally decomposing the polymer and effects adhesion through the formation of covalent bonds. Two of the adhesive resins used are furfuryl alcohol and phenolics. Often, these resins are filled with carbon powder to reduce shrinkage and improve the final bond strength. Several proprietary formulations are available, but all employ the same basic mechanisms.

Carbon and graphite and carbon–carbon composites are prepared for general purpose bonding by (1) abrading with emery cloth, (2) removing dust, and (3) solvent degreasing with acetone.

14.7 ACHIEVING THE MAXIMUM JOINT INTEGRITY BETWEEN COMPOSITES

As stated at the beginning of this chapter, the challenge in joining reinforced materials or composites comes from attempting to match the strength or other properties in the joint area with the base material or substrate. The degree of success that can be attained depends on the matrix type and reinforcement form and on the nature of the joining process.

For organic-matrix composites, adhesive bonding can come close to matching substrate properties in the actual joint area, in both thermosetting and thermoplastic types. This is often achieved by using laminated splices or doublers, usually external to the composite laminate. In thermoplastic types, welding, using either heat or solvents (i.e., solvent cementing), can usually produce nearly identical properties in the joint and substrate.

For metal-matrix composites, the integrity or continuity of reinforcement is almost always lost across the joint, unless splice plates or doublers are used. If the form of reinforcement is discrete particles or short fibers, or whiskers, braze fillers could, conceivably be designed to contain these particles, fibers or whiskers, imparting reinforcement to the joint area. By proper joint design, solid-phase welding, using resistance, capacitor-discharge, or diffusion bonding techniques, could produce a close match of properties across the joint.

For ceramic-matrix and carbon–carbon composites, there are some good possibilities of achieving close property match between the joint and the substrate, provided the form of reinforcement is not continuous and aligned. In ceramic–ceramic composites, inorganic adhesives or ceramic brazes could be formulated to contain some reinforcing particles or whiskers. In carbon–carbon composites, bonding by pyrolyzing a carbon-filled resinous adhesive can come close to matching properties.

With mechanical fastening, the integrity or continuity of reinforcement is always lost in all types of composites, organic-, metal-, ceramic-, or carbon-matrix.

There is, in conclusion, no perfect means of joining composites, any more than there is of joining monolithic materials, perhaps even less so.

SUMMARY

Composites are materials that are actually composed of two or more other materials: one in the form of a *reinforcement* for the other, called a *matrix.* The resulting properties of the combination, or composite, are superior to or unique from those attainable in any individual, or so-called monolithic, material. Composites are principally characterized by the material that makes up their matrix and can be organic- or polymer-matrix (i.e., either thermosetting or thermoplastic), metal-matrix, ceramic-matrix, intermetallic-matrix, or even carbon-matrix. The reinforcement in any of these matrices can be metallic, intermetallic, ceramic, glass, polymeric, or carbon-based (e.g., graphite). The form of the reinforcement, which also characterizes the composite secondarily, can be discrete particles, discontinuous or continuous fibers (either randomly oriented or aligned), random or oriented whiskers, or oriented laminates.

The joining of composites poses a special challenge, i.e., how to attempt to achieve joint strength or other properties anywhere close to the strength or other properties of the parent composite substrate. This is because it is difficult to impossible to achieve continuity of the reinforcement across the joint. The options for joining composites depend principally on the matrix but also on the nature of the reinforcement.

Organic-matrix composites or fiber-reinforced plastics (FRPs) are usually adhesively bonded or mechanically fastened. Thermoplastic types can also be fusion bonded or welded, however. The polymeric adhesive used generally matches the matrix, i.e., thermosetting adhesives for thermosetting matrices and thermoplastic adhesives for thermoplastic matrices. For joining thermosetting to thermoplastic composites (or monoliths), adhesive alloys are used. In mechanical fastening, fiber damage must be avoided if possible during hole preparation. This is usually accomplished by "softening" the hole region by eliminating (or dropping-off) reinforcement or by using reinforcement-free inserts. The same general principles used with monolithic plastics apply otherwise.

Metal-matrix composites, or MMCs, can be brazed, adhesively bonded using organic adhesives, or welded. In welding and brazing, it is the metallic matrices that are being joined, so choice of filler (if any) and process must consider compatibility with these matrices. At the same time, however, damage to the reinforcement (especially continuous fibers) by both elevated temperature and localized pressure or upset must be minimized. Solid-phase welding or resistance- or capacitor-discharge welding (which cause minimal melting) are good choices. The object is to minimize the peak temperature and time at temperature during heating, to minimize physically damaging reinforcements outright or chemically through reactions between the reinforcement and the metal of the matrix. In mechanical fastening, fiber damage must again be considered, and galvanic corrosion must be avoided.

Ceramic-matrix composites, or CMCs, are usually brazed or adhesively bonded, although ceramic–ceramic composites (or CCCs) can be fusion welded. Some CMCs can be diffusion bonded, just as can their monolithic counterparts. In brazing, achieving wetting is the major challenge, and the same techniques are used as with monolithic ceramics, i.e., active brazes or metallization of the ceramic substrate. Adhesive bonding usually uses inorganic adhesives or cements or mortars. Mechanical fastening is generally not used.

Carbon, graphite, and carbon–carbon composites can be mechanically fastened, brazed, or bonded. In brazing, wetting is again a major challenge but can be achieved by metallizing or by using strong carbide-forming additives in the metal filler. In bonding, a resinous adhesive is pyrolyzed to effect bonding. Resulting properties are excellent.

In both inherently brittle ceramic- and carbon-matrix composites, matching of C.T.E.s is critical for all forms of joining but especially for brazing or cementing.

PRACTICE QUESTIONS AND PROBLEMS

1. Define what is meant by a *composite material*. Differentiate between a natural and a synthetic composite. Give some examples of each.
2. What is the role of the matrix in a composite material? What are the various basic materials that can make up the matrix of a composite?
3. What are some of the particular advantages of composites versus monolithic materials?
4. In general, what are the special challenges posed by a composite material to joining? Explain how the matrix and the reinforcement influence joining.

5.   What are the options for joining organic- or polymer-matrix composites?

6.   What are the special considerations associated with mechanically fastening organic- or polymer-matrix composites? How are these particular problems overcome in practice?

7.   How are organic- or polymer-matrix composites joined by adhesives? Differentiate between the adhesive bonding of thermosetting- versus thermoplastic-matrix types.

8.   How can reinforcement be achieved across an adhesively bonded joint or can it not be achieved?

9.   Explain how *cocuring* can be used to join thermosetting-matrix composites or fiber-reinforced polymers. What is the advantage of cocuring as a means of adhesive bonding versus more conventional adhesive bonding?

10. Explain why and how thermoplastic-matrix composites, or FRPs, can be fusion bonded. How could solvent cementing of such composites be considered fusion bonding? Or can it?

11. What are the special properties of metal-matrix composites, or MMCs, that tend to drive joining process selection? What are the options for joining MMCs, assuming service temperature is paramount?

12. Describe the special challenges of welding metal-matrix composites, or MMCs, and give some examples of specific processes that will produce sound joints.

13. What are the special problems associated with mechanically fastening metal-matrix composites? How are these problems overcome in practice?

14. What are the special properties of ceramic-matrix composites, or CMCs, that tend to drive joining process selection? What are the options for joining CMCs?

15. Explain what is meant by *direct bonding* of ceramic-matrix composites. Give two specific process examples.

16. Describe how the brazing of metal- or ceramic-matrix composites differs from the brazing of the monolithic metals or ceramics, respectively. How is the brazing of the monolithic and reinforced forms of metals or ceramics the same?

17. What are the special properties of carbon, graphite, and carbon–carbon composites that tend to drive the selection of a joining process? What are some of the particular challenges of joining carbon, graphite, or carbon–carbon composites?

18. Describe the problems associated with mechanically fastening carbonaceous materials.

19. Describe the problems associated with brazing carbonaceous materials.

20. Explain how the adhesive bonding or simply bonding of carbonaceous materials like graphite or carbon–carbon composites is different from conventional adhesive bonding.

REFERENCES

1.   Devletian, J.H., "SiC/Al Metal Matrix Composite Welding by a Capacitor-Discharge Process," *Welding Journal,* Vol. 67, No. 6, 1987, pp. 33-39.

2.   Matthews, F.L., (ed.), *Joining Fibre-Reinforced Plastics,* London: Elsevier Applied Science, 1987.

3. Moorhead, A.J., and Kennedy, C.R., "Brazing of Carbon and Graphite," *Metals Handbook,* Volume 6, *Welding, Brazing and Soldering,* 9th ed., Metals Park: Ohio: ASM International, 1983, pp. 1061–1066.
4. Schwartz, Mel M., *Brazing,* 2nd ed., Metals Park, Ohio: ASM International, 1990.
5. Schwartz, Mel M., *Ceramic Joining,* Materials Park, Ohio: ASM International, 1990.
6. Strong, A. Brett, *Fundamentals of Composites Manufacturing: Materials, Methods and Applications,* Dearborn, Michigan: Society of Manufacturing Engineers, 1989.

15

Joining Dissimilar Material Combinations

15.1 INTRODUCTION

15.1.1 Need for Joining Dissimilar Materials

As described in the introductory chapter of this book, one of the major reasons or needs for joining is to permit dissimilar materials to be used in an assembly, creating so-called hybrid structures. *Dissimilar materials* as used here refers to materials of different fundamental types (e.g., metals, ceramics, polymers, composites), as opposed to materials with different compositions within a particular type. Dissimilar materials enable the achievement of function where design requirements call for diverse and, often, divergent properties unobtainable in single materials.

Dissimilar materials often enable the attainment of high structural efficiency in several ways. They do this, first, by minimizing weight by using the lowest density material with the appropriate strength (i.e., the highest specific strength) or modulus or stiffness (i.e., highest specific modulus or stiffness), for stiffness critical designs, in each area of the assembly or structure. Second, dissimilar materials provide damage tolerance to the overall structure by changing the material, and the elastic properties, along a potential crack path. This imparts damage tolerance to the structure without having to rely on the use of discrete parts that are simply fastened together. Third, dissimilar materials optimize the design by matching the correct material to the needed property or behavior (e.g., refractoriness, electrical conductivity, thermal or electrical insulation, corrosion, or wear resistance), rather than compromising in some areas by settling for a less than optimal material used to fabricate the entire structure.

In addition to these important property advantages, the use of dissimilar materials often allows the costs of fabrication and/or operation to be minimized by allowing optimum materials to be used in specific areas of the design. In this way, inherently expensive or difficult-to-fabricate (and, thus, expensive to fabricate) high-performance materials only need be used where they are absolutely required and provide real benefit.

The availability of more basic types (e.g., advanced metals and alloys, structural and electronic ceramics, specialty glasses, intermetallics, structural polymers, and a host of synthetic composites) and varieties within a type (e.g., thermosetting or thermoplastic, polymer-matrix, metal-matrix, ceramic-matrix, intermetallic-matrix, and carbon-matrix composites) *and* the increasing demands for higher quality, better performance, longer life and higher reliability, and lower cost lead designers to use more and more combinations of materials in their designs. Thus, joining of dissimilar materials is becoming increasingly important.

15.1.2 The Special Challenge of Joining Dissimilar Materials

The fundamental challenge of joining dissimilar materials is compatibility! In order to produce an acceptable joint in terms of structural strength, structural efficiency and structural integrity (i.e., quality and reliability), the different chemical, physical, and mechanical properties of the various materials being joined must be, or must be made to be, compatible. This becomes increasingly difficult as the basic nature (i.e., atomic structure, microstructure, and properties) of the various materials involved become more different.

Joining a crystalline ceramic to an amorphous glass is relatively simple, just as joining a thermoplastic polymer to a thermosetting-matrix composite is relatively simple. In both cases, there is reasonable chemical compatibility (i.e., bonding is the same) and physical properties (e.g., melting or glass-transition temperatures, C.T.E.s, electrochemical activity) are close. On the other hand, joining a metal to a glass or a metal to a ceramic can be much more difficult. For these combinations, fundamental structures and properties differ greatly.

The compatibility of chemical, physical, and mechanical properties is important, both during the actual process of attempting to create the joint and during the operation of the joint in service. If chemical incompatibilities are great enough, it will usually be impossible to create joints by fusion welding, since intermixing of the materials will not occur, or, if it occurs, adverse chemical reactions will lead to embrittlement and cracking. If joining processes are used that do not require mixing (such as brazing or mechanical fastening), bonding may be established, but adverse chemical reactions (e.g., formation of brittle intermetallics or galvanic corrosion) may occur with time, especially in certain aggravating environments (e.g., at elevated temperature or in the presence of water). If physical properties are drastically different, the same thing applies. It may be impossible to produce a satisfactory, crack-free joint by fusion welding, brazing, or even soldering if C.T.E.s are drastically different. If the differences are less drastic, such problems as distortion or cracking may occur with time in service, regardless of the process used to create the joint. Even mechanical properties can be incompatible, although this is less common. Drastic differences in strength or ductility could, for example, cause the joint to be of limited utility.

Achieving successful joints between dissimilar materials requires a sound understanding of the inherent nature of the materials to be joined *and* of the various joining options and their means of achieving bonding.

In Chapters 2 through 10, the various joining options were described. In Chapters 11 through 14, the basic material types were described in terms of how they can be joined to themselves. Now we need to look at combining different materials into sound structural assemblies using any or all of the joining options at our command.

15.2 LOGICAL AND ILLOGICAL COMBINATIONS OF MATERIALS

While every combination of metals, intermetallics, ceramics, glasses, polymers, and composites is conceivable, some combinations are logical and some are illogical in terms of their compatibility or their incompatibility. The more dissimilar two materials are, fundamentally, the more one must question the logic of their being combined. Of course, whether a combination makes sense depends on the application, i.e., the demands. Some combinations may not make sense for use in primary structural applications[1] but may be acceptable for secondary structural applications or for nonstructural applications.

Polymers and metals are fundamentally different—in strength, elasticity, plasticity, melting temperature, service temperature, electrical and thermal characteristics, chemical stability, etc. Polymers might logically be joined to metals, however, where there is a need to (1) provide corrosion protection of a surface, (2) reduce friction at a surface, (3) enhance damage tolerance by arresting crack propagation, or (4) reduce weight (in lower loaded areas. On the other hand, if the metal were also being used for its resistance to elevated or cryogenic temperature, then joining to a polymer would not be logical as the polymer would not perform well. Likewise, joining a ceramic to a polymer is generally illogical since the ceramic is almost always being used for its refractory properties, and it is inherently corrosion and wear resistant as well.

Just as some combinations of materials might generally be considered illogical because of the dramatically different structure and properties of each, the difficulty in joining such significantly different materials is also usually great. Further, the options for producing sound joints become more limited.

Table 15.1 summarizes the logical combinations of materials in joints for various primary service conditions and also quantifies the difficulty that can be expected in creating a suitable joint. Preferred joining options, in decreasing order, are also listed. This tabulation is not precise but can serve as a useful guide. The hybrid processes of weldbonding, weldbrazing, and rivet-bonding, would only be practical or possible if their fundamental, parent processes were possible.

In the following sections, methods for joining various material combinations will be described by combination.

---

[1]As used here, and as typically used, *primary structural applications* are those where the successful performance of the material is essential or critical to the performance of the structure. Loads are usually high and the principal load path runs through such material elements. Failure of a primary structural element or area will result in the complete and, often, immediate loss of the system. *Secondary structural applications* are those where failure of this structural element or area will not result in the complete and immediate failure or loss of the system.

Table 15.1 Summary of Logical Combinations of Materials in Joints, Relative Difficulty of Joining, and Joining Options

Material 1	Material 2	Difficulty [a]	Options [b]
Metal A	Metal A-matrix composite	2-3	B, W, M, A
	Metal B-matrix composite	3-4	B, M, A, W[c]
	Ceramic	3-5	B, A, M, S
	Ceramic-matrix composite	3-5	B, A, M
	Glass	3-4	S, M
	Intermetallic compound	3-5	B, M
	Thermoplastic polymer	2-3	A, M
	Thermosetting polymer	2-3	A, M
	Organic-matrix composite	2-3	A, M
Ceramic A	Glass	1-3	W, S, A
	Ceramic A-matrix composite	1-3	B, M, A, W
	Ceramic B-matrix composite	2-4	B, M, A
	Metal-matrix composite	3-5	B, M
	Intermetallic compound	3-5	B, M, W[c]
	Polymer	2	A
Polymer	Organic-matrix composite	1-2	A, M, W[c]
	Metal	2	A, M
	Metal-matrix composite	2	A, M
	Ceramic	2	A
	Ceramic-matrix composite	2	A

[a]Ranked as 1, no problem, to 5, extremely difficult or impossible.
[b]A, adhesive bonding; B, brazing; F, fusion welding; M, mechanical fastening; N, nonfusion welding; S, soldering; T, thermal spraying.
[c]If compatible.

15.3 JOINING METALS TO CERAMICS

15.3.1 General Comments

Ceramics are becoming increasingly important as engineering materials and, as such, are being considered for wider and more demanding applications. Frequently, it is desirable to combine the properties of a ceramic and a metal in a structure, so joining these fundamentally dissimilar materials becomes necessary.

Generally, ceramics are stronger (especially in compression), more refractory, less reactive, and less thermally expansive or conductive than metals. Some of these differences are shown in Table 12.1. Obviously, there are some exceptions to these generalities, such as the high thermal conductivity of BeO or the refractoriness of molybdenum, tantalum, and tungsten, but for the most part they apply.

These differences in properties arise from fundamental differences in atomic structure and microstructure. On an atomic scale, ceramics are bonded ionically or covalently or by mixed ionic–covalent bonds, while metals are bonded metallically. These different bond types give rise to different cohesive strengths and melting points (e.g., higher for the higher strength ionic and covalent bonds than for the metallic bonds), differences in plasticity or ductility (due to the ability of metals to slip without disrupting charge balance and bringing like charged ions together), and differences in electrical, thermal, optical, and magnetic properties (due to the difference in electron distribution). On a microstructural level, ceramics contain inherent microflaws or microcracks that limit tensile strength (compared to compressive strength) and toughness. All these factors mean that each material can meet different demands. Such differences in structure, chemistry, and properties, however, can make joining difficult or impossible.

It should be pointed out that ceramics and metals can be combined to produce joints for structural applications where strength and/or vacuum tightness (or hermeticity) are often the primary requirements or may be considered simply to form a nonstructural ceramic–metal interface for electrical and electronic components or for chemical catalysis or corrosion resistance. In many structural applications, specific characteristics of the ceramics have led to their selection for critical components, forming parts of a total system that is largely metallic. Here, the success of the system depends on the ability to form ceramic-to-metal joints of adequate quality and integrity.

As pointed out in Chapter 1, there are many reasons for wishing to join particular ceramic and metal components. The motives can usually be related to design, manufacturing, or economic factors. In a spark plug, an insulating ceramic must be bonded to a conductive metal electrode to function. Metal may be needed to support a ceramic and provide a degree of toughness as well as provide a sink for heat dissipation, as in a ceramic-tipped cutting tool. Metal may be used to reduce cost by changing from ceramic to metal as soon as the ceramic is no longer needed for its principal properties, such as refractoriness or wear resistance in furnace liners or in engines.

15.3.2 General Methods for Joining Metals to Ceramics

Macroscopic[2] metal–ceramic components can be joined by one of three major techniques: (1) mechanical joining by interlocking or with fasteners or fastening devices; (2) direct joining by nonfusion or fusion welding; and (3) indirect bonding by organic or inorganic adhesives, braze alloys, or solders. Whatever the process, aside from mechanical joining, the formation of successful joints, here as in other material combinations, depends on the achievement of intimate contact between the joint elements, the conversion of these contacting surfaces into an atomically bonded interface, and the ability of this interface to accommodate thermal expansion mismatch stresses generated during cooling after fabrication or temperature changes in service.

15.3.3 Mechanical Joining

The mechanical joining or attachment of ceramics to metals has been important in traditional applications and is still important in certain new applications. Traditional

---

[2]Microscopic bonding or bonding on the microscopic scale is an issue in composites for achieving effective load transfer between reinforcing fibers (for example) and the matrix. Bonding at this level will not be considered in this treatment.

applications include the "tying-in" or attachment of furnace wall or roof refractories with metal hooks, hangers, brackets, or interlocking dogbones. A newer application is the mechanical clamping of heat-resistant ceramic thermal protection tiles on the leading edges of NASA's space shuttle metal wing.

Less popular, but possible, is attachment by more conventional mechanical fasteners such as screws or bolts. To avoid damage to the brittle ceramic because of localized stress concentrations from bearing loads, bearing areas must be large. Oversized heads and/or feet on bolts or nuts or washers are typically used. Another means of overcoming localized points of high stress during mechanical joining is to use a compliant metal interlayer between the metal and the ceramic part. One example is the use of a platinum intermediate layer between a nickel-base superalloy turbine disc and an Si_3N_4 blade.

Fastener holes in ceramics can be difficult to produce unless they are molded and fired in. Inherently hard, abrasion-resistant ceramics will cause excessive tool wear and, often, will suffer damage themselves by chipping or fracture. Again, to avoid localized bearing, holes are often sleeved with metal inserts, especially when screws are used.

Mechanical joining provides fairly good mechanical integrity under moderate loads, and thermal expansion differences are relatively easy to accommodate by design. The single greatest advantage of mechanical joining, here as anywhere, is that disassembly can be readily accomplished, thereby facilitating service, maintenance, repair, expansion or, change.

15.3.4 Direct Joining by Welding

Metals and ceramics can be directly joined or bonded by using solid-phase, nonfusion, or to a lesser extent, fusion welding processes. Diffusion bonding is usually the solid-phase welding method of choice, while fusion welding has been accomplished using high energy density processes such as electron-beam, laser-beam, or imaging arc.

Diffusion Bonding

The success of diffusion bonding or solid-phase welding by hot pressing or pressing and sintering at elevated temperature depends critically on the achievement of adequate interfacial contact. This contact is achieved by pressing together very flat or contour-matching surfaces and relying on intimate contact between increasing numbers of microscopic asperities. Initially contact is increased by plastic deformation of the metal component, but subsequently it is increased by creep, grain growth, and diffusion, which spheroidize and seal any residual porosity.

As opposed to fusion welding, a similar refractoriness (i.e., melting point or range) of the ceramic and the metal is neither necessary nor usual in diffusion bonding. There are reports of successful bonds between such metals as Ag, Al, Au, Pd, Pt, Pb, Zn, Cu, Cr, Ni, Fe, and Nb to alumina. One practical example is the diffusion bonding of Nb to alumina sapphire (at 1700°C or 3092°F) in high-pressure sodium vapor lamps. Although diffusion does take time, bonding can occur surprisingly quickly, sometimes in seconds but usually in several minutes or hours.

The direct solid-state bonding of both noble and transition metals to ceramics without the use of any intermediate material or interlayer is believed to be due to some combination of surface and bulk chemical reactions. Usually, the joint is heated under pressure to near the melting point of the metal. The metal is usually in a sheet or foil form. For nonoxygen-active metals (e.g., the noble metals), strong bonds have been produced at pressures around 1 MPa (or about 150 psi). For oxygen-active metals, higher pressures of 10 MPa (or about 1500 psi) are required. Joint shear strengths can be quite high, up to 8-12 kg/mm^2 (or 10 to 15 ksi).

Diffusion bonding can also be accomplished using a *graded powder bonding technique* or functional gradient material (FGM) joint. Here, the metal-to-ceramic bond is formed by pressing layers of metal and ceramic powders together. By grading the composition of the powder layers from pure metal on the metal side of the joint to pure ceramic on the ceramic side of the joint through intermediate composite compositions, chemical compatibility as well as physical property matching is achieved. This technique is analogous to the graded seal or matched seal commonly employed in metal-to-glass sealing.

Achievement of the necessary contact and bonding for a particular application depends on the judicious selection of not only the materials but also the bonding parameters such as surface roughness, surface cleanliness, processing environment, pressure, time, and temperature. Some roughening of the metal surface is highly beneficial, but care must be exercised to maintain macroscopic contour and freedom from contaminants. Increasing the bonding pressure or time improves contact and bond quality but not nearly as much as increasing the temperature. Prolonged times or higher temperatures can lead to chemical reactions at the ceramic-metal interface that initially enhance bonding but often cause degradation, as their progressive growth can generate volume mismatch strains and stresses from the reaction product.

Usually, the environment should be inert, e.g., a vacuum or gas with a low oxygen activity. However, beneficial reactions can be induced by selecting the appropriate bonding environment. For example, oxygen enhances the bonding of Co, Cu, Au, Fe, Ni, Pd, Pt, or Ag to Al_2O_3, BeO, MgO, SiO_2 and ZrO_2. Table 15.2 lists some of the successful combinations of metals and ceramics that have been diffusion bonded.

A special variation of diffusion bonding for joining metals to ceramics is *electrostatic bonding*. Here, besides pressure, temperature, and time, voltage is applied across the components to effect a bond. The effect of the voltage is to cause ionic conduction and induce strong electrostatic attraction between components. The process has been used for joining metals and semiconductors to various glasses (e.g., borosilicates) or, to a far lesser extent, glass-containing ceramics (e.g., Al_2O_3).

Having achieved a bond, it is imperative to preserve it. Here, design is of crucial importance, especially for accommodating the strains and stresses generated during cooling after fabrication or from temperature excursions in service. This can be difficult in diffusion bonded joints since the pressure needed to cause bonding creates joints that are not suited to accommodating mismatches. Figure 15.1 shows that the effect of mismatch on bond strength can be significant.

Nonfusion and Fusion Welding

Nonfusion welds have been successfully produced between metals and ceramics using

Table 15.2   Some Successful Combinations of Metals and
Ceramics That Have Been Diffusion Bonded

Al_2O_3-Al	Al_2O_3-Nb
Al_2O_3-Au	Al_2O_3-Ni
Al_2O_3-Cr	Al_2O_3-Nichrome
Al_2O_3-Cu	Al_2O_3-Pb
Al_2O_3-Fe	Al_2O_3-Pt
Al_2O_3-Kovar	Al_2O_3Steel
Al_2O_3-Mo	Al_2O_3-Stainless steel
	Al_2O_3-W

B_4C-Si
BeO-Cu
NbC-Nb, Ta, Mo, W
SiC-Mild steel (using Co or Ni)
SiC-Nb
SiC-W or WC (using Pt or Ni)
SiO_2-Cu
ThO-W
TaC-Nb, Ta, Mo, W
UO_2-Stainless steel
ZrC-Nb, Ta, Mo, W
ZrO_2-Nb

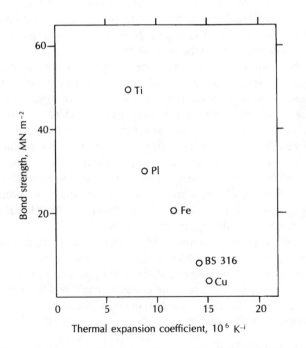

Figure 15.1 Plot of diffusion bond joint strength versus coefficient of thermal expansion for various metal-alumina combinations.

various friction processes, including ultrasonic welding. Conventional friction welding has been used to join aluminum cooling elements to alumina chip carriers, while ultrasonic welding has been used to join 94–96% Al_2O_3 to moly-manganese metallized Al_2O_3.

While rare, fusion welding, using high energy density processes, has been used to join metals to ceramics. Despite the difference in C.T.E.s, Al_2O_3 has been EB-welded to W, Mo, and Fe-Ni-Co alloys using very high accelerating voltages (90 kV) and low currents (2 mA). Laser welding has also successfully produced metal-to-ceramic welds.

15.3.5 Indirect Bonding

The most common method of achieving high integrity joints between metals and ceramics is to use a wide range of intermediate bonding materials. These intermediate materials can be organic adhesives, glasses or glass-ceramics, oxide mixtures (including cements and mortars), or metals. Actual bonding using some of these intermediates may be accomplished with the intermediate either in the liquid or solid state, depending on the intermediate and the process.

Organic adhesives and inorganic adhesive cements and mortars were discussed in Chapter 12, Section 12.3, as were glasses or glass-ceramics for joining ceramics to ceramics or to glass. Ceramic and metal brazes were also discussed in Chapter 12, Section 12.4. Much of what applied there applies here for joining metals to ceramics. The best joints are those developed after firing a suitable intermediate to fuse the joint elements together. Sometimes by oxidizing the metal component, bonding is facilitated with these inorganic adhesives or cements. Metal intermediates are used as either solid-state diffusion bonding agents (see Section 15.3.4) or as brazes or solders, with or without pretreatment of the ceramic surfaces to render them wettable. Often, the surface of the ceramic is metallized to facilitate wetting by the braze alloy or solder. The single greatest challenge of bonding ceramics to metals using additives or intermediates is, as usual, achieving wetting.

The special indirect bonding processes of *liquid-phase bonding* and *solid-phase bonding* will be discussed next.

Liquid-Phase Bonding

Both molten metals and glasses can be introduced between the mating surfaces of a metal and a ceramic to create joints, provided that both base materials are able to be wet and the filler adheres and remains bonded during postfabrication cooling. The process is either brazing or soldering, depending on the melting temperature of the intermediate (i.e., brazes above 450°C or 840°F, solders below).

The ability of metals and glasses to satisfy these requirements differs significantly. Metals do not wet oxides as readily as do glasses, but metals are far less sensitive to the detrimental effects of thermal mismatch stress generated during postfabrication cooling. Few ceramics are wetted by molten metals, but metals are wetted by molten glasses, in accordance with the usually large surface energy values for metals.

The usually poor wettability of ceramics is related to the nonmetallic character of their bonding. In ceramics, with their localized ionic or covalent or mixed bonding, electron movement is restricted compared to metals, with their delocalized bond character. There is some evidence that the wettability decreases with increasing ionicity. Thus, Pauling's electronegativity rules predict that oxides and fluorides are generally very ionic, and practice shows them to be the least wetted of ceramic families. Borides, phosphides, nitrides, and sulfides are less ionic and are often less nonwetted (i.e., more wetted). Carbides can display some metallic characteristics, such as electrical conductivity, and often can be wetted well.

One common approach to promoting wetting is to make the inherently different materials more similar. There are two approaches to making metals and ceramics more similar: (1) preoxidize the metal so that fabrication of a joint requires the easier formation of a similar oxide/oxide interface, and (2) metallize the ceramic surface using electroless plating or various vapor deposition processes to require the formation of a metal–metal interface. Both approaches are used in practice (see Chapter 12, Section 12.4).

Two examples of metallizing discussed earlier in reference to joining ceramics to ceramics using metallic braze fillers are worth repeating here. The first is a process known generally as the *sintered metal powder technique,* with the most common approach using Mo and debased Al_2O_3, in the process known as *moly-manganese.* Here, the alumina grains are held together by a glassy phase. When Mo is applied to the ceramic surface as a powder (often mixed with MnO_2, Mn, or Ti) and fired in a reducing atmosphere, the glassy phase migrates into the metal powder, binds the particles to each other and to the alumina, and thereby promotes wetting. Molybdenum and tungsten, with various metal oxides, are used for high temperature applications, while other approaches use mixtures with Rh, Fe, Ni, and Cr. Bond strengths can approach 70 MPa (or 10,000 psi) and higher.

The second approach uses TiH_2 activation. Here, titanium hydride powder is applied to the ceramic surface before vacuum brazing. During brazing, it dissociates (at 350-550°C or 662-1022°F) and forms a wettable Ti surface.

A third major approach is known as the *reactive-* or *refractory-metal salt technique.* Here, the ceramic substrate is painted with a solution of a refractory or reactive metal salt such as lithium molybdate. The ceramic substrate is dried and then treated at a high temperature to dissociate (i.e., reduce) the salt to a metal that bonds to the ceramic. Besides solutions in various solvents, molten salts can also be used. In one form of this approach, TiO_2 and KCl (or NaCl) react to form $TiCl_3$ and KO (or NaO), then the $TiCl_3$ reacts to form Ti, which bonds to the ceramic.

Other methods of metallizing a ceramic include sintering a finely divided mixture of metal and glass powder to the ceramic's surface to allow wetting of the metal particles by the braze or applying a metal layer by chemical or physical deposition processes. Examples include sublimation and vaporization (or vacuum metallizing), sputtering, ion plating, and thermal (e.g., plasma) spraying.

There is also the approach of using a braze filler that contains an active metal (such as titanium or zirconium) to react with the ceramic. These are called *active-metal brazes* (see Chapter 12, Section 12.4). The drawback of such brazes is their poor ductility due to the formation of complex microstructures containing intermetallics.

Besides brazing, metals and ceramics can be soldered together. Here, the ceramic joint element must definitely be metallized. Soft-soldered ceramic-to-metal joints are useful when the manufacturing and service temperatures must or can be kept low. Thermal expansion mismatch is not much of a concern since temperature excursions are limited and since the soft solder alloy can yield to accommodate strain. A common approach is to fire a layer of silver (Ag) onto the ceramic at the joint faying surface at about 750° to 850°C (or 1382° to 1562°F). Sometimes the Ag is mixed in a glaze to promote adherence. Silver-bearing Sn-Pb solders are used to avoid dissolving the silver metallized layer.

Solid-Phase Bonding

In principle, solid-phase bonding achieved by hot pressing using metal interlayers has two clear advantages: (1) it does not require that the bonding agent wet the ceramic, and (2) the ductility of the pure metal bonding interlayers can more easily accommodate thermal expansion mismatches. If low melting point metals are used, fabrication temperatures can also be kept low, minimizing any chances of thermal degradation of the workpiece. The process and mechanisms of solid-phase indirect bonding are the same as diffusion bonding (discussed in the previous section).

The most common interlayer to date is aluminum. It has been used to join titanium alloys to Al_2O_3, steel to Al_2O_3, and quartz to steel. For hermeticity or higher temperature service or caustic environments, however, aluminum is inadequate, and so copper, nickel, and silver have been considered.

Of special interest is the use of several interlayers of graduated metal–ceramic powder mixtures to produce a joint between severely mismatched workpieces. This technique, developed by the Japanese, was described in the previous subsection. It is another way of producing functionally gradient material joints.

15.4 JOINING METALS TO GLASSES

15.4.1 General Comments

Glass-to-metal seals or joints are important in electrical, electronic, vacuum, and some chemical applications. This combination of materials is important when optical transparency is required to allow direct observation of some function or operation occurring within a system, for chemical stability, for permitting the achievement of high vacuum (10^{-8} Torr or greater), including ease of removing contamination, and, occasionally, for providing unique electrical, photonic, or optoelectronic properties. The metal portion of the system is often there for structural purposes or to dissipate heat.

Glass-to-metal seals or joints can be assigned to one of four types. Type 1, matched seals, are those in which the metal is sealed directly to the glass, the resulting stress in which is kept to within safe limits by selecting a glass and a metal with coefficients of thermal expansion that are closely alike. Type 2, unmatched seals, are those in which the thermal expansion of the metal differs from the glass and the dangerously high stresses that would normally arise are avoided by using: (a) metal parts of very small diameters; (b) ductile metals, which by their yielding relieve some of the stresses in the glass; or (c) intermediate glasses and graded seals, the last (end) members of which match the substrates or joint elements. Type 3, soldered seals, are those in which the metal member is soldered to a layer of metal previously applied to the surface of the glass by one of several methods. Finally, Type 4, mechanical joints or seals, are those in which the joint between the glass and the metal components involves purely inter-locking and/or frictional (as opposed to chemical bonding) forces.

Often in metal-to-glass joining, hermeticity is more important than strength, pro-vided some strength is available to maintain joint integrity.

15.4.2 Properties of Metals Suitable for Glass-to-Metal Seals

A metal that can be attached to glass to produce a gas-tight seal should conform to the following general requirements: (1) its melting point must be higher than the working temperature[3] of the glass; (2) sufficient quantities of the metal should be available in a clean metallurgical state (i.e., free from nonmetallic inclusions); (3) it must be suffi-ciently ductile to enable it to be formed into wire or strip without mechanical defects (e.g., cracks); (4) the curves of thermal expansion of both the metal and the glass should, in the case of matched seals, follow one another closely over the same specific range of temperature; (5) no allotropic transformations, accompanied by marked changes in thermal expansion rate, should occur over the range of temperature to which the joint will be exposed (possibly –45 to 1200°C or –50 to 2000°F); (6) any layer of oxide formed in making the glass-to-metal seal should adhere firmly to both the metal and the glass; and (7) ease of joining to other metals by welding or brazing or soldering is desirable and, often, essential.

Lists of suitable metals and alloys for use with low thermal expansion, or *hard,* and high thermal expansion, or *soft,* glasses are given in Tables 15.3 and 15.4, re-spectively.

15.4.3 Glasses Used for Sealing to Metals

Many glasses, possessing quite different physical properties, are capable of wetting and, thus, fusing to metals. As a result, the only other requirement that needs to be con-sidered is C.T.E. matching. Table 15.5 lists some of the important physical properties of glasses and metals used successfully in seals. Notice that the C.T.E.s of the mating

---

[3]The *working temperature* of a glass is defined as the temperature at which the glass reaches a viscosity in the range of 10^3 to 10^5 Pa-sec (or 10^4 to 10^6 poise), at which level the glass can be shaped easily yet retain its newly worked shape. Because glasses do not melt at a specific temperature or, even, over a specific range (since they are amorphous) but rather soften continuously with increasing temperature, various viscosity ranges have been established for allowing various functions to be carried out. Other examples, besides *work-ing range* are: annealing range ($10^{11.5}$ to $10^{12.5}$ Pa-sec or $10^{12.5}$ to $10^{13.5}$ poise), softening point ($10^{6.6}$ Pa-sec or $10^{7.6}$ poise), and melting range (5 to 50 Pa-sec or 50 to 500 poise). The viscosity of water at 20°C is 0.001 Pa-sec or 0.01 poise, and 1 Pa-sec = 1 kg/mm/sec = 10 poise.

Table 15.3 Properties of Metals Suitable for Sealing to Low Thermal Expansion, or Hard Glasses

Metal	Melting Point (°C)	Maximum Operating Temp.		a × 10⁴ (20°C—350°)[a]	Ultimate Strength (tons/ sq. in.)	Yield Stress (tons/ sq. in.)	Elongation (% on 100 mm)	Spec. Elect. Resistance (ohm/cm)	Thermal Conductivity (cals/sq cm/cm/ °C/sec)
		In Vacuo	In Air						
Tungsten	3350	3000	300	4·4	99	85	4	5·6	0.38
Molybdenum	2450	2000	200	5·5	47	41	15—20	4·8	0.35
50% W, 50% Mo alloy	ca. 2800	2000	200	5·0	80	74	25—30	8·6	—
84% W, 12% Ni, 4% Co	—	—	—	6·8	—	—	—	—	—
Fernico I (54% Fe, 28% Ni, 18% Co)[b]	ca. 1450	ca. 1000	ca. 600	4·5	40	28—30	24	46	—
Kovar (54% Fe, 29% Ni, 17% Co)	ca. 1450	ca. 1000	ca. 600	4·7 (39)[c]	38—40 (27)[c]	25 (26c)[d]	32	44	0.04
Tantalum	2800	2500	—	6·5	—	—	—	15·5	0.13

[a]The usual symbol a is used throughout in referring to the linear coefficient of thermal expansion.
[b]*J. App. Physics*, 1941, **12**, 698.
[c]The figures in brackets were determined on specimens of iron–nickel–cobalt alloys made by pressing and sintering pure metal powders and subsequently fabricating into wire.
All temperatures, unless otherwise stated, are expressed in degrees Centigrade.
Data taken from T. Takamori, "Solder Glasses," *Treatise on Materials Science and Technology*, Vol. 17, Glass II, M. Tomozawa and R.H. Doremus (Eds.), Academic Press, New York, 1979, p. 186. Reproduced by kind permission of Society of Glass Technology.

Table 15.4 Properties of Metals Suitable for Sealing to High Thermal Expansion, or Soft Glasses

Metal	Melting Point (°C)	Maximum Operating Temp.		$a \times 10^4$ (20°C— 350°)[a]	Ultimate Strength (tons/ sq. in.)	Yield Stress (tons/ sq. in.)	Elongation (% on 100 mm)	Spec. Elect. Resistance (ohm/cm)	Thermal Conductivity (cals/sq cm/cm/ °C/sec)
		In Vacuo	In Air						
Platinum	1750	1600	1400	9·25	8—9	ca. 2	30—40	10·6	0.166
Copper	1083	400	150	17·8	16—17	9—10 (50—60)	30	1·75	0.920
Nickel	1452	900	400	14·5	34	16 (35)	25	7·5—10·0	0.14
Iron	1530	500	200	13·2	15—17	8 (40—50)	30	9·6	0.17
50% Ni, 50% Fe alloy	—	1000	—	9·5	35—36	22—25	25—28	49	0.025
26% Cr-Fe alloy	—	1000	1000	10·2	39—41	28—30 (35)	18—20	68	0.03

[a]The elongation figures in brackets wer determined on specimens of larger diameter over a gauge length = 4 $\sqrt{\text{area}}$.
Data taken from T. Takamori, "Solder Glasses," *Treatise on Materials Science and Technology*, Vol. 17, Glass II, M. Tomozawa and R.H. Doremus (Eds.), Academic Press, New York, 1979, p. 187. Reproduced by kind permission of Society of Glass Technology.

Table 15.5 Physical Properties of Glasses and Metals Used in Seals

Combination	Metal	$a \times 10^6$ (Metal)*	Glass	$a \times 10^6$ (Glass)*	Annealing Range (Glass)** (°C)		Color of Seal	State of Strain (annealed seals viewed at right angles to longitudinal axis)	Diameter of Wire (2a) (mm)	Diameter of Sheathed Single Wire Seal (2b) (mm)	Ratio b/a	Maximum Tensile Stress (after normal annealing) (kg/cm²)
1a	Tungsten	4·4	Corning 720MX	3·3	553°	510°	Straw to light brown	Severe compression	2·5	7	2·8	480 rl†
1b	"	"	Pyrex	3·2	—	—	"	"	1·0	4.1	4·1	520 rl†
2a	Molybdenum	5·5	Corning 705AJ	4·6	496	461	Light brown	Severe compression	2·5	7	2·8	215 rl†
2b	"	"	Corning G71	5·0	513	479	"	Slight tension	2·5	7	2·8	02 cl†
4a	Platinum	9·4	G.E.O.X4	9·6	520	450	Bright metallic	Severe tension	0·8	4·1	4·1	000 cl
5a	26% Cr-Fe	10·2	Corning G5	8·9	429	404	Greenish grey	Compression	2·5	7	2·8	128 rl†
6a	Fernico I	—	Corning 705AJ	4·6	496	461	Grey	Compression	2·5	7	2·8	118 rl†
6b	(64% Fe, 28% Ni, 18% Co)		Corning 705AO	5·0	495	463	"	Slight tension	2·5	7	2·8	59 cl†
7a	Fernico II (54% Fe, 31% Ni, 15% Co)	—	Corning 705AO	5·0	495	463	Grey	Strain free	2·5	7·5	3·0	About 10‡
8	British Kovar type alloys, e.g. "Nicosel," "Telcoseal No. 1" and "Darwin's F" alloys	4·5	B.T.H. C40	4·8	497	—	"	Slight compression	1·0	3·0	3·0	0—100 rl (according to metal)
9a	Fernichrome (37% Fe, 30% Ni, 25% Co, 8% Cr)	9·95	G.E.C. FON	5·1	500	440	"	Strain free	2·5	7	2·8	14 cl†
9a			Corning G5	8·9	429	404	"	Very slight tension	2·5	7	2·8	12 rl
9b		—	Corning G8	9·2	510	475	"	Compression	—	—	2·8	84 rl†
10	50/50 Ni-Fe Alloy	9·5	G.E.C. LI¶	9·1	410	350	"	Slight tension	1·04	3·75	3·6	34 cl
15	Copper	17·8	Many glasses if suitably shaped	3·5–10·2	—	—	Red to gold	Strain fee a fraction of millimetre from joint	—	—	—	—

rl. signifies in radial direction.

cl. signifies in circumferential direction.

* a = mean coefficient of linear thermal expansion between 20° and 350° except for Corning glasses where range is 0–310°.

† Values taken from Hull and Burger's paper.

‡ Values taken from Hull, Burger, and Navias' paper.

Data taken from T. Takamori, "Solder Glasses," *Treatise on Materials Science and Technology*, Vol. 17, Glass II, M. Tomozawa and R.H. Doremus (Eds.), Academic Press, New York, 1979, p. 187.

Reproduced by kind permission of Society of Glass Technology.

pairs of metals and glasses are close, generally within 1 to 1.5 x 10^{-6} mm/mm/°C (as was given as a guide for matching C.T.E.s earlier).

15.4.4 Methods for Producing Metal-to-Glass Joints and Seals

Various methods for producing joints or seals between metals and glasses are described in the following paragraphs for the four principal types: (a) matched seals, (b) unmatched seals, (c) soldered seals, and (d) mechanical seals.

Matched Seals

The construction used for the current leads of some lamps and vacuum tubes is typical of the matched type of glass-to-metal seal. The wires are surrounded by a flanged piece of glass tubing, which had been heated until thoroughly soft and, then, had been squeezed around the wires by pincer-like jaws. This type of construction is usually termed a *pinch seal.* It is the most common type of glass-to-metal seal.

In matched seals, alloys and glasses of similar thermal expansion are used, so that the joints are substantially free from strain. Metals or alloys in circular, disc, or tubular forms are typically used, to minimize stress concentrations from sharp features.

Unmatched Seals

Pinch seals can be made between alloys and glasses that are not matched in terms of their C.T.E.s, provided the expansion mismatch is dealt with in some other way. There are essentially three ways. First, metal elements of very small dimensions (i.e., diameter, or cross-sectional area, typically 0.8 mm or 0.035 in.) can be used. Second, ductile metal elements (e.g., of copper or molybdenum) can be used to distort under the stresses of mismatch, thereby relieving the stress in the glass. Third, glasses of intermediate expansion can be used to bridge the expansion difference between the metal and the final glass. Here, glass-to-glass bonds are made everywhere except between the final intermediate glass and the metal element.

Intermediate glasses, or *graded seals,* were quite common before the wide range of alloys and glasses listed earlier (in Tables 15.3 through 15.5) were available. Intermediate glasses were typically fluxes of easily fusible (i.e., low melting) glasses containing a moderate proportion of silica, lead oxide, boric oxide, and alkalis. The most popular is called *Pantin glass.* It contains 38.5% SiO_2, 53.23% PbO, 0.80% CaO, 0.38% MgO, 0.45% Na_2O, and 5.45% K_2O.

Cracking in seals made in this manner are reduced because stresses at the glass–metal boundary are lowered by (1) the lower temperatures required to make the seals, (2) the thin fusing layers (to reduce stresses), and (3) less oxidation of the metal element.

Soldered Seals

To overcome difficulties experienced in sealing metals directly to glass, joints can be successfully made by depositing a metal coating on the glass and, then, soldering this coating to the metal component of the seal. In the earliest and still most common method, the metal coating was applied by painting the glass surface with a suspension of fine metal powder or with such compounds as platinum chloride or silver oxide. This coating is then heated to deposit the metal by chemical reduction. Newer options for applying a metal coating include heating a film of silver or platinum in liquid suspension; evaporation or deposition in vacuum; cathodic sputtering; thermal spraying; deposition from an aerosol; or reduction of a metallic oxide. Subsequent soldering is fairly conventional (see Chapter 9).

Obviously, indium-based solders (Chapter 9, Section 9.4.10) can be used to join metals to glasses.

Mechanical Joints

Several clever mechanical methods of achieving glass-to-metal seals have been successful. In each, the joint or seal is obtained through mechanical interlocking as opposed to chemical bonding, often relying on extremely tight, even interference, fits. In one type of joint, called a *ground joint,* a metal pin (often low-expansion Invar) is precision ground to fit tightly into a tapered hole in the glass (typically fused silica). Sealing is assured by using mercury entrapped with a cement. Several possible designs are shown in Figure 15.2a and b. A second method involves producing a metal annulus with lead. The lead is first preplaced, then the glass is heated to collapse tightly around the metal component, and finally the lead is melted to create a tight seal. This approach is shown in Figure 15.2c.

Other methods for mechanically joining metals to glasses include (1) electroplating (e.g., with copper) to create a seal, and (2) thermally spraying molten metal to create a seal.

Some mechanical joints rely entirely on interlocking design features and use no filler to effect the seal. An example is shown in Figure 15.2d.

15.5 JOINING OF METALS TO POLYMERS

15.5.1 General Comments

The joining of metals to polymers is challenging because the atomic structures of the two material types differ greatly, and thus properties are quite different. The significant structural differences suggest joining methods that rely principally on mechanical forces and avoid intermixing, i.e., mechanical fastening and adhesive bonding. The viscoelastic behavior of polymers versus the elastic-plastic behavior of metals, however, means that the stresses imposed by the mechanical attachments must be kept low, otherwise severe deformation and loss of integrity can occur.

As in all joining of dissimilar materials, joining of polymers to metals can be facilitated by making the materials more alike. For this combination, the most common approach is to metallize the surface of the polymer, while another method is to fill the polymer with metal powder. Either of these allows soldering of the metallized layer to the metal component, often using fusible alloys.

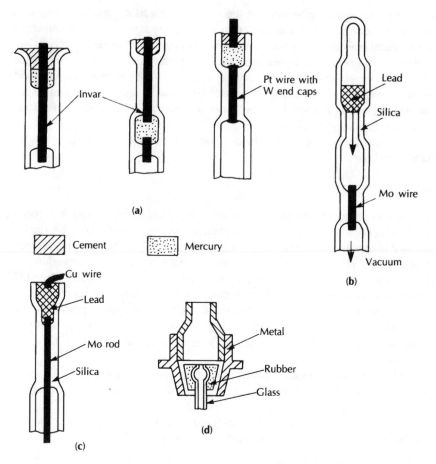

Figure 15.2 Ground-joint seal and lead-seal variations and mechanical methods for sealing metals to glasses. (a) Silica glass-metal; (b) sealing molybdenum to silica using lead; (c) lead-silica glass seal; (d) mechanical glass to metal seal.

15.5.2 Joining Methods

As mentioned earlier, there are three ways of joining metals to polymers: (1) mechanical fastening, (2) adhesive bonding, and (3) soldering of metallized polymers.

Mechanical Fastening

Polymers can be joined to metals by mechanical means using fasteners or interlocking design features. As in joining polymers to one another mechanically, joint and fastener design must consider the viscoelastic nature of the polymeric material. Suitable fasteners include screws, bolts, rivets, and pins. With each, to prevent viscoelastic deformation, or cold flow, broad bearing surfaces (e.g., heads, feet, and/or washers) are used.

Holes can be sleeved with threaded or unthreaded inserts. In any case, applied stresses should be kept low.

The most common approach to joining using design features is to take advantage of the polymer's elastic properties, using mechanical snap fits. Originally, this technique was seen in inexpensive assemblies, such as plastic toys, for example, with metal axles snapped into plastic clevises on cheap toy cars and trucks. Today, these same properties are used in snap-fit fasteners (see Chapter 3, Section 3.5) to facilitate assembly, as part of an aggressive design for assembly philosophy and approach to facilitate automated manufacturing.

Adhesive Bonding

Metals can be joined to both thermosetting and thermoplastic polymers through the use of adhesives. The bonding to the metal side of the joint results from some combination of mechanical locking and weak chemical bonds, with mechanical locking predominating. Obviously, on the polymer side, bonding is almost strictly chemical, of both primary and secondary types, as well as through chain entangling through interdiffusion.

In adhesively bonding metals to polymers, the adhesive should be matched primarily to the polymeric component, i.e., considering whether the polymer is thermosetting or thermoplastic but, obviously, the adhesive must also be suited to the metal adherend. As in adhesive bonding polymers to themselves or one another, thermosetting adhesives are generally used for joining thermosetting polymers; thermoplastic adhesives for joining thermoplastic polymers; and alloys of thermosetting and thermoplastic adhesives for joining thermosetting to thermoplastic polymers.

For metal-to-polymer bonding, epoxies are especially widely used, but cyanoacrylates are gaining popularity.

Adhesives should be selected carefully, considering the application environment, and preparation of both joint elements (or adherends) must be carried out properly and carefully.

Soldering

In electronic applications (e.g., printed wire boards), metals and polymers are joined by soldering, after the polymer surface has been locally metallized, usually with copper. The copper metallized *pad* is then soldered to the metal lead of a component, which itself may be a polymer or metal or ceramic.

Because metallization of the polymer is well described in references on electronics' manufacturing, it will not be discussed here. General methods include chemical or physical (e.g., vapor) deposition and thermal spraying. Metallized layers can be copper, tin, or tin-lead alloy, aluminum, chromium, or nickel. The primary mechanism for adhesion of the metal layer to the polymer substrate is largely mechanical locking. Adhesion between the metallized layer and the solder is partly mechanical and partly metallic bonding.

Solders are selected based on the application demands but are largely tin–lead compositions. The greatest demand placed on these joints is usually the stress and strain caused by mismatch of thermal expansion coefficients between the metal and the polymer. These mismatches can and do get very large, and joint failure can and does occur.

15.6 JOINING OF METALS TO COMPOSITES

15.6.1 General Comments

The need to join metals to composites is growing daily as composites are used more and more. While there are structures that are composed entirely of composites, many structures, especially for demanding applications, consist of a mix of monolithic metallic and composite elements. The combination of materials is often as much for economic as technical reasons and sometimes is to facilitate repair, because integral composite structures do not allow easy access for repair.

Because composites are almost always used for their superior mechanical properties (e.g., specific strength and/or specific modulus), the need for joints that can sustain high loads and stresses is extremely important. The special challenge of joining composites to metals or any other material arises largely from the variety of composites and, especially, their matrices. Organic, metallic, intermetallic, ceramic, and carbon types all pose different problems and, so, are treated in separate sections below.

15.6.2 Joining Metals to Organic-Matrix Composites

There are generally two options for joining metals to composites with organic or polymeric matrices: (1) mechanical fastening, and (2) adhesive bonding. A third option is rivet-bonding, which can offer the advantages of both approaches. Mechanical fastening is still the most common method, but adhesive bonding is gaining. Rivet-bonding is restricted primarily to high performance (e.g., aerospace) applications.

Mechanical Fastening

Advanced organic-matrix composite materials, which are increasingly being used in high performance structures such as aircraft have very different properties than the metals they typically replace. Although composites often reduce the number of structural components required in the assembly and offer alternative joining methods (e.g., adhesive bonding), mechanical fastening still plays a vital role.

Composite materials differ from monolithic metals by not being as ductile (i.e., exhibiting only elastic rather than elastic-plastic behavior) and by being anisotropic in many cases. Mechanical fastening of metals to organic-matrix composites usually uses fasteners. Mechanical fasteners used with organic-matrix composites carry or transfer shear loads through the joint. They typically develop clamping forces, however, and resist loads at the joint that act in the through-the-thickness direction, even if they are developed secondarily by bending or other complex loading. This turns out to be the weakest direction in most composites and in all laminated composites.

For the foregoing reasons, fasteners must be carefully designed and selected for metal-composite joining, considering several factors. First, galvanic compatibility must be considered. When metal fasteners are coupled with composites containing graphite fibers and are exposed to a corrosive environment, graphite's low electrical potential causes the fastener to act as an anode and corrode, often very rapidly. Current density is the best indicator of compatibility, as shown in Figure 15.3. Titanium and its alloys (e.g., 6Al-4V), Fe-Co-Cr multiphase alloys MP159 and MP35N, and Ni-base inconel alloys 600 and 718 are compatible with graphite fiber-reinforced composites, showing essentially no corrosion after 500 hours of 5% salt-spray testing. Corrosion-resistant steels A-286, PH13-8Mo, PH17-7, 301, 304, and 316 are also acceptable for use with graphite-fiber composites. Monel alloys, 400, 405, and 440 corrosion-resistant steels and aluminum are not compatible with graphite and should not be used. Even when fastener materials are compatible, sealants are generally applied to deny access to the corrosive medium.

Second, pull-through failures must be considered. When through-the-thickness shear forces act on a fastened composite joint, they can pull the fastener through the composite laminate. The failure load is influenced by bending moments, in-plane stresses, and dynamic effects that act in combination with the through-the-plane shear stress to lower the failure load. Structures joined with shear, flush-head, and blind fasteners are particularly susceptible to this kind of failure. Pull-through strength can be improved by using fasteners with larger bearing circumferences because they develop higher through-plane shear loads. Tension rather than shear-head fasteners should be used if pull-through strength is critical.

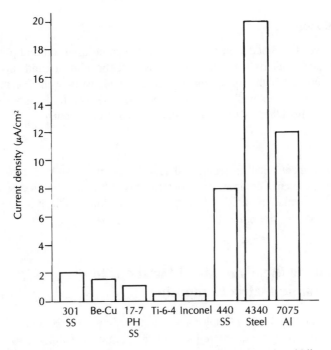

Figure 15.3 Plot of current density for metals coupled with graphite as an indicator of galvanic corrosion potential.

Third, loss of fastener preload must be considered. The viscoelastic nature of the matrix of organic-matrix composites can lead to the loss of the preload needed in a fastener in order for it to function properly. This phenomenon is called *stress relaxation*. Other problems with preload can arise from volume changes in the polymeric matrix due to environmental effects (e.g., swelling from moisture pick-up). Loss of preload can be minimized by using large bearing areas on fastener heads or nuts or by using washers.

Fourth, any tendency for fastener rotation must be considered. Fastener cocking, as opposed to turning under torque installation, in a joint is known as *fastener rotation* and leads to nonuniform bearing contact between the fastener shank and the fastener hole. At these more highly loaded points, the composite can fail. Fasteners with large bearing diameters resist cocking forces and retard undesirable rotation.

Fifth, galling of threaded fasteners must be considered. Galling or seizing can occur with some fasteners at nuts. This can lead to improper installation (e.g., cocking or insufficient preloading). To prevent galling, dry-film lubricants are used.

Sixth, damage to the composite during fastener installation must be considered. Some fasteners (especially metal fasteners) damage composites. Damage occurs when the clamping stress exceeds the compressive strength of the laminate; radial expansion of the fastener hole (from expansion of the fastener shank) delaminates or buckles composite plies; impact forces delaminate the structure; or turning fasteners abrade or splinter the composite surface. Current design practice is to avoid interference fits without using a protective sleeve, fasteners without adequate bearing area, or rivets and blind fasteners that expand radially. Rivet guns should also be avoided.

Adhesive Bonding

Many adhesives that bond well to the base resin can be used to bond polymers reinforced with such materials as glass fibers or synthetic high-strength fibers, such as Kevlar, to metals. For thermosetting matrices, epoxies work well. For thermoplastic matrices, hot-melts work well. Cyanoacrylates also work well for both types. The polymeric and metallic adherends must be properly prepared, of course.

Rivet-Bonding

The benefits of mechanical fastening and adhesive bonding can be combined by rivet-bonding organic-matrix composites to metals. This hybrid joining process is discussed in Chapter 10, Section 10.4.5. Let it be said here that the benefit of this process is the improvement to peel strength.

15.6.3 Joining Metals to Metal-Matrix or to Ceramic-Matrix Composites

Metals can be joined to metal-matrix and to ceramic-matrix composites, much as they can to their monolithic counterparts. Options, in decreasing order of preference, include brazing, adhesive bonding, and mechanical fastening. For low load applications, exposed to low service temperatures, soldering is another, less common, option.

For metals that are being joined to a metal-matrix composite (or MMC) where the alloys are the same or compatible, nonfusion and fusion welding are possible. Some diffusion welding is possible for metals to ceramic-matrix composites (or CMCs).

The major consideration in joining metals to MMCs or CMCs is minimizing damage to the reinforcement due to thermal degradation or chemical interaction or mechanical effects (e.g., breakage).

Joining to MMCs

Welding by solid-state DB, friction, resistance-spot, or high energy density fusion processes (e.g., EBW, LBW, and PAW) is possible between metals and MMCs where the matrix is compatible with the metal joint component. It is desirable to minimize the heat input to the joint to minimize adverse reactions between the reinforcement and the matrix in the heat-affected zone, as well as to minimize thermal degradation of the reinforcement itself in the fusion zone. For this reason, nonfusion processes, especially those that localize heating at the joint faying surfaces, are generally preferred. For non-fusion processes using pressure, pressure localization and plastic deformation or upsetting must be minimized to avoid damaging continuous reinforcing fibers. Fusion processes that produce minimal melting and involve rapid heating and cooling, such as resistance-spot or capacitor-discharge, are also viable options, however. The high energy density fusion welding processes (such as EBW, LBW, PAW, and imaging arc) have produced acceptable joints, provided the reinforcement is not totally destroyed by vaporization within the keyhole.

Brazing is probably the most common method for joining metals to MMCs. Normal procedures are followed, except particular care should be exercised to prevent thermal damage to the reinforcement or reaction between the reinforcement and the matrix. Essentially, any brazing process can be used, but localized heating methods may be preferred, depending on the particular composite (e.g., reinforcement and matrix). For particle or chopped fiber reinforcements versus continuous fiber reinforcements, heating damage is often less detrimental. Obviously, if mechanical loading will be low, or is of secondary importance, soldering can be used with little problem. The selection of filler for either brazing or soldering is based on the composition of the monolithic metal and the composite matrix.

Adhesive bonding works well in joining metals to MMCs, with little or no risk to the reinforcing phase; however, in-service loading must be as close to pure shear or pure tension as possible to avoid peel or cleavage failures. Furthermore, service temperature must be limited to prevent adhesive degradation, and other degrading environments must be avoided or considered carefully.

In mechanical fastening, hole preparation is a special challenge. Drilling can be difficult to accomplish at all or may result in severe tool wear. In addition, care must be exercised to avoid damaging reinforcing fibers that are continuous. So-called softened areas, devoid of reinforcement, are often designed into the joint area or inserts are used.

Two other excellent options for joining metals to MMCs are weldbonding and weldbrazing, particularly if the welding process is resistance-spot. Here, loading is spread by the adhesive or braze, respectively, and peel strength is enhanced by the spotwelds while heating is minimized.

*Joining to Ceramic-Matrix or Carbon-Matrix
Composites*

Brazing is by far the most common method used to join metals to ceramic-matrix or carbon-matrix composites. For the most part, the process is carried out exactly as it would be for monolithic metals to monolithic ceramics or carbon (see Section 15.3). There is usually little risk of thermal damage to the reinforcement in the ceramic if that reinforcement is a metal or ceramic. This is because the brazing temperature is usually sufficiently low to avoid thermal degradation or reactions, and often the reinforcement itself is refractory.

Adhesive bonding, using organic adhesives, works well with the same caveats, i.e., out-of-plane loading should be avoided, and environmental limitations should be kept in mind.

Nonfusion welding, using friction or diffusion bonding, is also possible. When these processes can be used, joint properties can be excellent.

For brazing and welding and, to a far lesser extent, for adhesive bonding, mismatch of C.T.E.s must be carefully considered and dealt with.

15.7 JOINING OF CERAMICS TO POLYMERS

The use of ceramics in combination with polymers is relatively rare except for electronic applications, and for these applications loading is usually relatively low. Therefore, except for the rare case in which a metallized ceramic would be soldered to a metallized polymer, the only really viable option for joining is adhesive bonding using organic adhesives. The choice of adhesive is dictated by the nature of the polymeric joint component: thermosetting adhesive for thermosetting polymer, thermoplastics adhesives for thermoplastic polymers. To deal with expected differential C.T.E.s the adhesive might contain an elastomeric component to facilitate strain accommodation.

An interesting development has recently been reported by Garth L. Wilkes of Virginia Polytechnic Institute that could change the relative importance and approach to joining polymers to ceramics. Wilkes has grafted ceramics onto polymer chains, creating a new class of materials he calls *ceramers*. They are considerable stronger, stiffer, and more solvent resistant than conventional polymers and could be useful for making lightweight structures. Joining of these materials will be a challenge, but the chemistry of their synthesis may also hold the key to better ways to bond polymers and ceramics macroscopically.

15.8 JOINING CERAMICS TO COMPOSITES

15.8.1 General Comments

Except for electronic applications, where ceramic chip packages or chip carriers are adhesively bonded to fiber-reinforced polymer (or FRP) substrates or boards, ceramics will likely only be joined to metal- or ceramic- or carbon-matrix composites. This is because elevated temperature service is an extremely likely requirement that leads to the choice of these combinations. This being the case, likely joining options are brazing, mechanical fastening, and welding (at least between ceramics and ceramic- or metal-reinforced ceramic-matrix composites.

Adhesive bonding is also possible between ceramics and metal-, ceramic-, or carbon-matrix composites. For joining to MMCs, organic adhesives could be used, but service temperatures would be severely restricted. For joining to CMCs, organic or inorganic adhesives (i.e., cements or mortars) could be used, with far less restriction on service temperatures for the inorganic adhesives. For joining to carbon-matrix composites, organic adhesives might be used and pyrolyzed.

15.8.2 Joining Methods

The preferred method of joining ceramics to composites depends strongly on the matrix of the composite.

Ceramics to MMCs

Brazing is almost certainly the process of choice for joining ceramics to metal-matrix composites. The techniques and procedures are the same as for joining monolithic ceramics and metals, relying on metallization of the ceramic, oxidation of the metal (matrix), or causing bonding through an interfacial reaction. Noble-metal, active-metal, and refractory metal brazes could all be considered. For the most severe temperature and loading environments, diffusion brazing should be considered.

Welding, especially by solid-state diffusion, with or without the aid of an intermediate, is possible, just as between monolithic metals and ceramics (Section 15.3.3). Adhesive bonding using organic adhesives can again be used, but with the same service temperature restrictions. Mechanical fastening is possible, but the brittleness of the ceramic severely limits loading, demands that loading not be concentrated and that bending and impact be avoided The same precautions must be taken to avoid damaging the reinforcement in the MMC as is usually the case in mechanically fastening composites.

Ceramics to CMCs

Once again, brazing is the process of choice in joining ceramics to CMCs, using either active metal braze alloys or ceramic braze compositions. Techniques are the same as for brazing ceramics to ceramics (see Chapter 12, Section 12.4, or Chapter 14, Section 14.5). Welding is possible, just as between monolithic ceramics, but with the added complexity created by the reinforcement (i.e., thermal degradation). In fact, thermal degradation of ceramic reinforcements and reaction between ceramic reinforcements and ceramic matrices is usually less serious than with metals and metals or metals and ceramics.

Adhesive bonding using inorganic adhesives is certainly a viable option, just as with monolithic ceramics. Mechanical fastening, while possible, would be limited, because loading would have to be limited and temperature serviceability would also be limited compared to most of the ceramics.

15.9 JOINING POLYMERS TO POLYMER-MATRIX COMPOSITES

15.9.1 General Comments

The combination of polymers and fiber-reinforced polymers or polymer-matrix composites is a common one, widely used in aerospace and likely to be used in automobiles. The problems encountered in attempting to join these materials are relatively minor, provided the resin matrix of the composite and the unreinforced polymer are of the same type, i.e, both thermosetting or both thermoplastic. Even if they are different, however, there are good joining options.

By far, the preferred method for joining polymers and FRPs is to use organic adhesives. For thermosetting types, use thermosetting adhesives; for thermoplastics, thermoplastic adhesives; and for thermosets to thermoplastics, adhesive alloys. For thermoplastics to thermoplastic-matrix composites, solvent cementing is an option, as is fusion bonding. They both offer excellent joint strength and are being used in the most advanced military aircraft.

While mechanical fasteners could be used, they generally have not been, largely because of viscoelastic distortion of the fastener hole or fastener pull-through. One attractive opportunity is to use fasteners that are themselves thermoplastics, possibly reinforced.

It is unlikely, although not impossible, that polymers would be joined to composites with metallic or ceramic or carbon matrices. In this case, adhesive bonding is again the preferred approach.

15.9.2 Joining Methods

As mentioned previously, the principal method for joining polymers to organic-matrix composites is by adhesive bonding, although, for the special case of thermoplastic polymers to thermoplastic-matrix composites, fusion bonding and welding are also possible.

Adhesive Bonding

Unreinforced plastics can be joined effectively to reinforced plastics by using adhesives of the appropriate type. Thermosetting polymers should be adhesively bonded to thermosetting-matrix composites using thermosetting adhesives, with the most popular being epoxies or modified epoxies. Either one- or two-component types can be used. It is possible to enhance the strength of the adhesive slightly by mixing in some reinforcement. This works best if the composite is not continuously reinforced.

Likewise, thermoplastic polymers should be bonded to thermoplastic-matrix composites using thermoplastic adhesives, especially hot-melts. Solvent cementing can also be accomplished, however, possibly with the need for additional resin at the bond line.

Dissimilar polymer types, whether reinforced or not, can also be adhesively joined using adhesive alloys containing mixtures of thermosetting and thermoplastic components, possibly with elastomerics added for flexibility. Details for adhesively bonding all of these materials are given in Chapter 13.

Welding or Fusion Bonding

Thermoplastics and thermoplastic-matrix composites can be welded or fusion bonded by causing local softening and applying pressure to cause interdiffusion or intermixing. Various friction or vibration methods, as well as hot-gas or other heating methods can be used. It may be necessary to add some resin of the same composition as the composite matrix to the faying surface to promote bonding. Otherwise, it is possible to have resin starvation and poor bond integrity.

15.10 LOGICAL AND ILLOGICAL COMBINATIONS REVISITED

As stated earlier, some combinations of dissimilar materials make little sense from a practical standpoint, particularly if the inherent properties of the component materials are drastically different. Examples include widely different strengths, widely different moduli, widely different expansion coefficients, and widely different temperature resistance (including melting point).

Illogical combinations should be avoided and some other combination used.

SUMMARY

Dissimilar materials must frequently be joined to create hybrid structures in which diverse or even divergent properties, unattainable in individual materials, are required. Hybrid structures enable (1) minimization of weight, (2) improved damage tolerance in the structure, (3) optimized matching of properties to design needs, and (4) improved economy in both basic material costs and the cost of fabricating those (especially difficult) materials. The challenge of joining dissimilar materials is obtaining sufficient compatibility to permit the materials to join and the joint to function in service. Chemical (i.e., atomic structure and bond type), physical (e.g., melting temperature and C.T.E. response to temperature), and mechanical (e.g., stress-strain response and temperature stability of properties) compatibility are all required, and the closer the match the better. While every combination of the basic material types (i.e., metals, intermetallics, ceramics, glasses, polymers, and composites) is conceivable, not every combination is practical based on drastic differences in fundamental structure and properties.

For joining metals to ceramics, the choices are (1) mechanical fastening by interlocking features or fasteners; (2) direct bonding by nonfusion or, to a lesser extent, fusion welding; and (3) indirect bonding using intermediates that include organic or inorganic adhesives (in adhesive bonding or cementing, respectively), or metallic fillers that are melted to enable brazing or soldering or remain solid for assisting in diffusion bonding. In mechanical fastening, stress concentration in the ceramic element must be minimized, and C.T.E.s must be matched or compensated for by joint and fastener design. In direct bonding, solid-state diffusion welding is possible and gives good results, but other welding processes are possible, including friction, ultrasonic and high-energy-density EBW, and LBW. In indirect bonding, liquid-phase and solid-phase bonding is possible using adhesives, braze or solder fillers, or intermediates for DB. One key to facilitating the indirect joining of ceramics and metals (or other dissimilar materials) is to make them appear less different. Oxidizing the surface of the metal or metallizing the surface of the ceramic accomplishes this task.

When metals are joined to glass, the principal need is hermeticity, with mechanical strength being secondary. The brittle nature of glasses demands that C.T.E.s be matched or the metal element be kept small or a compliant intermediate be used to accommodate strain. Joining methods predominantly include mechanical seals relying solely on interlocking or achieving sealing through the use of a soft or melted metal interlayer (e.g., lead-seal). Soldering of glasses with metallized surfaces to metals is also possible.

Metals can be joined to polymers simply by using organic adhesives, although mechanical fasteners are also used. With adhesives, here, as elsewhere, the adhesive and the substrates should be of the same polymer type, i.e., thermosetting or thermoplastic, otherwise the adhesive should be an alloy of both types. When fasteners are used, loading must not be permitted to concentrate to such high levels that cold flow by viscoelastic deformation occurs in the polymer. Once again, soldering can be accomplished by metallizing the surface of the polymer.

For joining metals to composites, one needs to consider the material making up the matrix of the composite and then select the joining process accordingly. For organic-matrix composites, mechanical fastening and adhesive bonding are the preferred processes, and the hybrid process of rivet-bonding may be the best of all. For metal-matrix composites, welding with processes that minimize heat input and melting (such as DB, resistance, capacitor-discharge, and friction) are preferred, brazing with metallic fillers is an excellent choice, and adhesive bonding and mechanical fastening are viable options. Weldbonding and weldbrazing are also good choices for some applications where out-of-plane loading is likely. For ceramic-matrix composites, brazing with either metallic or ceramic fillers is preferred, with adhesive bonding an option.

For joining ceramics to polymers, adhesive bonding using organic adhesives is the process of choice, with the selection of adhesive based on the polymer type. For joining ceramics to composites, organic adhesives are used to bond to organic-matrix types, brazing or solid-state welding is used to join to metal-matrix types, and ceramic or metal brazing or adhesive cementing is used to join to ceramic-matrix types.

Finally, for joining polymers to organic-matrix composites, adhesive bonding is clearly preferred, followed by mechanical fastening. Fusion bonding or welding is also possible for thermoplastic types.

PRACTICE QUESTIONS AND PROBLEMS

1. From a technical standpoint, why is it so important to be able to join dissimilar combinations of materials in modern designs? Why from an economic standpoint?

2. What makes the joining of dissimilar materials so challenging? Describe some of the special challenges.

3. Are some combinations of materials more difficult to join than others? If so, what combinations are especially difficult? Why? If not, why not?

4. Why are some combinations of materials considered illogical? Give some examples of illogical combinations.

5. Why is it so important to be able to join ceramics to metals? What properties of the ceramics can be enhanced by such joining? What properties of ceramics are normally being sought in such combinations?

6. What are some direct methods for joining ceramics to metals? What are some indirect methods for joining these materials?

7. One way to facilitate joining of metals and ceramics is to make them appear less different from one another. What are the two most common approaches taken to accomplish this?

8. Why is it so desirable to be able to join metals and glasses? Give some important application examples.

9. What are the four types of glass-to-metal seals? Describe each briefly.

10. What properties of metals make them suitable for joining to glasses? What properties of glasses make them suitable for joining to metals?

11. Describe what is meant by *matched seals*. What is an *unmatched seal?* Can unmatched seals be made to work? If so, how? If not, why not?

12. Describe some clever ways of sealing metals to glasses using purely mechanical means.

13. How are metals and polymers joined to one another by soldering?

14. What are some of the special problems associated with producing structurally sound joints between metals and polymer-matrix composites? What techniques are used in practice?

15. What kinds of problems arise during the mechanical fastening of metals to polymer-matrix composites using fasteners? How are these problems overcome in practice?

16. What are the preferred methods for joining metals to metal-matrix composites? Why are these methods preferred? What about for joining metals to ceramic-matrix composites? What about for joining metals to carbon–carbon composites?

17. How are ceramics usually joined to polymers? Give some examples of where such joining is important.

18. Differentiate between the preferred methods for joining ceramics to metal-, ceramic-, and carbon-matrix composites.

19. How challenging is it to join polymers to polymer-matrix composites? Explain your answer.

20. When can polymers be fusion bonded to polymer-matrix composites? What are the advantages of fusion bonding, if any?

REFERENCES

1. Allen, R.V., and Borbridge, W.E., "Solid-State Metal-Ceramic Bonding of Platinum to Alumina," *Journal of Materials Science,* 18, 1983, pp. 2835–2843.

2.   Eagers, Thomas W., "Ceramic-Metal Bonding Research in Japan," *Welding Journal,* Vol. 66, No. 11, pp. xx.

3.   Landt, Richard C., "Mechanical Fasteners for Advanced Composite Materials," *Joining Technologies for the 1990's,* John D. Buckley and Bland A. Stein, Park Ridge, N.J.: Noyes Data Corporation, 1986, pp. 54–63.

4.   Matthews, F.L., *Joining Fibre-Reinforced Plastics,* London: Elsevier Applied Science, 1987.

5.   Patridge, J.H., *Glass-to-Metal Seals,* Sheffield, England: Society of Glass Technology, 1967.

6.   Schwartz, Mel M., *Ceramic Joining,* Metals Park, Ohio: ASM International, 1990.

Index

539